Diagnostic Techniques in Public Health

Diagnostic Techniques in Public Health

Analytical and Chemometric Approaches in Disease Detection and Management

Edited by Jinghong Li and Yan Du

WILEY-VCH

Editors

Prof. Jinghong Li
Tsinghua University
Department of Chemistry
Beijing, 100084
P.R. China

Prof. Yan Du
Changchun Institute of Applied Chemistry
Chinese Academy of Sciences
Changchun 130022
P.R. China

Cover Design: Wiley
Cover Image: © Supapich Methaset/Shutterstock, © dizain/Shutterstock

Typesetting Straive, Chennai, India
Printing and Binding CPI Group (UK) Ltd, Croydon, CR0 4YY

Library of Congress Card No.: applied for

British Library Cataloguing-in-Publication Data A catalogue record for this book is available from the British Library.

Bibliographic information published by the Deutsche Nationalbibliothek
The Deutsche Nationalbibliothek lists this publication in the Deutsche Nationalbibliografie; detailed bibliographic data are available on the Internet at <http://dnb.d-nb.de>.

Print ISBN: 978-3-527-35409-2
ePDF ISBN: 978-3-527-84793-8
ePub ISBN: 978-3-527-84792-1
oBook ISBN: 978-3-527-84794-5

C9783527354092_270226

Contents

Preface

In an era defined by the increasing frequency and severity of public health emergencies—ranging from global pandemics and emerging zoonoses to the resurgence of antimicrobial resistance—the demand for precise, rapid, and scalable diagnostic technologies has never been more urgent. Scientific progress in this field is not merely driven by technological curiosity but by an existential imperative: to detect, understand, and contain threats to human health with unprecedented accuracy and speed.

This volume, *Diagnostic Techniques in Public Health: Analytical and Chemometric Approaches in Disease Detection and Management*, emerges as a timely and comprehensive response to this imperative. It brings together a series of meticulously curated chapters that bridge the fields of molecular biology, analytical chemistry, nanotechnology, biomedical engineering, and informatics to address one central question: how can diagnostic science evolve to meet the demands of a dynamic and interconnected world?

The book is structured to reflect both the conceptual foundations and cutting-edge frontiers of public health diagnostics. The early chapters establish the essential methodological basis, delving into **molecular diagnostic technologies**—from classical polymerase chain reaction (PCR) to next-generation isothermal amplification methods—and the enduring relevance of **ELISA-based immunoassays**, now enhanced by nanobody engineering, aptamer systems, and integration with microfluidic platforms. As the narrative progresses, the reader is introduced to **genomic sequencing** as a cornerstone of pathogen surveillance, enabling not only real-time outbreak tracking but also genomic epidemiology and predictive modeling.

Building upon these molecular pillars, subsequent chapters explore transformative technologies such as **CRISPR-based diagnostics**, whose programmability and field-deployability align with the World Health Organization's ASSURED criteria, and **high-throughput microfluidic systems**, which achieve unparalleled levels of automation, multiplexing, and reagent economy. The volume culminates in a forward-looking discussion of **plasmonic biosensors** and **flexible bioelectronics**, where the convergence of material science and biomedical instrumentation opens entirely new vistas for noninvasive, continuous, and personalized health monitoring.

What distinguishes this work is not only its breadth of technical coverage but also its clear commitment to translational relevance. Each chapter situates its scientific discussion within the broader context of public health preparedness, emphasizing point-of-care deployment, low-resource adaptability, regulatory considerations, and ethical dimensions. Moreover, the interplay between core diagnostic principles and enabling technologies—such as artificial intelligence, smartphone-based data acquisition, and digital epidemiology—is carefully woven throughout the text, underscoring the systemic integration required for next-generation diagnostic infrastructures.

This book is intended for a diverse readership: biomedical researchers seeking to advance assay performance; public health professionals aiming to translate bench technologies to field practice; engineers and data scientists developing the next wave of biosensing platforms; and policymakers navigating the complex landscape of technology adoption and health system resilience. It is our conviction that the chapters herein will not only inform and inspire but also serve as a catalyst for cross-disciplinary collaboration and innovation.

I would like to extend my deepest gratitude to the contributing authors for their intellectual rigor and commitment to excellence and to the editorial and production teams who brought this ambitious volume to fruition. As global health challenges continue to evolve, we hope this book will provide both a foundation and a framework for the development of diagnostic strategies that are not only scientifically robust but also globally equitable, operationally feasible, and ethically grounded.

Jinghong Li
Beijing

1

Analytical and Chemometric Approaches in Public Health Emergencies

Global Significance, Challenges, and Disease Detection Strategies

Yan Du[1,2] *and Jinghong Li*[3,4,5,6]

[1] *Chinese Academy of Sciences, Changchun Institute of Applied Chemistry, State key Laboratory of Electroanalytical Chemistry, Changchun 130000, China*
[2] *University of Science and Technology of China, School of Applied Chemistry and Engineering, Hefei 230026, China*
[3] *Beijing Life Science Academy, Beijing 102209, China*
[4] *Tsinghua University, Department of Chemistry, Key Laboratory of Bioorganic Phosphorus Chemistry & Chemical Biology, Beijing 100084, China*
[5] *New Cornerstone Science Laboratory, Shenzhen 518054, China*
[6] *University of Science and Technology of China, Hefei National Laboratory of Physical Science at Microscale, Center for BioAnalytical Chemistry, Hefei 230026, China*

CHAPTER MENU

Public health emergencies (PHEs) are situations that pose significant threats to the health of populations across regions, requiring rapid and coordinated responses. These emergencies are unpredictable, with wide-reaching impacts that can overwhelm health systems and disrupt societal functions [1, 2]. Causes include pandemics, bioterrorism, and severe environmental disasters, necessitating comprehensive preparedness strategies that include legal, ethical, and logistical considerations [2–4]. Effective PHE management relies on systems capable of mobilizing resources swiftly and adapting to evolving threats, underscoring the importance of health infrastructure resilience and international cooperation [2, 3, 5].

In the landscape of global health, the role of international organizations is crucial in orchestrating unified responses to PHEs. Organizations like the World Health Organization (WHO) not only coordinate these efforts but also foster cross-country collaborations to strengthen health system resilience and preparedness [6, 7]. This

Diagnostic Techniques in Public Health: Analytical and Chemometric Approaches in Disease Detection and Management, First Edition. Edited by Jinghong Li and Yan Du.

collective approach ensures rapid and effective responses to emerging health threats by leveraging shared resources, expertise, and surveillance systems [8, 9]. As health challenges become more interconnected globally, the emphasis on international cooperation and strategic partnerships has intensified, highlighting the need for a cohesive global health security framework [10–12].

Throughout history, these infectious diseases, such as smallpox, influenza, plague, and cholera, along with more recent outbreaks like severe acute respiratory syndrome (SARS), COVID-19, and monkeypox, have posed serious threats to human health. This chapter explores the multifaceted nature of PHEs, emphasizing their unpredictability and widespread impact. As global health threats continue to evolve, there is an increasing need for a cohesive and integrated global health security framework. More importantly, chemical measurements of disease biomarkers plays a critical role in responding to major PHEs, addressing the growing demands for higher accuracy, faster and simpler methods, and increased detection throughput.

1.1 Beyond Infectious Threats: Expanding the Horizons of Global Health Challenges

Global health challenges are increasingly shaped by factors beyond infectious diseases, including climate change, urbanization, and social disparities. These issues now play a significant role in disease emergence and management, broadening the scope of threats to global health.

1.1.1 Introduction to Global Health Challenges

Global health challenges now encompass noncommunicable diseases (NCDs) like cardiovascular disease, cancer, and diabetes, driven by urbanization, globalization, and environmental changes, contributing significantly to global mortality [13]. The resurgence of tuberculosis, ongoing pandemic threats, antimicrobial resistance from antibiotic overuse, and rising mental health burdens underscore the complexity of modern health threats, especially in low-resource settings [14–16]. Addressing these interconnected challenges requires an integrated approach that combines the management of infectious and noninfectious diseases [15, 17, 18].

1.1.2 Determinants Influencing the Spread and Management of Diseases

Disease spread and management are influenced by various biological, environmental, social, and economic factors. Biological aspects, such as pathogen virulence and human immunity, play a crucial role in transmission dynamics [19]. Environmental conditions, including climate change, geography, and urbanization, impact vector populations and human behaviors, expanding the reach of diseases such as malaria, dengue, and Zika to new regions [20–25]. Social determinants, such as healthcare access, education, and infrastructure, significantly affect disease control, especially in regions with inadequate systems that struggle to manage outbreaks [26, 27]. Globalization and increased mobility complicate containment, while economic factors, including funding and resource allocation, are critical for effective prevention and response strategies [28, 29]. A multidisciplinary approach is essential to address these interconnected determinants for both infectious and noninfectious disease management.

1.2 Tracing the Seven Public Health Emergencies of International Concern Announcements by the World Health Organization: From H1N1 to Monkeypox

Since implementing the International Health Regulations (IHR) in 2005, the WHO has declared seven PHEICs to address significant global health threats. These declarations mobilize international cooperation and resources, highlighting outbreak severity and guiding a coordinated response. Covering crises from the H1N1 influenza pandemic to the recent monkeypox outbreak, these PHEICs reveal evolving health security measures and challenges posed by emerging infectious diseases.

1.2.1 Overview of the World Health Organization's Criteria for Declaring a Public Health Emergency of International Concern

Under the IHR, the WHO declares a PHEIC based on criteria assessing a health threat's global significance. The event must pose a serious public health risk, have the potential for international spread, threaten cross-border transmission, and require a coordinated international response [30–33]. An Emergency Committee reviews the evidence, considering if the event is unexpected or poses long-term risks that could overwhelm health systems. The goal is to raise awareness, mobilize resources, and implement temporary health measures to contain the outbreaks [34–36].

1.2.2 List of Public Health Emergencies of International Concern Declarations with Dates and Brief Descriptions

1.2.2.1 H1N1 Influenza Pandemic (2009)

The H1N1, or swine flu, pandemic began in April 2009 in Mexico, leading the WHO to declare a PHEIC on April 25, 2009, as the virus spread globally. Unlike seasonal flu, it mainly affected younger people without pre-existing conditions. By June, it was declared a pandemic, infecting an estimated 700 million to 1.4 billion people, with a death toll between 151,700 and 575,400. The rapid spread strained health systems and required extensive vaccination, travel restrictions, and quarantine measures [32, 37].

1.2.2.2 Poliovirus (2014)

In May 2014, the WHO declared a PHEIC due to the international spread of wild poliovirus, primarily affecting Pakistan, Afghanistan, and Nigeria, where conflict and weak health systems hindered eradication efforts [32]. New cases in previously polio-free countries such as Syria and Cameroon heightened concerns about further spread. WHO responded with urgent vaccination campaigns and travel restrictions. While cases have declined, ongoing international cooperation and strengthened health systems are essential for complete eradication [37].

1.2.2.3 Ebola Virus Disease Outbreak in West Africa (2014)

Beginning in December 2013 in Guinea, the Ebola outbreak spread to Liberia and Sierra Leone, becoming the largest and deadliest to date. The WHO declared a PHEIC on August 8, 2014, as the virus spread rapidly with high mortality, compounded by weak health systems and cultural practices [32, 38, 39]. With over 28,000 cases and 11,000 deaths, the crisis underscored the need for global coordination. In response, treatment centers, diagnostic tools, and an experimental vaccine were introduced [37, 40].

1.2.2.4 Zika Virus (2016)

The 2016 Zika outbreak raised global concerns due to its rapid spread and association with severe birth defects, particularly microcephaly. Initially identified in Uganda, Zika spread rapidly across the Americas, leading the WHO to declare a PHEIC on February 1, 2016, in response to rising cases of neurological complications [37]. The virus's multiple transmission routes, including mosquito-borne and sexual transmission, posed unique challenges. Public health strategies involved mosquito control, awareness campaigns, and travel guidance. Despite these efforts, no effective vaccine was available, and Zika remains a long-term public health concern [40, 41].

1.2.2.5 Ebola Virus Disease Outbreak in the Democratic Republic of Congo (2019)

The 2019 Ebola outbreak in the Democratic Republic of Congo, the second-largest in history, was complicated by regional conflict. The WHO declared a PHEIC on July 17, 2019 [32, 37]. Unlike previous outbreaks, a ring vaccination strategy was implemented using an approved vaccine. However, insecurity, misinformation, and distrust hindered response efforts, and healthcare workers faced violence. The outbreak ended on June 25, 2020, with 3,481 cases and 2,299 deaths [39, 40].

1.2.2.6 COVID-19 Pandemic (2020)

Beginning in late 2019 in Wuhan, China, COVID-19 quickly became a global health crisis, prompting the WHO to declare a PHEIC on January 30, 2020 [40]. Transmitted through respiratory droplets, the virus caused widespread illness, severely impacting vulnerable populations and overwhelming health systems worldwide. Governments responded with lockdowns, social distancing, mask mandates, and travel restrictions [42]. Despite these measures, COVID-19 resulted in millions of deaths and significant economic disruption, highlighting both strengths and weaknesses in global health systems and advancing telemedicine, digital health, and vaccine technologies [43, 44].

1.2.2.7 Monkeypox (2022)

The 2022 monkeypox outbreak marked the first large-scale global spread of the virus, which had traditionally been endemic to Central and West Africa. Human-to-human transmission was reported in non-endemic countries, leading the WHO to declare a PHEIC on July 23, 2022, as cases spanned over 70 countries [37, 40]. Predominantly affecting men who have sex with men, the outbreak raised concerns about stigmatization [45]. Symptoms included fever, swollen lymph nodes, and rash. The response involved smallpox vaccination, contact tracing, isolation, and targeted vaccination efforts [32, 46].

1.2.3 Comparative Analysis of the Responses to Each Public Health Emergency of International Concern and the Evolution of Global Health Security Measures

Responses to the WHO's PHEICs, from H1N1 to COVID-19, highlight trends in global health security. The H1N1 outbreak revealed gaps in vaccine access, emphasizing the need for international coordination and equitable frameworks—lessons that informed the rapid vaccine development during COVID-19 [47, 48]. The 2014 Ebola crisis exposed vulnerabilities in local health systems, while the 2019 DRC Ebola response demonstrated improved vaccine use and faster mobilization, though challenges like conflict persisted [49]. The 2014 polio and 2016 Zika outbreaks underscored the importance of sustained immunization, surveillance, and strategies for vector-borne diseases [1, 32, 40]. The 2022 monkeypox outbreak stressed the need for zoonotic disease surveillance and preventive measures in non-endemic regions [42].

COVID-19, the most disruptive PHEIC, emphasized global health system weaknesses while accelerating advances in digital health. Together, these PHEICs highlight the need for timely cooperation, robust surveillance, and equitable resource allocation to mitigate future crises.

1.3 A Close Look at Smallpox, Severe Acute Respiratory Syndrome, and Influenza Variations

Beyond the seven PHEICs, diseases such as SARS, smallpox, and influenza have caused widespread outbreaks with high transmission and mortality rates. Despite the differences in origin and transmission, these respiratory syndromes highlight shared global challenges in containment, immunization, diagnosis, surveillance, vaccine development, and international cooperation. Studying these diseases offers insights into managing future pandemics and strengthening public health defenses.

1.3.1 Eradication Effort and Lessons Learned from Smallpox

Smallpox, with a 30% fatality rate, was eradicated through the WHO's 1967 Intensified Eradication Program, which focused on ring vaccination, surveillance, and containment [50–52]. Achieved in 1980, this success underscored the importance of political will, community engagement, and resource mobilization, setting a model for combating future health threats such as polio and COVID-19 [53, 54].

1.3.2 Etiology, Transmission, and the Global Response of Severe Acute Respiratory Syndrome

SARS, caused by the SARS-CoV virus, emerged in China in 2002, spreading to over 30 countries by 2003. Primarily transmitted via respiratory droplets, SARS is believed to have originated in bats, with transmission to humans through civet cats [43, 44, 55]. The outbreak underscored the need for global surveillance and rapid response, as health systems implemented strict isolation and hygiene protocols. Lessons from SARS informed pandemic management strategies, including COVID-19 preparedness [56].

1.3.3 Evolution of Influenza Strains and Their Implications for Global Pandemics

The continuous mutation of influenza strains, particularly subtype A, through antigenic drift and shift leads to seasonal outbreaks and potential pandemics, as exemplified by the H1N1 virus in 1918 and 2009 [57–59]. Rapid mutations complicates vaccine development, requiring frequent updates to match circulating strains. Past pandemics, such as the 1957 Asian Flu and 1968 Hong Kong Flu, highlight influenza's vast impact, and the need for robust surveillance, preparedness, and public health interventions to manage future outbreaks [60–62].

1.4 Analytical and Chemometric Approaches in Major Public Health Emergencies

The WHO provides critical guidelines on disease surveillance, containment, and response strategies for managing PHEICs, emphasizing early detection and rapid diagnostic measures to identify pathogens and contain outbreaks effectively [63, 64]. Advanced analytical and chemometric approaches are vital in this process, as they enable the precise pathogen identification and comprehensive pathogenomic analysis, which are essential for effective epidemic control.

Analytical and chemometric methods, such as molecular probes coupled with signal amplification and conversion techniques, achieve high sensitivity and specificity, allowing single-base resolution and single-molecule detection. These capabilities support rapid, accurate, and high-throughput diagnostics, critical for identifying new and variant pathogens with precision. Through these methods, chemical measurement has become a cornerstone in infectious disease control, improving our understanding of viral mechanisms and strengthening public health responses.

In response to recent public health crises, some important works have pioneered chemical measurement innovations to address these needs. These works include developing dual-labeled immunochromatographic technology for SARS antigen detection, single-base resolution nucleic acid detection tools for COVID-19, and studies on nucleic acid structural evolution in the monkeypox virus. These advancements not only aid epidemic prevention and control but also hold significant potential in fields such as vaccine development, detection of antibiotic-resistant microorganisms, and single-base drug target screening.

These contributions underscore the strategic importance of chemical measurement in preparing for future PHEs. Establishing robust research teams and technologies focused on rapid-response chemical measurement will be essential for proactive epidemic management, effectively integrating disease prevention demands with scientific innovation to strengthen the global health security.

1.4.1 Severe Acute Respiratory Syndrome Virus Antigen Immunoassay

The principle of colloidal gold-based immunochromatographic detection is that colloidal gold, which adsorbs the target protein, moves along the test strip via capillary action and interacts with the antigen–antibody complex on the test line, resulting in a visible color change [65]. This technique is known for its simplicity and speed. Since its introduction in the 1990s, it has been widely applied in protein detection across various fields. However, there have been some limitations: first, it can only achieve qualitative or semi-quantitative detection, making quantitative analysis challenging; second, it is susceptible to interference from other components in the sample, reducing robustness; and third, it is prone to false negatives when the target analyte is present at low concentrations [66].

To address these issues, as an example, a dual-labeled rapid immunochromatographic test strip was developed, combining colloidal gold and semiconductor fluorescent nanoparticles. Building upon the traditional colloidal gold immunochromatographic technique, CdTe semiconductor fluorescent nanoparticles were immobilized on the detection line [67, 68]. Due to the quenching effect of colloidal gold on fluorescent nanoparticles, this method enabled simultaneous colorimetric and fluorescence detection, significantly enhancing the sensitivity and extending the linear detection range, thus allowing for quantitative analysis. This method is capable of detecting various sample types, including whole blood, serum, plasma, and urine, with high sensitivity and accuracy in the detection of hepatitis B surface antigen, human chorionic gonadotropin, and human immunodeficiency virus (HIV). At the beginning of 2003, the sudden outbreak of SARS gripped China. Supported by the dual-labeled rapid immunochromatographic test strip technology, researchers successfully achieved rapid antigen detection of the SARS virus [69, 70].

1.4.2 Rapid Detection of COVID-19 and Its Variants

1.4.2.1 High-Sensitivity Rapid Detection of COVID-19

Respiratory viruses, such as COVID-19, pose significant risks due to their high transmissibility, long incubation period, and infectiousness even during the incubation phase. The winter–spring season is already a peak time for influenza, and the elderly and children—who are particularly vulnerable to COVID-19—are also at higher risk for influenza. The surge in patients with fever strains medical resources and increases the risk of cross-infection, posing great challenges to epidemic control. Therefore, it is crucial to rapidly differentiate and screen for influenza and COVID-19 viruses.

During the peak of the Wuhan COVID-19 outbreak in 2020, to facilitate the rapid screening of patients with fever, an ultra-sensitive colorimetric detection method was fabricated for bacteria and SARS-CoV-2 based on botulinum toxin sensitivity [71]. This method enabled highly sensitive detection of pathogenic microorganisms, allowing the rapid distinction between COVID-19 infection and seven influenza subtypes, including H1N1, H1N2, H3N2, H5N1, H7N9, H9N2, and H10N8. In October 2020, the Chinese Center for Disease Control and Prevention (CDC) reported the detection of live COVID-19 virus on the outer packaging of cold-chain food products in Qingdao. This new epidemic situation highlighted the need for COVID-19 detection in environmental samples, such as cold-chain logistics packaging. In response to this new demand, a novel method was reported for analyzing COVID-19 residues on cold-chain food and packaging, enabling high-sensitivity detection of the virus from environmental samples.

1.4.2.2 Single-Base Resolution Identification and Rapid Screening of COVID-19 Variants

Since December 2020, several COVID-19 variants, including Alpha, Beta, Delta, and Omicron, have been identified globally. These variants exhibit varying levels of transmissibility and pathogenicity, posing new challenges to epidemic control and demanding more stringent requirements for the molecular diagnostic methods of COVID-19.

Existing molecular diagnostic methods for viruses include polymerase chain reaction (PCR) and antigen-based enzyme-linked immunosorbent assay (ELISA). PCR, considered the gold standard for viral nucleic acid detection, involves amplifying viral nucleic acids using specific primers designed to target viral sequences under the action of nucleic acid polymerase. However, PCR is a time-consuming, labor-intensive process that requires highly skilled operation, as well as dependence on PCR instruments and professional laboratories, which limits its large-scale screening capacity. ELISA, on the other hand, relies on antibodies to recognize viral surface proteins, but its detection sensitivity and specificity are often inconsistent, leading to false positives and false negatives. Consequently, neither of these methods can effectively distinguish between viral variants.

In June 2021, faced with the new challenges posed by the Omicron variant, a novel, multiplex, amplification-free, single-base resolution method was quickly developed together with paper-based test strips and related instruments (Figure 1.1) [72]. A double-stranded DNA probe was designed with precise sequence configurations, in which C–C base mismatch sites chelate Ag^+. Only when the target nucleic acid molecule fully complements the recognition region of the probe does the double-stranded probe unwind, releasing Ag^+. Urease activity is inhibited by Ag^+, preventing the breakdown of urea into NH^{4+}, and the pH change in the

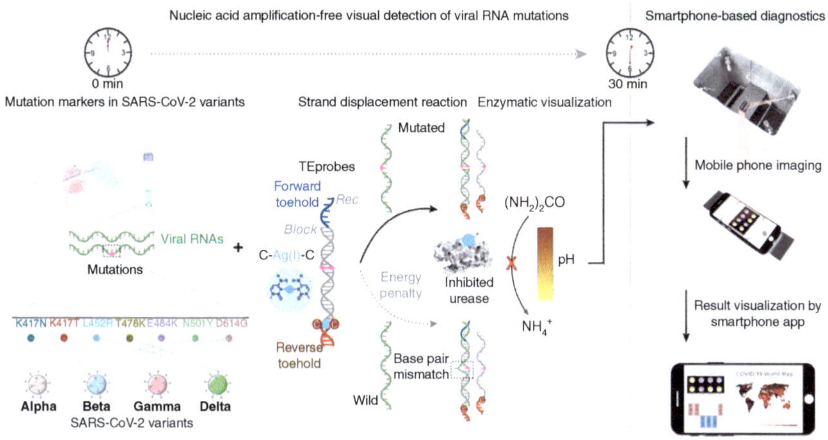

Figure 1.1 Single-base resolution nucleic acid rapid detection technology and instrument. *Source:* [72] / Springer Nature.

solution is easily indicated by a colorimetric indicator. Without the need for nucleic acid amplification, this method achieves a sensitivity of approximately 400 copies per sample and can effectively distinguish various COVID-19 variants, such as Delta and Omicron, that differ by only a single base. For newly emerging variants, the development of new detection probes can be completed within four days. The entire detection process can be performed within 30 minutes, without the need for large instruments. To further enhance throughput, a paper-based array chip and an integrated, enclosed detection instrument, along with a smartphone application, were integrated for reading the results, making it more suitable for clinical sample detection. This method enables rapid viral nucleic acid screening that can be applied in community hospitals and even at home. In July 2022, the US journal C&EN (Chemical & Engineering News) reported on this single-base resolution nucleic acid rapid detection technology, praising it as a simple and efficient method that produces visible signals on test strips.

1.4.2.3 Single-Base Resolution Nucleic Acid Rapid Detection Technology Assisting Vaccine Development and Drug Screening

The single-base resolution nucleic acid rapid detection technology and instrument offer several significant advantages: (i) the ability to distinguish single-base mutations, identifying specific subtypes of mutant pathogens without the need for the complex processes of library preparation, sequencing, and data analysis involved in sequencing technologies; (ii) high accuracy, with specificity greater than 95% and sensitivity over 90%; (iii) simplicity and speed, yielding results within 30 minutes through easy-to-learn procedures; and (iv) low cost, with a per-sample cost of less than 5 Chinese Yuan (CNY). This method holds great potential for playing a key role in scientific research and drug development within the biomedical field. Several biopharmaceutical companies, including Vanke and TusPharma, have already participated in the industrial application of this technology.

For instance, this technology can be used for detecting antibiotic-resistant microorganisms and screening drug targets, as well as for vaccine development [73–75]. The occurrence of microbial resistance and immune escape leads to the failure of existing drugs and vaccines. The root cause of resistance and immune escape lies in pathogen gene mutations. Therefore, developing screening methods that target mutant microorganisms rather than wild-type pathogens can significantly improve the efficiency of drug and vaccine screening. However, current analysis of mutant microorganisms relies solely on nucleic acid sequencing. Although sequencing offers high throughput, it requires two to three days for sample preparation, testing, and data analysis, resulting in poor timeliness. The single-base resolution nucleic acid rapid detection technology and instrument can quickly achieve quantitative analysis of mutant pathogens before and after drug treatment, assess the resistance of mutant pathogens, and, in combination with structural biology, explore the mechanisms behind resistance [76, 77]. This method can also facilitate high-throughput screening of highly active vaccines and potential antiviral drugs for mutant strains. The technology has the potential to become a universal platform for the rapid quantification of mutant pathogens and will be of great significance in the rapid screening of antibiotics and vaccine development.

1.4.3 Genetic Evolution Monitoring of the Monkeypox Virus

The monkeypox virus is an orthopoxvirus closely related to the smallpox virus. Historically, monkeypox was considered a low-contagion disease endemic to sub-Saharan Africa [78]. However, since May 2022, over 16,000 cases of monkeypox infection have been reported across 72 countries worldwide (as of July 26). The rapid outbreak of such a large number of cases in a short period was highly unusual. Given its high fatality rate (1–11%), the spread of monkeypox could exacerbate the global health crisis. Consequently, the sudden emergence of monkeypox, alongside the ongoing COVID-19 pandemic, has garnered widespread attention worldwide. Unfortunately, the reasons behind the abnormal increase in human-to-human transmission of the monkeypox virus remain unclear [79]. Therefore, monitoring the genetic characteristics of the monkeypox virus and exploring the biological mechanisms driving its accelerated transmission are crucial and urgent.

Nucleic acid secondary structures are key elements of viral genetic pathways. Intracellular nucleic acid macromolecules can form secondary structures such as G-quadruplexes. Studies have shown that these secondary structures regulate interactions between nucleic acids and proteins related to gene expression, ultimately determining cell fate [80]. As a result, nucleic acid secondary structures are considered potential therapeutic targets for diseases such as tumors and viral infections [81]. Among these structures, the RNA G-quadruplex (G4) is a topological structure of RNA molecules formed by stacking two to three guanine planes (G-planes) through Hoogsteen hydrogen bonds. RNA G-quadruplexes are present in various organisms and play a role in multiple RNA-related processes such as translation, RNA splicing, and RNA transport. RNA G-quadruplexes have also been identified in the genomes of many viruses, including HIV, human papillomavirus (HPV), herpes simplex virus (HSV), hepatitis C virus (HCV), Epstein–Barr virus (EBV), and SARS-CoV-2 [82]. These RNA G-quadruplexes can inhibit the expression of viral proteins and may play critical roles in the viral life cycle. As such, RNA G-quadruplexes are considered potential antiviral targets.

To study the folding status and biological regulatory mechanisms of virus-related RNA G-quadruplexes within cells, a single-molecule imaging technology was specifically developed to target G-quadruplex structures of certain genes in cells [83]. This technique utilizes a dual-modular probe system that recognizes both the RNA G-quadruplex structure and its adjacent RNA sequence, supported by rolling circle amplification, allowing for highly specific single-cell, single-molecule imaging of gene-specific G-quadruplexes.

In May 2022, when the first reports of monkeypox virus emerged, research on the evolutionary characteristics of monkeypox virus nucleic acids was initiated. For the first time, the presence of RNA G-quadruplex structures in the mRNA of the *C9L* gene of the monkeypox virus was discovered and a progressive evolution in this gene sequence was observed (Figure 1.2) [84]. As the RNA G-quadruplex sequence of the *C9L* gene evolved, its thermodynamic stability gradually decreased. All monkeypox virus strains identified globally in the past five years harbor the least stable RNA G-quadruplex variant, C9L-RG4-5. Using this single-molecule imaging technology for specific gene G-quadruplex structures within cells, the *C9L* RNA G-quadruplex can indeed form in mammalian cells at the single-molecule level. As evolution progresses, the reduced stability of the *C9L* RNA G-quadruplex led to an increase in C9L protein expression levels. Given the key role of the *C9L* gene in the monkeypox virus's evasion of host innate immune responses, the increased protein expression resulting from the reduced stability of the C9L RNA G-quadruplex may be one of the reasons why the currently circulating monkeypox virus strains spread more rapidly.

the G-quadruplex Sites in the C9L Gene

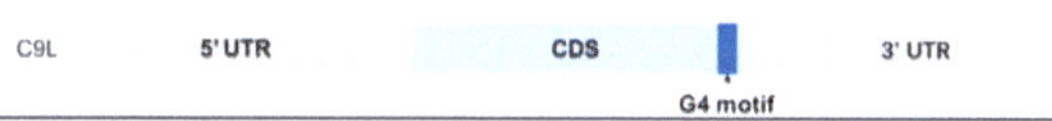

MKPV C9L PQS	PQS sequence	G-score	Overlaps
C9L-RG4-5	GGAATATATGGGATGGAATGGAATGGTAAATAATTTTGAAA	26	9
C9L-RG4-6	GGAATATATGGGATGGAATGGAATGGAATGGTAAATAATTT	26	25
C9L-RG4-7	GGAATATATGGAATGGAATGGAATGGAATGGAATGGTAAAT	26	25
C9L-RG4-7-1G3	GGAATATATGGGATGGAATGGAATGGAATGGAATGGTAAAT	26	41
C9L-RG4-7-2G3	GGAATATATGGAATGGGATGGGATGGAATGGAATGGTAAAT	26	66
C9L-RG4-8-1G3	GGAATATATGGGATGGAATGGAATGGAATGGAATGGAATGG	26	71
C9L-RG4-8-2G3	GGAATATATGGGATGGAATGGGATGGAATGGAATGGAATGG	26	110
C9L-RG4-8-3G3	GGAATATATGGAATGGGATGGGATGGGATGGAATGGAATGG	26	167
C9L-RG4-8-5G3	GGAATATATGGAATGGGATGGGATGGGATGGGATGGGATGG	52	341

Conservation of G-quadruplex in the C9L Gene

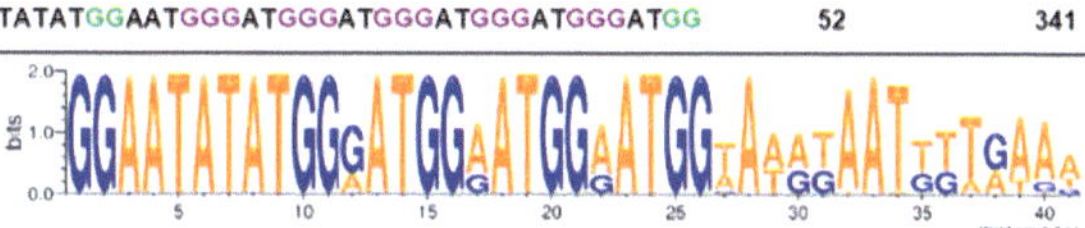

Figure 1.2 Variants of the G-quadruplex sites in the monkeypox virus *C9L* gene and their sequence conservation.

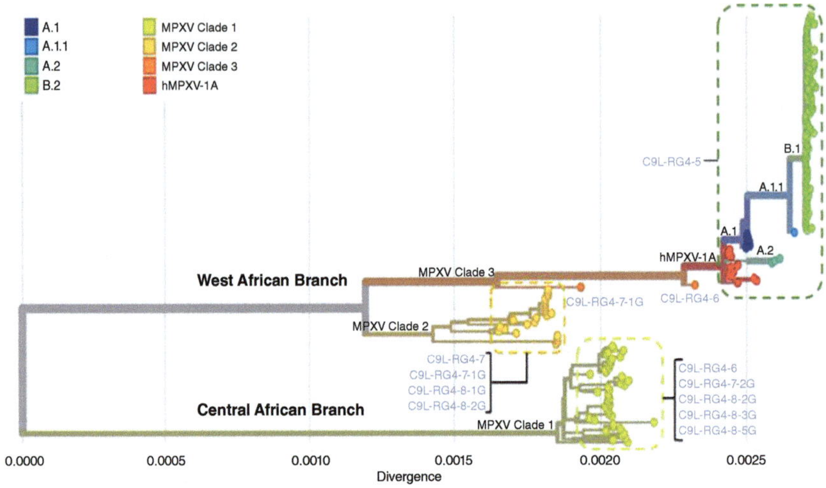

Figure 1.3 Phylogenetic tree of the monkeypox virus and the corresponding G-quadruplex types in the *C9L* gene. *Source:* Adapted from Nextstrain.

Moreover, the RNA G-quadruplex stabilizer, RGB-1 can stably bind to the RNA G-quadruplex variant C9L-RG4-5 present in the current strains and effectively reduces C9L protein expression levels (Figure 1.3). This research was also featured in a detailed report by the prominent UK-based biomedical media outlet, *NEWS Medical & Life Sciences*, and received high praise. On July 23, 2022, WHO declared that the monkeypox outbreaks in multiple countries constituted a "Public Health Emergency of International Concern." In light of the ongoing global spread of monkeypox, our research provides new directions for studying the transmission, pathogenesis, and antiviral targets of the monkeypox virus.

1.5 Conclusion and Discussion

The landscape of global health challenges has expanded beyond traditional infectious diseases to include a complex interplay of factors, such as NCDs, climate change, urbanization, and social disparities. These determinants significantly influence disease emergence and management, necessitating an integrated and multidisciplinary approach to health security. Biological factors like pathogen virulence and human immunity; environmental conditions, such as climate change and urbanization; and socio-economic influences, including healthcare access and globalization, collectively impact the spread and control of diseases. Addressing these interconnected challenges requires robust international collaboration and strategic partnerships.

Since the implementation of the IHR in 2005, the WHO has declared seven PHEIC, ranging from the H1N1 influenza pandemic to the recent monkeypox outbreak. These declarations underscore the evolving nature of global health threats and the critical importance of rapid, coordinated international responses. Comparative analyses of these PHEICs reveal advancements in global health security measures, such as improved surveillance systems and vaccine development, while also highlighting persistent gaps in preparedness, resource allocation, and equitable access to medical interventions.

Historical outbreaks like smallpox, SARS, and various influenza pandemics provide valuable lessons for future pandemic preparedness. The eradication of smallpox demonstrates the effectiveness of global vaccination campaigns, surveillance, and containment strategies. The SARS outbreak emphasized the need for global surveillance systems and rapid response mechanisms, informing strategies for managing subsequent pandemics, including COVID-19. The continuous evolution of influenza virus underscores the challenges in vaccine development and the necessity for ongoing vigilance in surveillance and public health interventions.

Advanced analytical and chemometric approaches have become indispensable in managing major PHEs. Precise pathogen identification and comprehensive pathogenomic analyses are essential for effective epidemic control. We have significantly contributed to this field by developing innovative chemical measurement technologies, such as dual-labeled immunochromatographic assays for SARS antigen detection, single-base resolution nucleic acid detection tools for COVID-19 variants, and studies on nucleic acid structural evolution in monkeypox virus. These technologies enhance the sensitivity, specificity, and speed of pathogen detection, facilitating the rapid and accurate diagnosis critical for controlling outbreaks.

The strategic importance of these chemical measurement innovations extends beyond immediate epidemic response. They hold substantial potential for applications in vaccine development against emerging pathogenic variants, detection of antibiotic-resistant microorganisms, and single-base drug target screening. Establishing robust research teams and investing in rapid-response capabilities are essential steps toward a resilient public health framework. Integrating these technological advancements with infectious disease prevention strategies will strengthen global health security and preparedness.

In conclusion, addressing the multifaceted challenges of global health requires a comprehensive and integrated approach that leverages advanced analytical technologies and fosters international collaboration. The lessons learned from past and present PHEs emphasize the critical need for timely cooperation, robust surveillance systems, and equitable resource distribution. Continued innovation in analytical and chemometric methods will play a pivotal role in enhancing our ability to detect, respond to, and prevent future health crises, ultimately contributing to a more resilient global health infrastructure.

References

1 Holroyd, T.A., Oloko, O.K., Salmon, D.A. et al. (2020). Communicating recommendations in public health emergencies: the role of public health authorities. *Health Security* 18 (1): 21–28.

2 Nelson, C., Lurie, N., Wasserman, J., and Zakowski, S. (2007). Conceptualizing and defining public health emergency preparedness. *American Journal of Public Health* 97 (Supplement_1): S9–S11.

3 Gostin, L.O., Friedman, E.A., and Wetter, S. (2020). A responding to Covid-19: how to navigate a public health emergency legally and ethically. *Hastings Center Report* 50 (2): 8–12.
4 Zhao, S., Cao, J., Shi, Q. et al. (2020). A quality evaluation of guidelines on five different viruses causing public health emergencies of international concern. *Annals of Translational Medicine* 8 (7): 500–513.
5 Wilder-Smith, A. and Osman, S. (2020). Public health emergencies of international concern: a historic overview. *Journal of Travel Medicine* 27 (8): taaa227.
6 Laaser, U. and Epstein, L. (2010). Threats to global health and opportunities for change: a New Global Health. *Public Health Reviews* 32 (1): 54–89.
7 Gushulak, B. and Weekers, J. (2009). Macpherson D migrants and emerging public health issues in a globalized world: threats, risks and challenges, an evidence-based framework. *Emerging Health Threats Journal* 2: e10.
8 Kirwan, D. (2009). Global health: current issues, future trends and foreign policy. *Clinical Medicine* 9 (3): 247–253.
9 Lautier, M. (2014). International trade of health services: global trends and local impact. *Health Policy* 118 (1): 105–113.
10 Fidler, D.P. (2003). Emerging trends in International Law Concerning Global Infectious Disease Control. *Emerging Infectious Disease Journal* 9 (3): 285.
11 Murray, C.J.L., Anderson, B., Burstein, R. et al. (2011). Development assistance for health: trends and prospects. *The Lancet* 378 (9785): 8–10.
12 Wilson, K., Brownstein, J.S., and Fidler, D.P. (2010). Strengthening the International Health Regulations: lessons from the H1N1 pandemic. *Health Policy and Planning* 25 (6): 505–509.
13 Ly, A. (2016). In global health, cancer challenges and control in African settings. *Cancer Therapy and Oncology International Journal* 2 (3): 555587.
14 Waldman, A.J. and Balskus, E.P. (2018). The human microbiota, infectious disease, and global health: challenges and opportunities. *ACS Infectious Diseases* 4 (1): 14–26.
15 Varmus, H., Klausner, R., Zerhouni, E. et al. (2003). Grand challenges in global health. *Science* 302 (5644): 398–399.
16 Salam, M.A., Al-Amin, M.Y., Salam, M.T. et al. (2023). Antimicrobial resistance: a growing serious threat for global public health. *Healthcare* 11 (13): 1946.
17 Holst, J. (2020). Global health – emergence, hegemonic trends and biomedical reductionism. *Globalization and Health* 16 (1): 42.
18 Liu, Y., Lee, J.M., and Lee, C. (2020). The challenges and opportunities of a global health crisis: the management and business implications of COVID-19 from an Asian perspective. *Asian Business & Management* 19 (3): 277–297.
19 Hsu, C.-H., Schambow, R., Montenegro, M. et al. (2023). Factors affecting the spread diagnosis, and control of African Swine Fever in the Philippines. *Pathogens* 12 (8): 1068.
20 Lakshmi Priyadarsini, S. and Suresh, M. (2020). Factors influencing the epidemiological characteristics of pandemic COVID 19: a TISM approach. *International Journal of Healthcare Management* 13 (2): 89–98.

21 Luz, S. and Masoodian, M. (2022). Exploring environmental and geographical factors influencing the spread of infectious diseases with interactive Maps. *Sustainability* 14 (16): 9990.

22 Asif, Z., Chen, Z., Stranges, S. et al. (2022). Dynamics of SARS-CoV-2 spreading under the influence of environmental factors and strategies to tackle the pandemic: a systematic review. *Sustainable Cities and Society* 81: 103840.

23 Dupraz, J. and Burnand, B. (2021). Role of health professionals regarding the impact of climate change on health—an exploratory review. *International Journal of Environmental Research and Public Health* 18 (6): 3222.

24 Myers, S.S. and Patz, J.A. (2009). Emerging threats to human health from global environmental change. *Annual Review of Environment and Resources* 34 (34): 223–252.

25 Papworth, A., Maslin, M., and Randalls, S. (2015). Is climate change the greatest threat to global health? *The Geographical Journal* 181 (4): 413–422.

26 Ebi, K.L. and Hess, J.J. (2020). Health risks due to climate change: inequity in causes and consequences. *Health Affairs* 39 (12): 2056–2062.

27 Yin, J.D.-C. and Lui, J.N.-M. (2024). Factors influencing risk perception during Public Health Emergencies of International Concern (PHEIC): a scoping review. *BMC Public Health* 24 (1): 1372.

28 Budreviciute, A., Damiati, S., Sabir, D.K. et al. (2020). Management and Prevention Strategies for Non-communicable Diseases (NCDs) and Their Risk Factors. *Frontiers in Public Health* 8: 574111.

29 Dorélien, A.M., Ramen, A., Swanson, I., and Hill, R. (2021). Analyzing the demographic, spatial, and temporal factors influencing social contact patterns in U.S. and implications for infectious disease spread. *BMC Infectious Diseases* 21 (1): 1009.

30 Schell, D. and Wang, M. (2020). Huynh T L D This time is indeed different: A study on global market reactions to public health crisis. *Journal of Behavioral and Experimental Finance* 27: 100349.

31 Jee, Y. (2020). WHO International Health Regulations Emergency Committee for the COVID-19 outbreak. *Epidemiology and Health* 42: e2020013.

32 Wenham, C. and Stout, L. (2024). A legal mapping of 48 WHO member states' inclusion of public health emergency of international concern, pandemic, and health emergency terminology within national emergency legislation in responding to health emergencies. *The Lancet* 403 (10435): 1504–1512.

33 Wenham, C., Kavanagh, M., Phelan, A. et al. (2021). Problems with traffic light approaches to public health emergencies of international concern. *The Lancet* 397 (10287): 1856–1858.

34 Durrheim, D.N., Gostin, L.O., and Moodley, K. (2020). When does a major outbreak become a Public Health Emergency of International Concern? *The Lancet Infectious Diseases* 20 (8): 887–889.

35 Eccleston-Turner, M. and Wenham, C. (2021). Case Studies on the PHEIC Declaration. In: *Declaring a Public Health Emergency of International Concern: Between International Law and Politics* (ed. M. Eccleston-Turner and C. Wenham), 73–127. Bristol University Press.

36 Hollmann, L. (2023). Declaring a public health emergency of international concern: between international law and politics. *International Affairs* 99 (1): 386–387.
37 Wilder-Smith, A. and Osman, S. (2020). Public health emergencies of international concern. *BMJ Global Health* 5 (7): e002273.
38 Gostin, L.O. and Friedman, E.A. (2014). Ebola: a crisis in global health leadership. *The Lancet* 384 (9951): 1323–1325.
39 Mohammed, A.S., Khwaja, M.I.S., and Khushhal, K.Z. (2015). Public Health Emergency of International Concern (PHEIC) has Declared Twice in 2014; Polio and Ebola at the Top. *AIMS Public Health* 2 (2): 218–222.
40 Giesecke, J., Bedford, J., Enria, D. et al. (2019). The truth about PHEICs. *The Lancet.*
41 Petersen Lyle, R., Jamieson Denise, J., Powers Ann, M., and Honein Margaret, A. Zika virus. *New England Journal of Medicine* 374 (16): 1552–1563.
42 Chen, S., Zhang, Z., Yang, J. et al. (2020). Fangcang shelter hospitals: a novel concept for responding to public health emergencies. *The Lancet* 395 (10232): 1305–1314.
43 Triggle, C.R., Bansal, D., Ding, H. et al. (2021). A comprehensive review of viral characteristics, transmission, pathophysiology, immune response, and management of SARS-CoV-2 and COVID-19 as a basis for controlling the pandemic. *Frontiers in Immunology* 12.
44 Guo, G., Ye, L., Pan, K. et al. (2020). New insights of emerging SARS-CoV-2: epidemiology, etiology, clinical features, clinical treatment, and prevention. *Frontiers in Cell and Developmental Biology* 8.
45 Wenham, C. and Eccleston-Turner, M. (2022). Monkeypox as a PHEIC: implications for global health governance. *The Lancet* 400 (10369): 2169–2171.
46 Mitjà, O., Ogoina, D., Titanji, B.K. et al. (2023). Monkeypox. *The Lancet* 401 (10370): 60–74.
47 Rajapaksha, R.M.N.U., Khatri, R.B., Abeysena, C. et al. (2022). Success and challenges of health systems resilience-enhancing strategies for managing Public Health Emergencies of International Concerns (PHEIC): a systematic review protocol. *BMJ Open* 12 (11): e067829.
48 Zarocostas, J. (2023). With the COVID-19 PHEIC over, what next? *The Lancet* 401 (10389): 1642–1643.
49 Chanda, E. (2020). Global Health Security. In: *Handbook of Global Health* (ed. R. Haring, I. Kickbusch, D. Ganten, and M. Moeti), 2315–2334. Springer International Publishing.
50 Heymann, D.L. and Wilder-Smith, A. (2020). Successful smallpox eradication: what can we learn to control COVID-19? *Journal of Travel Medicine* 27 (4): taaa090.
51 Maldonado, Y.A. (2023). Lessons from a house on fire—from smallpox to polio. *The Journal of Infectious Diseases* 227 (9): 1025–1027.
52 Breman, J.G. (2021). Smallpox eradication: African origin, African solutions, and relevance for COVID-19†. *The American Journal of Tropical Medicine and Hygiene* 104 (2): 416–421.

53 Javed, H., Rizvi, M.A., Fahim, Z. et al. (2023). Global polio eradication; can we replicate the smallpox success story? *Reviews in Medical Virology* 33 (3): e2409.

54 Gebre, T. (2021). Rethinking disease eradication: putting countries first. *International Health* 13 (3): 215–221.

55 Assefa, Y., Gilks, C.F., van de Pas, R. et al. (2021). Reimagining global health systems for the 21st century: lessons from the COVID-19 pandemic. *BMJ Global Health* 6 (4): e004882.

56 Ganesh, B., Rajakumar, T., Malathi, M. et al. (2021). Epidemiology and pathobiology of SARS-CoV-2 (COVID-19) in comparison with SARS, MERS: an updated overview of current knowledge and future perspectives. *Clinical Epidemiology and Global Health* 10: 100694.

57 Harrington, W.N., Kackos, C.M., and Webby, R.J. (2021). The evolution and future of influenza pandemic preparedness. *Experimental & Molecular Medicine* 53 (5): 737–749.

58 Monto, A.S. and Fukuda, K. (2020). Lessons from influenza pandemics of the last 100 years. *Clinical Infectious Diseases* 70 (5): 951–957.

59 Wille, M. and Holmes, E.C. (2020). The ecology and evolution of influenza viruses. *Cold Spring Harbor Perspectives in Medicine* 10 (7): a038489.

60 Petrova, V.N. and Russell, C.A. (2018). *The evolution of seasonal influenza viruses. Nature Reviews Microbiology* 16: 47–60.

61 Piret, J. (2022). Boivin G Pandemics throughout history. *Frontiers in Microbiology* 13.

62 Dhanasekaran, V., Sullivan, S., Edwards, K.M. et al. (2022). Human seasonal influenza under COVID-19 and the potential consequences of influenza lineage elimination. *Nature Communications* 13 (1): 1721.

63 Worsnop, C.Z., Nass, S., Grépin, K.A., and Lee, K. (2023). An analysis of WHO's Temporary Recommendations on international travel and trade measures during Public Health Emergencies of International Concern. *BMJ Global Health* 8 (7): e012615.

64 Villarreal, P.A. (2017). The World Health Organization's Governance Framework in Disease Outbreaks: A Legal Perspective. In: *The Governance of Disease Outbreaks: International Health Law: Lessons from the Ebola Crisis and Beyond*, 1e (ed. L. Vierck, P.A.P.D. Villarreal, and A.K.L.L.M. Weilert), 243–272. Baden-Baden: Nomos Verlagsgesellschaft mbH & Co. KG.

65 Quesada-González, D. (2015). Merkoçi A Nanoparticle-based lateral flow biosensors. *Biosensors and Bioelectronics* 73: 47–63.

66 Bahadır, E.B. and Sezgintürk, M.K. (2016). Lateral flow assays: principles, designs and labels. *TrAC Trends in Analytical Chemistry* 82: 286–306.

67 Wang, M.J., Ji, X.H., Wang, L.Y. et al. (2003). The assembly, hybridization and detection of DNA on gold nanoparticles with silver enhancement. *Acta Physico-Chimica Sinica* 19 (9): 879–882.

68 Li, Y., Ma, B., Fan, Y. et al. (2002). Electrochemical and Raman Studies of the Biointeraction between Escherichia coli and Mannose in Polydiacetylene Derivative Supported on the Self-Assembled Monolayers of Octadecanethiol on a Gold Electrode. *Analytical Chemistry* 74 (24): 6349–6354.

69 Wang, M., Wang, L., Wang, G. et al. (2004). Application of impedance spectroscopy for monitoring colloid Au-enhanced antibody immobilization and antibody–antigen reactions. *Biosensors and Bioelectronics* 19 (6): 575–582.

70 Hong, X., Li, J., Wang, M. et al. (2004). Fabrication of Magnetic Luminescent Nanocomposites by a Layer-by-Layer Self-assembly Approach. *Chemistry of Materials* 16 (21): 4022–4027.

71 Song, F., Shen, Y., Wei, Y. et al. (2021). Botulinum toxin as an ultrasensitive reporter for bacterial and SARS-CoV-2 nucleic acid diagnostics. *Biosensors and Bioelectronics* 176: 112953.

72 Zhang, T., Deng, R., Wang, Y. et al. (2022). A paper-based assay for the colorimetric detection of SARS-CoV-2 variants at single-nucleotide resolution. *Nature Biomedical Engineering* 6 (8): 957–967.

73 Gong, N., Zhang, Y., Teng, X. et al. (2020). Proton-driven transformable nanovaccine for cancer immunotherapy. *Nature Nanotechnology* 15 (12): 1053–1064.

74 Gong, N., Ma, X., Ye, X. et al. (2019). Carbon-dot-supported atomically dispersed gold as a mitochondrial oxidative stress amplifier for cancer treatment. *Nature Nanotechnology* 14 (4): 379–387.

75 Zhang, K., Qin, S., Wu, S. et al. (2020). Microfluidic systems for rapid antibiotic susceptibility tests (ASTs) at the single-cell level. *Chemical Science* 11 (25): 6352–6361.

76 Huang, J., Zhong, Y., Li, W. et al. (2021). Fluorescent and opt-electric recording bacterial identification device for ultrasensitive and specific detection of microbials. *ACS Sensors* 6 (2): 443–449.

77 Wei, Y., Yang, Z., Zong, C. et al. (2021). Trans single-stranded DNA cleavage via CRISPR/Cas14a1 activated by target RNA without destruction. *Angewandte Chemie International Edition* 60 (45): 24241–24247.

78 McCollum, A.M. and Damon, K. I, (2014). Human Monkeypox. *Clinical Infectious Diseases* 58 (2): 260–267.

79 Harris, E. (2022). What to know about Monkeypox. *JAMA* 327 (23): 2278–2279.

80 Varshney, D., Spiegel, J., Zyner, K. et al. (2020). The regulation and functions of DNA and RNA G-quadruplexes. *Nature Reviews Molecular Cell Biology* 21 (8): 459–474.

81 Dumas, L., Herviou, P., Dassi, E., and Cammas, A. (2021). Millevoi S G-quadruplexes in RNA biology: recent advances and future directions. *Trends in Biochemical Sciences* 46 (4): 270–283.

82 Métifiot, M., Amrane, S., and Litvak, S. (2014). Andreola M-L G-quadruplexes in viruses: function and potential therapeutic applications. *Nucleic Acids Research* 42 (20): 12352–12366.

83 Dai, Y., Teng, X., and Li, J. (2022). Single-cell visualization of monogenic RNA G-quadruplex and occupied G-quadruplex ratio through a Module-Assembled Multifunctional Probes Assay (MAMPA). *Angewandte Chemie-International Edition* 61 (7): e202111132.

84 Dai, Y., Teng, X., Hu, D. et al. (2024). Evolution of Monkeypox C9L RNA G-quadruplex elevates the translation of its only Kelch-like protein by accelerating G-quadruplex unfolding kinetics. *CCS Chemistry* 6: 953–963.

2

Mastering Clinical Diagnostics

Modern Strategies for PHEIC Responses

Xinxin Shen and Xuejun Ma

Chinese Center for Disease Control and Prevention, National Key Laboratory of Intelligent Tracking and Forecasting for Infectious Diseases, NHC Key Laboratory of Medical Virology and Viral Diseases, National Institute for Viral Disease Control and Prevention, Changping District, Changbai Road 155, Beijing, 102206, China

CHAPTER MENU

2.1 Clinical Evaluation Comprehensive Pathway: From Cutting-edge Laboratory Analysis

Laboratory services are an essential and fundamental part of public health systems. With the occurrence of infectious disease outbreaks such as the COVID-19 pandemic or other public health events, Centers for disease control and prevention (CDCs) are at the heart of public health investigation and response mechanisms. The pathogen analysis laboratory is the core and primary department of CDCs at different levels in the response to infectious disease outbreaks.

Nucleic acid detection (NAT) technology is widely used for the early detection of infectious disease pathogens in the CDCs, owing to its advantages of high sensitivity and specificity. Thus, NAT is crucial for the control of infectious disease outbreaks. During the COVID-19 pandemic, polymerase chain reaction (PCR) has become the main technique and gold standard for the detection of SARS-CoV-2 RNA worldwide. However, PCR-based nucleic acid detection technology remains relatively time-consuming, requires specialized instrumentation and technical personnel, is difficult to adopt in primary department of CDCs of rural areas, and is not yet able to meet the needs of bedside examinations for simple and rapid testing. Serum-specific IgG and IgM antibody detection, in contrast, has limitations because of the late appearance of antibodies, thus making early diagnosis difficult. Antigen

Diagnostic Techniques in Public Health: Analytical and Chemometric Approaches in Disease Detection and Management, First Edition. Edited by Jinghong Li and Yan Du.

detection, though simple and rapid, has the drawbacks of insufficient sensitivity and a high false-negative rate.

The emerging nucleic acid isothermal amplification technique (IAT) overcomes the shortcomings of PCR technology, through its advantages of simple operation, rapid detection, and accurate results. Therefore, IAT is more suitable for bedside examinations (on-site tests) in primary CDCs at the county and rural levels, which may lack technical expertise and have limited human resources. Currently, IAT still cannot replace PCR in pathogen detection, but IAT technology is expected to move toward the development of digitalization, miniaturization, simplification, field use, and automation.

2.2 Recombinase-aided Isothermal Amplification of Nucleic Acids: A New Dawn in Detection and Analysis

2.2.1 Introduction

Nucleic acids, fundamental components of genetic material, are ubiquitous across all forms of life, including plants, animals, and microorganisms. The advent of techniques for the rapid in vitro amplification and analysis of nucleic acids has revolutionized life sciences, establishing nucleic acid amplification as a cornerstone technology in this field [1]. Since the initial breakthrough in 1983, when in vitro nucleic acid amplification was first achieved, the technology has seen exponential development and diversification [2].

The journey began with the polymerase chain reaction (PCR), a method that quickly became indispensable in molecular biology. Over the years, nucleic acid amplification technologies have evolved, introducing a range of methodologies that cater to different scientific needs. Among these, nucleic acid isothermal amplification technologies have garnered significant attention due to their operational advantages and reduced dependency on sophisticated instrumentation [3, 4].

Isothermal amplification technologies offer several benefits over traditional PCR, particularly in terms of simplicity and practicality [5]. Unlike PCR, which requires precise thermal cycling, isothermal techniques maintain a constant temperature throughout the amplification process. This feature makes them more suitable for point-of-care testing and field diagnostics, where access to advanced laboratory equipment is limited. Common isothermal amplification methods include the following:

A) Loop-mediated isothermal amplification (LAMP)
B) Helicase-dependent amplification (HDA)
C) Single-primer isothermal amplification (SPIA)
D) Recombinase polymerase amplification (RPA)
E) Rolling circle amplification (RCA)
F) Recombinase-aided amplification (RAA)

Among these, RAA stands out as a particularly promising technology, especially in the context of its development and application in China [6]. Nucleic acid amplification technologies have undergone significant evolution since their inception, with isothermal methods like RAA leading the charge in recent advancements. RAA's development underscores the importance of technological independence and innovation in the life sciences. As RAA continues to progress, it promises to play a crucial role in the future of rapid diagnostics and molecular biology research. This section introduces the principle, application, and progress of RAA technology with Chinese independent intellectual property rights.

2.2.2 Principle of the Recombinase-aided Amplification

Recombinase-aided amplification (RAA) is a novel isothermal nucleic acid amplification technology with Chinese intellectual property rights. RAA technology can realize the exponential amplification of target gene fragments in 5–20 minutes under isothermal conditions (generally 37–42 °C) [7]. The principle is to use the recombinant enzyme obtained from bacteria or fungi; under constant temperature, the recombinant enzyme can be tightly bound to the two pairs of specific primers for detecting the target gene, forming a polymer of the enzyme and primer; when the primer searches for sequences that are fully complementary to it on the template DNA, with the help of the single-stranded DNA binding (SSB), it will unwind the template DNA and convert the template DNA into single strands, and then the template DNA will be amplified within 5–20 minutes. When the template DNA is searched for a sequence that is fully complementary to the template DNA, with the help of SSB, the template DNA is unwound into single strands, and a new complementary strand of DNA is formed under the action of DNA polymerase. The end product of the reaction can be detected by conventional nucleic acid detection methods such as agarose gel electrophoresis or fluorescence intensity (Figure 2.1).

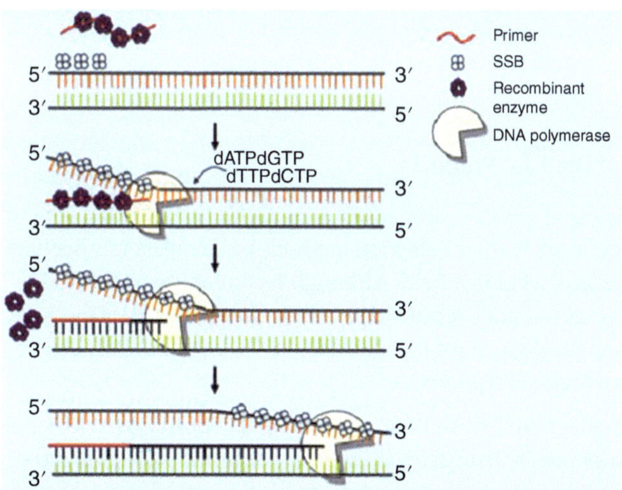

Figure 2.1 Reaction principle of recombinase-aided isothermal nucleic acid amplification technique. *Source:* Figure from Practical Diagnostic Virology, ISBN 978-7-83005-306-2, 2023.9.

The key aspect of RAA technology is to ensure that the recombinase proteins all bind to the DNA. In an in vitro nucleic acid amplification system, the recombinase proteins assist in the unwinding of the DNA template and induce strand substitution between the template and the primer, which is then catalyzed by the DNA polymerase to amplify the new DNA fragments. This substitution process requires ATP for energy, and the DNA polymerase can then specifically extend the target DNA after the strand substitution with the assistance of single-stranded DNA-binding proteins. Because ATP can be regenerated using phosphocreatine catalyzed by creatine kinase, the balance of chemical reactions during this reaction favors product synthesis. This process can be repeated, leading to efficient amplification of nucleic acids. In addition to recombinase, 3–4 enzymes are required for isothermal nucleic acid amplification, so a reasonable ratio between different enzymes is also an important determinant for the completion of isothermal nucleic acid amplification.

Compared with traditional PCR and other nucleic acid isothermal amplification methods (e.g. LAMP method), the advantages of the RAA amplification method include the following:

A) Good specificity, high sensitivity, independent intellectual property rights, and the application of RAA technology in the laboratory are not subject to the restrictions of patents in developed countries.
B) Fast detection time (within 5–20 minutes), low cost of reagents (half of fluorescent PCR), low equipment requirements, requiring only a thermostat or a simple water bath to complete amplification and detection, and it is suitable for rapid detection of a large number of samples and for the promotion of basic health facilities.
C) The RAA method can be used in a variety of ways, either using a portable thermostatic fluorescence detection system or combined with the flow test strip method of naked eye detection, which is very suitable for on-site rapid detection.
D) Simple operation, only need to add nucleic acid templates to launch the reaction. The lyophilized RAA reaction system contains primers and probes, which are stable and easy to transport.

2.2.3 The Detection of the RAA Products

In recent years, with the rapid development of RAA methods in the field of rapid nucleic acid detection, the study of their detection methods has increasingly become the focus of research by scientists in this field. Although recombinase-aided isothermal detection of nucleic acids has not been available a long time, a variety of methods for detecting amplification products have been developed [8, 9], which can be generally classified into three major categories:

A) The basic method of detection: The final product of the reaction can be detected by agarose gel electrophoresis. Because the two pairs of primers amplify different lengths of the target bands on the template strand, the amplification results can be judged based on the presence or absence of the bands in the electrophoresis profile and their size. However, this method requires an open-cap operation, which will produce aerosol contamination and affect the judgment of the results.

B) Real-time fluorescence detection is achieved by a specific fluorescent probe. The fluorescent probe is labeled with a tetrahydrofuran (THF) deletion site in the middle of the probe, flanked by a fluorescent group and a quenching group with a spacing of 1–5 nt, and a modifying group that inhibits polymerase elongation or amplification is labeled at the 3′ end of the fluorescent probe. The tetrahydrofuran site on the fluorescent probe in the double-stranded state is recognized by nucleic acid exonuclease III, which enzymatically cleaves the fluorescent and quenching groups to separate them, thus releasing fluorescence, see Figure 2.2. The fluorescence signal can be read in real-time by a portable

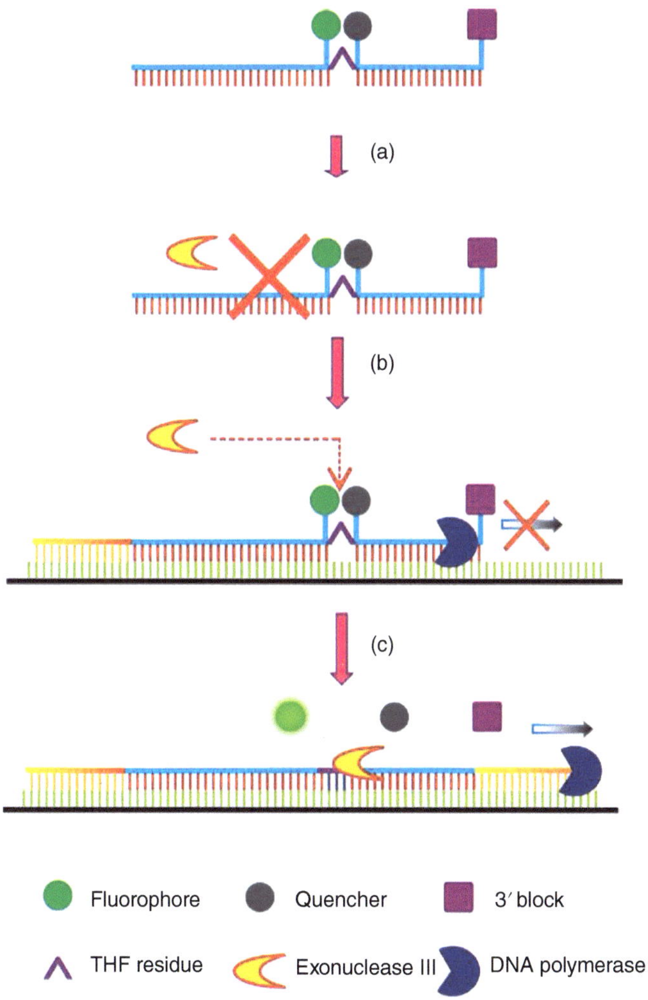

Figure 2.2 Principles of real-time fluorescence RAA technology. (a) When the probe is single-stranded, it is not recognized by the nucleic acid exonuclease. (b) Upon binding the probe to the target DNA, the nucleic acid exonuclease recognizes the nick, excises the tetrahydrofuran abasic site, and unblocks the 3′ end. (c) The fluorescent and quenching groups separate and fluoresce, while the polymerase completes the chain extension. *Source:* Figure from Practical Diagnostic Virology, ISBN 978-7-83005-306-2, 2023.9.

instant fluorescence scanner. Using a portable thermostatic fluorescence detection system, the results are interpreted by detecting the presence or absence and strength of the fluorescence signal.

C) Lateral flow dipstick (LFD): RAA combined with LFD is a method of detecting amplification products by adding endonuclease IV (nfo) and an nfo probe to the RAA base reaction unit. The nfo probe is labeled with an antigenic moiety at the 5′ end (usually a FAM moiety) and a blocker at the 3′ end, with a THF molecule in the probe sequence, and markers such as biotin or digoxin labeled on the reverse primer. The cut probe and the downstream primer form a double-labeled amplicon with both the probe fluorescent group marker and the primer-specific marker. On the lateral flow dipstick test strip, the corresponding antibody is used at the detection line to capture the markers carried by the primer, and the colloidal gold particle-labeled antibody combined with the fluorescent group carried by the amplicon can be observed to detect the product, as shown in Figure 2.3. The isothermal nucleic acid rapid detection technology

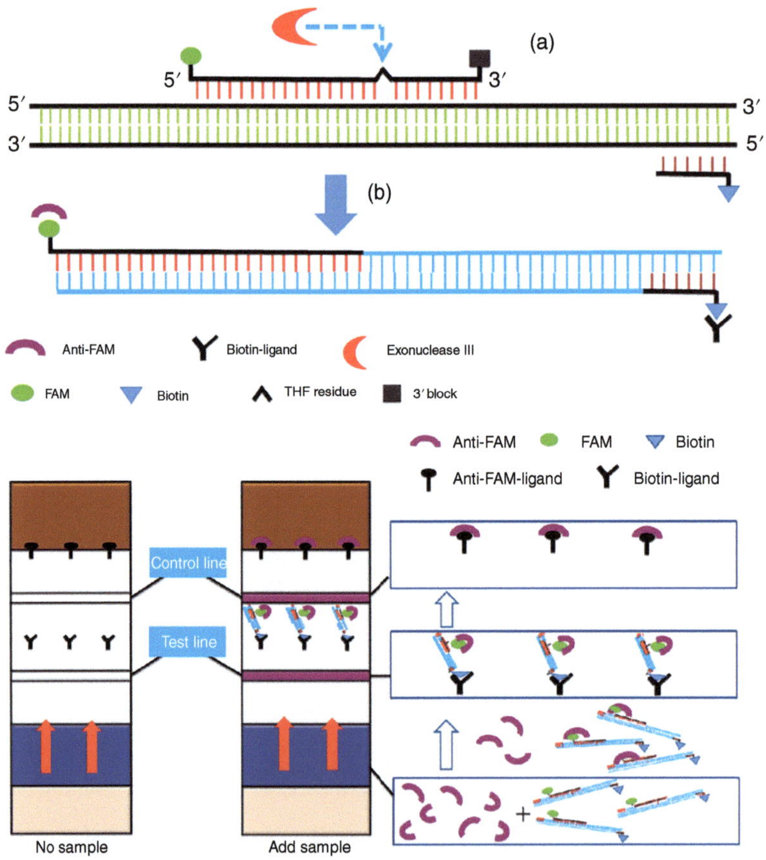

Figure 2.3 Principle of RAA for the lateral flow dipstick test strip method. (a) Endonuclease recognizes the THF site and cleaves it, and the polymerase extends from the cleavage site. (b) Generation of double-labeled amplicons containing FAM and biotin modifications. *Source:* Figure from Practical Diagnostic Virology, ISBN 978-7-83005-306-2, 2023.9.

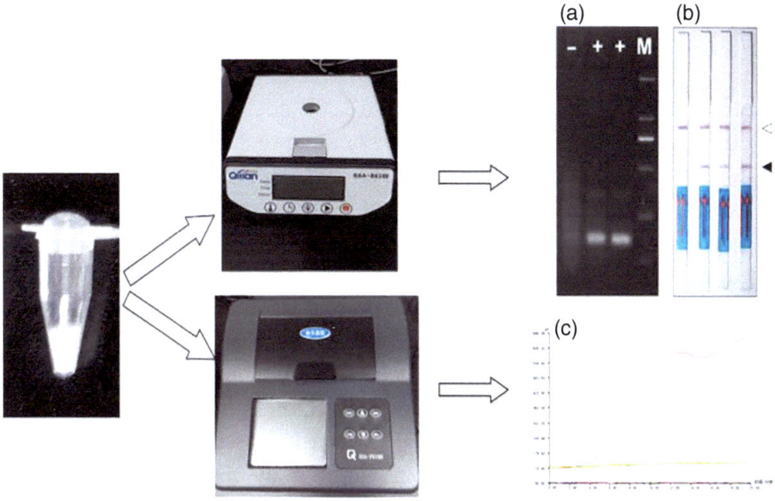

Figure 2.4 Interpretation of results of recombinase-aided amplification (RAA) method. (a) Agarose gel electrophoresis method. (b) Lateral flow dipstick test strip method. (c) Real-time fluorescence RAA method. *Source:* Figure from Practical Diagnostic Virology, ISBN 978-7-83005-306-2, 2023.9

combined with the lateral flow test strip method can be detected by the naked eye, which is very suitable for on-site rapid detection. Figure 2.4 shows the effect of the three amplification product detection methods.

2.2.4 Applications of Recombinase-aided Amplification

RAA detection technology is mainly applied to the detection of pathogens such as bacteria, viruses, and parasites. Among them, virus detection is mainly divided into five categories:

A) Respiratory tract transmitted viruses such as respiratory syncytial virus, respiratory adenovirus, influenza A virus and so on.
B) Blood-borne viruses such as hepatitis B virus, hepatitis C virus, human immunodeficiency virus, etc.
C) Intestinal viruses such as Coxsackie virus, enterovirus 71, etc.
D) Diarrhoeal viruses such as rotavirus and norovirus.
E) Mosquito-borne arboviruses such as yellow fever virus, West Nile virus, dengue virus, etc.

Bacterial detection is mainly for Mycobacterium tuberculosis, non-tuberculosis mycobacteria, *Mycobacterium pertussis*, and food-borne salmonella. Parasite testing is mainly for *Schistosoma japonicum*.

RAA detection technology is fast, easy to operate, inexpensive, and portable and has a good application prospect in the field of pathogen detection. Zhao et al. [10] established an isothermal nucleic acid detection method for *S. japonicum* recombinant plasmid with a sensitivity of 10 copies/μL and no cross-reactions were observed with *S. mansoni, S. japonicum, Ancylostoma duodenale, Clonorchis sinensis, Echinococcus granulosus, or Ascaris lumbricoides.* This test can be conducted at 39 °C, and the whole RAA reaction can be completed within 20 minutes. The validation of the RAA assay showed that it had 100% consistency with urine-egg microscopy, as it does not require an elaborate reading tool, is simple to use, and should be useful for field diagnostics and point-of-care applications. Balea et al. [11] developed a sensitive and specific ZIKV diagnostic using reverse transcription recombinase-aided amplification (RT-RAA) coupled with lateral flow detection (LFD) targeting a highly conserved region of the ZIKV NS1 gene. Their rapid, isothermal-ZIKV-diagnostic (Iso-ZIKV-Dx) can detect 500 copies of synthetic ZIKV RNA/μL in less than 30 minutes at a constant 39 °C. Using simulated urine samples, Iso-ZIKV-Dx also detects as low as 34.28 RNA copies/reaction of ZIKV (MR766 strain). Specificity testing confirmed that their test does not detect any cocirculating flaviviruses or chikungunya virus. Furthermore, lateral flow strips testing demonstrates positive diagnoses in as little as five minutes in a running buffer. Cui et al. [12] developed a real-time reverse transcription recombinase-aided amplification assay of influenza A viruses (IAVs). The assay can be completed within 20 minutes at 42 °C. The sensitivity of the real-time reverse transcription recombinase-aided amplification (RT-RAA) assay was 142 copies/reaction at 95% probability, which was comparable to the sensitivity of the RT-qPCR assay. The specificity assay showed that the real-time RT-RAA assay was specific to IAVs, and there was no cross-reactivity with other important viruses. In addition, 100%concordance between the real-time RT-RAA and RT-qPCR assays was achieved after testing 120 clinical specimens. Liu et al. [13] established a real-time RAA assay targeting conserved positions in the IS1081 gene of MTB, which is successfully established to detect MTB. The intact workflow was completed within 30 minutes at 42 °C with no cross-reactivity observed for non-tuberculous mycobacteria and other clinical bacteria, and the detection limit for recombinant plasmid of MTB IS1081 was 163 copies/reaction at 95% probability, which was an approximately 1.5-fold increase in analytical sensitivity for the detection of MTB, compared to conventional quantitative real-time PCR (qPCR; 244 copies/reaction). Furthermore, the result of clinical performance evaluation revealed an increased sensitivity of the RAA assay relative to qPCR was majorly noted in the specimens with low bacteria loads. Zhang et al. [14] present a RAA-mediated Salmonella detection, which achieved a rapid detection of food-borne Salmonella in 20 minutes. Salmonella was specifically detected in both the basic and the real-time fluorescence RAA detection systems. The basic detection system could detect down to 105 copy/μL plasmid DNA while the real-time fluorescence RAA detection system was more sensitive, recording specific detection at 103copy/μL plasmid DNA. This offers great potential in food-borne Salmonella control and its clinical diagnosis.

Xuejun Ma's team at the Central Laboratory of the Institute for the Prevention and Control of Viral Diseases, Chinese Center for Disease Control and Prevention (CDC), has been working on the application of isothermal rapid nucleic acid assays in recent years, which has led to a further expansion of the scope of the application of RAA. Chen et al. [7] established a rapid assay for the detection of two subtypes of respiratory syncytial viruses (RSV-A and RSV-B), with a limit of detection of up to 38 and 35 copies/reaction, respectively, and no cross reactions with other related respiratory viruses were observed. The RT-RAA assay was further utilized to detect and subgroup 306 clinical specimens and the results showed that 79 (25.82%, 79/306) samples were positive for RSV, of those 16 (20.25%, 16/79) were identified as RSVA and 63 (79.75%, 63/79) were RSVB, which is completely consistent with the results obtained by respiratory syncytial virus (RSV) RT-qPCR assay; Yan et al. [15] developed rapid methods for the detection of Coxsackievirus 10 (CA10) and Coxsackievirus 6 (CA6) with detection limits of 35 and 38 copies/reaction, respectively, with 100% specificity. Compared with commercial RT-qPCR assays, when testing 455 fecal specimens, the kappa value of the RT-RAA assay for CVA10 and CVA6 was 0.920 ($p < 0.001$) and 0.952 ($p < 0.001$), respectively; Shen et al. [16] developed a rapid method for the detection of hepatitis B Virus (HBV), which for the first time enabled amplification of nucleic acids by nonextractable nucleic acids. A total of 130 archived suspected HBV-infected serum samples were detected by commercial qPCR with DNA extraction and RAA assay without DNA extraction (heat treatment). Compared with the qPCR assay as a reference, the RAA assay obtained 95.7% sensitivity and 100% specificity and a kappa value of 0.818; Qi et al. [17] developed a rapid method for the detection of RSV, which includes dual real-time fluorescent reverse transcriptase-mediated isothermal nucleic acid amplification with internal reference for the first time, as well as a rapid method combined with side stream chromatography strips. The sensitivity of this assay for RSV was determined as 4.4 copies/reaction, and the specificity was 100%. Next, a duplex-rtRAA assay with an internal control was established. The sensitivity of the duplex-rtRAA assay approached 5.0 copies/reaction, and no cross-reaction with other common respiratory viruses was observed. The two detection methods (singleplex-rtRAA and duplex-rtRAA) developed in this study were used to test 278 clinical specimens, and the results showed absolute consistency with RSV RT-qPCR analysis, demonstrating 100% diagnostic sensitivity and specificity; Wang et al. [18] developed a dual rapid method for the detection of adenovirus 3 (ADV-3) and adenovirus 7 (ADV-7) by incorporation of internal reference. The analytical sensitivities of the duplex RAA assays for HAdV 3 and HAdV 7 were 5.0 and 14.8 copies/reaction, respectively (at 95% probability by probit regression analysis). No cross-reaction was observed with other types of HAdV or other common respiratory viruses. The duplex RAA assays were used to detect 152 previously defined HAdV-positive samples. These results agreed with those obtained using a published triplex quantitative real-time PCR protocol; Zhang et al. [19] established a dual rapid detection method for *Bordetella pertussis*, and the detection limit can reach 10 CFU/mL with no extraction of nucleic acid. The sensitivity of the internally controlled RAA assay was 10 copies or 10 CFU/mL per reaction in detecting plasmid DNA or *B. pertussis* strain. The optimum concentration of the IAC plasmid was determined to be 100 copies, and the introduction of IAC effectively reduced the occurrence of false negatives. Compared to the RT-PCR, RAA results with DNA extraction obtained 100% sensitivity and specificity, and the RAA results with heat-treated DNA showed 85.96% sensitivity and 100% specificity. RAA technology has also been researched in the detection of single-nucleotide polymorphism (SNP) and mutant strains of new coronaviruses, etc. For example, Duan et al. [20] established a rapid detection method for the 5,10-methylenetetrahydrofolate reductase (MTHFR) A1298C polymorphism associated with congenital heart disease; in this proof-of-concept study, a probe-directed recombinase amplification (PDRA) assay was developed to detect the A1298C polymorphism of MTHFR. The PDRA assay included two real-time reactions to detect the A and C nucleotides of A1298C polymorphism. Each reaction contained only one primer and one probe and was finished at 39 °C within 35 minutes. The results of the genotyping of 150 clinical samples using PDRA were completely consistent with those by direct sequencing. Additionally, when the 1000 Genomes Project HCB frequencies were used as the control group, MTHFR A1298C was found to be associated with congenital heart disease. Chen et al. [21] established a rapid detection method for the new coronavirus mutant strain D614G; with good reproducibility and specificity, the PDRA assay worked well with a concentration in the range of 10^3–10^7 copies/reaction. Compared with direct sequencing as a reference, the RAA assay obtained 100% sensitivity and 100% specificity using clinical samples. A particularly important contribution is that the fluorescent RAA technology rapid detection of SARS-CoV-2 nucleic acid kit and supporting instruments developed by Xuejun Ma's team were awarded the registration certificate of Class III of medical devices (kit and supporting instruments) by the State Drug Administration of China (SDA) and the CE certification of the European Union in September 2021, which achieves the breakthrough of 0 to 1 in the industrialization of RAA technology. The detection sensitivity of the kit is 384 copies/mL, and it can complete the whole process of 16 clinical samples in 43 minutes, with a total compliance rate of more than 98% with the quantitative PCR results from three clinical institutions.

2.2.5 Advantages and Limitations of Recombinase-aided Amplification

RAA has demonstrated significant advantages in the field of nucleic acid detection, including the following aspects:

A) High sensitivity: RAA technology can detect nucleic acid molecules at very low concentrations. This gives it great potential for use in the early detection of pathogens and the analysis of low-abundance samples. For example, in the early detection of infectious diseases, RAA technology can achieve positive detection before the number of pathogens reaches the lower limit of traditional detection methods, thus improving the timeliness and effectiveness of disease prevention and control.
B) High specificity: RAA technology relies on the high specificity of primers and probes to identify target sequences, which can effectively differentiate between similar sequences and reduce false-positive results. This is particularly important in the context of complex samples, such as environmental and clinical samples containing multiple microorganisms, where RAA can still accurately identify and detect the target microorganism.

C) Rapid response: The reaction time of RAA technology is usually less than 30 minutes, which is much shorter than traditional PCR technology. This is important for clinical emergencies and on-site testing applications, such as rapid screening during outbreaks and immediate diagnosis of acute infectious diseases.

Despite the above advantages of RAA technology, it has some limitations that need to be overcome:

A) Increased difficulty in primer and probe design: The primer and probe lengths required for RAA technology are usually longer than those for conventional PCR reactions. This adds to the complexity of designing suitable primers and probes, especially when the target nucleic acid sequence is conserved or highly variable.
B) Insufficient quantitative analysis capability: Currently, RAA technology is mainly used for qualitative testing and is not yet capable of accurate quantitative analysis. This limits its use in applications that require accurate measurement of target nucleic acid concentrations, such as in drug efficacy monitoring and disease progression assessment.
C) Low throughput for multiple detection: RAA technology has certain technical difficulties when performing multiple detection (i.e., simultaneous detection of multiple targets), especially when the number of targets is more than two, resulting in low detection throughput, which makes it difficult to meet the demand for high-throughput detection. This poses a challenge for application scenarios that require simultaneous detection of multiple pathogens or multiple gene mutations.

Overall, RAA technology has notable advantages in terms of sensitivity, specificity and reaction speed, offering an effective solution for nucleic acid detection. However, regarding primer and probe design, quantitative analysis, and multiplexed detection throughput, further technical optimization and improvement are necessary to enhance its extensive applicability and practicality.

2.2.6 Recent Advances of Recombinase-aided Amplification

In response to the limitations of traditional recombinase-mediated isothermal amplification of nucleic acids, Xuejun Ma's team developed the RAP technology and the fluorescent dye-based RAA technology, which effectively addressed the technical challenges associated with the design of the RAA long probes and the low throughput.

A) Recombinase-aided PCR assay research, development, and applications.

Fan et al. [22] established a novel N-PCR approach named Recombinase-aided PCR (RAP) by integrating the advantages of RAA and quantitative PCR (qPCR) detection techniques. RAP is a robust detection method that is more rapid and sensitive than conventional qPCR. The principle of RAP involves a first-round amplification step using one pair of RAA primers (outer primers), followed by a second round of amplification using one pair of qPCR primers (inner primers) and a qPCR probe. The entire process is performed in a single tube. The RAA

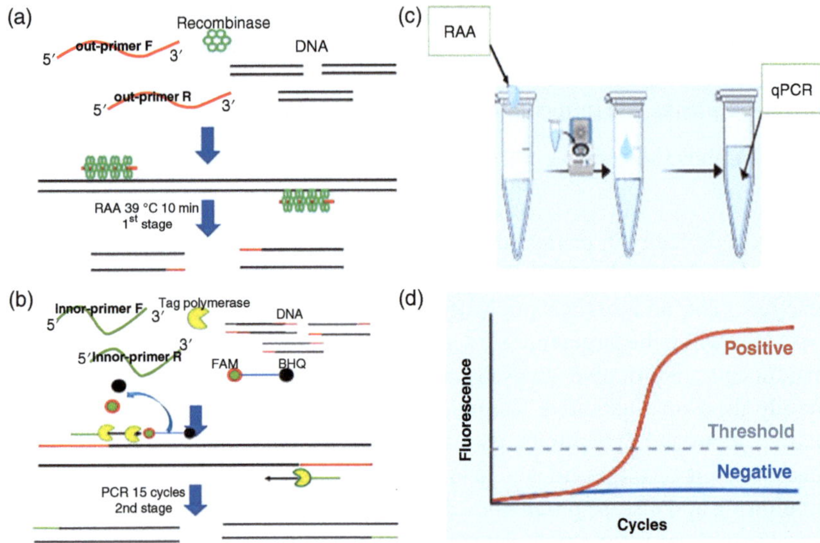

Figure 2.5 The working principle of recombinase-aided PCR (RAP). (a) RAA for 10 minutes at 39 °C (first stage of RAP). (b) qPCR 15 cycles (second stage of RAP). (c). The reaction tube comprises two compartments. The tube lid (inner surface) compartment contains the RAA mix and the tube bottom compartment houses the qPCR mix. (d) The results can be exhibited by real-time fluorescence curves with qPCR machine. *Source:* Ref. [22] / Frontiers / CC BY 4.0.

products are placed in the lid of the tube and briefly centrifuged to mix with the qPCR buffer at the bottom of the tube. The qPCR occurs at the bottom of the tube after heat activation of a hot-start Taq DNA polymerase. At this point, the RAA components are inactivated as shown in Figure 2.5.

RAP capitalizes on the rapid and high amplification efficiency of RAA and the flexibility of the TaqMan probe in qPCR to detect pathogens, thereby avoiding the need for designing a complex RAA probe. In assays for respiratory DNA viruses (HADV3 and HADV7), RAP demonstrated detection limits of 6 and 17 copies/reaction for HADV3 and HADV7, respectively (probit analysis, $p \leq 0.05$), which are lower than those of the corresponding qPCR assays. The clinical application of RAP in testing 200 samples yielded results consistent with those obtained from qPCR. Notably, no cross-reaction was observed with other common respiratory virus-positive samples. The testing results using serially diluted clinical samples also indicated that RAP possesses higher sensitivity.

The rapidity of RAP (approximately one hour and single-tube assay) reduces the risk of contamination and saves running time compared to qPCR. RAP does not depend on an isothermal detection device, as it can be performed on a conventional real-time thermal cycler, indicating its applicability in general laboratories. Preliminary studies explored overlapping qPCR primers with RAA primers to simplify the design of RAP primers, and no adverse impact on RAP amplification efficiency was observed.

However, RAP has some limitations. Although the RAP analysis result is shown by the CT value of qPCR, the CT value cannot quantify the virus titer in the original samples. Due to the extremely high amplification efficiency of RAA, most respiratory clinical samples generate no or very low CT values in the second stage of the qPCR, suggesting that the amplification for most samples reaches the plateau phase after RAA. Further improvements to RAP will aim to: (i) use a Taq enzyme that can function in the RAA reaction mix containing a high Mg^{2+} concentration to allow the use of only one mix in the tube; (ii) develop multiplex detection of pathogens; and (iii) reduce the RAA reaction time and increase qPCR cycles for quantitative detection of pathogens.

Moreover, the RAP method required a higher level of operational maintenance than the RAA method, as the RAA occurred on the inner surface of the tube cap. This necessitated the maintenance of the circular liquid drop on the surface to ensure the efficiency of RAA. Furthermore, the preadded magnesium could also impact the efficacy of RAA. To address these limitations of the RAP method, Tian et al. [23] optimized the reverse transcription RAP (RT-RAP) using a single-tube temperature-adjustable block, enabling one-pot and two-stage amplification for the detection of HIV-1.

Docosane exists as an insoluble solid at room temperature. However, when the temperature is above 44 °C, it melts, forming a molten docosane wax that can be used for sealing the qPCR mix. Through multiple comparisons, they explored the optimal amount of docosane (ranging from 10 to 45 μL) for separating the first-stage and second-stage reactions of the RAP assay.
The RT-RAP assay was performed in two stages within a single tube. The tube was divided into two parts using a heat-removable docosane barrier. Briefly, the docosane was molten at 60 °C, and 25 μL of liquid docosane was added into the qPCR tube containing 40 μL of the qPCR mix. The docosane was then solidified to form a block sealing on the sample. Thereafter, 8 μL of RT-RAA mix and 2 μL of the templates were added to the block surface, and 1 μL 140 mM magnesium acetate was suspended on the inner surface of the tube lid before covering the tubes. The mixture was short-centrifuged to initiate the RT-RAA, and the RT-RAA products were then used as templates for the subsequent qPCR. The scheme of the one-pot two-stage RT-RAP is shown in Figure 2.6.

The turnaround time of the assay is only 50 minutes and can be performed with commonly available laboratory equipment, the qPCR devices. The RT-RAP assay could detect approximately 5 and 14 copies/reaction of HIV-1 DNA and RNA using recombinant plasmids and standard reference strains, respectively. Additionally, they found that the clinical performance of RT-RAP (detected 169 samples out of 170 specimens) was consistent with that of qRT-PCR. The sensitivity and specificity of RT-RAP were 100.00% (99/99) and 98.59% (70/71), respectively, while its positive and negative predictive values were 99.00% (99/100) and 100.00% (70/70), respectively. The total coincidence rate of the RT-RAP was 99.41% (169/170), with a kappa value of 0.988 ($p < 0.05$). In summary, TIAN FENGYU et al. developed a one-pot and two-stage RT-RAP assay for detecting

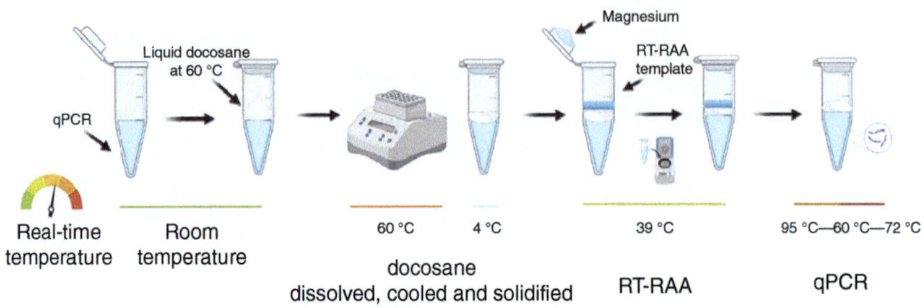

Figure 2.6 An overview of the one-pot two-stage RT-RAP. A 40 μL qPCR mix was added in a qPCR tube, followed by 25 μL of liquid docosane for wax sealing upon cooling. Magnesium, template, and RT-RAA mix were briefly-centrifuged to activate RT-RAA assay, and the RT-RAA products were used as templates for the subsequent qPCR. *Source:* Ref. [23] / MDPI / CC BY 4.0.

different HIV-1 strains prevalent in China. The assay can be completed within 50 minutes using conventional qPCR devices available in most molecular laboratories. Additionally, since this RT-RAP assay combines RAA's rapidity with the qPCR probe's simplicity, it exhibits significant sensitivity and specificity, which is crucial for ensuring HIV prevention and treatment efficacy in China.

Despite these interesting findings, there were some limitations to this study. The RT-RAP assay could not accurately quantify HIV-1 viral copy numbers because of the rapidity and qualitative nature of the RT-RAA stage. Furthermore, the current RT-RAP assay only supports 2 μL of the templates, which is one-tenth the volume of qRT-PCR though the detection sensitivity of RT-RAP is equivalent to qRT-PCR in this case. There was no exogenous internal control to monitor the whole reaction in the present study, and this may also serve as a direction for subsequent optimization. Additionally, more clinical samples are needed to further evaluate the feasibility of RT-RAP assay for more diverse HIV-1 subtypes across China. To enhance the convenience and efficiency of this detection method, integrating the microfluidic chip into RT-RAP may also be a viable option. This would facilitate rapid, sensitive, and high throughput HIV-specific detection at the point-of-care settings.

Hua et al. [24] successfully developed the multiplex recombinase-aided qPCR (mRAP) assay for susceptible and rapid detecting *Klebsiella pneumoniae* and the presence of two drug-resistant genes, blaKPC-2, and blaNDM-1, which can be accomplished in a closed test tube using conventional PCR equipment.

The mRAP assay was performed in two steps, with the RAA reaction in the first step and the qPCR reaction in the second step, the products of the amplification of the RAA reaction provided a large amount of DNA template for the qPCR reaction. The RAA reaction mixture included magnesium acetate, out-primer for each of the three genes, template, and RAA reaction buffer. The qPCR reaction mixture included inner primer and probe for each of the three genes, enzyme-free water, dNTP, MgC12, Taq enzyme, and 4× buffer. Next, 30 μL of docosane was added to the preprepared qPCR system and then kept in a thermostat at 60 °C for 1 minute and at 4 °C for 30 seconds, causing the docosane to be a barrier to load on top of the qPCR mixture. In addition, magnesium acetate was added to the reaction cap as the initiator of the RAA reaction, and brief centrifugation was carried out instantly before the reaction to ensure that magnesium ions fell into the RAA system to start the RAA reaction. The three genes are amplified in a single test tube, and depending on the genes contained in the template, the corresponding fluorescence curves are displayed. The single-tube, two-step mRAP assay is shown schematically in Figure 2.7.

Currently, the gold standard detection methods of CRKP are based on the culture (biochemical identification and MALDI-TOFMS) and drug sensitivity testing in the clinic, which takes 24–72 hours for assays and requires specialized instruments. The mRAP assay can achieve high detection sensitivity for the khe, blaKPC-2, and blaNDM-1 genes in *K. pneumoniae*, and its accuracy can be comparable to the gold standard method. In this study, after optimization of the working conditions the sensitivities of mRAP for khe, blaKPC-2 and blaNDM-1 gene were 10, 8, and 14 copies/reaction, respectively, using recombinant plasmids, which are 10–100 times higher than those of qPCR assays. This superior sensitivity of mRAP to qPCR was also confirmed by the comparative testing of nucleic acids of the reference strains. The Kappa value of qPCR and mRAP for detecting khe, blaKPC-2, and blaNDM-1 genes was 1, 0.855, and 1, respectively ($p < 0.05$). Regarding the clinical performance of mRAP, two clinical samples in the study showed CRKP results by mRAP assay and confirmed by phenotypic assay but missed using the qPCR assay. This might be the higher sensitivity of the mRAP method, suggesting that mRAP is indeed a method capable of highly sensitive in diagnosing the presence of blaKPC-2 and blaNDM-1 in *K. pneumoniae*, and it provides a valuable screening tool for large-scale testing and controlling the spread of blaKPC-2 and blaNDM-1.

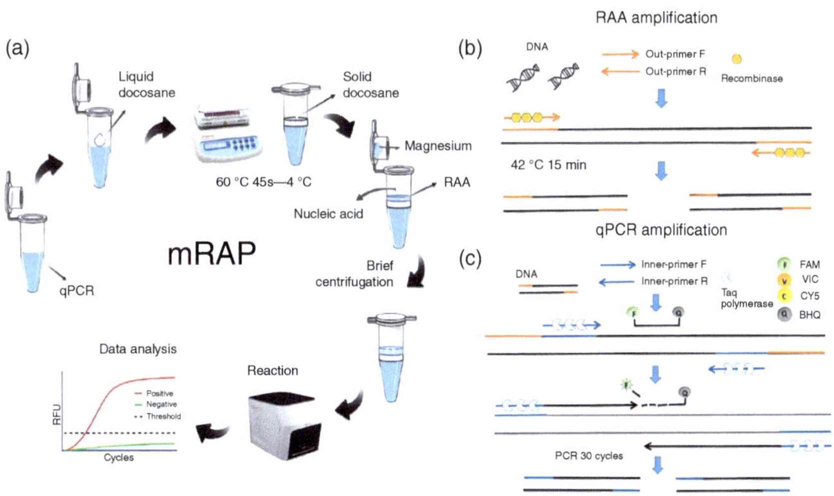

Figure 2.7 Working principle of mRAP. (a) The mRAP assay protocol. (b) RAA reaction at 42 °C for 15 min. (c) qPCR reaction at 95 °C for 10 minutes (1 cycle), 95 °C for 10 s and 65 °C for 15 s (30 cycles). *Source:* Ref. [24] / John Wiley & Sons / CC BY 4.1.

The mRAP assay can meet the clinical needs of the most common drug-resistant K. *pneumoniae*, but the detection of uncommon resistance genes other than blaKPC-2 and blaNDM-1 may be missed. Besides, this method cannot be used for absolute quantification of pathogen nucleic acid detection, rather being used as a rapid and sensitive, qualitative assay. Third, only 56 sputum specimens infected with *K. pneumoniae* were tested in this study, and more carbapenem-resistant sputum specimens are needed to validate the method. In the future, mRAP will be further improved in terms of simplicity by combining it with microfluidic and automation techniques, thus facilitating more practical diagnosis.

RAA coupled with enrichment using M1 protein-conjugated magnetic beads (M1 beads) is a rapid and highly sensitive assay for detecting Epstein-Barr virus (EBV) in the blood [25]. *Staphylococcus aureus* (SA), *Pseudomonas aeruginosa* (PA), and *Acinetobacter baumannii* (AB) account for a substantial proportion of bloodstream infections (BSI) caused by bacteria and are common causes of death. Blood culture remains the gold standard for separating and detecting BSI in clinical practice; however, its extended detection cycle and low positivity rate hinder its ability to fulfill the requirements for early rapid diagnosis and treatment of BSI in clinical practice. Zhao et al. [26]. developed mRAP combined with the M1 bead enrichment method to detect SA, PA, and AB in patients with BSI and compared it with qPCR to evaluate its clinical performance. These polysaccharide residues recognized by the M1 protein exist widely on the surface of various pathogens, including bacteria, fungi, viruses, etc. M1 beads have good enrichment effects on common pathogenic bacteria in BSI. Owing to the low concentration of pathogenic bacteria in the patient's blood, conventional molecular biology methods have limited sensitivity and cannot effectively detect them. Therefore, M1 beads should be used for the pretreatment of the plasma sample to capture pathogenic bacteria. After enrichment, the relative concentration of pathogenic bacteria can be increased, and other interfering components in the plasma can be removed, thereby enhancing amplification efficiency and sensitivity. Enriched live, intact bacteria can also be used for drug sensitivity testing.

This study aimed to develop an assay that can be used for the early diagnosis of three types of bacteremia caused by SA, PA, and AB in the bloodstream based on recombinant human manna binding lectin protein (M1 protein)-conjugated magnetic bead (M1 bead) enrichment of pathogens combined with RAP. The probes for SA, PA, and AB were labeled with FAM, VIC, and ROX fluorophores, respectively, so that they could be detected simultaneously in the same RAP reaction system for blood-infected bacteria. Simulated and clinical plasma samples were enriched with M1 beads and then subjected to multiple recombinase-aided PCR (M-RAP) and qPCR assays. Kappa analysis was used to evaluate the consistency between the two assays.

Simulated plasma samples (2, 4, 6, 8, and 10 CFU) were introduced to a 14-mL round-bottomed tube, and M1 beads (1 mg) were added. Subsequently, the system was replenished to 10 mL with PBS. A final concentration of 4 mmol/L $CaCl_2$ was added, and the mixture was thoroughly mixed for 1 hour. After placing it on a magnetic rack to separate magnetically, the supernatant was discarded. The system was then washed three times using PBS to remove the plasma components and pathogenic bacteria that had not been adsorbed onto the M1 beads. Finally, the solution was resuspended in 400 μL of PBS. The PBS-simulated samples (2, 4, 6, 8, and 10 CFU) were also enriched using the same method; after overnight cultivation on the plate, the number of colonies was counted. A Nucleic Acid Extraction Kit was used to extract DNA from both unenriched simulated plasma specimens and enriched M1 bead suspensions as templates for MRAP and qPCR. The preparation and enrichment processes for the M1 beads are shown in Figure 2.8a. After nucleic acid extraction from the simulated plasma samples with different concentrations of SA, PA, and AB bacteria before and after enrichment, MRAP testing and qPCR testing were performed separately, and the results were compared. These experiments were performed in triplicate. For clinical sample testing, 30 heparin anticoagulant plasma samples were collected from patients with positive results on blood culture tests, including 10 with SA, 10 with PA, and 10 with AB, as well as 30 blood culture-negative samples. After enrichment with M1 beads, M-RAP and

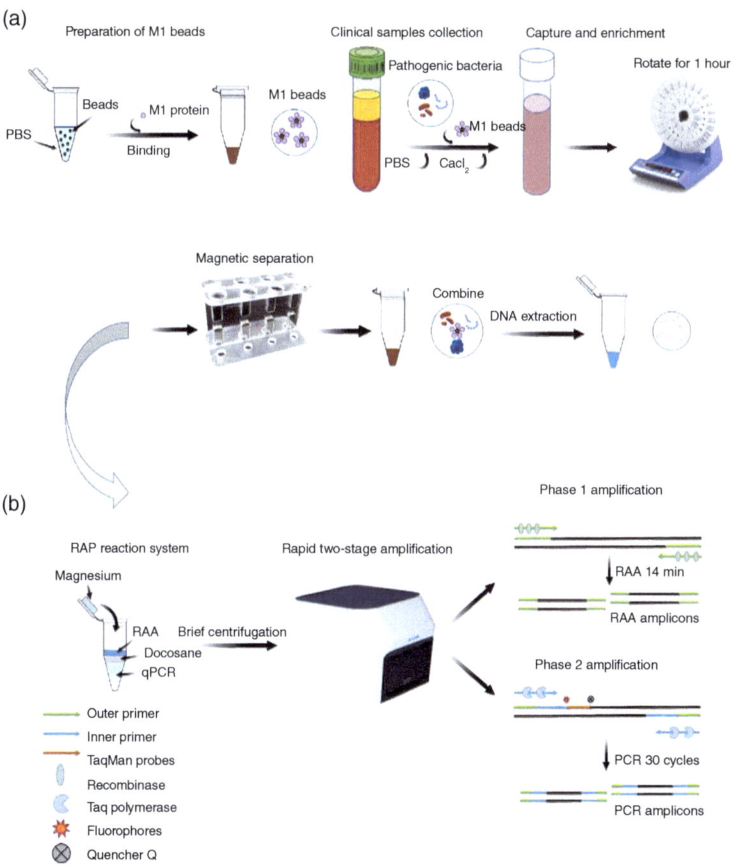

Figure 2.8 The enrichment process of M1 beads (a) and the detection principle of M-RAP (b). M1 protein, recombinant human mannan-binding lectin protein; M1 beads, M1 protein-conjugated magnetic beads; RAP, recombinase-aided PCR; PCR, polymerase chain reaction. *Source:* [26] / with permission of Elsevier.

qPCR assays were performed. The consistency of the two detection methods was compared to evaluate the clinical performance of the M-RAP detection method. The M-RAP detection principle is shown in Figure 2.8b.

The M-RAP method had sensitivity rates of 1, 10, and 1 copies/μL for the detection of SA, PA, and AB plasmids, respectively, without cross-reaction to other bacterial species. The M-RAP assay obtained results for < 10 CFU/mL pathogens in the blood within four hours, with higher sensitivity than qPCR. A total of 60 plasma samples, of which 30 plasma specimens from blood culture-positive patients and 30 plasma specimens from blood culture-negative patients were enriched by M1 beads to assess the clinical manifestations of M-RAP. qPCR assays of both sets of specimens were performed, and the results were compared. The Kappa values of 0.839, 0.815, and 0.856 were obtained, all exceeding 0.75, with a *P* value of <0.05 indicating a high level of consistency in the detection results of the two methods. Although it was unable to collect clinical blood samples that were infected with multiple pathogenic bacteria, M-RAP successfully established a multiple reaction system, demonstrating the feasibility of the detection method. They established M-RAP assay revealed higher sensitivity and accuracy. It could directly detect plasma from patients with <10 CFU/mL bacteremia without the need for blood specimen culture and could obtain valid results within four hours. This presents a novel approach for detecting bacteremia in patients

However, the M-RAP assay has some limitations. Due to the extremely high amplification efficiency of RAA, some samples already reached the threshold of qPCR detection during the RAA amplification period; hence, no CT value was generated in the second stage. Consequently, M-RAP cannot accurately quantify the copy number in samples; it serves as a qualitative detection method. Additionally, although the M-RAP was performed in a single tube, the two reaction stages were separated by physical barriers, increasing the complexity of the experiment.

In summary, the integration of M-RAP with magnetic bead separation technology enables the direct detection of clinical plasma samples. This approach facilitates the rapid detection of trace amounts (<10 CFU/mL) of common bacteria in the blood, eliminates the time-consuming microbiological culture process, shortens the detection time, and provides a new method for the early detection of trace pathogens in the blood of patients with BSI, which is of great significance and advances the rapid diagnosis of BSI in the early stages. Although the M-RAP assay has some limitations, it is a stable, fast, highly sensitive, and specific detection method. M-RAP technology can not only be used to detect bacteria, fungi, viruses, and other pathogens but can also be combined with other technologies, such as microfluidics technology, to achieve high throughput detection in different scenarios with broad application potential in clinical practice.

Zhang et al. [27] developed a multiplex LNA probe-based RAP method (MLP-RAP) for the first time to detect multiple point mutations of rifampicin-resistant tuberculosis. Furthermore, the docosane-based physical isolation strategy allows both RAA and qPCR reactions carried out in a single tube within one hour. The schematic diagram of the MLP-RAP assay for detecting rifampicin-resistant *Mycobacterium tuberculosis* is shown in Figure 2.9.

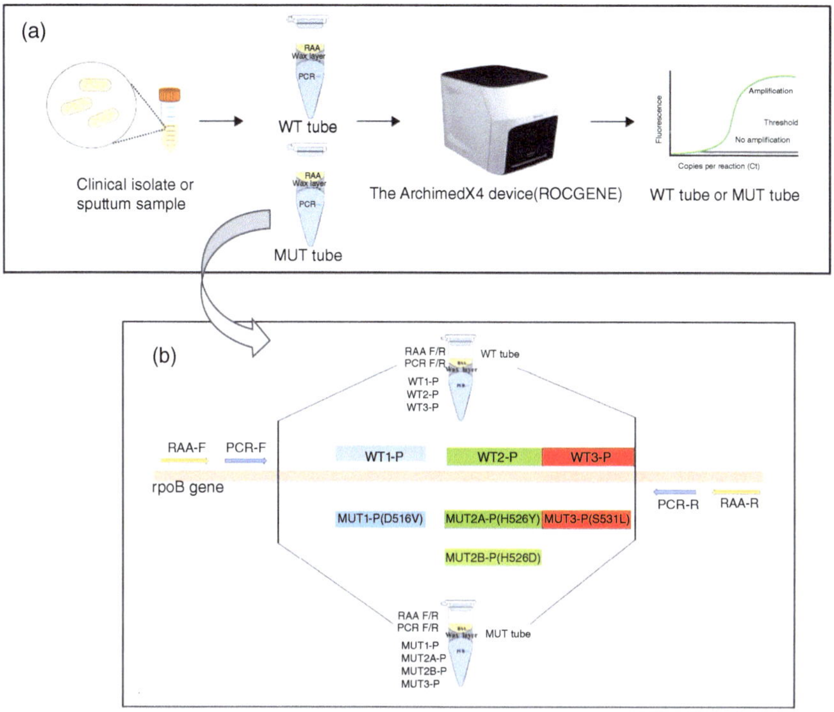

Figure 2.9 (a) Flow chart of the MLP-RAP method. (b) The schematic diagram of the MLP-RAP method for detection of RR-TB. RAA primers, qPCR primers, and qPCR LNA probes in WT and MUT tubes. Fluorescent dye RAA technology development and application. *Source:* Ref. [27] / Frontiers / CC BY 4.0.

The primer and probe set for MLP-RAP assay of RR-TB included RAA-F, RAA-R, PCR-F, PCR-R, WT1-P, WT2-P, WT3-P, MUT1-P, MUT2A-P, MUT2B-P, and MUT3-P. They designed two tubes, referred to as wild-type (WT) tube and mutant-type (MUT) tube, for simultaneous detection of rifampicin-susceptible and rifampicin-resistant TB in each sample. The RAA primers (RAA-F and RAA-R) and PCR primers (PCR-F and PCR-R) were the same for each tube, but the LNA probes were base-specific (different) in WT and MUT tubes. The WT tube had three LNA probes labeled with different fluorescence, namely WT1-P with HEX label covering codon 516, WT2-P with Cy5 label covering codon 526, and WT3-P with FAM label covering codons 531 and 533. In the MUT tube, there were four differently labeled fluorescent LNA probes (MUT1-P with HEX label, MUT2A-P with Cy5 label. MUT2B-P with ROX label, and MUT3-P with FAM label). The four LNA probes were exactly matching four mutant rpoB gene sequences (D516V, H526Y, H526D, and S531L), respectively, which could accurately identify the presence of mutation types and cover the base mutation positions in mutant rpoB gene sequences. Multiplex LNA probe-based RAP method assay was conducted in two tubes (WT and MUT).

In each tube, the RAA reaction was performed at 39 °C in the first stage, docosane formed a wax layer to perfectly separate the RAA (upper phase) from the qPCR systems (lower phase). When the RAP turned into the qPCR stage at 95 °C, the wax layer with a melting point of 43–46 °C melted and moved to the top layer of the tube. Meanwhile, the RAA components were inactivated at 9 °C and mixed with the PCR system. The RAA-amplified DNA fragment was therefore used as a template for qPCR. The sensitivity of the MLP-RAP assay was 5 copies/mL in detecting DNA plasmids, which is 20 times more sensitive than qPCR (100 copies/mL). In addition, the MLP-RAP results using 78 boiled sputum samples achieved 100% concordance with the GeneXpert results. Meanwhile, the MLP-RAP can be performed on a conventional real-time PCR device that is more suitable for general common laboratories. Of note, after the addition of the extracted nucleic acid, the whole process of 96 samples could be completed within one hour. The MLP-RAP offers a shorter turn-around time (one hour) to generate diagnostic results than GeneXpert (2.5 hour). In addition, the MLP-RAP assay (4.5 $ per reaction) provides the lower test price for detecting RIF resistance–a tenth that of GeneXpert (45 $ per reaction). Therefore, the MLP-RAP assay is also a cost-effective method and has important potential for improving the diagnosis and control of RR-TB. The MLP-RAP assay was designed to detect four common mutations associated with RIF resistance in the MUT tube, including D516V (GAC→GTC), H526D (CAC→GAC), H526Y (CAC→TAC), and S531L (TCG→TTG) and was able to detect 5% mutations in the genomic template containing 5×10^2 genomes/mL. Analysis of clinical isolates confirmed that the MLP-RAP assay detected more heteroresistant samples than the Xpert MTB/RIF assay. Hence, given the practically high prevalence of heteroresistance in China, the better capability of MLP-RAP to detect mixed infection may facilitate laboratory staff to identify more RIF-heteroresistant TB cases at the early stage of the TB.

Therefore, MLP-RAP enabled rapid and highly sensitive detection of rifampicin-resistant *M. tuberculosis*. Nonetheless, the MPL-RAP assay has some limitations. First, the MPL-RAP analysis result cannot quantify the bacteria titer in the original samples. A special software needs to be developed to accurately analyze the results of rifampicin heteroresistance and give the percentage of drug resistance. Second, this assay only detects the mutation types at codons 516, 526, 531, and 533, not all codons of RRDR. Third, this study only tested 41 rifampicin-sensitive sputum samples, and more rifampicin-resistant sputum samples are needed to verify this method. With the distinctive advantages of high sensitivity, rapidity, cost-effectiveness, and generality, they intend to extend the MLP-RAP strategy to other mutations associated with other drug resistance in TB. Ultimately, they attempt to integrate the MPL-RAP assay with microfluidic devices to achieve multiple and simultaneous screening of multidrug-resistant (MDR) and extensively drug-resistant tuberculosis (XDR-TB) patients from sputum samples.

Fluorescent RAA assays are increasingly being used in the detection of a variety of pathogens and have the advantages of rapidity and simplicity and similar sensitivity and specificity, compared with real-time PCR assays, but they require a complex probe design. The real-time quantitative Eva Green-RAA method for detecting SA and *Listeria monocytogenes* (LM) was developed by ZHAO ZIJIN et al, which eliminates the need to design complex RAA probes. The method combines the advantages of the self-avoiding molecular recognition system (SAMRS) primer and the Eva Green fluorescent dye to minimize the generation of primer dimers while avoiding the need for complex probe design, making the assay more convenient and economical.

The real-time Eva Green-dye-based RAA assay with SAMRS primers (Eva Green-RAA assay) for the detection of SA and LM established in this study could obtain results in less than 30 minutes with good sensitivity. The instrument is also portable and simple to use, making this method very suitable as a supplement to the standard method for the detection of psychrophilic bacteria and widely applicable under various conditions. Because the products are free of other DNA double-stranded components, all the products can be regarded as target products, and then relative quantification can be achieved by plotting the standard curves with the results obtained for the recombinant plasmids.

The SAMRS uses the nucleotide analogs A*, G*, C*, and T* to replace A, G, C, and T, respectively, at certain positions in a primer, where A* pairs with T, T* pairs with A, G* pairs with C, and C* pairs with G. The SAMRS nucleotide will form two hydrogen bonds with the normal nucleotides and the A*–T* pair and the C*–G* pair will only form 1 or 1.5 hydrogen bonds, which do not contribute substantially to the stability of the double strand. Therefore, the SAMRS nucleotides can bind normally to the DNA single strand, but SAMRS nucleotides cannot interact with each other, and the generation of primer dimers can be minimized or even avoided. Previous studies have used SAMRS in isothermal amplification techniques, such as recombinant polymerase amplification (RPA) and helicase-dependent amplification (HDA), and shown that placement of SAMRS nucleotides at the 3′ end of the primer-binding site avoided the generation of most primer dimers. In the SAMRS, the natural bases (A, G, C, and T) are replaced with unnatural bases (A*, G*, C*, and T*), which have strategically adjusted hydrogen bonding to retain the ability to correctly pair in the Watson–Crick model, while the bonding between the SAMRS pairs is weaker and does not contribute substantially to the stability of the double strand, as shown in Figure 2.10.

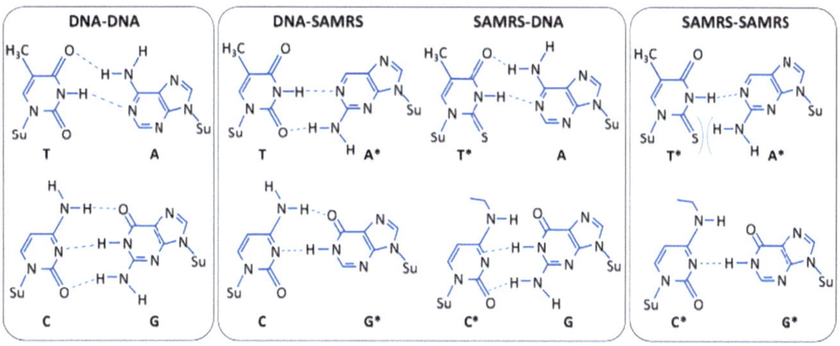

Figure 2.10 Chemical structures of the self-avoiding molecular recognition system (SAMRS). Pairs between standard nucleobases (left, DNA-DNA). Pairs between standard nucleobases and their SAMRS complements (middle, DNA-SAMRS and SAMRS-DNA). Pairs between SAMRS nucleobases and their formal SAMRS complements (right, SAMRS-SAMRS). Su, sugar backbone. *Source:* Ref. [28] / John Wiley & Sons / CC BY 4.1.

Zhao et al. [28] using SA and LM, as examples, Eva Green dye-based RAA assays with SAMRS primers (Eva Green-RAA) were developed. As a reference and comparison, a traditional fluorescence probe RAA method and RAA with SAMRS primers (SAMRS RAA) for detecting SA and LM were also investigated. Serial dilutions of recombinant plasmids were used to evaluate the sensitivity of the assays. Unenriched and enriched simulated milk samples were used to evaluate the limits of detection (LOD) of these methods. Simulated milk samples were used to explore the sensitivity of the dual Eva Green RAA assay. The sensitivity of the fluorescent RAA method for detecting SA and LM was 10 copies/μL using plasmids and the sensitivity of the SAMRS RAA and Eva Green-RAA for detecting SA and LM plasmids was 1 copies/μL. The LOD values of the Eva Green-RAA for SA and LM in unenriched simulated milk samples were 100 and 50 CFU/mL, respectively, and the LOD value for both SA and LM using enriched simulated milk samples was 10 CFU/mL. The Eva Green RAA had linear amplification in real-time in the range of 1–105 copies/μL of the plasmids of SA and LM. The sensitivity of the dual Eva Green-RAA assay for SA and LM was estimated to be 102 CFU/mL.

Compared with the corresponding RAA and SAMRS RAA, their results showed that the Eva Green-RAA assays for SA and LM had enhanced sensitivity, as evidenced by investigations of recombinant plasmids, standard strain nucleic acids, and enriched and unenriched simulated milk samples. The enhanced sensitivity was attributed to the complete elimination of the generation of primer dimers in RAA assays, which was demonstrated in the results of AGE and further confirmed by HRM, suggesting that the readout of the Eva Green-RAA detection results could be directly achieved from the fluorescence curves without the need for performing HRM. Thus, the Eva Green-RAA method that they have developed in the present study is a reliable quantitative assay and was further improved to be a dual assay for SA and LM. However, at present, only dual HRM detection methods have been explored. The presence of multiple sets of primers in a reaction system increases the possibility of producing primer dimers, which makes it difficult to realize multiple detection using Eva Green RAA. More effort is needed to further optimize the structure of SAMRS nucleotides to investigate the feasibility of multiplex Eva Green RAA.

The Eva Green-based RAA assay, utilizing SAMRS primers and employing real-time detection, has been developed for the specific purpose of identifying SA and LM. In this study, it was demonstrated that the Eva Green-RAA assay can produce results in less than 30 minutes, while also demonstrating high levels of sensitivity. The instrument is also portable and simple to use, making this method very suitable as a supplement to the standard method for the detection of psychrophilic bacteria and widely applicable under various conditions. Furthermore, the combination of the Eva Green-RAA assay with HRM enables the realization of multiple detection. Despite the synthesis of the SAMRS primer being a more challenging and costly process and the inability to mass produce it at this time, the Eva Green RAA assay offers an innovative approach to implementing multiple RAA detection using a melting curve.

2.2.7 Summary

RAA technology is an emerging isothermal nucleic acid amplification technique. Compared with traditional PCR technology, RAA technology has many significant advantages. First, it requires less equipment and reaction conditions, and usually only requires a constant temperature of 37 °C for amplification. This feature makes RAA technology more suitable for resource-limited environments, such as field operations or primary care units. Second, the reaction time of RAA technology is very short, usually within 20 minutes to complete the amplification, which greatly improves the detection efficiency.

RAA technology has been widely used for the rapid diagnosis of a wide range of bacterial and viral infections, such as new coronavirus, influenza virus, enterovirus, Salmonella, SA, and LM etc. Its high sensitivity and specificity make it an important tool for rapid screening of pathogenic pathogens in clinical settings and food safety testing. In addition, RAA technology can rapidly identify genetic mutations and gene polymorphisms, supporting early diagnosis and personalized treatment of genetic diseases.

Although RAA technology has a wide range of applications in many fields, its limitations in long-fragment nucleic acid amplification still need to be overcome. Currently, RAA technology is mainly used to amplify shorter nucleic acid fragments (usually no more than 500 base pairs), which limits its application in complex genetic manipulation. However, with the continuous development of technology, scientists are actively exploring and optimizing RAA technology to improve its amplification capacity and applicability.

2.3 Summary

In this chapter, we detail the principle, applications, advantages, limitations, and the recent advances of the RAA. We conlude that RAA technology is becoming an important tool in the field of nucleic acid detection by its low equipment requirements, fast response time, and high accuracy. Although it still has some limitations in the field of gene manipulation, with the continuous improvement and optimization of the technology, the application of RAA technology in life science research is expected to be more promising, especially in the fields of gene manipulation and complex genome analysis. In addition, microfluidic chips, a new reaction and detection vehicle, can be combined with RAA to develop a compact, portable platform for bedside detection or instant detection. Microfluidic chips can also integrate multiple reactions to achieve simultaneous detection of multiple samples or multiple targets. The major challenge faced by microfluidic chips is avoiding compromising clinical performance in the integration of nucleic acid extraction, amplification, and detection steps in a single chip. Fabrication cost is another major concern, although recently emerging mobile phone- and paper-based microfluidic devices provide attractive alternatives in resource-poor settings, real-world field applications of RAA also require consideration of miniaturization of the instrument or visual interpretation without instrumentation, affordable assay costs, smart data transfer (e.g., with smartphones), and battery-powered devices [29].

References

1 Caetano-Anollés, G. (1996). Scanning of nucleic acids by in vitro amplification: new developments and applications. *Nature Biotechnology* 14 (13): 1668–1674.

2 Gingeras, T.R., Richman, D.D., Kwoh, D.Y. et al. (1990). Methodologies for in vitro nucleic acid amplification and their applications. *Veterinary Microbiology* 24 (3-4): 235–251.

3 Analytical Methods Committee, AMCTB No 59 (2013). PCR – the polymerase chain reaction. *Analytical Methods: Advancing Methods and Applications* 6 (2): 333–336.

4 Schluger, N.W. and Rom, W.N. (1995). The polymerase chain reaction in the diagnosis and evaluation of pulmonary infections. *American Journal of Respiratory and Critical Care Medicine* 152 (1): 11–16.

5 Zhao, Y., Chen, F., Li, Q. et al. (2015). Isothermal amplification of nucleic acids. *Chemical Reviews* 115 (22): 12491–12545.

6 Munawar, M.A. (2022). Critical insight into recombinase polymerase amplification technology. *Expert Review of Molecular Diagnostics* 22 (7): 725–737.

7 Chen, C., Li, X., Li, G. et al. (2018). Use of a rapid reverse-transcription recombinase aided amplification assay for respiratory syncytial virus detection. *Diagnostic Microbiology and Infectious Disease* 90 (2): 90–95.

8 Lin, H., Zhao, S., Ye, Y. et al. (2022). A Fluorescent Recombinase Aided Amplification Assay for Detection of Babesia microti. *The Korean Journal of Parasitology* 60 (3): 201–205.

9 Wu, K., Zhang, Y., Zeng, S. et al. (2021). Development and application of RAA Nucleic Acid Test Strip Assay and Double RAA gel electrophoresis detection methods for ASFV and CSFV. *Frontiers in Molecular Biosciences* 8: 811824.

10 Zhao, S., Zhang, Q., Wang, X. et al. (2023). Development and performance of recombinase-aided amplification (RAA) assay for detecting Schistosoma haematobium DNA in urine samples. *Heliyon* 9 (12): e23031.

11 Balea, R., Pollak, N.M., Hobson-Peters, J. et al. (2023). Development and pre-clinical evaluation of a Zika virus diagnostic for low resource settings. *Frontiers in Microbiology* 14: 1214148.

12 Cui, H., Zhang, C., Tu, F. et al. (2022). Rapid detection of influenza A viruses using a real-time reverse transcription recombinase-aided amplification assay. *Frontiers in Cellular and Infection Microbiology* 12: 1071288.

13 Liu, Y., Ren, W., Xue, Z. et al. (2023). Real-time recombinase-aided amplification assay for rapid amplification of the IS1081 gene of Mycobacterium tuberculosis. *European Journal of Clinical Microbiology & Infectious Diseases: Official Publication of the European Society of Clinical Microbiology* 42 (8): 963–972.

14 Zhang, X., Guo, L., Ma, R. et al. (2017). Rapid detection of Salmonella with recombinase aided amplification. *Journal of Microbiological Methods* 139: 202–204.

15 Yan, T., Li, X., Wang, L. et al. (2018). Development of a reverse transcription recombinase-aided amplification assay for the detection of coxsackievirus A10 and coxsackievirus A6 RNA. *Archives of Virology* 163 (6): 1455–1461.

16 Shen, X., Qiu, F., Shen, L. et al. (2019). A rapid and sensitive recombinase aided amplification assay to detect hepatitis B virus without DNA extraction. *BMC Infectious Diseases* 19 (1): 229.

17 Qi, J., Li, X., Zhang, Y. et al. (2019). Development of a duplex reverse transcription recombinase-aided amplification assay for respiratory syncytial virus incorporating an internal control. *Archives of Virology* 164 (7): 1843–1850.

18 Wang, R., Zhang, H., Zhang, Y. et al. (2019). Development and evaluation of recombinase-aided amplification assays incorporating competitive internal controls for detection of human adenovirus serotypes 3 and 7. *Virology Journal* 16 (1): 86.

19 Zhang, R., Li, G., Li, X. et al. (2019). A rapid and sensitive recombinase aided amplification assay incorporating competitive internal control to detect Bordetella pertussis using the DNA obtained by boiling. *International Journal of Infectious Diseases : IJID: Official Publication of the International Society for Infectious Diseases* 86: 108–113.

20 Duan, S., Li, G., Li, X. et al. (2018). A probe directed recombinase amplification assay for detection of MTHFR A1298C polymorphism associated with congenital heart disease. *BioTechniques* 64 (5): 211–217.

21 Chen, Z., Shen, X., Wang, J. et al. (2021). Development of a PDRA method for detection of the D614G mutation in COVID-19 virus – Worldwide, 2021. *China CDC Weekly* 3 (21): 448–453.

22 Fan, G., Zhang, R., He, X. et al. (2021). RAP: a novel approach to the rapid and highly sensitive detection of respiratory viruses. *Frontiers in Bioengineering and Biotechnology* 9: 766411.

23 Tian, F., Jin, C., Ji, S. et al. (2023). A rapid and sensitive detection of HIV-1 with a one-pot two-stage reverse transcription recombinase aided real-time PCR assay. *Tropical Medicine and Infectious Disease* 8 (2).

24 Hua, S., Wang, J., Zhao, Z. et al. (2024). A multiplex recombinase-aided qPCR Assay for highly sensitive and rapid detection of khe, bla(KPC) (-2), and bla(NDM) (-1) genes in Klebsiella pneumoniae. *Journal of Clinical Laboratory Analysis* 38 (9): e25038.

25 Li, J., Chen, X., Tie, Y. et al. (2022). Detection of low-load Epstein-Barr virus in blood samples by enriched recombinase aided amplification assay. *AMB Express* 12 (1): 71.

26 Zhao, Z.J., Chen, X.P., Hua, S.W. et al. (2024). Establishment of a multiplex detection method for common bacteria in blood based on human mannan-binding lectin protein-conjugated magnetic bead enrichment combined with recombinase-aided PCR technology. *Biomedical and Environmental Sciences : BES* 37 (4): 387–398.

27 Zhang, R., Ou, X., Sun, X. et al. (2023). Multiplex LNA probe-based RAP assay for rapid and highly sensitive detection of rifampicin-resistant Mycobacterium tuberculosis. *Frontiers in Microbiology* 14: 1141424.

28 Zhao, Z., You, Y., Hua, S. et al. (2024). A novel dye-based recombinase aided amplification assay with enhanced sensitivity and specificity. *iLABMED* 2 (4): 294–306.

29 Ma, X.J. (2022). Isothermal amplification technology for diagnosis of COVID-19: current status and future prospects. *Zoonoses* 2: 5.

3

Diving into the ELISA Technique

Its Role and Innovations in Public Health

Xin Wang[1,2], Nan Cheng[1], and Juewen Liu[2]

[1] *China Agricultural University, College of Food Science and Nutritional Engineering, Beijing Laboratory for Food Quality and Safety, No.17 Qinghua East Road, Haidian District, Beijing 100083, China*
[2] *University of Waterloo, Waterloo Institute for Nanotechnology, Department of Chemistry, 200 University Avenue West, Waterloo, ON N2L 3G1, Canada*

CHAPTER MENU

3.1 Introduction to ELISA and Its Components

Enzyme-linked immunosorbent assay (ELISA) is one of the most widely used diagnostic tools in public health due to its simplicity and the ability to provide results with high specificity, sensitivity, and reproducibility. ELISA techniques were originally derived from radioimmunoassay (RIA). Due to safety concerns regarding the use of radioisotopes, subsequently, the procedures were modified to use enzymes as labels for signal generation [1]. ELISA was first used to detect immunoglobulin G (IgG) in rabbit serum. Since then, it has been continuously expanded to a variety of detection applications, such as human chorionic gonadotropin (hCG) for the detection of pregnancy, human immunodeficiency virus (HIV), and the most recent severe acute respiratory syndrome coronavirus 2 (SARS-CoV-2). ELISA has continued to innovate and develop to date.

As an analytical methodology, ELISA relies on the basic immunologic concept of antigen binding to specific antibodies, allowing the detection of various antigen targets like proteins, peptides, viruses, and bacteria. ELISA mainly consists of two key

Diagnostic Techniques in Public Health: Analytical and Chemometric Approaches in Disease Detection and Management, First Edition. Edited by Jinghong Li and Yan Du.

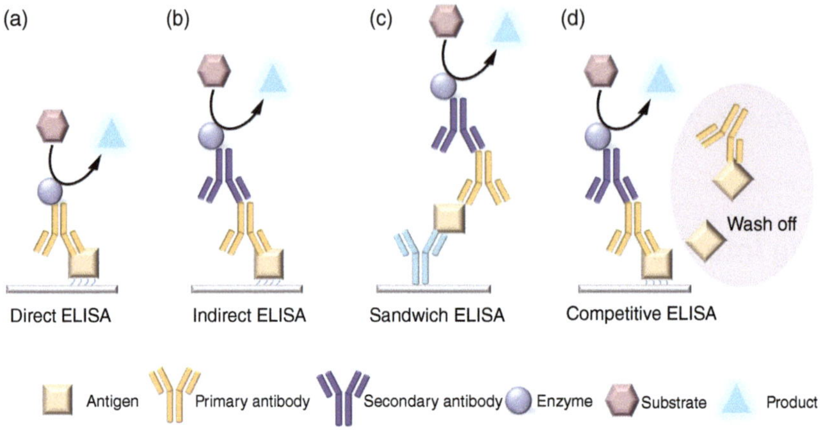

Figure 3.1 Schematic of four typical ELISA formats. (a) Direct ELISA, (b) indirect ELISA, (c) sandwich ELISA, (d) competitive ELISA. The first three methods (a–c) are suitable for the detection of protein targets. The last one, competitive ELISA, is typically used for the detection of small molecules, where small-molecule targets are first immobilized on the plate.

components: an immune-based binding system for target recognition, and a detection system. The target recognition system is based on antibodies and antigens forming a complex, which determines the assay's specificity. Meanwhile, ELISA can be performed in different formats depending on different targets or assay requirements, mainly including direct, indirect, sandwich, and competitive ELISA (Figure 3.1) [2, 3]. When two monoclonal antibodies with specificity for different epitopes of an antigen are available, the "sandwich"-type ELISA method is often employed. When only one specific epitope is available, the antigen coating method is commonly used. The detection system primarily focuses on enzymes and their roles in signaling. Traditionally, signal generation in this context depends on the color change induced by an enzyme-labeled antibody or antigen acting on a chromogenic substrate. The amount of colored product formed is directly proportional to the concentrations of the standards used in the assay. The most commonly used enzymes include alkaline phosphatase (ALP), horseradish peroxidase (HRP), and glucose oxidase [4].

Compared to the "gold standard" polymerase chain reaction (PCR) technique for the detection of viruses and bacterial cells, ELISA possesses many advantages. PCR testing involves complex sample pre-treatment and nucleic acid extraction, and then proceeds to PCR amplification with fluorescence monitoring. For an RNA virus, an additional reverse transcription step is required. Moreover, specialized equipment is often needed to carry out these procedures. In contrast, ELISA testing is much simpler. Commercial kits only require basic sample pre-treatment, followed by sample incubation, plate washing, and colorimetric measurement to complete the process. Certainly, this is under the premise of considerable maturity in the product development of ELISA. Traditional ELISA involves numerous manual steps, consumes large amounts of reagents (100–200 μL), and has limited detection performance and multiplexing capabilities [5, 6]. When facing urgent demands for rapid and precise testing, such as during a pandemic like coronavirus disease 2019 (COVID-19), these limitations become particularly evident.

Despite these limitations, ELISA remains an important tool and a workhorse in basic research and public health. Over the past few decades, significant advancements in chemistry, substrates, nanomaterials, and engineering design have improved the analytical performance of ELISA methods or their variants, making them more stable, faster, more sensitive, and higher throughput. This chapter will provide an overview of ELISA technology, its recent developments and innovations, and applications in public health.

3.2 Recognition Molecules

One of the key aspects of detection is the ability to precisely identify target molecules. The selectivity of a traditional ELISA system primarily depends on the specificity of the antibody and the purity of the antigen. Among them, recently developed nanobodies deserve special discussion. Nanobodies are also known as single-domain antibodies or VHH antibodies (variable domain of heavy-chain antibodies), which are initially derived from camelid animals and possess unique characteristics such as a smaller molecular size and higher stability [7]. The use of nanobodies in ELISA has been explored due to these advantages [8, 9]. They offer higher sensitivity and specificity, especially in complex sample matrices, and are more amenable to engineering modifications, which can enhance the performance of the detection system. In addition, aptamers, also called chemical antibodies, are short oligonucleotide sequences proven to selectively bind to proteins, viruses, metal ions, small organic molecules, and even cells. Aptamers have been utilized as alternatives to antibodies or engineered antibodies in various biological applications. Methods utilizing aptamers as capture molecules instead of primary or secondary antibodies in ELISA detection have been developed and employed for the detection of various biomolecules. This is known as aptamer-ELISA [10, 11]. Finally, molecularly imprinted polymers (MIPs, also called plastic antibodies) have also been employed as recognition probes in place of antibodies in ELISA [12, 13]. These polymers are prepared using the molecular imprinting technology to achieve specific molecular recognition properties. MIPs are synthesized by introducing a target molecule (template molecule) into a polymer network during the polymerization process, followed by removal of the template molecule, leaving behind cavities that match the shape of the template molecule. These cavities exhibit high selectivity toward the target molecule. Currently, the recognition molecules used in ELISA technology mainly consist of antibodies (nanobodies), aptamers, and MIPs. Figure 3.2a illustrates the different properties of these probes in terms of specificity, stability, affinity, and cost.

3.2.1 Antibodies

Antibodies, also known as immunoglobulins, are "Y"-shaped proteins produced by the immune system to help combat foreign substances (see Figure 3.2b for the

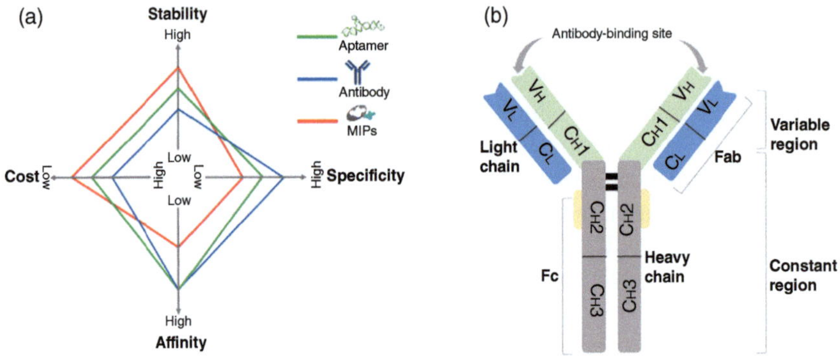

Figure 3.2 (a) A comparison of different properties of aptamers, antibodies, and MIPs. *Source:* Reproduced from [2] / with permission of Elsevier. (b) The structure of an IgG antibody. Target binding happens in the Fab chains, and the Fc chain is relatively stable, can resist protease degradation, and is often used as a site for functional modification.

structure of an IgG antibody). Any substance that induces the immune system to produce antibodies against it is considered an antigen. Antigens encompass a diverse array of entities such as viruses, bacteria, fungi, chemicals, and food allergens. Some antigens are endogenous, produced within the body, such as those from cancer cells. Additionally, viral and bacterial infections can stimulate immune responses. Organ transplants may sometimes be rejected because proteins on the surface of transplanted organs can act as antigens in the recipient's body. The immune response involves five different antibodies, each with distinct structures and functions: IgG, IgA, IgM, IgD, and IgE.

IgG: This antibody appears in the gamma band during protein separation and is the most prevalent type, comprising approximately 75% of serum antibodies in humans. It is the most commonly used antibody in immunoassays.

IgA: During protein separation, IgA is observed in both the beta and gamma bands.

IgM: Known as macroglobulin due to its faster sedimentation rate compared to IgG, IgM predominates in the early stages of an infection and is subsequently replaced by IgG as the illness progresses.

IgD: In protein separation, IgD appears after the beta and gamma bands.

IgE: This antibody is produced only after exposure to specific allergic antigens, which can cause allergic conditions such as erythema.

In ELISA, the quality of both antigen and antibody is crucial for the success of the experiment. A high purity of antigen and its high potency in stimulating an immune response, and a strong affinity of the antibody, are required. Both monoclonal and polyclonal antibodies can be used as capture and detection antibodies in ELISA. Monoclonal antibodies exhibit inherent specificity for a single epitope, allowing for the precise detection and quantification of even minor differences in antigens. Polyclonal antibodies are typically used as capture antibodies, while monoclonal antibodies are used as detection antibodies in sandwich ELISA to provide higher specificity. In addition to traditional monoclonal antibodies, recombinant monoclonal antibodies can also be used in ELISA. For example, rabbit recombinant antibodies are derived from cell lines that produce antibodies and are engineered to express specific antibody heavy and light chains. Compared to traditional monoclonal antibodies derived from hybridomas, recombinant antibodies are less susceptible to cell line drift or batch-to-batch variability, thus achieving optimal antigen specificity. Another consideration when selecting antibodies is cost, producing polyclonal antibodies is typically about five times cheaper than monoclonal antibodies. Therefore, a compromise often needs to be made between the specificity achieved by monoclonal antibodies and the time and cost required.

Secondary antibodies are antibodies used to target and bind to the primary antibody. Many secondary antibodies are labeled with tags such as Alexa Fluor fluorescent dyes or HRP, allowing the signal from the secondary antibody to be detected. When selecting secondary antibodies, considerations typically include species specificity and reactivity, cross-adsorption, purification methods, antibody type and subtype, and labeling.

Here, we list some key points related to antibodies for ELISA [1, 6]:

1) Antibody-specific information: It is essential to gather detailed information about the antibodies involved, such as whether they are monoclonal or polyclonal antibodies. Polyclonal antibodies, produced by different B-cell clones, recognize multiple epitopes, potentially increasing sensitivity but with the risk of cross-reactivity. Affinity: High-affinity antibodies bind more tightly to their target antigens, enhancing assay sensitivity. Epitope recognition: Understanding the specific epitope recognized by the antibody ensures that the ELISA can detect the target antigen accurately, even in the presence of similar molecules.
2) Cross-reactivity information: Cross-reactivity refers to the ability of an antibody to bind to non-target antigens that share structural similarities with the target antigen. This can lead to false-positive results, undermining the specificity of the assay.
3) Matrices information: Different sample matrices (e.g., serum, plasma, urine, saliva) can affect the binding properties of antibodies and the overall signal of the assay. Components such as proteins, lipids, and salts can interfere with antibody–antigen interactions. It is crucial to validate the ELISA for each specific matrix to ensure accurate and consistent results. This involves spiking known amounts of antigen into the matrix and verifying recovery rates and linearity.

3.2.2 Aptamers

Aptamers are short, single-stranded DNA or RNA that exhibit high affinity and specificity for target molecules. Aptamers fold into unique three-dimensional (3D) structures, enabling them to bind to targets with precision akin to antibodies [14]. Over the past decade, aptamers have demonstrated potential in applications ranging from biosensors, therapy, targeted drug delivery, to diagnostic tools [15]. Unlike conventional antibodies acquired through in vivo immunization pathways, aptamers offer several distinct advantages, including ease of synthesis, modification, and stability, making them attractive candidates for incorporation into ELISA assays. In particular, aptamers can be obtained for molecules without immunogenicity and under non-physiological conditions. Once the sequence is known, aptamers can be chemically synthesized using automated synthesizers, and there are no apparent batch-to-batch differences. The selection protocol was reported in 1990 independently by two research groups: Szostak [16] and Tuerk [17]. The general process of aptamer selection is called Systematic Evolution of Ligands by Exponential Enrichment (SELEX). Using SELEX, aptamers applicable to a variety of targets, including metal ions [18], small organic molecules [19, 20], antibiotics [21], and even whole cells [22], have been successfully isolated. A typical SELEX starts with 10^{14}–10^{15} random DNA or RNA sequences forming a library. After incubation with target molecules, unbound sequences are removed, and the bound ones are eluted and amplified by PCR for further selection rounds. After the binding affinity reaches saturation, the sequences that survived SELEX are subsequently sequenced and characterized. Figure 3.3 illustrates the main steps of SELEX.

In an ELISA assay using aptamers, aptamers specific to the target molecule are immobilized on a solid support, such as a microplate well or a sensor surface. When a sample containing the target molecule is added, the target molecule selectively binds to the immobilized aptamers. After washing away unbound molecules, an

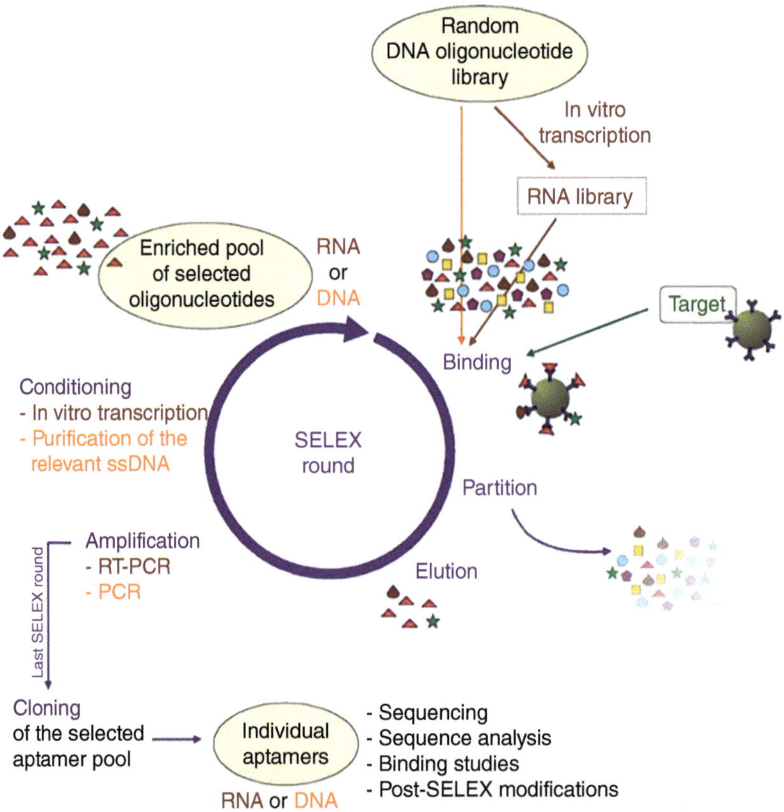

Figure 3.3 Flow diagram illustrating the in vitro selection of target-specific aptamers using SELEX technology. *Source:* Ref. [23] / with permission of Elsevier.

enzyme-conjugated secondary aptamer or detection probe is introduced, which binds specifically to the target molecule bound to the immobilized aptamers. Finally, a substrate for the enzyme is added, and the resulting enzymatic reaction produces a detectable signal, typically a color change or fluorescence, which can be quantified using a spectrophotometer or a fluorescence reader.

3.2.3 Molecularly Imprinted Polymers

Molecular imprinting is a general molecular recognition technique used to create materials with selective binding properties for specific molecules. As such, MIPs are synthetic materials designed to mimic natural antibodies and receptors in their ability to recognize and bind target molecules with high selectivity [24, 25]. For a given target or template molecule, the process typically involves the following steps. First, functional monomers are selected based on their capacity to interact with the template molecule via non-covalent interactions, including hydrogen bonding, electrostatic interactions, or van der Waals forces. The second step is the formation of pre-polymerization complexes, in which the template molecule and functional monomers are mixed to form a complex through non-covalent interactions. Third, polymerization is initiated to develop a polymer network encasing the template molecule, thus forming cavities or imprints that correspond to the template molecule's size, shape, and functional groups. Fourth, the template molecules are extracted from the polymer matrix, leaving behind cavities or imprints that are complementary to the template molecule in size, shape, and chemical functionality. Finally, the resulting MIP exhibits selective recognition and binding toward the template molecule and structurally related compounds.

In ELISA, MIPs serve as synthetic recognition elements that mimic the binding specificity of antibodies [12]. MIPs offer several advantages over antibodies, including enhanced stability, lower cost, and ease of production. In an ELISA assay using MIPs, the MIPs are immobilized on a solid support, such as a microplate well or a sensor surface. When a sample containing the target molecule is added, the target molecule selectively binds to the cavities or imprints within the MIPs due to their complementary molecular recognition sites. After washing away unbound molecules, an enzyme-conjugated secondary antibody or detection probe is introduced, which binds specifically to the target molecule bound to the MIPs. Finally, a substrate for the enzyme is added, and the resulting enzymatic reaction produces a detectable signal [26, 27]. Overall, MIPs offer a promising alternative to antibodies in ELISA assays, providing robust and selective detection of target molecules with potential applications in diagnostics and biomedical research. However, the selectivity and affinity of MIPs are typically not as good as antibodies or aptamers.

3.3 Signaling Molecules, Enzymes, and Nanozymes

In ELISA reactions, signaling molecules and enzymes play crucial roles in detecting and quantifying the presence of target antigens or antibodies. Signal molecules,

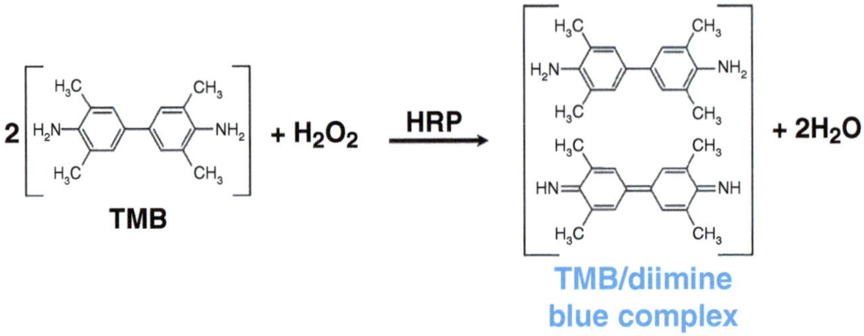

Figure 3.4 The oxidation reaction of 3,3′,5,5′-TMB with H_2O_2, catalyzed by HRP, generates a blue oxidation product. *Source:* Reproduced from [28] / with permission of Elsevier.

often referred to as reporter molecules, are typically linked to secondary antibodies and are used to generate a detectable signal. These signals can be colorimetric, fluorescent, luminescent, or others, depending on the type of enzyme and substrate used.

One of the most used enzymes in ELISA is HRP. HRP catalyzes the oxidation of substrates such as 3,3′,5,5′-tetramethylbenzidine (TMB), leading to a color change that can be detected spectrophotometrically (Figure 3.4). Another widely used enzyme is ALP, which catalyzes the hydrolysis of substrates like *p*-nitrophenyl phosphate (*p*NPP) to produce a yellow color that can also be measured [29]. Additionally, fluorescent signals can be generated using enzymes such as HRP in combination with fluorogenic substrates. For instance, HRP can oxidize Amplex Red to produce a fluorescent product that emits light at a specific wavelength. Luminescent signals are generated using enzymes like luciferase or HRP with chemiluminescent substrates, such as luminol, which emit light as a result of the enzyme-substrate reaction [1].

3.3.1 Enzyme/Antibody Conjugates

Enzyme/antibody conjugates are integral to the function of the ELISA, which are formed by linking an enzyme to an antibody. The enzyme serves as a reporter molecule that produces a measurable signal, while the antibody provides specificity by binding to the target antigen. Enzyme/antibody conjugates facilitate signal amplification through enzyme-substrate reactions. For instance, HRP can convert multiple molecules of TMB into a colored product, amplifying the signal generated by each antigen–antibody interaction. This allows for the detection of very low concentrations of the target molecule, enhancing the sensitivity and specificity of the assay. The enzyme/antibody conjugation process involves meticulous optimization to maintain antibody affinity and enzymatic activity. Various coupling chemistries, such as amine-reactive cross-linkers or sulfhydryl-based conjugation, offer flexibility in conjugate design. Moreover, advancements in recombinant antibody technology have spurred the development of engineered antibody variants with enhanced stability and specificity, further augmenting ELISA sensitivity. Here are some common methods of conjugation:

i) ***Chemical Crosslinking:***
   Enzyme/antibody conjugation involves the use of bifunctional reagents that form covalent bonds between the enzyme and the antibody. Some widely used cross-linkers include glutaraldehyde, which forms Schiff base linkages, and carbodiimides, which facilitate the formation of amide bonds between carboxyl and amine groups [30].
   a) Glutaraldehyde crosslinking: glutaraldehyde is a commonly used cross-linker that reacts with amino groups on both the enzyme and the antibody, forming stable covalent bonds. The reaction conditions, such as pH and temperature, must be carefully controlled to prevent over cross-linking, which can lead to loss of activity.
   b) Periodate oxidation: periodate oxidation is another method used, particularly for enzymes with carbohydrate moieties such as HRP. Sodium periodate oxidizes the sugar residues to form aldehyde groups, which can then react with amino groups on antibodies, forming Schiff bases that are subsequently reduced to stable secondary amines.
   c) Carbodiimide coupling: carbodiimide coupling, using reagents like EDC (1-ethyl-3-[3-dimethylaminopropyl] carbodiimide), is another popular approach. This method involves the activation of carboxyl groups on enzymes or antibodies, which then react with primary amines to form stable amide bonds.
ii) ***Avidin-biotin Interactions:***
   A non-covalent method of conjugation involves the use of avidin or streptavidin and biotin [31]. The high-affinity interaction between biotin and avidin/streptavidin allows for strong and specific binding. Enzymes can be biotinylated and then bound to avidin/streptavidin-conjugated antibodies, or vice versa. This method offers flexibility and simplicity but can introduce additional steps and potential non-specific binding.

The performance of enzyme/antibody conjugates in ELISA is influenced by several factors, including pH, temperature, and ionic strength. Enzymes have optimal pH and temperature conditions at which their activity is maximized. Deviation from these conditions can reduce enzyme activity and stability, impacting the assay's sensitivity and accuracy. Similarly, the ionic strength of the buffer can affect antibody binding and enzyme activity, necessitating careful optimization as well.

3.3.2 Nanozymes and Nanozyme/Antibody Conjugates

Traditional ELISA relies on natural enzymes like HRP or ALP to catalyze a substrate reaction that signals the presence of the target analyte. However, natural enzymes are limited by high preparation costs and are susceptible to inactivation. So, the search for more robust, efficient, and cost-effective alternatives has led to an intensive exploration of nanozymes and nanozyme/antibody conjugates.

Nanozymes are nanomaterials that can catalyze the conversion of an enzyme's substrate to product [32–34]. Although their definition is still controversial [33–35], the initial discovery was ground-breaking. In 2007, Gao et al. discovered that ferromagnetic nanoparticles possessed intrinsic peroxidase-like activity [36], and the field rapidly expanded after that. Researchers began to investigate a variety of nanomaterials, including metals, metal oxides, and carbon-based nanomaterials, for their enzyme-mimicking properties. Nanozymes exhibit catalytic properties akin to natural enzymes but offer advantages such as greater stability, lower cost, and tunable activity. Nanozymes can mimic various enzymatic activities, including oxidase, peroxidase, catalase, and superoxide dismutase, making them ideal candidates for signal amplification in ELISA.

The creation of nanozyme/antibody conjugates involves the functionalization of nanozymes with antibodies specific to the target analyte. This conjugation can be achieved through various chemical linkers or physical adsorption techniques, and most methods are similar to antibody conjugation [30]. The goal is to retain the bio-recognition ability of antibodies while leveraging the catalytic prowess of nanozymes. Nanozymes often exhibit lower catalytic turnover rates compared to natural enzymes, and there is potential for further improving the catalytic activity of nanozymes in ELISA. Their surface properties can be engineered to improve interactions with substrates and increase assay sensitivity [37].

3.3.3 Some Common Substrates in ELISA

The choice of substrate also greatly influences the sensitivity and detection limits of ELISA assays [38]. TMB and ABTS (2,2′-azino-bis(3-ethylbenzothiazoline-6-sulfonic acid diammonium salt) represent two of the most widely employed chromogenic substrates, each offering unique advantages in signal generation. TMB, upon enzymatic oxidation by HRP, yields a blue-colored product, the absorbance of which is related to analyte concentration. In some assays, a strong acid is added to the TMB reaction product to stop the reaction and convert the solution to a yellow color. Conversely, ABTS undergoes oxidative coupling in the presence of HRP, producing a green-colored product with distinct spectral properties. The selection of substrate depends on several factors, including assay format, detection instrumentation, and desired sensitivity. TMB exhibits high sensitivity and compatibility with commonly used ELISA plate readers, making it a preferred choice for routine assays. In contrast, ABTS offers enhanced signal stability and broader dynamic range, making it suitable for applications requiring prolonged incubation times or multiplexed detection [39]. Additionally, *p*NPP is a chromogenic substrate used with ALP. ALP hydrolyzes *p*NPP to *p*-nitrophenol, producing a yellow color with an absorbance maximum at 405 nm. Its simplicity and reliability make it suitable for diagnostic assays, including detecting various diseases and monitoring immune responses. Moreover, the advent of chemiluminescent substrates (such as luminol for HRP) and fluorescent substrates (such as 4-methylumbelliferyl phosphate [MUP]) has expanded the horizons of ELISA. High-sensitivity substrates enable the detection of low-abundance biomarkers, facilitating early diagnosis and treatment of diseases such as cancer, HIV, and other infectious diseases. These substrates, coupled with advances in detection instrumentation, empower researchers to push the limits of ELISA sensitivity.

3.4 Different Assay Formats

ELISA have diverse formats, each tailored to meet specific diagnostic or research objectives. In the realm of in vitro diagnosis, from direct detection of pathogen antigens to the quantification of host immune responses, ELISA assays offer a comprehensive toolkit for surveillance, outbreak investigation, and therapeutic monitoring [40, 41]. This part delves into the intricacies of different ELISA formats, elucidating their principles.

Traditional ELISA includes direct, indirect, sandwich, and competitive methods (Figure 3.1). Direct ELISA represents a straightforward and robust approach for the detection of bacterial and viral antigens. In this format, an analyte of interest is directly immobilized on a solid support. Subsequent incubation with a detection antibody conjugated to an enzyme, such as HRP or ALP, enables the visualization of antigen–antibody complexes through enzymatic conversion of chromogenic or fluorogenic substrates. A blocking step before adding enzyme-labeled antibodies is crucial. Otherwise, false positives might occur [42]. Direct ELISA offers rapid detection without the need for complex sample preparation, making it well-suited for point-of-care diagnostics and surveillance of acute infections.

Indirect ELISA enhances sensitivity by employing an additional antibody layer for signal amplification. Following the binding of immobilized antigen by a primary detection antibody, a secondary antibody, conjugated to an enzyme, recognizes and binds to the primary antibody. The enzymatic reaction catalyzed by the conjugated enzyme generates a detectable signal, indicative of antigen presence in the sample (Figure 3.1b). Indirect ELISA offers enhanced sensitivity, allowing for the detection of low-abundance antigens and the differentiation of closely related strains or serotypes [43].

Sandwich ELISA, also known as capture ELISA, epitomizes the epitope specificity and signal amplification capabilities inherent to ELISA techniques [29]. In this format, the target antigen present in the sample serves as the linchpin between two distinct antibodies, known as capture and detection antibodies, respectively. A solid-phase surface, typically a microtiter plate, is coated with capture antibodies, which specifically bind to epitopes on the target antigen. Subsequent incubation with a sample leads to the capture of the target antigen by the immobilized antibodies. Following antigen capture, a detection antibody, conjugated to an enzyme such as HRP or ALP, is introduced. This detection antibody recognizes a different epitope on the target antigen, forming a "sandwich" complex comprising the antigen bound between the capture and detection antibodies. Oftentimes, the capture antibody is not modified with an enzyme, and the signal is generated by introducing an enzyme-labeled secondary antibody specific for the capture antibody (Figure 3.1c). This way, the same enzyme-linked secondary antibody can be used for signaling a range of targets by binding to the same Fc region of the capture antibody. The conjugated enzyme catalyzes a chromogenic or fluorogenic substrate, generating a measurable signal directly proportional to the concentration of the target antigen in the sample. Sandwich ELISA offers exceptional sensitivity, specificity, and versatility [44].

Competitive ELISA finds utility in quantifying the concentration of antibodies or small molecules, such as toxins or metabolites, in biological samples [45]. In this format, antigen-coated wells compete with the target analyte present in a sample for binding to a limited supply of enzyme-conjugated antibodies. The signal intensity inversely correlates with the concentration of the analyte in the sample, as higher analyte concentrations would inhibit the formation of antigen–antibody complexes and subsequent enzymatic reactions (Figure 3.1d). Competitive ELISA offers precise quantification over a wide dynamic range, although the detection relies on the decrease of signal, which is analytically less desirable.

While traditional ELISA formats above continue to play pivotal roles in public health, other forms of ELISAs have emerged, capitalizing on technological advancements to enhance sensitivity, throughput, and multiplexing capabilities, thereby addressing the evolving needs of public health initiatives. For example, multiplex ELISA enables simultaneous detection and quantification of multiple analytes within a single assay. This format utilizes the microarray technology or a bead-based platform, wherein capture antibodies specific to different target antigens are immobilized at discrete locations or on distinct microspheres, respectively [46–48]. Following sample incubation and detection of antibody binding, the signal generated from each analyte is quantified using fluorescence, chemiluminescence, or colorimetric readouts. Multiplex ELISA offers high throughput, reduced sample volume requirements, and enhanced cost-effectiveness, facilitating comprehensive pathogen profiling.

Digital ELISA represents a significant advancement over traditional ELISA techniques. This method leverages digital technology to provide quantifiable results at the single-molecule level, making it an essential tool in diagnostics, particularly for detecting trace amounts of biomarkers [49, 50]. Digital ELISA operates by leveraging microscopic beads, each coated with capture antibodies specific to the target antigen. These beads are mixed with a sample, allowing the target antigens to bind to the capture antibodies. Subsequently, detection antibodies conjugated with enzymes (such as HRP) are added, forming a sandwich complex with the antigen between the capture and detection antibodies. The beads are then distributed into thousands of tiny wells or droplets, ensuring each well contains at most one bead, effectively isolating individual antigen molecules. A substrate is then introduced to react with the enzyme, producing a detectable signal (fluorescent or colorimetric) in wells containing the target antigen. The wells are scanned and analyzed digitally, with signals indicating the presence of antigen, allowing for precise counting and quantification of individual molecules. Despite its higher cost and complexity, the benefits of early and precise detection of pathogens make digital ELISA a powerful tool in modern diagnostics. Digital ELISA is still a sandwich ELISA, which requires the target to be combined with a pair of antibodies. The background signal caused

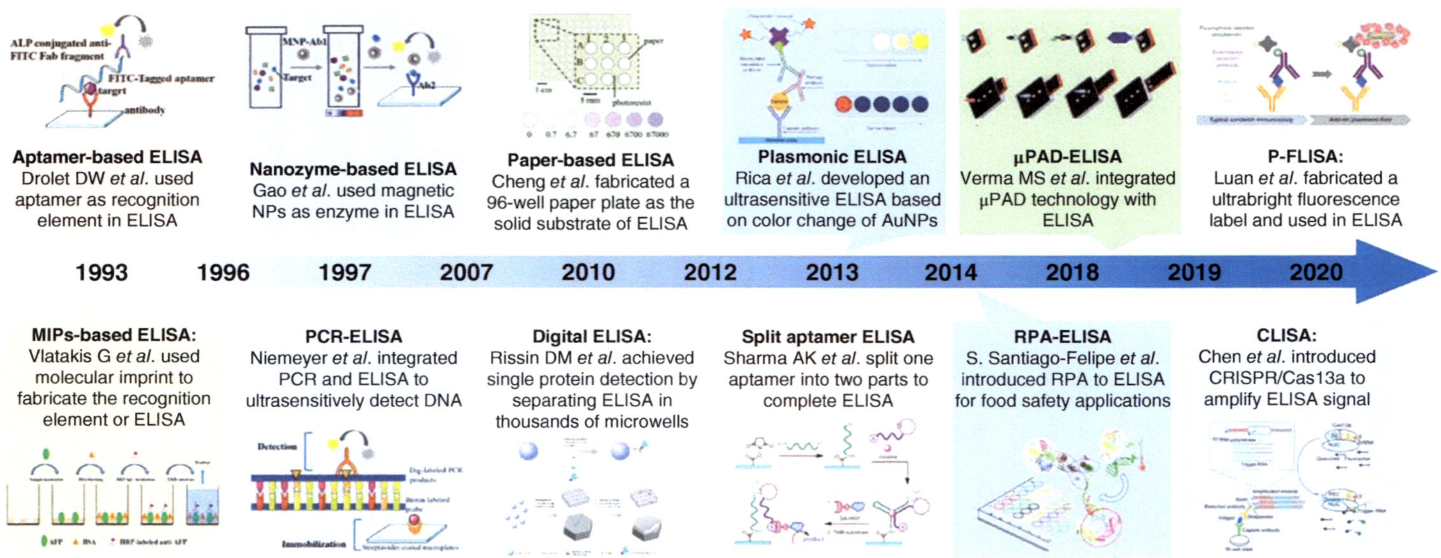

Figure 3.5 An overview of the evolution of representative ELISA format-derived technologies in recent decades. *Source:* Reproduced from Ref. [2] / with permission of Elsevier.

by non-specific adsorption of microbeads, enzyme labels, surfaces, and other reagents cannot be ignored.

Electrochemical ELISA harnesses the principles of electrochemistry to detect and quantify antigen–antibody interactions, offering rapid, sensitive, and label-free detection of bacterial and viral pathogens [51]. In this format, electrodes functionalized with capture antibodies selectively bind to target antigens, modulating electrochemical signals upon antigen binding. The magnitude of the electrochemical response correlates with the concentration of the target antigen in the sample, enabling quantitative analysis with high sensitivity and specificity. Electrochemical ELISA holds promise for point-of-care diagnostics, environmental monitoring, and surveillance of infectious diseases in resource-limited settings. Figure 3.5 shows more ELISA formats not yet discussed, including paper-ELISA, plasmonic ELISA, PCR-ELISA, and so on. By harnessing the principles of immunology, enzymology, microfluidics, and electrochemistry, these diverse ELISA methods empower researchers and clinicians to combat infectious threats with precision, efficiency, and innovation.

3.5 ELISA for Viral Detection: SARS-CoV-2 as an Example

After the general description of ELISA, a few examples of applications will be discussed. In late 2019, an acute respiratory disease pandemic rapidly spread worldwide, significantly impacting people's lives for several years. The virus responsible for this pandemic was reported in December and initially named 2019 Novel Coronavirus (2019-nCoV). It was later renamed SARS-CoV-2, and the viral pneumonia it caused was named COVID-19. SARS-CoV-2 is the seventh coronavirus known to infect humans and belongs to the *Betacoronavirus* genus and *Sarbecovirus* subgenus. It is a positive-sense single-stranded RNA virus [52]. Its genome shares 79% sequence identity with SARS-CoV and 50% with MERS-CoV [53]. The virus comprises four proteins: spike protein (S), nucleocapsid (N), small membrane envelope protein (E), and a membrane protein (M) (Figure 3.6). SARS-CoV-2 posed a significant threat to global public health, surpassing both SARS and MERS in terms of infection population and spread. The experiences and challenges encountered during the fight against SARS-CoV-2 remain crucial today. This part introduces the ELISA method used in SARS-CoV-2 diagnosis to highlight the methodology of ELISA-based viral diagnostics.

3.5.1 SARS-CoV-2 Antigen Detection

The target proteins commonly used for SARS-CoV-2 antigen detection are the spike protein (S), including the receptor-binding domain (RBD), and the nucleocapsid (N) protein. These proteins contain conserved, specific peptide sequences. Compared to antibody detection, antigen detection can be performed before clinical symptoms appear or during the acute infection phase.

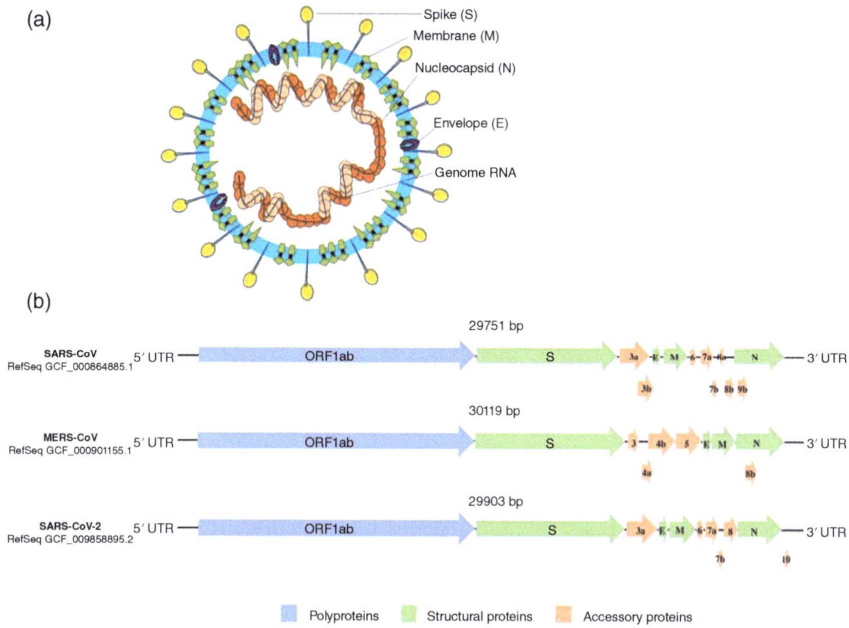

Figure 3.6 (a) Schematic diagram of the SARS-CoV-2 virus structure. (b) Comparison of the genome structures of SARS-CoV, MERS-CoV, and SARS-CoV-2. *Source:* Ref. [54] / with permission of American Chemical Society.

Adnan and co-workers illustrate the classic sandwich ELISA method for detecting SARS-CoV-2 antigens as follows [55]:

Preparation procedures: monoclonal antibodies (capture antibodies) against the antigen are diluted to the appropriate concentration in a coating buffer (pH > 9 bicarbonate buffer). 100 μL/well is used to coat a 96-well plate, incubated overnight at 4 °C (or one hour at 37 °C), and unbound antibodies are discarded. The wells are then blocked with 150 μL/well of blocking buffer (PBS, 0.1% Tween-20, 2% BSA) at 37 °C for one hour. After washing three times with washing buffer (50 mM Tris, 0.05% Tween-20, 0.1% SDS, 0.8% NaCl), the detection process begins.

Detection procedures: samples from the nasopharyngeal area are diluted to the appropriate concentration with dilution buffer (PBS, 0.1% Tween-20, 1% BSA) and added to the wells, incubated at 37 °C for 15 minutes, followed by five washes with washing buffer. Then, HRP-labeled detection antibodies are added, incubated at 37 °C for 10 minutes, followed by the washing steps, and then color development. Alternatively, unlabeled monoclonal antibodies can be used, followed by incubation and washing. Then, HRP-labeled secondary antibodies were added. After 5 minutes of incubation and washing for five times, 100 μL of TMB substrate was added. The reaction is incubated in the dark at room temperature for another five minutes before adding 100 μL of stop solution (1.5 M H_2SO_4) and measuring absorbance at 450 nm for quantification. As described here, this is quite a lengthy process and requires an experienced technician and equipment for the detection. So, this method is applicable in hospitals and other centralized labs.

Mladenovic Stokanic et al. [56] employed polyclonal antibodies from two species for nucleocapsid protein detection, achieving a detection limit of 0.93 ng/mL and a linear range of 1.52–48.83 ng/mL. Despite its lower sensitivity, this is a rapid, simple, and cost-effective antigen detection method. Apart from the classical ELISA mode, Girt et al. [57] utilized nanobodies against the spike protein in place of monoclonal antibodies in ELISA, achieving a detection limit of 21 $TCID_{50}$/mL for pseudoviruses. It could detect 305 pfu/mL of X-ray-irradiated (12.19 kGy) inactivated SARS-CoV-2, and 103 ffu/mL of Empigen (a zwitterionic detergent, 0.05%) and heat (60 °C, 30 minutes) inactivated SARS-CoV-2. Using site-selective biotinylated nanobodies further improved the detection limit to 16 $TCID_{50}$/mL for pseudoviruses and 16 ffu/mL for Empigen-inactivated viruses, significantly simplifying the detection process for human samples. Di Domenico et al. [58] used polyvinylidene fluoride (PVDF) membranes in place of ELISA plates for detection, achieving a detection limit of 5 ng/mL for spike protein and nucleocapsid proteins. The detection time is within 30 minutes, making it suitable for point-of-care and large-scale screening.

There is also a report of not using antibodies at all. Pomplun et al. [59] utilized a high-affinity peptide screened against the spike protein RBD to detect the RBD in a fetal bovine serum mixture (Figure 3.7). The principle involves adding the mixture to an ELISA plate for adsorption, followed by binding with a biotin-labeled affinity peptide and the addition of streptavidin-conjugated HRP and TMB substrate. This method can detect concentrations between 100 pM and 100 nM with a signal significantly stronger than the background.

One effective method for evaluating antibody performance is to construct an ELISA targeting SARS-CoV-2 antigens, and detecting inactivated SARS-CoV-2 improves detection safety and simplifies procedures. In addition, ELISA can also be used to measure protein subunit-based SARS-CoV-2 vaccines [60].

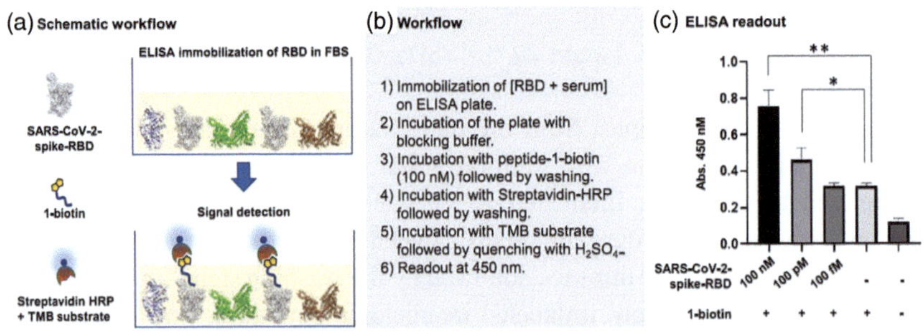

Figure 3.7 Detection of SARS-CoV-2 spike RBD by ELISA. (a) Schematic diagram of the ELISA assay. (b) Assay workflow. (c) Relationship between ELISA absorbance reading and RBD concentration. *Source:* Ref. [59] / American Chemical Society / CC BY 4.0.

3.5.2 SARS-CoV-2 Antibody Detection

During the COVID-19 pandemic, another application of ELISA is serological testing. This involves the detection of the presence of SARS-CoV-2 antibodies generated through adaptive immune responses to determine whether an individual has been previously infected with COVID-19. During the early stages of SARS-CoV-2 infection, there is a significant expansion of memory B cells and plasma cells. These cells secrete serum IgM and IgA antibodies around days 5–7 after the onset of symptoms, and IgG antibodies around days 7–10. Typically, the titers of IgM and IgA decrease approximately 28 days later, while the IgG titer peaks around 49 days [61]. Detecting IgM and IgA antibodies in serum can aid in the early diagnosis of SARS-CoV-2 infection, whereas IgG measurement can assess the immune status of the body after infection or vaccination [62], and dried blood spot sampling can also meet the testing needs [63].

Classical ELISA serological testing involves using a 96-well polystyrene plate coated with purified recombinant SARS-CoV-2 proteins, followed by the addition of serum samples and HRP-conjugated anti-human IgG antibodies. The procedures are like antigen detection and can be referenced from the method described in Klumpp-Thomas et al.'s published work [64]. Here are some additional examples of novel detection strategies. Fernandez et al. [65] utilized chimeric recombinant proteins designed from the S1 subunit of the S protein and the N protein as antigens in an ELISA to detect SARS-CoV-2-specific IgG. They achieved better sensitivity (95.7%) and specificity (95.5%) than the traditional mode. Wouters et al. [66] developed a competition ELISA based on the competition between neutralizing antibodies (all isotypes) in convalescent plasma and a single commercially available neutralizing mouse monoclonal antibody (mAb). This assay was used for the rapid quantification of SARS-CoV-2 neutralizing antibodies in convalescent plasma. The competition ELISA screened neutralizing antibodies in COVID-19 convalescent plasma by competing for a single epitope, achieving a sensitivity and specificity of up to 61% and 100%, respectively. Kasetsirikul et al. [67] utilized Whatman chromatography paper pressed into a paper-based platform, replacing the ELISA plate, to achieve paper-based ELISA detection of SARS-CoV-2 humanized antibodies. This assay can detect the SARS-CoV-2 humanized antibody at 10 ng/μL with RSD around 10%, and it can complete the detection within 30 minutes, offering advantages over traditional ELISA.

Compared to the gold standard RT-qPCR for the detection of SARS-CoV-2, ELISA has both advantages and limitations. The abundance of different targets during the infection process indicates the utility of these two methods at different time points (Figure 3.8). Antigen ELISA and RT-qPCR samples are taken from the upper respiratory tract, whereas antibody ELISA samples are taken from saliva and blood, leading to less consistency. The diagnostic time for the antibody ELISA is also later. However, ELISA is less expensive than RT-qPCR, making it a more convenient method for large-scale screening in resource-limited areas.

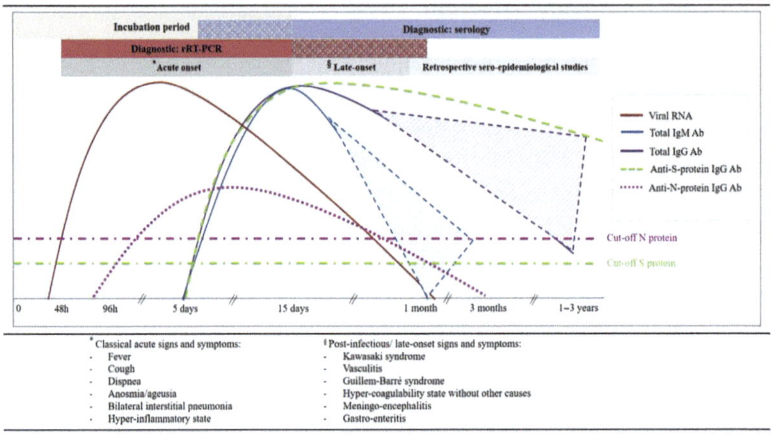

Figure 3.8 Dynamic changes in the production of SARS-CoV-2 markers during infection and laboratory diagnosis. *Source:* Ref. [68] / with permission of Elsevier.

3.6 ELISA for Bacterial Identification

ELISA is widely used in the field of microbiology for the identification and quantification of bacterial species. This technique leverages the specificity of antibodies to detect and measure bacterial antigens, making it an essential method in both clinical diagnostics and research laboratories. Common target molecules include surface antigens like proteins, polysaccharides, and lipopolysaccharides. Additionally, toxins and enzymes produced by bacteria can also serve as detection targets. ELISA can be performed using various sample types: whole bacterial cells, which are straightforward but sometimes less sensitive; bacterial lysates, which enhance sensitivity by releasing intracellular markers; and purified antigens, which offer high specificity but require complex preparation. This versatility makes ELISA an indispensable tool in both clinical diagnostics and research, enabling the precise identification of diverse bacterial species. This section mainly introduces detection of bacteria and their biomarkers.

For the detection of whole bacterial cells, Wang et al. [44] developed a portable, highly sensitive detection platform for *Staphylococcus aureus*, a significant foodborne pathogen (Figure 3.9). The study utilized mesoporous core-shell Pd@Pt nanozymes as signal amplifiers in a colorimetric nanozyme-linked immunosorbent assay (N-ELISA). Three immunoassay types were compared: conventional ELISA (Type I), Pd@Pt nanozyme-labeled antibodies (Type II), and Pd@Pt-HRP bioprobes combining Pd@Pt nanozymes with HRP (Type III). The Pd@Pt-HRP bioprobes demonstrated the highest sensitivity, achieving a limit of detection (LOD) of 1×10^3 CFU/mL, significantly lower than the conventional ELISA's 1×10^5 CFU/mL. This enhanced sensitivity was attributed to the superior catalytic activity of the Pd@Pt nanozymes, which outperformed natural HRP. The portable

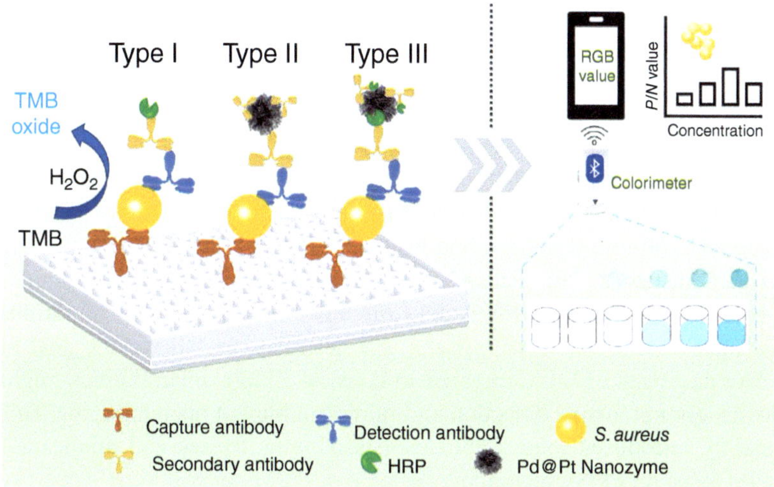

Figure 3.9 Schematic diagram illustrating the principle of enhancing ELISA for portable detection of *Staphylococcus aureus* using Pd@Pt nanozymes. *Source:* Ref. [44] / with permission of Elsevier.

colorimetric readout, captured using a smartphone, facilitates rapid, on-site pathogen detection and reflects the advantages of nanozyme applications.

For the detection of bacterial biomarkers, Montagut et al. [69] explored the use of the Pseudomonas quinolone signal (PQS), a key molecule in the quorum-sensing system of *Pseudomonas aeruginosa*, as a biomarker for infection. The researchers developed antibodies against PQS and created a microplate-based ELISA capable of detecting PQS at low nanomolar concentrations in complex biological media. The assay demonstrated an LOD of 0.36 ± 0.14 nM in culture broth media. This method allowed for the monitoring of PQS levels in clinical isolates from patients with different infection statuses, showing a clear correlation between PQS levels and the chronic or acute nature of the infection. The innovative aspect of this study lies in its development of a reliable and sensitive immunoassay for PQS, offering potential for improved diagnostics and patient management in *Pseudomonas aeruginosa* infections by enabling precise pathogen identification and disease progression monitoring. Additionally, Cui et al. [70] developed an indirect competitive ELISA (ic-ELISA) to quantify enterobactin (Ent), a siderophore produced by Enterobacteriaceae, which serves as a biomarker for gut health. The method involved immunizing mice with an Ent conjugate vaccine, producing monoclonal antibodies with high affinity (3.1×10^{-10} M) and specificity. The ic-ELISA demonstrated an LOD of 0.39 μg/mL and high accuracy, with intra- and inter-assay recovery rates up to 94.6% and coefficients of variation between 4.0% and 12.3%. This assay was validated for detecting Ent in bacterial cultures, proving effective for monitoring gut inflammation and microbiota dysbiosis. This work features highly specific mAbs and a sensitive, reproducible ELISA, providing a robust tool for gut health diagnostics.

3.7 ELISA in Microfluidic Devices and Other Advanced Applications

Microfluidics refers to systems featuring integrated channels on the microscale, ranging from tens to hundreds of micrometers. These channels allow the controlled and systematic manipulation of small fluid quantities [71]. The advantages of microfluidics include an increased surface-to-volume ratio, which enhances mass transport in immunoreactions and leads to faster analysis. Additionally, the miniaturized dimensions of the microchannels reduce the consumption of samples and reagents. The technology also enables the automated integration of various functions, such as valves, pumps, mixers, and detectors, to achieve the goal of point-of-care diagnostics [72]. Compared to classical ELISA, microfluidics might be able to achieve automated detection and minimize human operated steps. This section mainly introduces some improved microfluidic devices and application technologies based on ELISA technology.

3.7.1 Microfluidic-Integrated ELISA

Jose and co-workers introduced a 3D reconfigurable microfluidic platform with planarized through-hole valves for pipette-free ELISA, significantly advancing point-of-care diagnostics [73]. This innovative system integrates latching valves and compression-based sealing to ensure leak-free operation and automation of ELISA procedures. The platform successfully detected TNF-α (tumor necrosis factor-alpha) with an LOD of 4.21 pg/mL, achieving results comparable to traditional 96-well plate methods. The use of smartphone-based colorimetric analysis further validated the platform's accuracy and reliability, highlighting its potential for portable, efficient, and decentralized diagnostic testing.

Zhou et al. presented a novel sample-to-answer robotic ELISA platform that integrates a robotic-microfluidic interface (RoMI) with a modular hybrid microfluidic chip incorporating a highly sensitive nanofibrous membrane, referred to as the robotic ELISA [74]. This system automates the entire ELISA process, including high-precision pipetting, efficient mixing, washing, and colorimetric readout via machine vision (Figure 3.10). The platform demonstrated a low LOD of 0.1 ng/mL for chloramphenicol in a 15 μL sample within 20 minutes, significantly faster than traditional ELISA methods. The use of pneumatically driven high-precision metering and back-and-forth flows for mixing and enrichment enhanced antibody–antigen binding and reduced incubation time. These features highlight the platform's potential for rapid, reliable, and decentralized diagnostic testing in resource-limited settings or high-risk environments.

Yang et al. [75] introduced a microfluidic magnetic analyte delivery (μMAD) technique designed for ultrasensitive detection of trace biomarkers. This method utilizes a polydimethylsiloxane (PDMS)-glass hybrid microchip and functionalized magnetic beads (MBs) to isolate and preconcentrate analytes from complex samples within microchannels, facilitating efficient fluorescence detection. Compared to

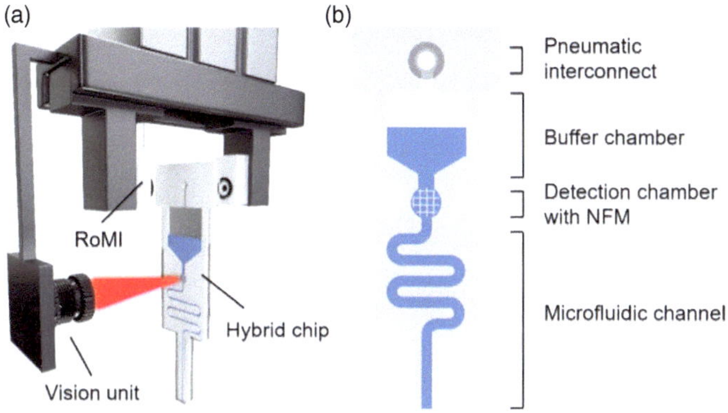

Figure 3.10 (a) Schematic representation of the robotic-assisted ELISA platform; (b) exploded view of the hybrid microfluidic chip containing the nanofibrous membrane (NFM). *Source:* Reproduced from Ref. [74] / with permission of American Chemical Society.

traditional ELISA, the μMAD-ELISA method significantly enhances sensitivity, achieving an LOD of 0.29 pg/mL for carcinoembryonic antigen (CEA), 0.047 pg/mL for prostate-specific antigen (PSA), and 0.021 pg/mL for interleukin-6 (IL-6), representing over a 200-fold sensitivity improvement. This integration of magnetic bead technology with ELISA principles highlights substantial advancements in selectivity, reproducibility, and operational simplicity.

He et al. [76] developed a cPlate (composable microfluidic plates) system designed to address the limitations of conventional ELISA, such as high reagent consumption and lengthy procedures. The cPlate system employs a fluid manipulation method using miniaturized 96-well plates and corresponding channel plates. This design allows for the simultaneous execution of 96 independent ELISA reactions without external equipment. The method demonstrated high sensitivity and specificity (~pg/mL), shorter assay time (~1 hour), and low sample consumption (~5 μL/well). The system was validated by quantifying four protein biomarkers in serum samples, showing comparable performance to commercial ELISA kits. The innovative aspect lies in its simplicity, scalability, and instrument-free operation, making it suitable for multiplexed point-of-care testing in resource-limited settings.

Van Nguyen et al. [77] developed a centrifugal microfluidic device to perform ELISA assays in a high-throughput and automated manner. Using a rapid dispenser structure with aliquoting microchannels, ultra-fast reagent loading, and precise handling of small volumes, a miniaturized diagnostic system was achieved (Figure 3.11). This system can process 40 samples simultaneously within 75 minutes, significantly reducing the time, cost, and labor compared to conventional methods. The device demonstrated an LOD of 10.2 ng/mL for SARS-CoV-2, successfully identifying various clinical samples and SARS-CoV-2 variants. This breakthrough enhances ELISA's applicability for point-of-care testing, addressing the critical need for rapid, high-throughput diagnostics during pandemics.

Iria et al. [78] developed a microfluidic ELISA (mELISA) to quantify Infliximab in plasma samples from inflammatory bowel disease patients. This method utilizes

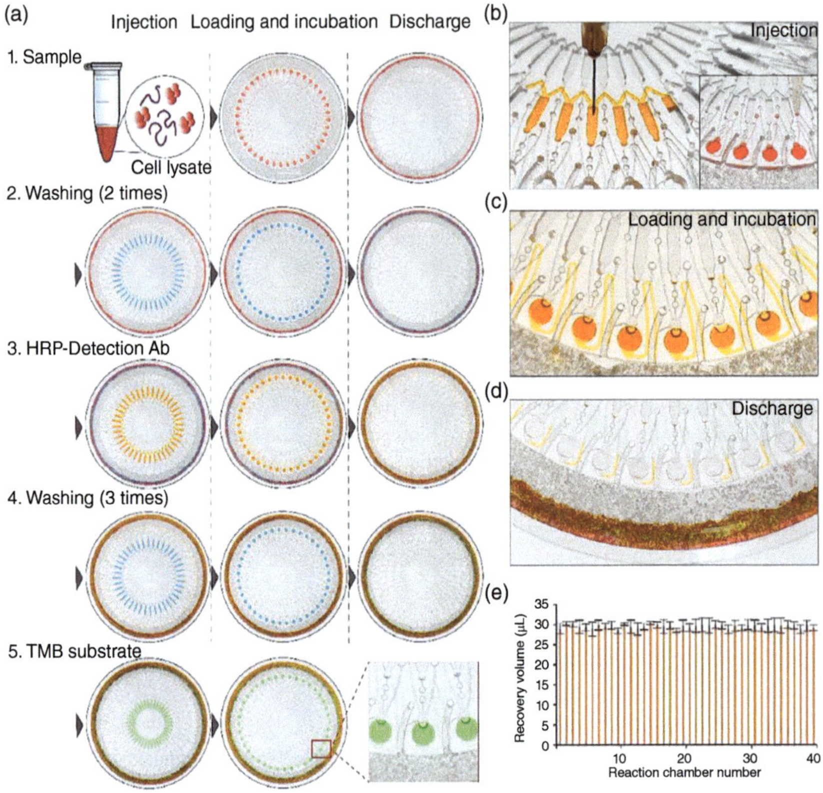

Figure 3.11 ELISA centrifugal microfluidic device for detecting SARS-CoV-2. (a) Workflow of the microfluidic device: 40 cell lysate samples are loaded into 40 antibody-coated reaction chambers and discharged into the waste chamber by centrifugal force at a speed of 5,000 rpm. (b) The ELISA analyzer can be programmed to load designated solutions into the dispensing chambers via syringe. (c) The spin drives the reagents into the antibody-coated reaction chambers without overflowing into the waste chamber due to the siphon valve. (d) The spin drives the reagents into the waste chamber. (e) The recovery volume of the reaction chamber demonstrates the high accuracy of the dispensing. *Source:* Reproduced from Ref. [77] / with permission of Elsevier.

a microfluidic structure fabricated through standard photolithography and PDMS molding techniques. The process involves coating microchannels with TNF-α, blocking it with casein, and sequentially introducing patient samples and detection antibodies using a 12-channel syringe pump. Chemiluminescent detection was performed using a fluorescence microscope, achieving results in just 24 minutes, significantly faster than conventional ELISA. The mELISA demonstrated a lower LOD of 0.026 μg/mL, which is 5.8 times more sensitive than commercial point-of-care devices. The study's results showed high correlation with conventional ELISA ($r = 0.998$; $R^2 = 0.996$; $p < 0.0001$), validating its accuracy. By integrating a simplified microfluidic design with rapid and sensitive detection, this method offers a robust, user-friendly platform for point-of-care applications.

3.7.2 Paper-Based Microfluidic ELISA

Paper-based mELISA is an innovative approach that integrates ELISA onto a paper-based microfluidic platform. This technique combines the simplicity and portability of paper-based devices with the sensitivity and specificity of traditional ELISA. This method typically involves patterning microfluidic channels on paper, using wax or other hydrophobic materials to define the flow paths for reagents and samples. Functionalized paper regions capture target analytes, while subsequent reagent flows facilitate the ELISA process, including antibody binding, washing, and colorimetric detection. This approach is particularly advantageous for point-of-care testing, offering low-cost, rapid, and easy-to-use diagnostic solutions without the need for sophisticated laboratory equipment. For instance, Gong et al. [79] developed a cost-effective, portable microfluidic platform for SARS-CoV-2 diagnosis. This system uses a spinning top for rapid serum separation and paper-based mELISA (μPAD) for antibody detection. Serum is isolated within five minutes and applied to wax-printed μPADs, where anti-RBD IgA, IgM, and IgG antibodies are measured using a colorimetric reaction and quantified via smartphone. The method showed high sensitivity (99.7% for early-stage, 100% for late-stage patients) and specificity (100% and 94.3%, respectively). This low-cost (~$5) and quick (~1 hour) platform is ideal for point-of-care testing, particularly in resource-limited settings. Verma et al. [80] introduced a 3D microfluidic μPAD for conducting ELISA assays. This device comprises a sliding strip containing the active sensing area and a surrounding structure that stores and distributes reagents. The device facilitates ELISA by manually moving the sliding strip to bring samples and reagents into contact in a sequential manner. The method demonstrated the ability to detect C-reactive protein in a concentration range of 1–100 ng/mL in 1,000-fold diluted blood. The integration of stored reagents within the device and a sliding mechanism eliminate the need for external equipment, making the device suitable for resource-limited places.

3.8 ELISA Using Aptamers

The emergence of aptamers as alternative recognition elements has opened new avenues for ELISA assays. This part explores the integration of aptamers into ELISA assays, highlighting their unique properties and applications in viral pathogen diagnostics and surveillance.

Park et al. developed an aptamer-based ELISA (ELASA) for quantifying infectious hepatitis C virus (HCV), which utilizes two aptamers that recognize different regions of HCV E2 envelope protein. In the ELASA system, one aptamer (capture aptamer) was immobilized on a 96-well plate, while the other aptamer (detection aptamer) was linked to HRP via a biotin-streptavidin interaction (Figure 3.12). This system demonstrated high sensitivity, detecting HCV at concentrations as low as 3.13×10^2 FFU/mL. The ELASA readout values linearly correlated with the

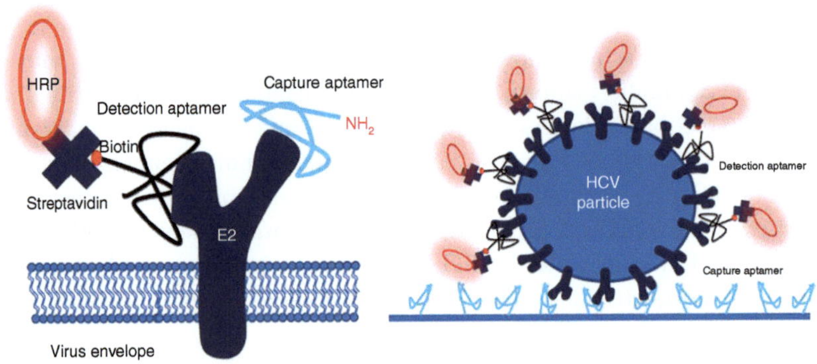

Figure 3.12 Schematic diagram of aptamer-based ELASA. A pair of DNA aptamers specifically interacting with different parts of the E2 protein on the HCV virion particle is depicted. *Source:* Ref. [81] / with permission of Elsevier.

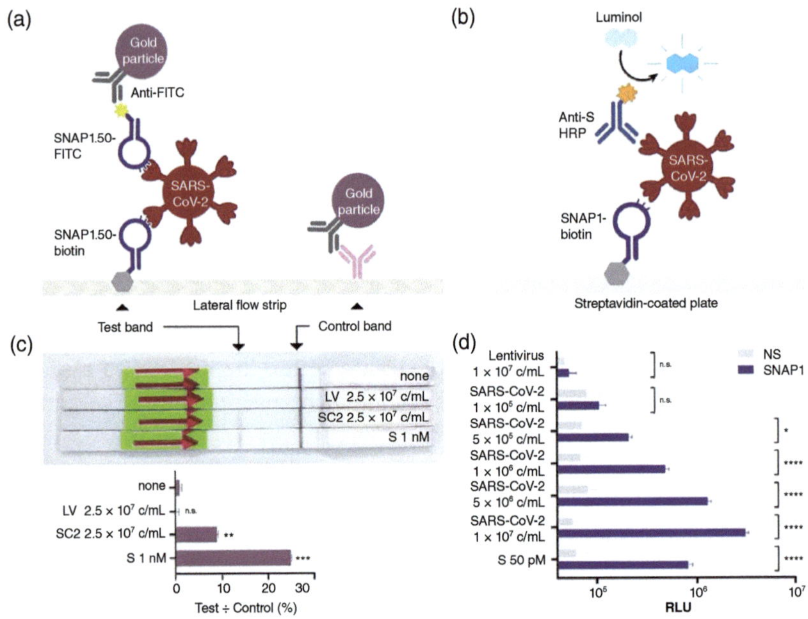

Figure 3.13 Aptamer-based detection of inactivated SARS-CoV-2 virus. (a) Schematic of a lateral flow strip for detection. (b) Schematic of aptamer-antibody sandwich ELISA. (c) Lateral flow strip detection results for S protein (S), control lentivirus (LV), and inactivated SARS-CoV-2 virus (SC2). (d) ELISA detection results for various concentrations of inactivated SARS-CoV-2 virus. *Source:* Reproduced from Ref. [82] / with permission of John Wiley & Sons.

infectious dose of HCV particles, but not with the amount of HCV RNA, providing a more accurate measure of infectivity. Additionally, this ELASA system proved effective in monitoring the activity of anti-HCV drugs, demonstrating its potential utility in both diagnostic and therapeutic contexts. The use of aptamers in place of antibodies and the application of advanced SELEX technology highlight significant potential in the field of viral infectivity assays.

Kacherovsky et al. [82] presented the discovery of DNA aptamers that specifically bind to the N-terminal domain of the SARS-CoV-2 spike protein (Figure 3.13). The researchers identified high-affinity aptamers by SELEX technology, particularly SNAP1 (SARS-CoV-2 spike protein N-terminal domain-binding aptamer 1), with a dissociation constant (K_d) of approximately 39 nM. This aptamer demonstrated exceptional specificity for the SARS-CoV-2 spike protein, with no binding observed to the related spike proteins of SARS-CoV and MERS-CoV. High-resolution cryo-electron microscopy confirmed the binding site of SNAP1 on the NTD of the spike protein. The aptamers were employed in lateral flow assays (LFA) and ELISA to detect UV-inactivated SARS-CoV-2 at concentrations as low as 2.5×10^7 and 5×10^5 copies/mL, respectively. This work highlights the potential of the SNAP1 aptamer as a rapid diagnostic tool for SARS-CoV-2, offering a cheaper and easier-to-produce alternative to antibodies. Similar strategies have also been reported to be used in the detection of the Zika virus [83].

3.9 Conclusions and Future Directions

ELISA, with its high efficiency, sensitivity, and specificity, has become an indispensable tool in public health. This chapter has reviewed the development of ELISA technology and its applications in the detection of viruses, bacteria, and other biomarkers related to public health. The performance of ELISA relies on the use of recognition molecules from traditional antibodies to emerging molecules and materials, including aptamers and MIPs, along with signal amplification techniques. Additionally, microfluidic devices have further promoted the application of ELISA in on-site rapid testing and high-throughput screening. Looking to the future, several directions can be anticipated for the advancement of ELISA technology:

- Integration with advanced technologies: Combining ELISA with advanced technologies such as digital assays and electrochemical detection can further enhance sensitivity and specificity. These integrations could facilitate ultrasensitive detection and allow for rapid and more accurate detection, although these assays have higher requirements for hardware.
- Microfluidic and portable ELISA systems: The evolution of microfluidic technology will likely lead to the development of more compact, portable, and automated ELISA systems. These systems can provide rapid and accurate diagnostics at the point of care, which is crucial during outbreaks and in remote areas.
- Multiplexed and high-throughput assays: Advances in multiplexed ELISA will enable the simultaneous detection of multiple analytes in a single assay, improving diagnostic efficiency and reducing sample volume and costs. This is particularly valuable for comprehensive pathogen profiling and monitoring complex disease states.

In conclusion, ELISA will continue to evolve with technological advancements, expanding its applications and maintaining its critical role in public health diagnostics. The ongoing innovation in ELISA technology promises enhanced detection capabilities, contributing significantly to global health and disease management efforts.

References

1 Hosseini, S., Vázquez-Villegas, P., Rito-Palomares, M., and Martinez-Chapa, S.O. (2018). *Enzyme-Linked Immunosorbent Assay (ELISA): From a to Z*. Springer.

2 Peng, P., Liu, C., Li, Z. et al. (2022). Emerging ELISA derived technologies for in vitro diagnostics. *TrAC Trends in Analytical Chemistry* 152: 116605.

3 Aydin, S. (2015). A short history, principles, and types of ELISA, and our laboratory experience with peptide/protein analyses using ELISA. *Peptides* 72: 4–15.

4 Dita, M.C. (2021). Enzyme-linked immunosorbent assay (ELISA): a narrative literature review. *Natural Sciences Engineering and Technology Journal.* 1 (2): 29–38.

5 Ahirwar, R., Bhattacharya, A., and Kumar, S. (2022). Unveiling the underpinnings of various non-conventional ELISA variants: a review article. *Expert Review of Molecular Diagnostics.* 22 (7): 761–774.

6 Reen, D.J. (1994). Enzyme-linked immunosorbent assay (ELISA). *Basic Protein Peptide Protocols.* 461–466.

7 De Meyer, T., Muyldermans, S., and Depicker, A. (2014). Nanobody-based products as research and diagnostic tools. *Trends in Biotechnology.* 32 (5): 263–270.

8 Li, D., Morisseau, C., McReynolds, C.B. et al. (2020). Development of improved double-Nanobody Sandwich ELISAs for human soluble epoxide hydrolase detection in peripheral blood mononuclear cells of diabetic patients and the prefrontal cortex of multiple sclerosis patients. *Analytical Chemistry.* 92 (10): 7334–7342.

9 Zhang, Y., Wang, T., Zhang, P. et al. (2024). Facile construction of sandwich ELISA based on double-nanobody for specific detection of alpha-hemolysin in food samples. *Talanta.* 274: 126021.

10 Drolet, D.W., Moon-McDermott, L., and Romig, T.S. (1996). An enzyme-linked oligonucleotide assay. *Nature Biotechnology.* 14 (8): 1021–1025.

11 Moreno, M., García-Sacristán, A., Martín, M.E., and González, V.M. (2022). Enzyme-Linked Oligonucleotide Assay (ELONA). In: *Nucleic Acid Aptamers: Selection, Characterization and Application*, 235–242. Springer.

12 Chen, C., Luo, J., Li, C. et al. (2018). Molecularly imprinted polymer as an antibody substitution in pseudo-immunoassays for chemical contaminants in food and environmental samples. *Journal of Agricultural and Food Chemistry.* 66 (11): 2561–2571.

13 Yadav, A.K., Verma, D., Dalal, N. et al. (2022). Molecularly imprinted polymer-based nanodiagnostics for clinically pertinent bacteria and virus detection for future pandemics. *Biosensors and Bioelectronics: X.* 12: 100257.

14 Ellington, A.D.S.J.W. (1992). Selection in vitro of single-stranded DNA molecules that fold into specific ligand-binding structures. *Nature.* 355 (6363): 850–852.
15 Dunn, M.R., Jimenez, R.M., and Chaput, J.C. (2017). Analysis of aptamer discovery and technology. *Nature Reviews Chemistry.* 1 (10): 0076.
16 Szostak, A.D.E.J.W. (1990). in vitro selection of RNA molecules that bind specific ligands. *Nature.* 346 (6287): 818–822.
17 Tuerk, C. and Gold, L. (1990). Systematic evolution of ligands by exponential enrichment: RNA ligands to bacteriophage T4 DNA polymerase. *Science.* 249 (4968): 505–510.
18 Guo, W., Zhang, C., Ma, T. et al. (2021). Advances in aptamer screening and aptasensors' detection of heavy metal ions. *Journal of Nanobiotechnology.* 19 (1): 166.
19 Gu, L., Ding, Y., Zhou, Y. et al. (2024). Selective Hemin binding by a non-G-quadruplex Aptamer with higher affinity and better peroxidase-like activity. *Angewandte Chemie International Edition.* 63 (6): e202314450.
20 Niu, C., Zhang, C., and Liu, J. (2022). Capture-SELEX of DNA Aptamers for estradiol specifically and estrogenic compounds collectively. *Environmental Science & Technology.* 56 (24): 17702–17711.
21 Zhao, Y., Ong, S., Chen, Y. et al. (2022). Label-free and dye-free fluorescent sensing of tetracyclines using a capture-selected DNA Aptamer. *Analytical Chemistry.* 94 (28): 10175–10182.
22 Tan, W., Donovan, M.J., and Jiang, J. (2013). Aptamers from cell-based selection for bioanalytical applications. *Chemical Reviews.* 113 (4): 2842–2862.
23 Stoltenburg, R., Reinemann, C., and Strehlitz, B. (2007). SELEX--a (r)evolutionary method to generate high-affinity nucleic acid ligands. *Biomolecular Engineering.* 24 (4): 381–403.
24 Hoshino, Y., Kodama, T., Okahata, Y., and Shea, K.J. (2008). Peptide imprinted polymer nanoparticles: a plastic antibody. *Journal of the American Chemical Society.* 130 (46): 15242–15243.
25 Poma, A., Turner, A.P., and Piletsky, S.A. (2010). Advances in the manufacture of MIP nanoparticles. *Trends in Biotechnology.* 28 (12): 629–637.
26 Shutov, R.V., Guerreiro, A., Moczko, E. et al. (2014). Introducing MINA–the molecularly imprinted nanoparticle assay. *Small.* 10 (6): 1086–1089.
27 Villa, C.C., Sánchez, L.T., Valencia, G.A. et al. (2021). Molecularly imprinted polymers for food applications: a review. *Trends in Food Science and Technology.* 111: 642–669.
28 Tabatabaei, M.S. and Ahmed, M. (2022). Enzyme-linked immunosorbent assay (ELISA). In: *Cancer Cell Biology: Methods and Protocols* (ed. S.L. Christian), 115–134. Springer.
29 Busa, L.S.A., Maeki, M., Ishida, A. et al. (2016). Simple and sensitive colorimetric assay system for horseradish peroxidase using microfluidic paper-based devices. *Sensors and Actuators B: Chemical.* 236: 433–441.
30 Tao, X., Wang, X., Liu, B., and Liu, J. (2020). Conjugation of antibodies and aptamers on nanozymes for developing biosensors. *Biosensors and Bioelectronics.* 168: 112537.

31 Bu, D., Zhuang, H., Zhou, X., and Yang, G. (2014). Biotin–streptavidin enzyme-linked immunosorbent assay for detecting Tetrabromobisphenol A in electronic waste. *Talanta.* 120: 40–46.

32 Wei, H. and Wang, E. (2013). Nanomaterials with enzyme-like characteristics (nanozymes): next-generation artificial enzymes. *Chemical Society Reviews.* 42 (14): 6060–6093.

33 Zandieh, M. and Liu, J. (2024). Nanozymes: definition, activity, and mechanisms. *Advanced Materials.* 36 (10): 2211041.

34 Wei, H., Gao, L., Fan, K. et al. (2021). Nanozymes: a clear definition with fuzzy edges. *Nano Today.* 40: 101269.

35 Scott, S., Zhao, H., Dey, A., and Gunnoe, T.B. (2020). Nano-apples and orange-zymes. *ACS Catalysis.* 10 (23): 14315–14317.

36 Gao, L., Zhuang, J., Nie, L. et al. (2007). Intrinsic peroxidase-like activity of ferromagnetic nanoparticles. *Nature Nanotechnology.* 2 (9): 577–583.

37 Panferov, V.G. and Liu, J. (2024). Nanozyme catalysis in a crowded environment: the impact of diffusion and surface shielding. *Nano Research.* 17 (7): 5795–5800.

38 Gibbs, J. and Kennebunk, M. (2001). Selecting the detection system—colorimetric, fluorescent, luminescent methods. Corning life sciences ELISA technical. *Bulletin.* 14.

39 Ornelas-González, A., Rito-Palomares, M., and González-González, M. (2022). TMB vs ABTS: comparison of multi-enzyme-based approaches for the colorimetric quantification of salivary glucose. *Journal of Chemical Technology & Biotechnology.* 97 (10): 2720–2727.

40 Ran, X., Yang, Z., Bai, M. et al. (2019). Development and validation of a competitive ELISA based on bacterium-original virus-like particles of serotype O foot-and-mouth disease virus for detecting serum antibodies. *Applied Microbiology and Biotechnology.* 103 (7): 3015–3024.

41 Yohans, H. (2021). Detection of microbial antigens by using immunodiagnostic techniques for the diagnosis of microbial diseases: a review. *Journal of Immunology Research Reports.* 1 (2): 1–9.

42 Halat, M.J. and Gravens, E. (2023). Screening for antibody specificity and sensitivity with ELISA. *ELISA: Methods and Protocols: Springer* 45–58.

43 Okda, F., Liu, X., Singrey, A. et al. (2015). Development of an indirect ELISA, blocking ELISA, fluorescent microsphere immunoassay and fluorescent focus neutralization assay for serologic evaluation of exposure to north American strains of porcine epidemic Diarrhea virus. *BMC Veterinary Research.* 11: 1–14.

44 Wang, X., Zhang, M., Pang, X. et al. (2022). Comparative study of Pd@Pt nanozyme improved colorimetric N-ELISA for the paper-output portable detection of *Staphylococcus aureus. Talanta.* 247: 123503.

45 Meyer, T. (2010). *Competitive ELISA Antibody Engineering*, 739–742. Springer.

46 Tighe, P.J., Ryder, R.R., Todd, I., and Fairclough, L.C. (2015). ELISA in the multiplex era: potentials and pitfalls. *PROTEOMICS–Clinical Applications.* 9 (3–4): 406–422.

47 Yanagisawa, N., Mecham, J.O., Corcoran, R.C., and Dutta, D. (2011). Multiplex ELISA in a single microfluidic channel. *Analytical and Bioanalytical Chemistry.* 401 (4): 1173–1181.

48 Wang, D., Zheng, Y., Kang, X. et al. (2015). A multiplex ELISA-based protein array for screening diagnostic antigens and diagnosis of Flaviviridae infection. *European Journal of Clinical Microbiology & Infectious Diseases.* 34 (7): 1327–1336.

49 Rissin, D.M., Kan, C.W., Campbell, T.G. et al. (2010). Single-molecule enzyme-linked immunosorbent assay detects serum proteins at subfemtomolar concentrations. *Nature Biotechnology.* 28 (6): 595–599.

50 Perez-Ruiz, E., Decrop, D., Ven, K. et al. (2018). Digital ELISA for the quantification of attomolar concentrations of Alzheimer's disease biomarker protein tau in biological samples. *Analytical Chimica Acta.* 1015: 74–81.

51 Arya, S.K. and Estrela, P. (2018). Recent advances in enhancement strategies for electrochemical ELISA-based immunoassays for cancer biomarker detection. *Sensors.* 18 (7): 2010.

52 Hu, B., Guo, H., Zhou, P., and Shi, Z.-L. (2021). Characteristics of SARS-CoV-2 and COVID-19. *Nature Reviews Microbiology.* 19 (3): 141–154.

53 Lu, R., Zhao, X., Li, J. et al. (2020). Genomic characterisation and epidemiology of 2019 novel coronavirus: implications for virus origins and receptor binding. *The Lancet.* 395 (10224): 565–574.

54 Mattioli, I.A., Hassan, A., Oliveira, O.N. Jr., and Crespilho, F.N. (2020). On the challenges for the diagnosis of SARS-CoV-2 based on a review of current methodologies. *ACS Sensors.* 5 (12): 3655–3677.

55 Adnan, N., Khandker, S.S., Haq, A. et al. (2022). Detection of SARS-CoV-2 by antigen ELISA test is highly swayed by viral load and sample storage condition. *Expert Review of Anti-infective Therapy.* 20 (3): 473–481.

56 Mladenovic Stokanic, M., Simovic, A., Jovanovic, V. et al. (2023). Sandwich ELISA for the quantification of Nucleocapsid protein of SARS-CoV-2 based on polyclonal antibodies from two different species. *International Journal of Molecular Sciences.* 25 (1): 333.

57 Girt, G.C., Lakshminarayanan, A., Huo, J. et al. (2021). The use of nanobodies in a sensitive ELISA test for SARS-CoV-2 spike 1 protein. *Royal Society Open Science.* 8 (9): 211016.

58 Di Domenico, M., De Rosa, A., and Boccellino, M. (2021). Detection of SARS-COV-2 proteins using an ELISA test. *Diagnostics (Basel).* 11 (4): 698.

59 Pomplun, S., Jbara, M., Quartararo, A.J. et al. (2021). De novo discovery of high-affinity peptide binders for the SARS-CoV-2 spike protein. *ACS Central Science.* 7 (1): 156–163.

60 Han, L., An, C., Liu, D. et al. (2022). Development of an ELISA Assay for the determination of SARS-CoV-2 protein subunit vaccine antigen content. *Viruses.* 15 (1): 62.

61 Stephens, D.S. and McElrath, M.J. (2020). COVID-19 and the path to immunity. *JAMA.* 324 (13): 1279–1281.

62 Mendoza, R., Silver, M., Zuretti, A.R. et al. (2021). Correlation of automated chemiluminescent method with enzyme-linked Immunosorbent Assay (ELISA) antibody titers in convalescent COVID-19 plasma samples: development of rapid, cost-effective semi-quantitative diagnostic methods. *Journal of Blood Medicine.* 12: 157–164.

63 Meyers, E., Coen, A., De Sutter, A. et al. (2022). Diagnostic performance of the SARS-CoV-2 S1RBD IgG ELISA (ImmunoDiagnostics) for the quantitative detection of SARS-CoV-2 antibodies on dried blood spots. *Journal of Clinical Virology.* 155: 105270.

64 Klumpp-Thomas, C., Kalish, H., Drew, M. et al. (2021). Standardization of ELISA protocols for serosurveys of the SARS-CoV-2 pandemic using clinical and at-home blood sampling. *Nature Communication.* 12 (1): 113.

65 Fernandez, Z., de Arruda, R.R., Torres, J.M. et al. (2023). Development and validity assessment of ELISA test with recombinant chimeric protein of SARS-CoV-2. *Journal of Immunological Methods.* 519: 113489.

66 Wouters, E., Verbrugghe, C., Devloo, R. et al. (2021). A novel competition ELISA for the rapid quantification of SARS-CoV-2 neutralizing antibodies in convalescent plasma. *Transfusion.* 61 (10): 2981–2990.

67 Kasetsirikul, S., Umer, M., Soda, N. et al. (2020). Detection of the SARS-CoV-2 humanized antibody with paper-based ELISA. *Analyst.* 145 (23): 7680–7686.

68 Caruana, G., Croxatto, A., Coste, A.T. et al. (2020). Diagnostic strategies for SARS-CoV-2 infection and interpretation of microbiological results. *Clinical Microbiology and Infection.* 26 (9): 1178–1182.

69 Montagut, E.J., Martin-Gomez, M.T., and Marco, M.P. (2021). An immunochemical approach to quantify and assess the potential value of the pseudomonas quinolone signal as a biomarker of infection. *Analytical Chemistry.* 93 (11): 4859–4866.

70 Cui, Y., Wang, H., Guo, F. et al. (2022). Monoclonal antibody-based indirect competitive ELISA for quantitative detection of Enterobacteriaceae siderophore enterobactin. *Food Chemistry.* 391: 133241.

71 Shang, L., Cheng, Y., and Zhao, Y. (2017). Emerging droplet microfluidics. *Chemical Reviews.* 117 (12): 7964–8040.

72 Han, K.N., Li, C.A., and Seong, G.H. (2013). Microfluidic chips for immunoassays. *Annual Review of Analytical Chemistry.* 6: 119–141.

73 Lizama, J.H., Chen, C.-J., Chang, W.-C. et al. (2024). Planarized through-hole valves enabling multilayered microfluidic architecture towards pipette-free ELISA. *Microchemical Journal.* 200: 110373.

74 Zhou, C., Fang, Z., Zhao, C. et al. (2021). Sample-to-answer robotic ELISA. *Analytical Chemistry.* 93 (33): 11424–11432.

75 Yang, J., Xiao, X., Xia, L. et al. (2021). Microfluidic magnetic analyte delivery technique for separation, enrichment, and fluorescence detection of Ultratrace biomarkers. *Analytical Chemistry.* 93 (23): 8273–8280.

76 He, Z., Huffman, J., Curtin, K. et al. (2021). Composable microfluidic plates (cPlate): a simple and scalable fluid manipulation system for multiplexed enzyme-linked Immunosorbent Assay (ELISA). *Analytical Chemistry.* 93 (3): 1489–1497.

77 Van Nguyen, H., Yang, J., Van Nguyen, H. et al. (2023). Development of a high-throughput centrifugal microsystem for enzyme-linked immunosorbent assay to detect SARS-CoV-2. *Chemical Engineering Journal.* 472: 144808.

78 Iria, I., Soares, R.R.G., Bras, E.J.S. et al. (2022). Accurate and rapid microfluidic ELISA to monitor infliximab titers in patients with inflammatory bowel diseases. *Analyst.* 147 (3): 480–488.

79 Gong, F., Wei, H.X., Qi, J. et al. (2021). Pulling-force spinning top for serum separation combined with paper-based microfluidic devices in COVID-19 ELISA diagnosis. *ACS Sensors.* 6 (7): 2709–2719.

80 Verma, M.S., Tsaloglou, M.N., Sisley, T. et al. (2018). Sliding-strip microfluidic device enables ELISA on paper. *Biosensors and Bioelectronics.* 99: 77–84.

81 Park, J.H., Jee, M.H., Kwon, O.S. et al. (2013). Infectivity of hepatitis C virus correlates with the amount of envelope protein E2: development of a new aptamer-based assay system suitable for measuring the infectious titer of HCV. *Virology.* 439 (1): 13–22.

82 Kacherovsky, N., Yang, L.F., Dang, H.V. et al. (2021). Discovery and characterization of spike N-terminal domain-binding Aptamers for rapid SARS-CoV-2 detection. *Angewandte Chemie International Edition.* 60 (39): 21211–21215.

83 Lee, K.H. and Zeng, H. (2017). Aptamer-based ELISA Assay for highly specific and sensitive detection of Zika NS1 protein. *Analytical Chemistry.* 89 (23): 12743–12748.

4

Metallic Nanomaterials for Plasmonic Biosensors and Flexible Bioelectronics

Heng Zhang[1,2], *Yi Chen*[1,2], *Cheng Yang*[1], *Yiyi Liu*[3], *and Wenlong Cheng*[4]

[1] *Southeast University, School of Biological Science and Medical Engineering, Jiangsu Key Laboratory for Biomaterials and Devices, State Key Laboratory of Digital Medical Engineering, Nanjing 210096, China*
[2] *Southeast University-Monash University Joint Graduate School, Southeast University Suzhou Research Institute, Suzhou 215123, China*
[3] *Nanjing Medical University, School of Biomedical Engineering and Informatics, Key Laboratory for Bio-Electromagnetic Environment and Advanced Medical Theranostics, Nanjing 211166, China*
[4] *School of Biomedical Engineering, Faculty of Engineering, University of Sydney, Darlington, NSW 2008, Australia*

CHAPTER MENU

4.1 Introduction

The rapid advancement of nanoengineering has revolutionized the field of biodiagnostics [1–4], paving the way for the development of highly sensitive and selective biosensing platforms. Among these cutting-edge technologies, metallic nanomaterial-based plasmonic biosensors [1, 2, 5–7] and flexible bioelectronics [8–14] have emerged as promising approaches, offering unprecedented capabilities for the detection and monitoring of biomolecules and disease biomarkers in real time. These cutting-edge techniques leverage the unique optical and electronic properties of nanomaterials and nanoarchitectures, enabling the creation of advanced biodiagnostic tools with remarkable capabilities.

Plasmonic biosensors exploit the collective oscillation of free electrons, known as surface plasmons, in metallic nanostructures when excited by electromagnetic radiation. This phenomenon, termed localized surface plasmon resonance (LSPR), results in extraordinary light–matter interactions and the generation of enhanced

Heng Zhang and Yi Chen contributed equally to this work.

Diagnostic Techniques in Public Health: Analytical and Chemometric Approaches in Disease Detection and Management, First Edition. Edited by Jinghong Li and Yan Du.

electromagnetic fields at the nanoscale [15]. The amplification of signals makes them ideal for real-time, label-free biodetection. By functionalizing these plasmonic nanostructures with biomolecular recognition elements, such as antibodies, aptamers, or nucleic acids, these sensors can selectively detect and quantify specific target analytes with excellent sensitivity and specificity, facilitating early disease diagnosis and therapeutic monitoring [7, 16–18].

Bioelectronics, on the other hand, represents a multidisciplinary field that integrates biomolecules, biological systems, or biological principles with electronic devices, creating hybrid bioelectronic interfaces [13, 19, 20]. These interfaces can convert biological signals into electrical signals or vice versa, enabling the development of implantable biosensors [21, 22], neural interfaces [23], and bio-inspired electronic devices [24, 25]. Bioelectronics has the potential to revolutionize personalized medicine, connected healthcare, prosthetics, and human–machine interactions by bridging the gap between biological and electronic systems [9, 26].

The synergistic convergence of plasmonics and bioelectronics has opened up new horizons in biodiagnostics, offering improved detection limits, multiplexing capabilities, and real-time monitoring of biological processes [9, 27–29]. Moreover, the integration of plasmonic biosensors and bioelectronics has the potential to create innovative biodiagnostic platforms that combine the advantages of both technologies, providing unprecedented sensitivity, selectivity, and versatility in the detection and analysis of biomarkers and biological processes [30–32]. As the field of nanotechnology continues to advance, the development of plasmonic biosensors and flexible bioelectronics is poised to make significant contributions to the field of biodiagnostics, offering innovative solutions to address emerging challenges in healthcare, environmental monitoring, and scientific research.

In light of the rapid advancements and promising future of this field, we aim to provide a broad overview of the technical developments in the integration of "plasmonics" and "biosensors." We seek to offer readers an in-depth understanding of the emerging field of plasmonic bioelectronics, focusing on the design principles, fabrication techniques, and mechanisms that enhance sensitivity and selectivity in biodetection. Furthermore, we discuss the integration of these materials into bioelectronic platforms, illustrating how they drive advancements in next-generation biosensing technologies. This chapter will cover the following key areas: (i) an overview of plasmonic nanostructures, (ii) plasmonic two-dimensional nanoassemblies, (iii) plasmonic biosensors, (iv) plasmonic bioelectronics, and (v) technical challenges and future outlooks.

4.2 Overview of Plasmonic Nanostructures

In the early 19th century, Michael Faraday observed that when gold particles are reduced to nanoscale sizes, their colloidal solutions exhibit a red color. Additionally, the color of these colloidal solutions can be further tuned by altering the particle size [33]. In 1908, Gustav Mie provided a theoretical explanation for this phenomenon and developed a method to calculate the extinction spectra (absorption and scattering spectra) for spherical metallic colloids of arbitrary size [34]. Since then, the study of interactions between light and metallic nanostructures has evolved into an interdisciplinary field known as plasmonics. Plasmonic nanostructures generally refer to metallic nanostructures that exhibit plasmonic effects, including those made of gold (Au), silver (Ag), platinum (Pt), and copper (Cu). Unlike traditional bulk materials, plasmonic nanostructures exhibit unique optical, chemical, and electrical properties due to their reduced size. These properties have led to widespread applications in various research fields, such as biomedicine, catalysis, energy, and sensing [35–40].

4.2.1 Optical Properties of Plasmonic Nanoparticles

Plasmonic nanoparticles can be considered as consisting of a positively charged metal core and a negatively charged electron cloud on their surface. Under the influence of an optical electric field, the electron cloud undergoes physical displacement relative to the metal core. However, this displacement is temporary, as the Coulomb attraction exerted by the positively charged metal core pulls the electron cloud back to its initial position. Therefore, under the continuous influence of these two driving forces (the optical electric field and Coulomb force), the interaction between the electron cloud and the metal core can be described as a collective coherent oscillation of the electron cloud relative to the metal core, known as plasmon oscillation, as illustrated in Figure 4.1a [15]. When the frequency of the oscillating electric field of the incident light matches the oscillation frequency of the plasmonic nanoparticles, LSPR occurs. This resonance represents a unique coupling effect between plasmonic nanoparticles and light and is the fundamental reason why Faraday's gold nanoparticles (AuNPs) solution exhibits a red color. At the LSPR resonance frequency, the absorption of electromagnetic radiation by the plasmonic nanoparticles reaches its maximum, and a locally enhanced electromagnetic field is generated on their surface, as depicted in Figure 4.1b. This phenomenon is one of the mechanisms behind the enhancement of molecular Raman signals [42]. Additionally, the strong electric fields, electron–hole pairs, and heat generated during the excitation and decay of LSPR can serve as excitation sources to induce chemical reactions, thereby holding significant potential for applications in photocatalytic reactions [37].

The resonance frequency of incident light with plasmonic nanoparticles, an important parameter of LSPR, is primarily determined by the inherent properties of the charges within the plasmonic nanoparticles, such as electron density and charge distribution [43]. By altering the size, morphology, and the dielectric constant of the surrounding medium of the plasmonic nanoparticles [44], the electron density and charge distribution on the particle surface can be adjusted, thus enabling the tuning of the LSPR resonance frequency.

Moreover, the LSPR resonance frequency typically lies within the ultraviolet to near-infrared spectral range. It can be characterized using UV-Vis-NIR spectroscopy, and the position of the LSPR absorption band can be used to infer the size of the plasmonic nanoparticles [45]. This optical property also facilitates the widespread

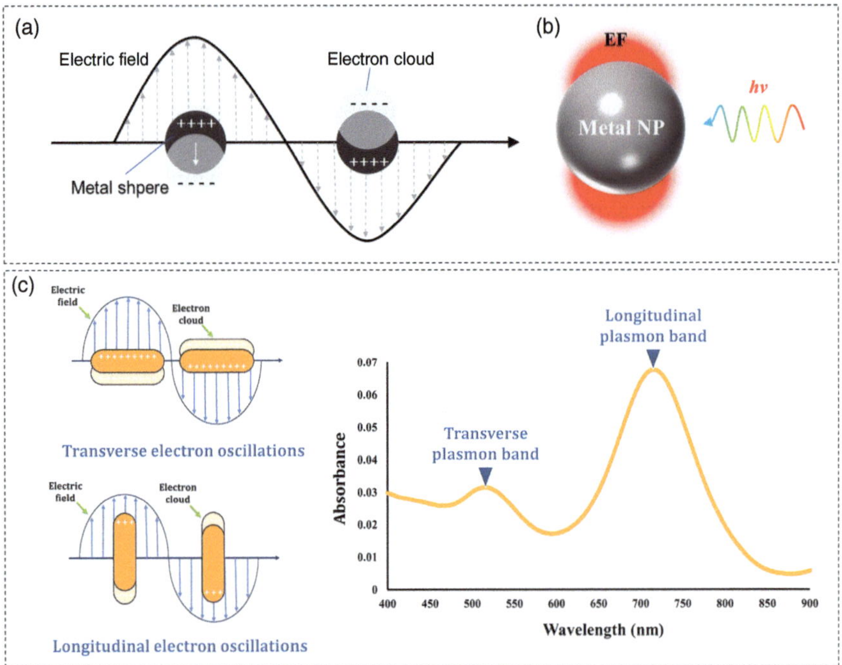

Figure 4.1 (a) Localized surface plasmon resonance *Source:* Ref. [15] / with permission of Annual Reviews. (b) Schematic of the localized electromagnetic field. (c) Schematic of the LSPR of gold nanorods (left) and their UV-vis absorption spectra (right). *Source:* Ref. [41] / with permission of Elsevier.

application of plasmonic nanoparticles in sensing. Various wet chemistry methods have been developed to synthesize plasmonic nanoparticles with precise control over their sizes and shapes, enabling the formulation of the so-called artificial periodic table of "meta-atoms" [46]. Further assembly of them into "plasmonic molecules" and "plasmonic crystals" allows for program plasmonic properties for various sensing applications [47].

4.2.2 Nanoplasmonic Building Blocks ("Meta-atoms")

Nanoplasmonic building blocks may be isotropic or anisotropic in overall morphologies. Isotropic nanostructures, such as gold nanospheres, are constrained by uniformity in three dimensions, resulting in a single mode of LSPR resonance. By contrast, anisotropic nanostructures exhibit more complex LSPR resonance modes [48]. For example, gold nanorods (AuNRs) display two distinct LSPR resonance modes: [41] transverse LSPR (TLSPR) and longitudinal LSPR (LLSPR), as shown in Figure 4.1c (left). The corresponding LSPR absorption bands are depicted in Figure 4.1c (right), where the TLSPR absorption peak is around 550 nm, and the LLSPR absorption peak lies in the near-infrared range (650–900 nm). Furthermore, the LLSPR can be tuned by adjusting the aspect ratio of the AuNRs [49]. Moreover, the LSPR absorption bands of anisotropic nanostructures are more sensitive to changes in particle morphology; even minor changes can lead to significant blue shifts or red shifts in the LSPR absorption bands, much more than the LSPR changes caused by changes in the size of isotropic structures [50]. Furthermore, by controlling the morphology of anisotropic nanostructures, it is possible to achieve a shift in LSPR absorption bands from the visible light range to the near-infrared range [51]. Since biological tissues weaken photon scattering and absorption, near-infrared light can penetrate deep into biological tissues [52], making it possible to excite nanoparticles' LSPR within the biological body, a characteristic application of plasmonic photothermal therapy [53]. In summary, anisotropic nanostructures possess superior LSPR performance, further expanding the application of plasmonic nanomaterials in biomedicine.

The controlled synthesis of anisotropic plasmonic nanostructures remains a significant and continuously evolving direction within the field of inorganic materials chemistry. Based on the use of seeds during the synthesis process, the methods for synthesizing anisotropic nanostructures can be divided into seed-mediated and seedless methods [54]. With the ongoing advancements in synthesis techniques and characterization technologies, various anisotropic nanostructures with well-defined morphologies, sizes, and structures have been synthesized. These can be broadly categorized into zero-dimensional (0D), one-dimensional (1D), and two-dimensional (2D) structures based on their shapes, as illustrated in Figure 4.2.

The 0D structures mainly include nanostars, nanourchins, nanoflowers, nanocubes, and nanocages [55–57, 64, 65]. The 1D structures primarily consist of nanorods, nanowires, nanotubes, and nanobelts [58–60, 66]. The 2D structures are predominantly composed of nanosheets with various morphologies [61–63].

4.2.3 Plasmonic "Meta-molecules"

Finite-number plasmonic nanoparticles may be assembled to form so-called plasmonic "meta-molecules." Compared to single-component plasmonic nanostructures, heterogeneous hybrid structures not only retain the functions of individual component nanostructures but also enable synergistic interactions between components, potentially exhibiting superior performance [67–69]. For example, the hybridization of noble metals with magnetic metals enhances magneto-optical coupling properties [67]; the hybridization of plasmonic metals with catalytic metals greatly enhances the photocatalytic efficiency of materials [68]; and the hybridization of dielectric material SiO_2 with noble metals can generate numerous "electromagnetic hotspots" [69]. Additionally, changing the relevant parameters of single-component nanostructures and overall heterogeneous hybrid structures allows for the adjustment of plasmonic heterogeneous hybrid structures, making them more suitable for practical applications. According to heterogeneity and hybridization methods, plasmonic heterogeneous hybrid nanostructures can be divided into three categories: alloy structures [70–76], core-shell structures [77–84], and heterodimers [85–88].

Plasmonic alloy structures are formed by doping two or more metals in certain proportions. Bimetallic Au/Ag is one of the most common alloy structures, as shown in Figure 4.3a. Under the excitation of the same incident light, discrete Au and Ag show

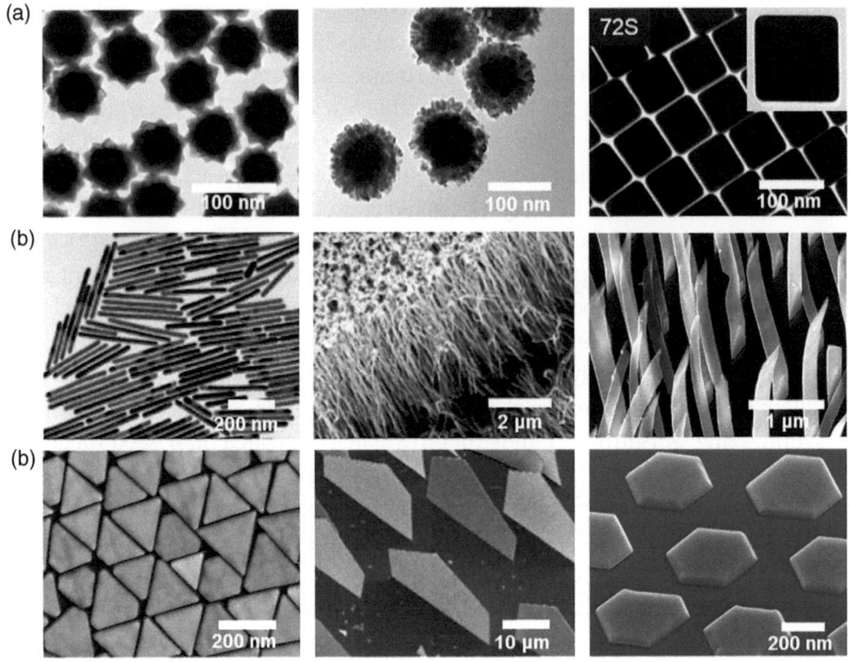

Figure 4.2 Plasmonic nanostructures with different morphologies. (a) 0D structures, from left to right: nanostar (*Source:* Ref. [55] / American Chemical Society), nanoflower (*Source:* Ref. [56] / with permission of Elsevier), and nanocube (*Source:* Ref. [57] / American Chemical Society); (b) 1D structures, from left to right: nanorod (*Source:* Ref. [58] / American Chemical Society), nanowire (*Source:* Ref. [59] / American Chemical Society), and nanobelt (*Source:* Ref. [60] / American Chemical Society); (c) 2D structures, from left to right: triangular (*Source:* Ref. [61] / American Chemical Society), pentagonal (*Source:* Ref. [62] / The Royal Society of Chemistry), and hexagonal nanosheets (*Source:* Ref. [63] / American Chemical Society).

two characteristic extinction peaks, while Au/Ag bimetallic alloy shows a single extinction peak [72]. Changing the proportion of Au and Ag allows for the adjustment of the optical properties of alloy structures, as shown in Figure 4.3b, where increasing the proportion of Au results in a gradual red shift of the extinction peak of Au/Ag bimetallic alloy. Furthermore, Au/Ag bimetallic alloys are a feasible strategy to enhance the stability of Ag; according to recent literature [74], doping 4.9 at% Au into Au/Ag bimetallic alloys can largely retain the superior LSPR performance of Ag, with stability comparable to pure Au. In addition to this, Au or Ag is also commonly hybridized with other metals such as Cu, Ni, Co, and metal oxides [71, 90–92], with widespread applications. Additionally, plasmonic hollow structures are another prominent type of alloy nanostructures [89, 93–95]. These hollow structures are typically driven by a galvanic reaction, where silver acts as a reducing agent and is oxidized, while the gold ions in chloroauric acid ($HAuCl_4$) serve as the oxidizing agent and are reduced (Figure 4.3c) [89]. By controlling the concentration of $HAuCl_4$ and the reaction time, the morphology of the hollow structures can be precisely regulated.

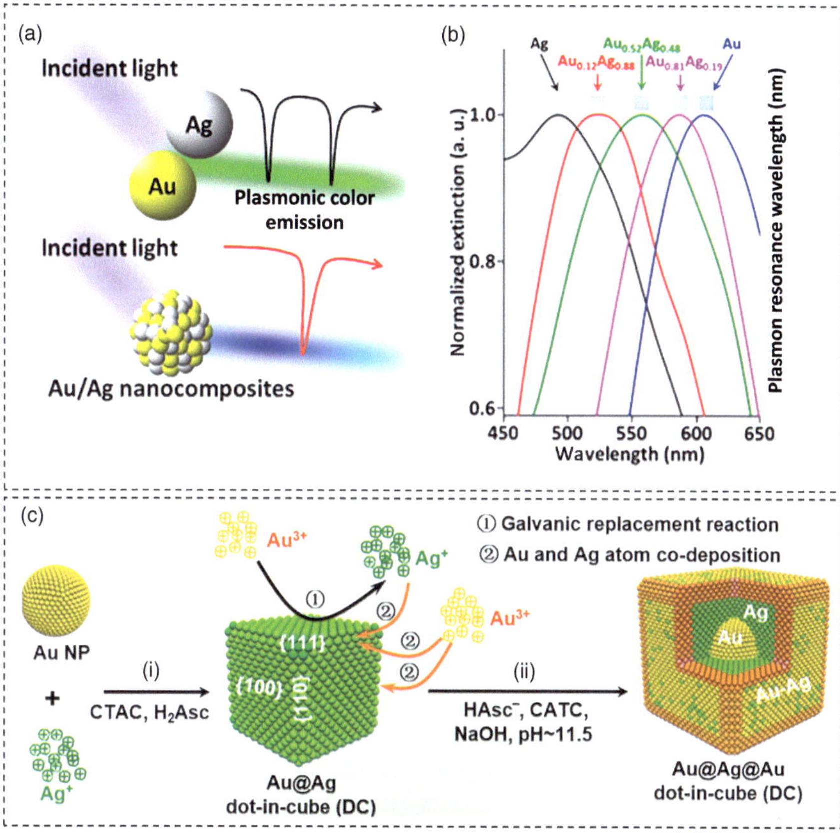

Figure 4.3 Plasmonic alloy structures. (a) Extinction peaks of discrete metal and Au/Ag bimetallic alloy structures *Source:* Ref. [72] / with permission of American Chemical Society. (b) Changes in the extinction spectrum and color due to variations in the metal composition ratio of the alloy structures. (c) Schematic illustrating the synthesis process of a plasmonic hollow structure. *Source:* Ref. [89] / American Chemical Society / CC BY 4.0.

Plasmonic core-shell structures [84], as the name suggests, grow a shell layer on the template of a core particle. This is also the most common heterogeneous hybrid structure. By changing the material, size of the core particle, and material and thickness of the shell layer, easy control over core-shell structures and their optical properties can be achieved. Since the concept was proposed, a large number of dual-component core-shell structures have been synthesized controllably, roughly classified into metal–metal and metal–nonmetal core-shell structure types; metal–metal core-shell structures mainly include Au@Ag [96], Ag@Au [97], Fe_3O_4@ noble metal [98], and noble metal @ metal oxide, etc. [99], while metal–nonmetal core-shell structures mainly include SiO_2@ noble metal [100], noble metal @ SiO_2 [101], and noble metal @ graphene, etc. [102]. Moreover, with the development of nanotechnology, more complex multicomponent core-shell structures have also been controllably prepared. Among them, the Langhammer group [81] reported a strategy for preparing a three-layer nanocomposite structure

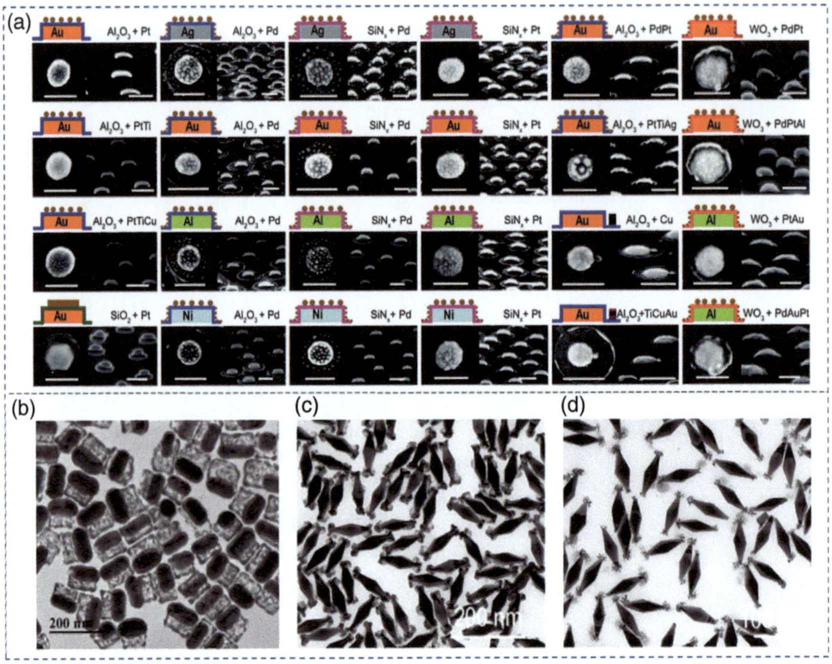

Figure 4.4 Plasmonic core-shell nanostructures and heterostructures. (a) Library of three-layer core-shell nanostructures (core-shell-nanoparticles). *Source:* Ref. [81] / The Royal Society of Chemistry. The top section of each image shows a schematic diagram of the structure, the bottom left shows a top-view SEM image, and the bottom right shows a side-view SEM image. Color codes for the metal cores: Ag (gray), Au (orange), Al (light green), and Ni (light blue); color codes for the shell layers: AlO_2 (purple), SiO_2 (gray-green), SiNx (magenta), and WO_3 (red). Scale bars: except for the last column, which is 100 nm, all others are 200 nm. (b) AuCuAu hetero-nanorod (*Source:* Ref. [103] / John Wiley & Sons), (c) Au@AgPd bicone (*Source:* Ref. [88] / American Chemical Society), (d) Au@AgPt bicone. *Source:* Ref. [88] / American Chemical Society.

(core-shell-nanoparticle), constructing a library of three-layer nanocomposite structures composed of various metal cores, oxide/nitride shells, and several nanosized metal particles, as shown in Figure 4.4a.

Plasmonic hybrid nanostructures are typically composed of two or more types of nanostructures with different shapes or sizes, including binary, ternary, and complex heterostructures [87]. Seed-mediated synthesis is one of the most commonly used methods in the synthesis of heterostructures [86]. Typically, a nanostructure with a well-defined morphology acts as a seed, and selective heterogeneous nucleation and growth occur at specific sites on its surface. Surfactants play a crucial role in this process because they not only act as stabilizers to prevent aggregation of the formed structures but also selectively adsorb on specific crystal facets of nanocrystals, thereby guiding the generation of nanostructures with different morphologies [104]. For example, Xu coworkers [85] utilized hexadecylamine to guide the selective growth of Cu on the surface of AuNRs, thereby synthesizing a special chiral AuCuAu hetero-nanorod, as shown in Figure 4.4b. Similarly, using AuNRs as seeds, the team further designed a spiky nanorod using polydopamine [103]. In addition, seed-mediated synthesis can be combined with galvanic replacement to achieve the synthesis of plasmonic heterogeneous hybrid nanostructures, such as Au@AgPd and Au@AgPt bicone structures [88], as shown in Figure 4.4c, d.

4.3 Plasmonic "Meta-crystals"

Monodisperse nanoplasmonic building blocks can also be assembled into extended arrays forming so-called plasmonic "meta-crystals." Specifically, the creation of ordered arrays and superlattices of plasmonic nanostructures can open up new possibilities for applications in areas such as plasmonic sensing, photocatalysis, and nanoscale optical manipulation. Among these, 2D plasmonic superlattice thin films represent a unique class of structures wherein plasmonic nanoparticles serve as building blocks, densely arranged at the nanoscale. This close-packed arrangement induces strong interparticle coupling among the plasmonic nanoparticles, leading to the formation of localized electromagnetic field enhancements, often referred to as "hotspots." These hotspots significantly amplify the Raman signal intensity of target molecules [42], making the films highly effective for applications such as biomolecule analysis, environmental pollutant detection, and the production of anti-counterfeiting labels [51, 105, 106].

The preparation of 2D superlattice thin films typically employs two main strategies: top-down and bottom-up approaches. The top-down approach, based on photolithography techniques, offers high resolution and reproducibility [107] but is costly and time-consuming. By contrast, the bottom-up approach relies on adjusting the interparticle forces to allow nanoparticles to spontaneously form 2D assemblies, known as self-assembly [108]. This method is simpler, more cost-effective, and widely used in the fabrication of 2D plasmonic superlattice thin films.

One common self-assembly method is droplet evaporation, where concentrated nanoparticle suspensions are dropped onto a substrate and allowed to naturally dry, forming ordered nanoparticle assemblies. However, achieving continuous arrangement at the micrometer scale is challenging [112], requiring strict control over environmental factors such as humidity, temperature, and evaporation rate of the solution [113]. These limitations restrict its application in constructing 2D superlattice thin films. Addressing these challenges, Cheng and co-workers [109] demonstrated a two-step drying-mediated droplet evaporation method, achieving the construction of centimeter-scale 2D films. As depicted in Figure 4.5a, b, this method involves dropping a concentrated chloroform solution of metal nanoparticles onto the surface of a water droplet at room temperature. As chloroform and the water droplet naturally evaporate, the metal nanoparticles self-assemble into a densely packed monolayer superlattice film. This approach allows the fabrication of superlattice films composed of various morphologies of metal nanoparticles, including nanospheres [106], nanopolyhedra [106], nanostars [106], nanocubes [110], and nanobipyramids [111] (Figure 4.5c). Moreover, this technique can be combined with acoustic levitation technology to achieve the preparation of 2D superlattice films without a substrate [114].

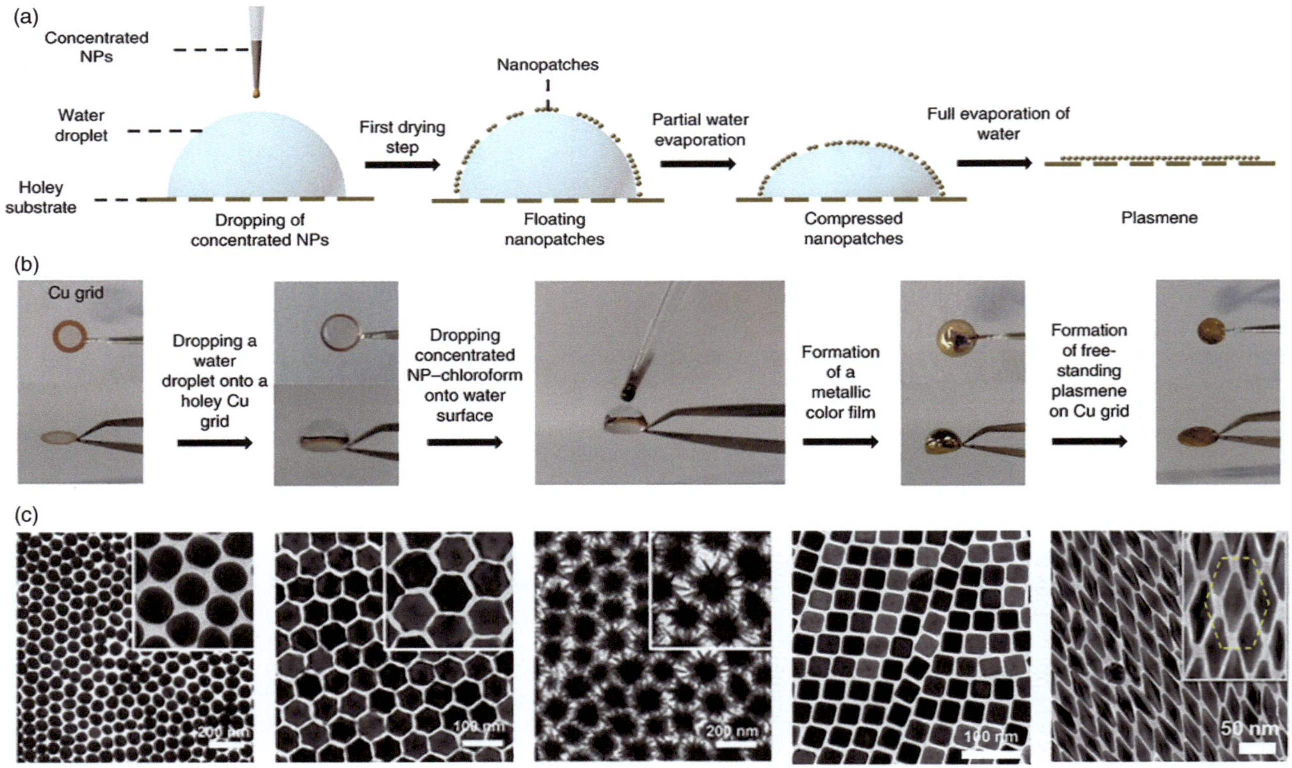

Figure 4.5 Preparation of a 2D superlattice film via a two-step drying-mediated droplet evaporation method. *Source:* Ref. [109] / Springer Nature. (a) Scheme of superlattice film fabrication. (b) Photographs of Cu grid, procedure for superlattice film fabrication and the final superlattice film on a Cu grid. (*Source:* [109] / Springer Nature) (c) Superlattice films constructed from nanoparticles with different morphologies, including nanospheres (*Source:* Ref. [106] / John Wiley & Sons), nanopolyhedra [106] / John Wiley & Sons, nanostars [106] / John Wiley & Sons, nanocubes (*Source:* Ref. [110] / with permission of Elsevier), and nanoshuttles (*Source:* Ref. [111] / American Chemical Society).

Interface assembly is another effective method for nanoparticle self-assembly, commonly used for constructing large-area 2D films. The key parameter is reducing the surface charge density of metal nanoparticles so that they can freely migrate to the interface between two phases to form an array structure. Traditional interface assembly methods primarily utilize ethanol as a solvent with a low dielectric constant to overcome energy barriers and reduce the surface charge density of AuNPs [118]. However, due to the insufficient hydrophobicity of AuNPs (contact angle less than 90°), the preparation time is often prolonged (greater than 20 minutes), and the resulting films may have a limited area (a few square centimeters) and uneven arrangement. Huang coworkers reported a rapid interface assembly technique that can achieve the construction of large-area (95 cm^2) 2D superlattice films within five seconds, with a particle assembly efficiency as high as 97.5%, representing a significant breakthrough compared to traditional interface self-assembly techniques [119]. The innovation lies in the introduction of perfluorinated molecules (perfluorododecanethiol, PFT), which have highly hydrophobic groups ($-CF_3$) on their surface, enabling metal nanoparticles to easily overcome energy barriers and reach the interface.

The evaporation of colloidal dispersions on a liquid substrate is another commonly employed method for interfacial assembly. In this approach, colloidal particles suspended in a liquid medium are introduced onto a secondary liquid substrate. As the solvent evaporates, the colloidal particles are driven to assemble at the three-phase contact line, resulting in the formation of an ordered colloidal crystal structure (Figure 4.6a) [115]. This method is simple, cost-effective, and can be performed at room temperature, making it widely applicable for assembling various functional nanoparticles. However, this self-assembly method also faces challenges, such as difficulties in precisely controlling the assembly process, the presence of defects in particle arrangements, and the challenging nature of large-scale fabrication of ordered structures. Exploring self-assembly conditions, including the shape, size, dispersing solvent, and capping ligands of the nanoparticles, is a common approach to overcoming the aforementioned challenges [115, 117, 119, 120]. For example, Ding et al. achieved control over this process by carefully tuning the amounts of free ligands introduced during the self-assembly [116]. By manipulating the ligand concentration, they could finely adjust the tip arrangements of Au nanobipyramids, leading to the formation of 2D superlattice films with various configurations, including tilted superlattice films, tip-on-tip superlattice films, and tip-to-tip superlattice films (Figure 4.6b–d). Similarly, the Stemmer group and colleagues obtained large-area (~1 cm^2) self-assembled AuNPs superlattices via interfacial self-assembly on diethylene glycol (DEG), with oleylamine (OAm) as the capping ligand [117]. They introduced an intermediate subphase exchange process to mitigate the effects of the low volatility of DEG during the NP membrane transfer process (Figure 4.6e). After forming the AuNPs superlattices, they exchanged the nonvolatile DEG subphase with volatile acetonitrile through a fluidic system. Subsequently, by introducing an excess of the target molecules into the acetonitrile, they performed ligand exchange from OAm to different capping ligands within the AuNPs superlattices (Figure 4.6f). The ligand exchange reaction was terminated by

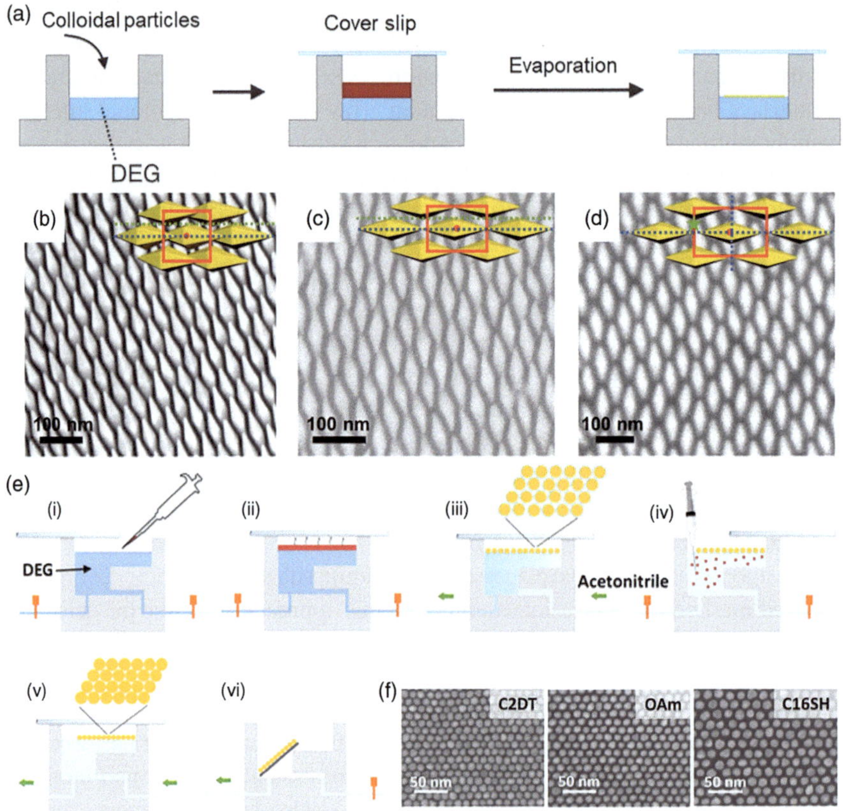

Figure 4.6 Preparation of 2D superlattice film via the evaporation of colloidal dispersions on a liquid substrate. (a) Scheme of superlattice preparation. First, colloidal particles in toluene are pipetted onto a liquid subphase (DEG); then, the well is covered with a glass slide to slow down the evaporation of the toluene *Source:* Ref. [115] / with permission of John Wiley & Sons. (b–d) SEM images of tilted superlattice films, tip-on-tip superlattice films, and tip-to-tip superlattice films (from left to right). *Source:* Ref. [116] / American Association for the Advancement of Science - AAAS. (e) A schematic representation of (i) and (ii) the growth of AuNPs superlattices, (iii) and (v) the subphase exchange, (iv) the in situ free-floating ligand exchange, and (vi) the film transfer process. *Source:* Ref. [117] / Springer Nature / CC BY 4.0. (f) High-resolution SEM images of AuNPs superlattices with different capping ligands. *Source:* Ref. [117] / Springer Nature / CC BY 4.0.

exchanging the subphase once again with clean acetonitrile. Utilizing acetonitrile as the subphase, they obtained large-scale ($\sim$1 cm^2) high-quality AuNPs superlattices with subnanometer interparticle gaps. These superlattices could be easily transferred and deposited onto solid substrates without the need for prolonged drying processes.

4.4 Plasmonic Biosensors

Plasmonic biosensors are advanced analytical devices that exploit the phenomenon of LSPR to detect and analyze various chemical and biological analytes with high sensitivity. LSPR occurs when the collective oscillation of free electrons on the surface of noble metal nanostructures, such as gold or silver, resonates with incident light at specific wavelengths. This resonance leads to a significant enhancement of the electromagnetic fields at the nanoparticle surface, amplifying the optical, electrical, and magnetic properties of the surrounding medium.

The specific properties of noble metal nanoparticles, including their size, shape, composition, and surface chemistry, can be meticulously tailored to optimize their performance in various sensing applications. For example, core-shell nanostructures, particularly gold–silver (Au–Ag) combinations, are favored for their ability to enhance sensitivity [120–122]. Beyond simple shape and size, plasmonic biosensors also exploit more complex nanostructures, such as shape-engineered AuNPs that exhibit quantum tunneling effects or interlocking gold nanowires (AuNWs) that provide unique conductive properties [123, 124].

These specialized features enable plasmonic biosensors to detect a broad spectrum of analytes, including metal ions, DNA, proteins, and volatile organic compounds (VOCs). The key to their versatility lies in the fine structure–property relationship established by the nanostructures, which can be precisely engineered to interact with specific target molecules.

Moreover, the surface chemistry of these nanoparticles can be functionalized with various ligands, antibodies, or receptors, allowing the biosensor to selectively bind to particular analytes. Upon binding, these interactions can induce shifts in the LSPR signal, change the refractive index, or alter the electrical conductivity, which are then measured and correlated to the concentration of the analyte. This multifaceted approach enables plasmonic biosensors to provide rapid, label-free detection with high sensitivity and specificity, making them invaluable tools in medical diagnostics, environmental monitoring, and other fields where precise detection of biological or chemical species is essential.

4.4.1 Plasmonic-based SERS Sensors

Surface-enhanced Raman scattering (SERS) is a phenomenon where the Raman scattering signals of molecules are significantly amplified when the molecules are adsorbed onto or in proximity to nanostructured metallic surfaces, typically composed of noble metals like silver or gold. Unlike Rayleigh scattering, which is elastic and involves no change in photon energy, Raman scattering is inelastic, resulting in scattered photons with different energies corresponding to the vibrational states of the molecule.

The enhancement in SERS arises from two primary mechanisms: the electromagnetic mechanism, which involves the excitation of localized surface plasmons (LSPs) on the metal nanostructures, leading to an intense local electromagnetic field, and the chemical mechanism, which involves charge transfer between the molecule and the metal surface. The electromagnetic enhancement is generally considered the dominant factor, with the local field amplifying the Raman signal by several orders of magnitude.

SERS was first observed in 1974 by Fleischmann et al. [125], who reported an unusually strong Raman signal from pyridine molecules on a roughened silver electrode. Subsequent studies by Jeanmaire, Van Duyne [126], Albrecht, Creighton [127], and Moskovits [128] further elucidated the underlying mechanisms, identifying the crucial role of surface plasmons in this process. The enhancement effect of SERS makes it a powerful tool for molecular detection, allowing for the identification and analysis of substances at extremely low concentrations, often down to the single-molecule level.

For an analyte to manifest SERS, three critical prerequisites must be satisfied: (i) the analyte must be adsorbed onto a SERS-active substrate, typically composed of nanostructured noble metals; (ii) the substrate must be illuminated with monochromatic radiation, most commonly from a coherent laser source; and (iii) the resultant inelastically scattered photons must be precisely analyzed using a high-resolution Raman spectrometer [129].

The unique features of Raman spectroscopy, such as its ability to provide fingerprint recognition, nondestructive analysis, and multiplexing, paired with the powerful enhancement potential of plasmonic nanostructures, make SERS-based platforms particularly promising for biosensing applications. The ability to detect Raman signals at extremely low analyte concentrations, even down to the single-molecule level, is motivating scientists to create advanced nanostructured substrates. These substrates are being tailored for integration into handheld, portable, and smart Raman devices, offering new possibilities in the field of biosensing.

An ideal SERS substrate should exhibit high sensitivity for accurate and rapid analysis. It should also ensure uniformity, eliminating the need to scan multiple spots, and reproducibility, where consistent results are obtained across different substrates, which is crucial for quantitative measurements. Stability is important, with substrates remaining effective over time and under varying environmental conditions. Large-area substrates are beneficial for portable instruments not coupled to microscopes. Additionally, ease of fabrication and cost-effectiveness are essential for integrating SERS technology into mainstream diagnostics, allowing for scalable and widespread use.

Among various plasmonic structures, gold- and silver-based nanostructures are the most commonly used due to their higher enhancement factors and plasmon resonances in the visible and near-infrared regions. Silver has a broader plasmon resonance range (300–1200 nm) compared to gold (500–1200 nm), making it more effective at amplifying Raman signals. However, silver is reactive to its surrounding environment and can form surface layers of oxides, chlorides, or other complex organic agents. It is also prone to oxidation and reacts with sulfur compounds in the atmosphere, making it less suitable for environmental or biological applications. By contrast, gold is more suitable for biomedical applications due to its excellent chemical stability and low toxicity.

Recent studies have made improvements on gold and silver substrates [130–138]. For example, in the modification of silver substrates, Li et al. anchored arrays of silver nanocubes (Ag NCs) on a flexible and transparent polydimethylsiloxane (PDMS) mask to fabricate SERS substrates [130]. As shown in Figure 4.7a, they employed a "zero-transfer" method to directly retrieve the plasmonic arrays from the interface.

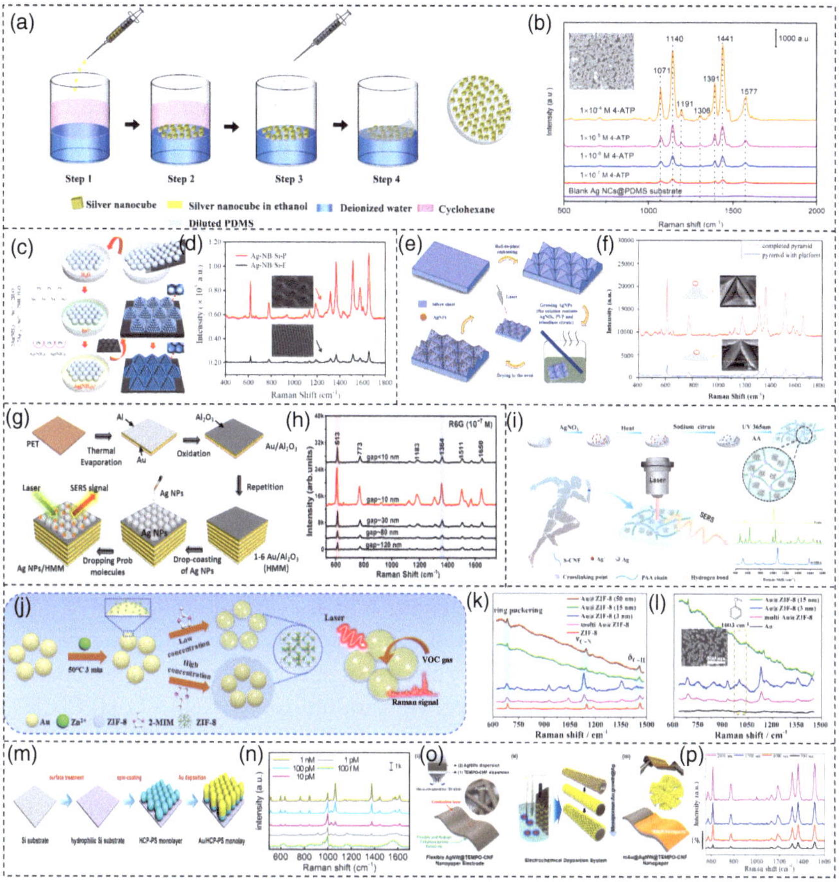

Figure 4.7 SERS substrates based on Au and Ag nanostructures. (a) Ag NCs@PDMS film fabrication via the "zero-transfer" method. *Source:* Ref. [130] / with permission of American Chemical Society. (b) High EF and reproducibility of Ag NCs@PDMS SERS substrate for 4-ATP detection. *Source:* Ref. [130] / American Chemical Society. (c) Silver substrate on silicon micro-pyramids via anisotropic wet etching. (d) SERS performance of silver-silicon substrate (*Source:* Ref. [131] / American Chemical Society), EF ~ 10^9. (e) Silver substrate fabrication using roll-to-plate embossing technology. *Source:* Ref. [132] / with permission of American Chemical Society. (f) SERS performance of embossed silver substrate (*Source:* Ref. [132] / American Chemical Society), EF ~ 10^9. (g) Sulfonated cellulose nanocomposite hydrogels for enhanced SERS signal. *Source:* Ref. [133] / with permission of Elsevier. (h) Ag NPs/HMM substrate SERS performance with superior signal enhancement. (i) Comparison of various substrates' SERS enhancement capabilities. *Source:* Ref. [134] / with permission of Elsevier. (j) Au@ZIF-8 core-shell NP synthesis with controlled shell thickness. *Source:* Ref. [135] / with permission of American Chemical Society. (k) Raman enhancement of toluene on Au@ZIF-8 NPs with 3 nm shell. *Source:* Ref. [135] / with permission of American Chemical Society. (l) SPR effect and gas adsorption of Au@ZIF-8. *Source:* Ref. [135] / American Chemical Society. (m) Fabrication of Au/HCP-PS nanosphere monolayer substrate. *Source:* Ref. [136] / with permission of American Chemical Society. (n) Enhanced SERS performance of Au/HCP-PS nanosphere substrate. *Source:* Ref. [136] / with permission of American Chemical Society. (o) Electrochemical deposition of mAu on AgNW@TEMPO-CNF nanopaper [137]. (p) High EF (8.72×10^{11}) for femtomolar detection on mAu@AgNW@TEMPO-CNF. *Source:* [137] / with permission of Elsevier.

This involved gently introducing a mixed PDMS solution onto the top of the Ag NC array and curing it at 65 °C for two hours. The successfully formed Ag NCs@PDMS film, resembling a shiny golden mirror atop the underlying DI water layer, could be easily lifted as a standalone film. Compared to other polymer-based SERS substrates, the resulting Ag NCs@PDMS film exhibited sufficient flexibility to withstand stretching and twisting. Furthermore, the prepared SERS substrate demonstrated a high EF (3.43×10^6) and excellent reproducibility (~12%) when detecting 4-ATP molecules (Figure 4.7b). Additionally, the substrate demonstrated high sensitivity and effective selective discrimination toward various analyte molecules, including rhodamine 6G (R6G), methylene blue (MB), and crystal violet (CV).

In another study, Jin et al. and Zhang et al. successfully constructed silver substrates on ordered micro-pyramid arrays of silicon and silver, respectively [131, 132]. Figure 4.7c and e illustrate the substrate fabrication processes, which, respectively, involved anisotropic wet etching and roll-to-plate embossing technology. These substrates demonstrated exceptional SERS performance, with enhancement factors reaching the order of 10^9 (Figure 4.7d and f). Other modifications, such as sulfonated cellulose nanocomposite hydrogels or Ag NPs/Hyperbolic metamaterials (HMM) SERS substrates [133, 134], have also demonstrated superior signal enhancement capabilities, as shown in Figure 4.7g–i.

Kim et al. reported the preparation of a high-performance cellulose nanofiber (CNF)–based flexible SERS substrate by electrochemically depositing a mesoporous gold (mAu) film onto 2,2,6,6-tetramethylpiperidine-1-oxyl CNF (TEMPO-CNF) nanopaper coated with silver nanowires (AgNWs) [137]. The fabrication process is shown in Figure 4.7o. The electrochemical deposition was performed using an electrochemical analyzer with a three-electrode system, where the AgNW@TEMPO-CNF nanopaper served as the working electrode (with Au@Si wafer as a control), an Ag/AgCl electrode as the reference, and a Pt wire as the counter electrode. The developed mAu@AgNW@TEMPO-CNF substrate enabled femtomolar-level detection and provided an enhancement factor of 8.72×10^{11} (Figure 4.7p).

Beyond silver-based substrates, gold-based improvements have also become a significant area of interest. For example, Chen et al. proposed an Au@ZIF-8 core-shell nanoparticle architecture [135], enabling the controlled synthesis of uniform single-core and multi-core nanoparticles, with precise control over the ZIF-8 shell thickness (ranging from 3 to 50 nm) to adjust the distance between metal nanoparticles and the electromagnetic field (Figure 4.7j). Due to the excellent gas adsorption properties of the ZIF-8 shell and the outstanding SPR effect of the Au core, the synthesized Au@ZIF-8 NPs with a 3 -nm shell thickness serve as a highly attractive core-shell structure, significantly enhancing the Raman peak of the ring-breathing vibration of toluene (Figure 4.7k and l). Additionally, well-ordered, gold-decorated hexagonally close-packed polystyrene (Au/HCP-PS) nanosphere monolayer substrates have also been shown to exhibit excellent enhancement performance [136], as demonstrated in Figure 4.7m and n.

The local enhancement of the electric field intensity produced by anisotropic plasmonic nanostructures is generally much stronger than that of isotropic structures,

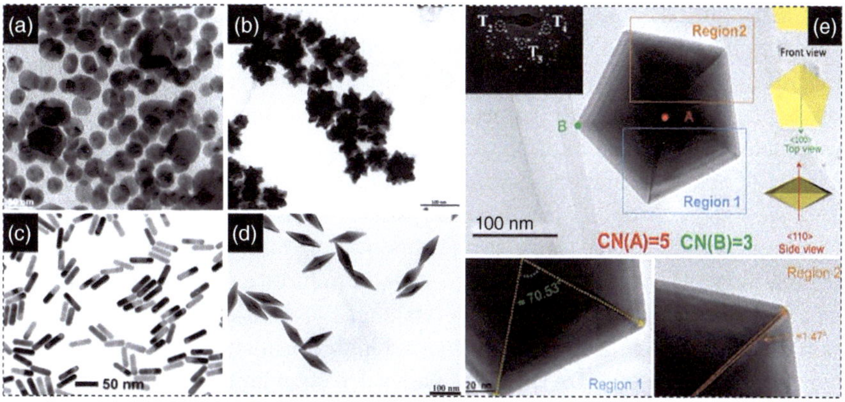

Figure 4.8 Plasmonic nanostructures. (a) Spherical (*Source:* Ref. [139] / with permission of Elsevier). (b) Star-shaped (*Source:* Ref. [140] / with permission of Elsevier). (c) Rod-shaped (*Source:* Ref. [141] / American Chemical Society). (d) Bipyramidal (*Source:* Ref. [142] / with permission of Elsevier). (e) Decahedral. *Source:* Ref. [143] / with permission of Elsevier.

particularly at sharp surface features like vertices and edges. Consequently, in addition to spherical plasmonic nanoparticles [139], various anisotropic shapes such as stars, rods, bipyramids, and decahedrons have also been utilized as SERS substrates (Figure 4.8) [140–143].

Thanks to their high sensitivity, plasmonic-based SERS sensors are widely applied in various biochemical scenarios. In biomarker monitoring, label-free SERS is primarily utilized for detecting small molecules, including metabolites in bacteria or eukaryotic cells, various neurotransmitters, and drugs, as well as pathogens such as hepatitis A virus [144], endogenous neurotransmitters [145], oral cancer-related DNA [146], and more.

While large biomolecules can be detected using SERS, their typical size (around 10 nm) and composition of similar building blocks result in highly comparable SERS spectra within the same class, making it difficult to differentiate between them based on spectral fingerprints alone. Therefore, specific labels are necessary for accurate SERS identification. This has given rise to indirect methods, where the detected SERS signal comes from a probe or label, often involving traditional immunoassays or oligonucleotide microarrays, allowing for the detection of proteins and nucleic acids. For indirect sensors, the simplest SERS probe must include a noble metal nanoparticle (typically gold or silver) and a monolayer of reporter molecules that act as fingerprint labels. By carefully designing the nanoprobe, such as adding a protective outer shell, and incorporating washing steps, background noise from biological fluids can be systematically reduced. As a result, only minimal modifications to traditional assays are needed, such as replacing fluorescent dyes with SERS probes, bringing indirect sensors closer to integration into commercial diagnostic tools [147].

SERS sensors play a crucial role in live-cell imaging. Raman scattering provides spectral information for all molecules that exhibit changes in polarizability. By analyzing the Raman peaks in the acquired spectra, it is possible to obtain Raman images of live cells. Furthermore, a wide range of excitation wavelengths, from ultraviolet (UV) to near-infrared (NIR), can be utilized for Raman spectroscopy. Due to the significant signal enhancement provided by SERS, the exposure time for each pixel can be greatly reduced, thereby substantially shortening the imaging time. This enables real-time monitoring of molecular signal changes. Additionally, the improved temporal resolution achieved by increasing the number of pixels in the field of view also contributes to enhanced spatial resolution.

Plasmonic-based SERS sensors also offer new avenues for noninvasive sampling in flexible wearable devices. For instance, Li et al. introduced a flexible wearable plasmonic paper-based microfluidic device leveraging SERS [148]. This device, equipped with a portable Raman spectrometer, achieves highly sensitive detection and quantitative analysis of uric acid (UA) in sweat, with a detection limit as low as 10 μM, and pH changes within the range of 4–8. The paper-based microfluidic device adheres to the skin, capturing and transporting sweat through capillary action in microchannels and pressure exerted by sweat glands (with a maximum hydrostatic pressure of 70 kN/m^2) for in situ SERS analysis (Figure 4.9a). Experimental results demonstrate that the SERS sensor exhibits excellent signal reproducibility (with an enhancement factor of up to 1×10^7) and effective molecular penetration capability.

He and colleagues developed a wearable microfluidic plasmonic sensor capable of identifying fingerprint information of target biomarkers like urea, lactate, and pH in sweat, in a portable and refreshable manner [149]. They designed a miniature plasmonic metasurface with uniformly distributed mushroom-shaped hotspots that exhibit high SERS activity and integrated it into a microfluidic platform. The integrated microfluidic SERS platform is composed of four layers (as shown in Figure 4.9b) from bottom to top: (i) a double-sided medical adhesive tape serving as the epidermal adhesion layer; (ii) a microfluidic PDMS layer; (iii) two ultrathin SERS chips embedded within the microfluidic layer's microcavities; and (iv) a Kapton tape encapsulation layer. Under the pressure of subcutaneous sweat glands, biofluids are drawn into the microchannels via capillary action. When sweat finally reaches the plasmonic metasurface, the SERS effect, generated by the orderly silver metasurface, allows for in situ Raman analysis, enabling the extraction of molecular characteristics of target analytes in sweat.

Additionally, Wang and colleagues developed a novel method for fabricating ultrathin, flexible, wearable SERS substrates and introduced a flexible substrate named silk fibroin-anodic aluminum oxide-Au nanoparticles (SF-AAO-Au) as an outstanding wearable glucose sensor for sweat [151]. This substrate, as shown in Figure 4.9d, consists of an SF film serving as a flexible support, an ultrathin AAO cavity as a template, and plasmonic Au NPs. By employing simple chemical etching and physical deposition, they created an innovative ultrathin layered structure with a 3D plasmonic array as the SF-AAO-Au SERS substrate, which exhibits excellent mechanical flexibility, uniformity, and sensitivity. The developed SF-AAO-Au SERS sensor demonstrated a detection limit (LOD) of 1.68×10^{-7} M within a concentration range of 10^{-7} to 10^{-3} M, and effectively detected glucose in human sweat on the skin.

Lee et al. developed a SERS material for contact lenses (SERS-LM) designed for the selective detection of glucose in human tears [150]. The SERS-LM is composed

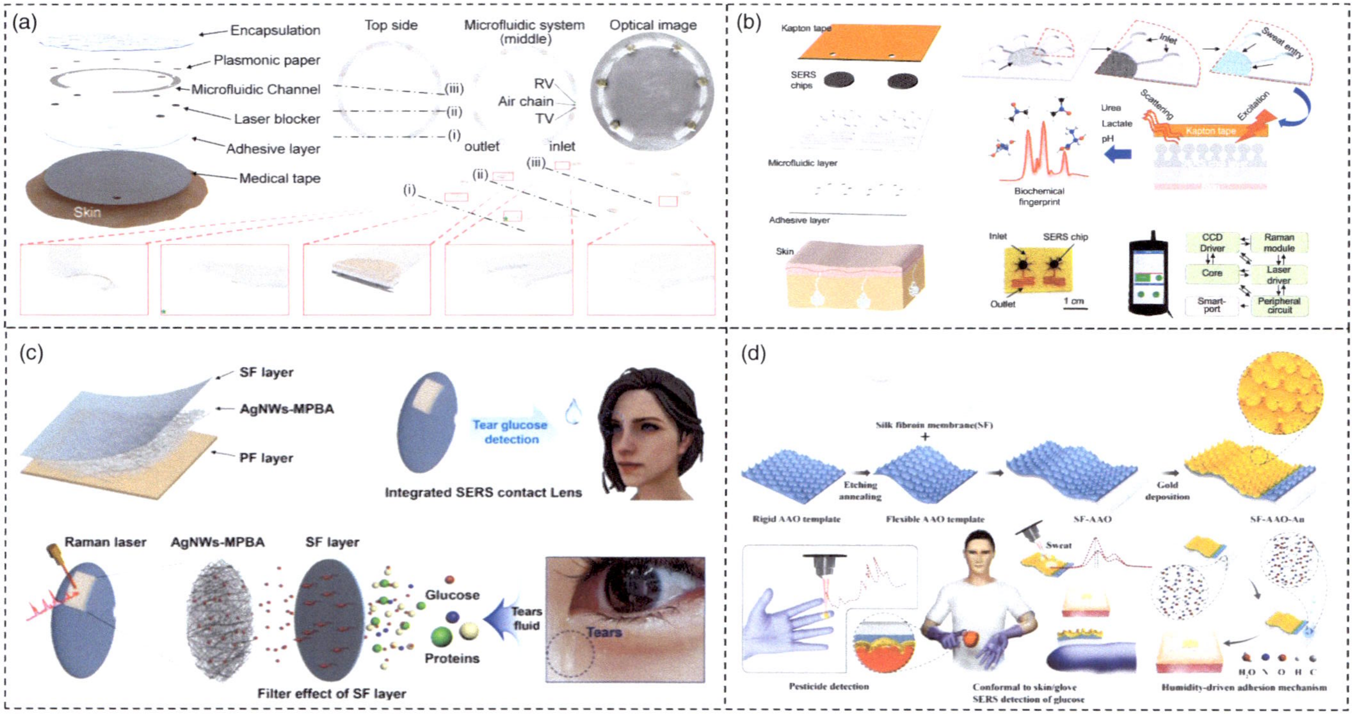

Figure 4.9 Flexible SERS sensors for noninvasive detecting. (a) Wearable plasmonic paper-based microfluidic device for SERS analysis of sweat. *Source:* Ref. [148] / American Chemical Society. (b) Microfluidic SERS sensor with mushroom-shaped hotspots for biomarker detection in sweat. *Source:* Ref. [149] / Springer Nature / CC BY 4.0. (c) SERS-LM for glucose detection in tears with a layered SF-AgNW structure. *Source:* Ref. [150] / with permission of Elsevier. (d) SF-AAO-Au flexible SERS substrate for glucose detection in sweat. *Source:* [151] / with permission of Elsevier.

of a layered structure, as illustrated in Figure 4.9c, consisting of a silk fibroin (SF) layer, a AgNW layer coated with 4-mercaptophenylboronic acid (MPBA), and a protective film (PF) layer. After material characterization, the SERS performance of the SERS-LM was evaluated, demonstrating uniform Raman signal generation with detection limits as low as 198 pM for MB and 211 nM for glucose.

Overall, the continued refinement of SERS substrates and the exploration of new applications underscore the growing importance of SERS in analytical science and its potential to become a key tool in various fields, including medical diagnostics, environmental monitoring, and biochemical research.

4.4.2 LSPR Colorimetric Sensors

The LSPR effect causes plasmonic nanomaterials to exhibit strong extinction in the visible light region, often accompanied by distinct yet strong colors. When surface etching or anisotropic growth occurs in the presence of an analyte, changes in the position or line shape of the LSPR peak can lead to noticeable color changes. Therefore, LSPR materials can be used as colorimetric sensors, where visible color changes indicate the presence or concentration of specific analytes.

LSPR is highly dependent on the size, composition, morphology, interparticle distance, and orientation of the plasmonic nanostructures [41, 152, 153]. Zeng et al. have highlighted that anisotropic plasmonic nanostructures hold greater potential for use in colorimetric sensing compared to isotropic ones [153]. For example, gold nanomaterials with anisotropic structures, such as nanorods, nanoplates, nanoflowers, nanoframes, nanocubes, and nanoprisms, exhibit higher sensitivity to changes in refractive index than lower-dimensional spherical AuNPs. Generally, as the length of these structures increases and their vertices become sharper, their refractive index sensitivity tends to enhance.

In the realm of anisotropic plasmonic nanostructures, AuNRs are among the most commonly used building blocks for designing colorimetric sensors with enhanced visual resolution. Electrons oscillating along the short and long axes of AuNRs give rise to TLSPR and LLSPR peaks, respectively [41]. The TLSPR peak of gold nanomaterials is relatively insensitive to changes in nanoparticle size or the refractive index of the surrounding medium. By contrast, the LLSPR undergoes a redshift as the aspect ratio of the AuNPs increases, making it highly sensitive to any refractive index changes in many sensor designs. For instance, as the aspect ratio increases from 1.7 to 5.2, the longitudinal plasmon peak shifts from 590 to 935 nm [153]. This shift is accompanied by a notable color change in the gold nanorod solution.

Building on the aforementioned characteristics of LSPR, the detection mechanism of LSPR-based colorimetric sensors primarily hinges on altering the shape and size of the plasmons, the interparticle distance, the composition, and the alignment of the particles (Figure 4.10). These factors allow for the design of various colorimetric sensing strategies tailored to specific applications.

Methods for altering the size and shape of plasmons include etching and growth. Etching refers to the oxidation process of nanoparticles, which can lead to changes in their aspect ratio, shape, LSPR peak position, and overall appearance. For

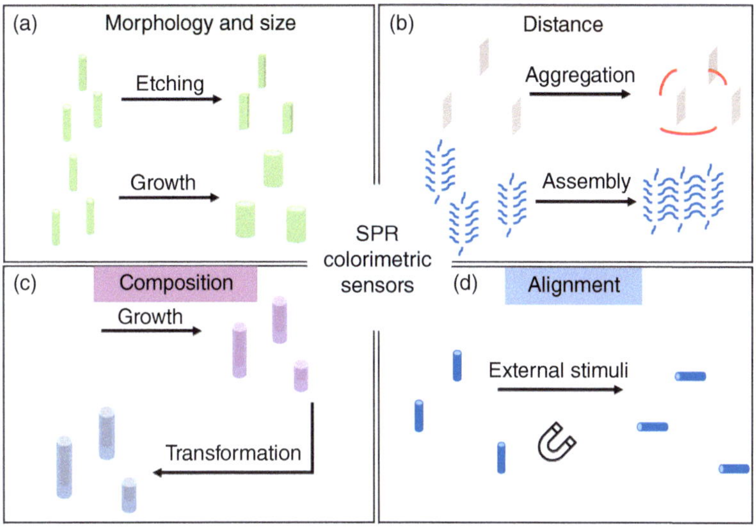

Figure 4.10 Detection mechanism of LSPR-based colorimetric sensors. (a) Shape and size. (b) Interparticle distance. (c) Composition. (d) Alignment.

example, in Au@FeOOH nanoparticles, the FeOOH shell is etched by the reducing agent ascorbic acid (AA), leading to changes in the surrounding environment of the AuNPs. This process results in a shift in the scattering color and a blue shift in the wavelength [154]. Similarly, in the case of passivated aluminum nanopillars on a wafer-scale flexible plasmonic metasurface, a significant decrease in spectral wavelength is observed with etching times of 425 and 300 seconds [155].

Growth-based methods rely on the reduction of precursor metal ions by adding analytes, which results in the formation of a new metal layer on the original plasmonic nanoparticles. For example, gold nanorods can serve as chiral seeds that grow into quadruply twisted nanoparticles with plasmonic optical activity [156]. During this process, the transverse LSPR mode gradually redshifts from 508 to 650 nm, while the longitudinal LSPR mode initially blueshifts from 1098 to 810 nm, and then gradually redshifts to 915 nm. Similarly, the growth of silver nanoparticles (AgNPs) on gold nanostars (AuNSs) leads to a significant blueshift in the LSPR peak [157].

In LSPR-based biosensors, one of the most common strategies is analyte-induced aggregation [5–7]. In this approach, the distance between plasmonic nanomaterials decreases due to aggregation. As a result, the electromagnetic fields of the aggregated plasmonic nanostructures couple due to their proximity, leading to a redshift in the plasmonic peak, with the solution's color changing proportionally to the analyte concentration (typically from pink to blue). Conversely, antiaggregation can occur, where the presence of the analyte prevents the plasmonic nanostructures from aggregating [41]. This effect is mainly due to the strong cross-linking and complexation capabilities of polyamine or thiol-functionalized compounds with target molecules.

Another method to alter the distance between particles is through assembly. Assembly refers to the formation of plasmonic nanostructures composed of multiple nanoparticles [158]. This includes all plasmonic nanoparticle assemblies as well as multicomponent nanostructures formed by coupling a plasmonic nanoparticle with another functional material. Common approaches include the creation of gap-tunable nanoparticle dimers and trimers through chemical growth, and the synthesis of core-shell assemblies using template nanoparticle synthesis. After assembly, the distance between nanoparticles decreases, leading to a redshift in the plasmonic peak.

In transformation-based colorimetric sensors, anisotropic core-shell nanoparticles are commonly used. The analyte modifies the composition of the nanoparticle shell through a transformation-based approach. For example, sulfide salts can be detected using Au@Ag core-shell nanoparticles, where Ag is converted into Ag_2S, resulting in a transformation of the plasmonic nanoparticles and a change in their optical properties [158]. In a specific type of growth-based method, hybridization occurs when nanoparticles grow in conjunction with a metal different from that of the original nanoparticles. Typically, the deposited metal shares a similar lattice structure with the original plasmonic nanostructure. Therefore, metals with compatible lattices, such as silver, copper, and gold, can be used together.

One of the unique characteristics of anisotropic plasmonic nanoparticles is their orientation-dependent optical properties. When these anisotropic nanoparticles are attached to counterparts with different functionalities, they can respond to various physical external stimuli, such as electric and magnetic fields, heat, mechanical pressure, and light.

Leveraging these properties, LSPR colorimetric sensors are widely used in various applications, including the detection of biomarker molecules and ions [5–7, 16–18].

Amirjani et al. successfully synthesized L-tryptophan–functionalized AgNPs using a one-pot synthesis method (Figure 4.11a) [5]. These L-tryptophan AgNPs exhibited a significant response to Mg^{2+} ions, with a response time of less than 10 minutes. By contrast, no response was observed for Ca^{2+} and other metal ions at concentrations up to 300 μM. Further studies revealed that these nanoparticles tend to aggregate at pH levels below 7 or above 11, while the response to 150 μM Mg^{2+} was maximized when the solution's pH was adjusted to 10–11. Additionally, the nanoparticles maintained their stability even under high salt conditions (60 mM NaCl).

UV-Vis spectroscopy indicated the appearance of a new peak at longer wavelengths. This phenomenon is attributed to the presence of carboxylic acid, amine, and amide groups in the L-tryptophan structure, with the amine group being responsible for AgNP functionalization. In the presence of Mg^{2+} ions and at pH levels above 9, the amine group becomes protonated and interacts with the magnesium cations due to its high affinity. This interaction, which loosens the surface amine bonds, significantly affects the stability of the AgNPs, leading to nanoparticle aggregation.

The team further advanced their work by immobilizing L-tyrosine–functionalized AgNPs onto a polyethylene terephthalate (PET) substrate, developing a recyclable optical nanosensor for the rapid detection of Pb^{2+} ions (Figure 4.11b) [6]. The mechanism relies on the carboxylic acid, hydroxyl, and amine functional groups of

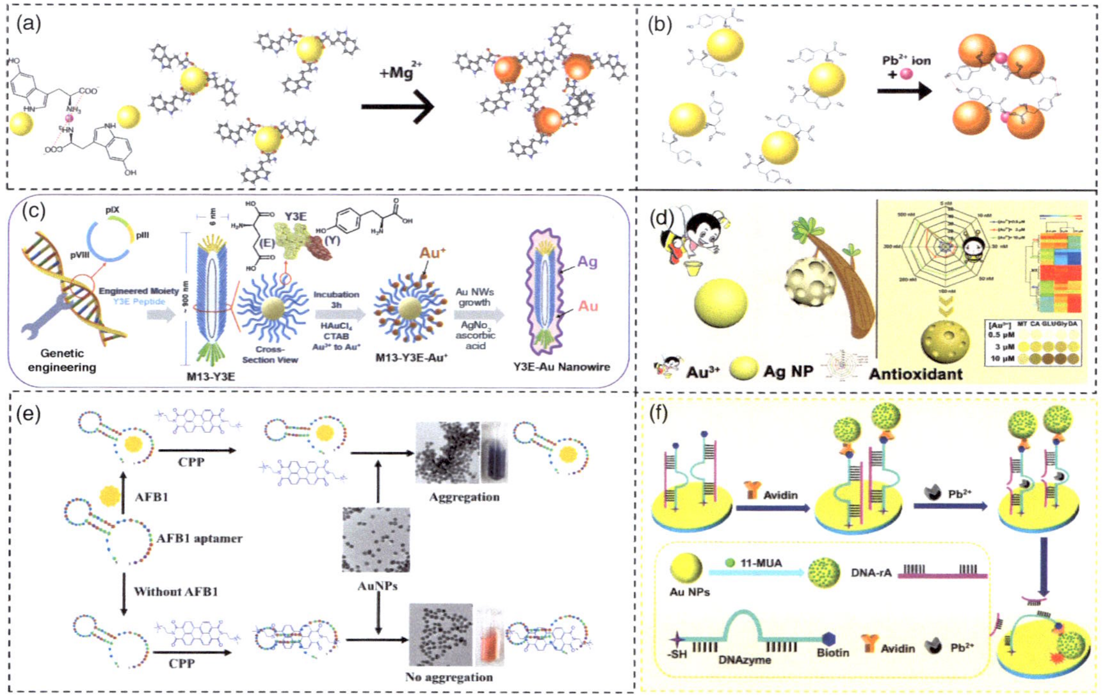

Figure 4.11 SPR colorimetric sensors. (a) L-tryptophan-functionalized AgNPs aggregate via Mg^{2+}-induced amine interaction. *Source:* Ref. [5] / with permission of Elsevier. (b) L-tyrosine-functionalized AgNPs aggregate through Pb^{2+}-induced carboxylic acid interaction. *Source:* Ref. [6] / with permission of Elsevier. (c) Y3E-Au NWs sensor for Hg(II) detection with nonaggregative sensing. *Source:* Ref. [16] / with permission of Elsevier. (d) Honeycomb-like AgNPs sensor array for antioxidant identification. *Source:* Ref. [159] / with permission of Springer Nature. (e) Colorimetric aptasensor using citrate-AuNPs for AFB1 detection. *Source:* Ref. [7] / with permission of Elsevier. (f) TFBG-SPR sensor for lead ion detection with DNAzyme and AuNPs. *Source:* Ref. [18] / with permission of Elsevier.

L-tyrosine, which act as metal chelating agents on the surface of the AgNPs. The carboxylic acid group interacts with Pb^{2+} ions, reducing the interparticle distance between the L-tyrosine–functionalized nanoparticles. Additionally, the hydroxyl group forms hydrogen bonds, facilitating this process. This interaction triggers the aggregation of the L-tyrosine–functionalized AgNPs. The sensor demonstrates a response time of less than 15 seconds and achieves a detection limit below 9 nM within a dynamic range of 1–1,000 nM.

Manivannan and colleagues developed an optical and colorimetric biosensor using a biomineralization process mediated by silver nitrate and surfactants (Figure 4.11c) [16]. This sensor is composed of AuNWs templated by transgenic M13 bacteriophages that express a specific gold-binding peptide, tyrosine-glutamate-glutamate-glutamate (Y3E). The flexible Au NWs form enriched Hg(II) binding sites on their surface, with the surface-coated AgNPs serving as more effective receptors for Hg(II) binding.

The results demonstrated that the Y3E–Au NWs sensor probes could efficiently preconcentrate the added Hg(II) and induce chemical interactions between the Au NWs and Hg(II), exhibiting nonaggregative sensing performance. This is attributed to the partially deposited Ag NPs on the Y3E–Au NWs, which likely react with the added Hg(II), leading to the deposition of elemental mercury or the formation of a mercury alloy between Hg and Au.

Zhang et al. developed a visual sensor array using honeycomb-like AgNPs as colorimetric probes to accurately identify antioxidants (Figure 4.11d) [159]. They selected three concentrations of $HAuCl_4$ solution (0.5, 3, and 10 μM) as sensor receptors. In the presence of $AuCl_4^-$, the AgNPs react with $HAuCl_4$, where the AgNPs are oxidized to Ag^+, and Au^{3+} is reduced to Au(0), which then deposits on the AgNPs surface. The presence of different antioxidants effectively inhibits the oxidation of AgNPs to varying degrees, resulting in distinct colorimetric responses. By employing this array, they were able to successfully identify five different antioxidants: dopamine, glucose, citric acid, glycine, and melatonin, both quantitatively and qualitatively at concentrations as low as 10 nM.

Lerdsri and colleagues successfully developed a label-free colorimetric aptasensor for the rapid and direct detection of aflatoxin B1 (AFB1) using a G-rich specific aptamer, cationic perylene probe (CPP), and unmodified citrate-stabilized AuNPs (Figure 4.11e) [7]. Under citrate conditions, the presence of CPP causes the negatively charged AuNPs to aggregate, resulting in a noticeable color change from red to blue. In the absence of AFB1, the negatively charged aptamer interacts with CPP through electrostatic interactions, forming an aptamer-CPP complex due to the random coiling of the aptamer. This interaction reduces the amount of free CPP in the solution, allowing the AuNPs to remain well-dispersed and the solution to retain its red color. When the specific aptamer binds to AFB1, its conformation changes from a random coil to a folded structure. This leaves enough CPP in the solution to interact with the negatively charged AuNPs, causing them to aggregate and turn the solution blue.

In addition to traditional LSPR sensors operating in the visible light range, SPR sensors based on fiber Bragg gratings (FBGs) offer an effective colorimetric sensing approach due to their high-quality factor, operation in the near-infrared region (which enhances penetration depth), and minimal interference from external environmental factors such as temperature and stress. This method detects refractive index changes caused by biochemical reactions on the sensor surface by monitoring the spectral response (wavelength and amplitude) of certain higher-order cladding modes.

For example, Wang and colleagues combined the efficient titled fiber Bragg grating (TFBG) SPR mechanism with the specific catalytic reaction between DNAzyme and lead (II) ions, along with the "hotspot" effect of AuNPs on a gold film, to propose a highly sensitive and selective method for detecting heavy metal lead ions (Figure 4.11f) [18]. When lead ions with high catalytic activity cleave the substrate strand containing rA, the double helix structure transitions to a single strand, causing the AuNPs linked to the DNAzyme to settle onto the sensor surface. The narrow gap (<2 nm) between the AuNPs and the gold film generates a "hotspot" effect, thereby enhancing the SPR signal. By measuring the TFBG-SPR signal enhancement, the detection of heavy metal lead ions (Pb^{2+}) is achieved.

These studies collectively highlight significant advancements in the development of highly sensitive and selective colorimetric sensors for detecting a variety of analytes, ranging from metal ions to antioxidants and toxins. By leveraging the unique properties of nanoparticles, such as their size, shape, and surface functionalization, researchers have created innovative sensing platforms that exhibit rapid response times, high stability, and low detection limits. The use of specific interactions between target molecules and functionalized nanoparticles, as well as the integration of sophisticated mechanisms like fiber Bragg grating-based SPR, has enabled the detection of even trace amounts of analytes with great accuracy. These developments pave the way for more reliable and efficient sensors that can be applied in diverse fields such as environmental monitoring, food safety, and medical diagnostics.

4.5 Soft Bioelectronics

Soft electronics, also known as flexible or stretchable electronics, represent a cutting-edge field in technology that combines the principles of electronic engineering with materials science. Unlike traditional rigid electronics, soft electronics are designed to be flexible, stretchable, and conformable to various shapes and surfaces. This innovative approach opens up a plethora of new applications and possibilities that were previously unattainable with conventional electronics.

However, the mismatch in Young's modulus between materials and human skin and organs has hindered the further development of soft electronics. To overcome this limitation, the development and innovation of soft materials are crucial. These innovations should focus on creating flexible, stretchable, and biocompatible materials that conform to the human body, enabling continuous and noninvasive monitoring of biological vital signs in complex and dynamic real-world environments. Among the various nanomaterials, metallic nanomaterials such as gold and silver stand out due to their excellent electrical and thermal conductivity, biocompatibility, and mechanical flexibility. These precious metals are the mainstream nanomaterials in the advanced field of soft electronics.

4.5.1 Materials and Design

A common strategy to impart flexibility or stretchability to conductive metallic films is through the design of specialized structures, such as micro-cracked [160–162], serpentine [163, 164], mesh-like [165–167], and wavy patterns [168–170].

Although gold films lack intrinsic stretchability, under specific conditions, they exhibit stretchability through microcracks when deposited on elastic substrates [160–162]. Recent reports have indicated that inert gold atoms can penetrate the surface of elastic polymers and diffuse throughout the free volume during the thermal deposition process of gold films, forming deformable microcracked gold interconnections (Figure 4.12a) [162]. By controlling the deposition rate of gold atoms, it is possible to regulate the morphology of the microcracks and, consequently, their electrical properties.

The serpentine structure achieves stretchability through its meandering design [163, 164]. In this configuration, a layer of conductive metal is evaporated onto a serpentine polymer (e.g., polyimide), which is typically firmly attached to soft substrates such as human skin and elastomers [164]. Figure 4.12b, c show a microneedle electrode based on a serpentine interconnect structure, which fractures at 80% strain. Its fracture limit is generally constrained by the strain limit of the polyimide film.

Mesh structures feature pores of various shapes [165–167]. For example, Fan et al. used flexible leaf veins as a self-assembly template, introducing conductive materials such as AgNWs and 2D-layered Mxene to construct a highly transparent conductive network (Figure 4.12d) [165]. This universal "weaved" method is also applicable to copper nanowires (CuNWs)/graphene oxide (GO). Furthermore, this mesh structure can be applied in implantable electronic systems. Kim group integrated Ag-Au-Pt NW/Pt NP nanocomposites with an elastic mesh substrate to develop an intrinsically flexible network electrode (Figure 4.12e) [166]. This electrode can securely adhere to the beating heart of a rat, allowing clear recognition of not only the QRS complex but also the P and T waves.

The wavy or wrinkle structure is created by deforming the elastomeric substrate either before or after the formation of the conductive film [168–170]. For instance, before forming the conductive film, metal nanostructures are first deposited onto a prestretched elastic substrate. When the strain is released, a wavy periodic structure forms [170]. This wavy configuration allows for reversible stretching and contraction, while preventing destructive strain on the metal nanostructures during the stretching process. The ultimate tensile strain of this structure largely depends on the initial prestrain intensity of the elastic substrate. An improved semiembedded method was recently reported (Figure 4.12f) [169]. Ag@Au nanowires were spray-coated onto a prestrained PDMS film. Subsequently, a prepolymerized PDMS (P-PDMS) solution was spin-coated over the Ag@Au NW network and cured to immobilize the nanowires. Upon releasing the prestrain, a wrinkled, semiembedded

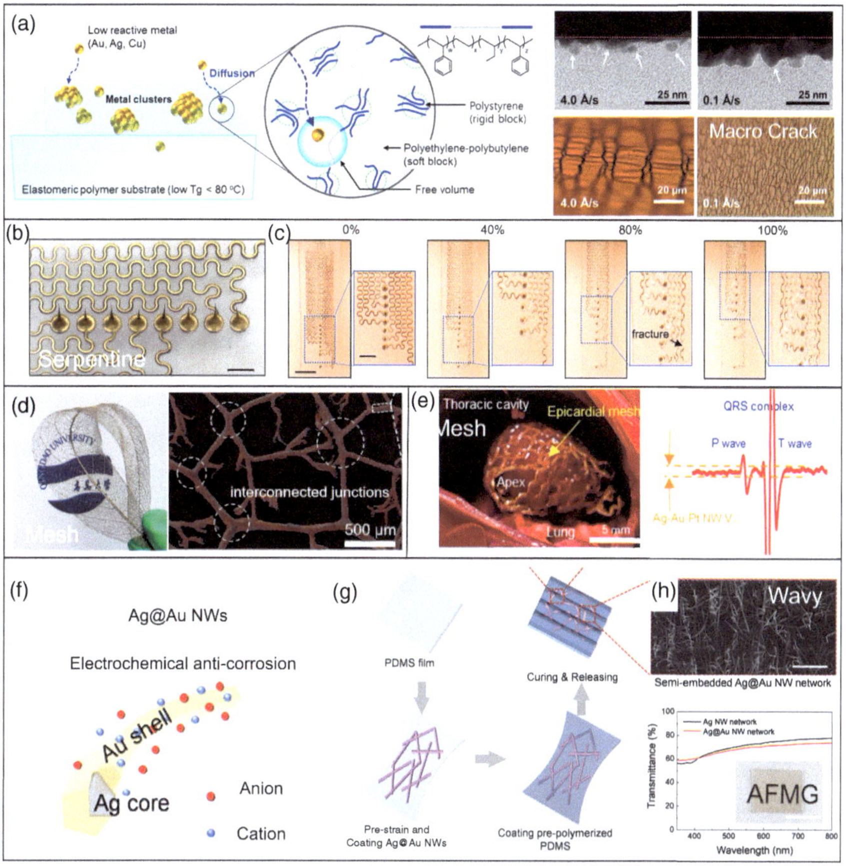

Figure 4.12 Structure-based soft electronics. (a) Macro-crack structure. *Source:* Ref. [162] / American Association for the Advancement of Science / CC BY 4.0. (b, c) Serpentine structure. *Source:* Ref. [164] / American Association for the Advancement of Science – AAAS. (d–e) Mesh structures. *Source:* Adapted from Refs. [165, 166]. (f–h) Wavy or wrinkle structure. *Source:* Ref. [169] / John Wiley & Sons. (i) Transmittance spectra of the semiembedded Ag@Au nanowire network and the Ag nanowire network [169]. The inset includes a digital photograph of the semiembedded Ag@Au NW network, with a scale bar of 1 cm.

Ag@Au NW network was formed (Figure 4.12g, h). The semiembedded Ag@Au nanowire network, with a sheet resistance of approximately 8 Ω/sq, achieved a transmittance of 69.0% at 550 nm. This value was only marginally lower than that of the semiembedded Ag NW network (Figure 4.12i), suggesting that the Au shell layer deposited on the Ag NWs had minimal impact on the transmittance.

Another widely used strategy is depositing metal nanomaterials onto or inside an elastic substrate to form a percolation network that integrates stretchability and conductivity [171–177]. The nanomaterials can be 0D [174–176], 1D [171–173], or 2D [177]. Specifically, 1D metallic nanowires exhibit unique advantages. For instance, AgNWs exhibit high conductivity and a low percolation threshold. These properties make AgNWs a primary choice for soft electronics. Jung et al. introduced a new method for making stretchable conductive films by floating various metal nanomaterials at the oil–water interface [178]. These materials include AgNWs, gold–silver core-shell nanowires (Au@Ag shell-core NWs), AgNPs, and AuNPs. They are embedded in a thin elastic film, combining stretchability with conductivity. For instance, AgNWs form a densely packed layer at the nanoscale (Figure 4.13a, b). The nanomembrane is about 250 nm thick and consists of a single

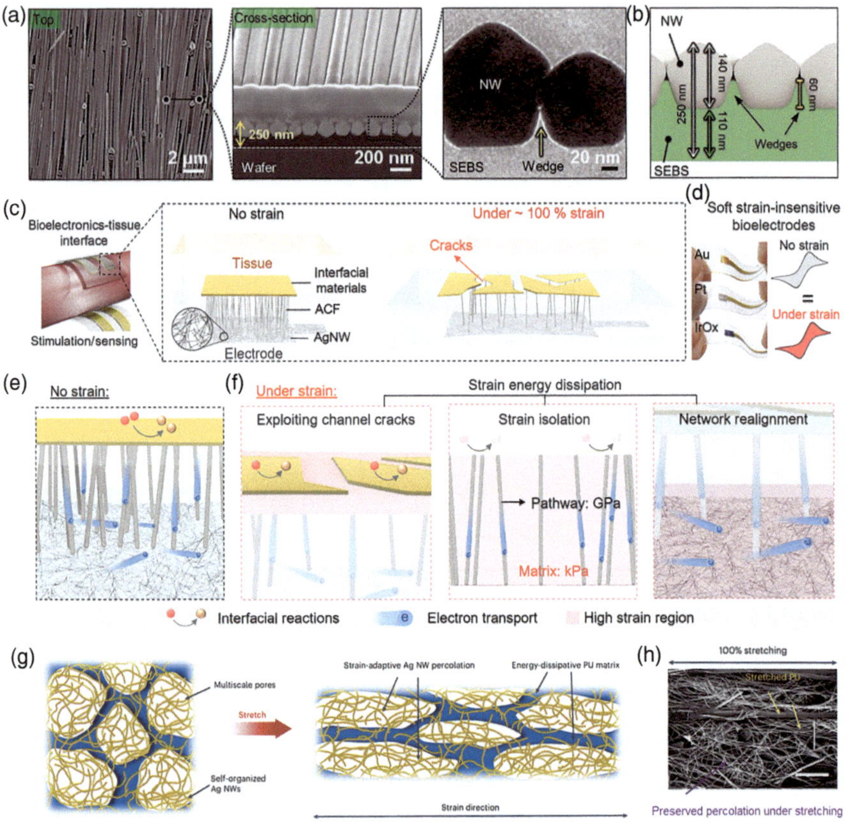

Figure 4.13 AgNWs-based percolation network. (a) SEM and higher-magnification TEM images of closely packed AgNWs in the nanomembrane. *Source:* Ref. [178] / American Association for the Advancement of Science – AAAS. (b) Schematic illustration of the nanomembrane cross-section that describes its detailed structure and dimensions [178]. (c) Illustration of the strain-insensitive bioelectrodes (SIB) as a bioelectronics–tissue interface under strain [179]. (d) Brittle noble metal-enabled SIBs for high fidelity stimulation and sensing under strain. *Source:* Ref. [179] / American Association for the Advancement of Science – AAAS. (e) Schematic presentation of the SIB's underlying electron transfer and out-of-plane and in-plane electron transport processes [179]. (f) Strain energy dissipation mechanisms within each layer of the SIB [179]. (g) Schematic of porous Ag NW nanocomposites in original (left) and stretched (right) states [171]. (h) SEM images of porous Ag NW nanocomposite after 100% stretching. *Source:* Ref. [171] / Springer Nature. Scale bars, 5 μm.

layer of nanowires approximately 140 nm in diameter, with an elastomer layer about 110 nm thick. Elastomer wedges about 60 nm high support the structure between the nanowires. As the film stretches, the AgNWs separate but stay interconnected even at 500% strain, maintaining conductivity under high tension. With specific structural design, AgNWs can achieve strain insensitivity [179]. The strategy (Figure 4.13c, d) involves: (i) creating a thin film for tissue interfacing, (ii) using stretchable AgNW traces in a poly(urethane acrylate) (PUA) matrix, and (iii) integrating these with an adhesive anisotropically conductive film (ACF). Various thin-film materials like Au, Pt, and iridium oxide (IrOx) can be deposited on the ACF-AgNW-PUA layers. The strain insensitivity is controlled by three mechanisms (Figure 4.13e, f): (i) crack channeling in the brittle thin film to release tensile strain energy, (ii) strain isolation in the ACF layer ensuring out-of-plane electrical conduction remains unaffected, and (iii) realignment of AgNW networks in the PUA layer to dissipate strain energy. The brittle film cracks to minimize strain, while the ACF maintains electrical connections, and the AgNW networks realign to manage strain. Similarly, Xu et al. described an electrically conductive phase-separated porous nanocomposite (PSPN) using Ag NWs with an exceptionally low percolation threshold and electrical properties unaffected by strain [171]. The combination of energy-dissipative porous polyurethane (PU) microstructures and adaptable Ag NW pathways ensures efficient electron transport even under significant strain (Figure 4.13g). As the PU elastomer undergoes structural changes to dissipate local stress during stretching, the Ag NWs maintain their percolation networks with minimal alteration in orientation (Figure 4.13h).

Although AgNWs have shown remarkable performance in stretchable electronics, their susceptibility to oxidation in air poses limitations. Additionally, direct exposure of human tissues to AgNWs can be harmful due to the potential leaching of silver ions from the nanowires. By contrast, AuNWs possess excellent mechanical compliance and robustness, chemical resistance, and outstanding biocompatibility [13]. Cheng's group developed a number of strategies to construct soft electronics via vertically aligned AuNW [59, 180–184]. The vertically aligned AuNWs can be directly grown on the surface of elastic substrates (such as PDMS and Ecoflex) without being restricted by the substrate topology, achieving conformal coating. The preparation process is straightforward: First, plasma treatment is used to introduce hydrophilic groups to the substrate. Next, the substrate undergoes aminization treatment using 3-aminopropyltriethoxysilane / 3-Aminopropyltrimethoxysilane (APTES/APTMS). Gold nanoseeds are then anchored to the substrate surface via electrostatic attraction forces. Finally, the substrate with anchored gold seeds is immersed in a growth solution containing gold precursors, strong surface binding ligands, and reducing agents. This leads to the spontaneous growth of vertically aligned AuNWs within minutes. Essentially, these vertically aligned AuNWs exhibit a Janus structure: the top layer ("head") is composed of densely packed AuNPs, while the bottom layer ("tail") consists of nanowires perpendicular to the substrate. This multifunctional head/tail configuration and multilayer nanostructure (Figure 4.14a, b) can even be transferred onto soft/hard substrates while maintaining overall structural integrity and conductivity.

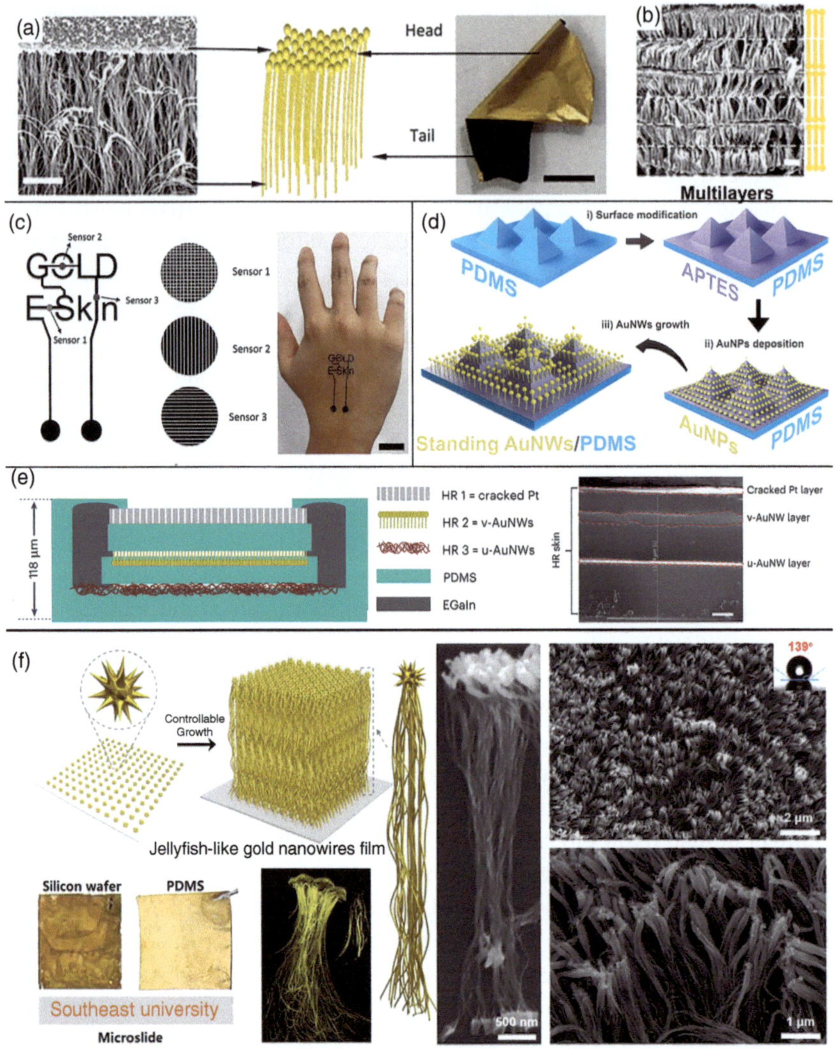

Figure 4.14 AuNWs-based percolation network. (a) SEM image (scale bar: 500 nm), schematic illustration and photographic image (scale bar: 1 cm) of the vertically aligned AuNWs. *Source:* Ref. [180] / American Chemical Society. (b) Layer-by-layer SEM images of vertically aligned AuNWs. *Source:* Ref. [180] / American Chemical Society. (c) Diagram showing the positions and shapes of three crack sensors embedded in a "gold E-skin" pattern and image of the tattoo-like "gold E-skin" circuit applied to the back of the hand. *Source:* Ref. [185] / John Wiley & Sons. (d) The fabrication process of the vertically aligned AuNWs on the pyramid structures. *Source:* Ref. [32] / with permission of American Chemical Society. (e) Diagram showing the structure of the hierarchically resistive (HR) skin, where EGaIn stands for Eutectic Gallium–Indium, and a cross-sectional scanning electron microscope (SEM) image of the HR skin. *Source:* Ref. [186] / Springer Nature. (f) Diagram, optical images, and SEM images of jellyfish-like AuNWs. *Source:* [187] / with permission of John Wiley & Sons.

By integrating local crack programming technology, wearable electronic tattoos based on vertically aligned AuNWs have been successfully fabricated [185]. These electronic tattoos require only a single type of material, thereby eliminating the need for multiple active material interfaces and avoiding welding or adhesive bonding. This allows for the seamless integration of strain/pressure sensors, anisotropic orientation-specific sensors, and strain-insensitive stretchable interconnects (Figure 4.14c). This strategy opens a new universal route for the design of next-generation wearable electronic tattoos. When combined with rational structural engineering (such as a hierarchical structure based on pyramid microarrays), vertically aligned AuNWs can also be used to design high-performance wearable pressure sensors (Figure 4.14d). Following the aforementioned preparation strategy, vertically aligned AuNWs can uniformly grow on the pyramid structures [32]. This configuration is particularly sensitive to pressure, resulting in pressure sensors with a low operating voltage of 0.1 V, high sensitivity under low pressure conditions, a rapid response time of less than 10 ms, and high durability, maintaining stable signals during 10,000 cycle tests.

Building on these foundations, Cheng group and colleagues recently presented a hierarchically resistive skin sensor composed of a laminated cracked platinum film, vertically aligned AuNWs, and a percolated AuNWs film, all integrated into a single sensor (Figure 4.14e) [186]. Consequently, this hierarchically resistive skin exhibits a staircase-shaped resistive response to tensile strain, with distinct sensing regimes corresponding to specific active materials. Through this single resistive signal, the sensor can identify up to five physical or physiological activities related to human throat speech: heartbeats, breathing, touch, and neck movement, thus functioning as a multimodal sensor. Additionally, this strategy for fabricating vertically aligned AuNWs has been further extended. Zhang et al. utilized ligand-free AuNSs instead of gold nanospheres as seeds to create a novel jellyfish-shaped AuNWs array (Figure 4.14f) [187]. In this array, the head is composed of AuNSs, and the tail consists of bundles of AuNWs. Compared to the previously developed vertically aligned AuNWs, this new array has a larger specific surface area and can also be directly grown on glass and PDMS. Building on the jellyfish-shaped AuNWs, Zhang et al. further investigated their multifunctionality in sensing applications. They found that these nanowires could be utilized for electrochemical sensing, humidity sensing, pressure sensing, strain sensing, and temperature sensing [30].

4.5.2 Flexible Biosensors

Flexible biosensors can seamlessly conform to the soft human skin with attributes such as lightweight, small size, and biocompatibility, making them ideal for real-time monitoring of biological systems [9]. Metallic nanomaterials can combine conductivity with flexibility and stretchability, which have been widely used in the fabrication of wearable flexible biosensors.

Common flexible biosensors include pulse sensors [30–32, 188, 189], temperature sensors [190–192], motion sensors [31, 192–198], and sweat sensors [134, 149, 151, 191, 199–203]. These wearable biosensors can be broadly categorized into physical biomarkers such as pulses, temperature, and motion, and chemical biomarkers, including glucose, lactate, cationic ions, and cortisol.

4.5.2.1 Chemical Biomarkers

Among various biomarkers, glucose concentration detection has been a major research focus due to its direct connection to diabetes. Currently, the global number of diabetes patients has exceeded 420 million, and it is projected to reach 642 million by 2040 [204]. Continuous noninvasive monitoring of glucose is therefore of great importance, as it can significantly reduce the risk of diabetes-related complications and eliminate the pain and stress associated with traditional finger-prick blood tests.

The glucose concentration in sweat has a certain correlation with blood glucose levels, making it possible to achieve noninvasive continuous glucose monitoring through sweat. However, compared to blood glucose concentrations (2×10^{-3} M to 40×10^{-3} M), the glucose concentration in sweat is much lower, ranging from 0.01×10^{-3} M to 1.11×10^{-3} M [205].

Methods for measuring glucose concentration include electrochemical detection, SERS, and colorimetry. Among these, electrochemistry is a powerful biosensing tool that can convert information from specific bioanalytes into measurable electrochemical signals. It is known for its high sensitivity, low detection limits, and ease of integration into electronic systems, making it commonly used in glucose sensing.

Imamura et al. developed a stretchable enzyme-free electrochemical electrode by depositing an Au film onto a thermoplastic shape-memory polymer surface [206], as shown in Figure 4.15a. This electrode demonstrated high glucose detection sensitivity even under a 210% tensile strain, with a detection limit as low as 2.22×10^{-8} M. However, the limited selectivity of enzyme-free sensors for glucose poses a challenge, restricting their practical application in glucose monitoring.

Combining enzymatic glucose sensors with plasmonics can enhance electron transfer rates. These nanomaterials include noble metals, transition metal nanoparticles, and nanostructured metal oxides [208].

Zhai et al. developed a dual-modal stretchable electrochemical glucose sensor by modifying a stretchable AuNWs film with glucose oxidase (Gox) [207]

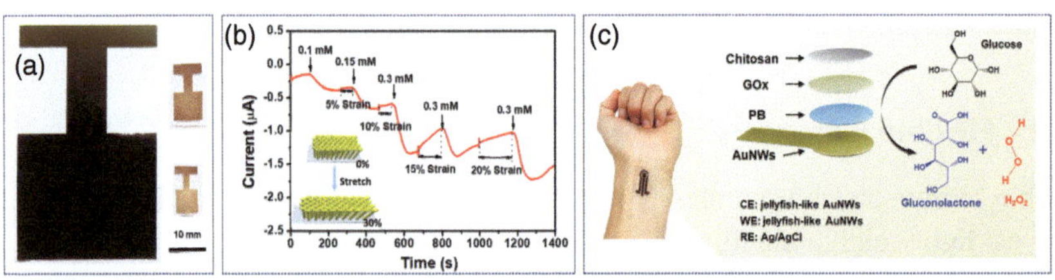

Figure 4.15 Electrochemical glucose sensor. (a) Stretchable enzyme-free Au electrode on thermoplastic polymer for glucose detection. *Source:* [206] / with permission of John Wiley & Sons. (b) Dual-modal AuNWs film glucose sensor with GOx for strain and glucose detection. *Source:* Ref. [207] / with permission of American Chemical Society. (c) Jellyfish-shaped stretchable AuNW film glucose biosensor with high sensitivity. *Source:* Ref. [30] / with permission of Elsevier.

(Figure 4.15b). This sensor can simultaneously respond to changes in glucose concentration and tensile strain, while maintaining high selectivity. Zhang et al. designed a stretchable electrochemical glucose biosensor based on an inherently stretchable jellyfish-shaped AuNWs film [30] (Figure 4.15c). This sensor exhibits high sensitivity (45.40 μA/mM/cm^2) and a low detection limit (10 μM).

In addition to glucose, electrochemical detection methods have also been reported for other substances such as acetaminophen [199], urea, lactate, pH [149], H_2O_2 [209], Na^+, and K^+ [203]. Monitoring these physiological parameters is crucial for the prevention and diagnosis of conditions such as liver failure and kidney damage.

4.5.2.2 Physical Biomarkers

Arterial pulse contains valuable cardiovascular information, including diastolic and systolic pressure, which are commonly used in diagnosing cardiovascular diseases and assessing overall health. Compared to traditional empirical pulse diagnosis in traditional Chinese medicine, real-time monitoring of pulse signals provides a more comprehensive and accurate reflection of health status, enabling early prevention and diagnosis of cardiovascular diseases. In recent years, the rapid development of flexible biosensors has made continuous real-time monitoring of pulse signals possible. Various pressure sensors have been developed for this purpose, including piezoresistive, capacitive, optical, triboelectric, piezoelectric, and magnetoelastic pressure sensors [210].

Zhu et al. developed a high-performance wearable pressure sensor based on a vertically aligned AuNW design [32], utilizing a hierarchical structure of pyramid microarrays (Figure 4.16a). The resulting sensor features a low-operating voltage of 0.1 V, high sensitivity under low-pressure conditions, a rapid response time of less than 10 ms, and excellent durability, with stable signals maintained after 10,000 cycles. It can wirelessly detect high-quality arterial pulses.

Li et al. designed an ultra-thin, flexible piezoresistive sensor with high sensitivity and a wide detection range, using pressure-sensitive materials and nanonetwork electrodes with a layered nanonetwork structure (Figure 4.16b) [188]. The layered nanonetwork is composed of AgNWs, graphene (GR), and polyamide nanofibers (PANF), achieving a sensitivity of 134 kPa^{-1} (0–1.5 kPa) and a low detection limit of 3.7 Pa, enabling accurate measurement of physiological signals such as pulse.

Yang et al. reported an ultra-sensitive capacitive pressure sensor based on a porous pyramid dielectric layer [211], which was used to measure the wrist pulse of a 26-year-old male, producing a waveform consistent with the subject's age. By further integrating the porous pyramid structure with PDMS coated with AgNWs and embedding the entire sensor into the elastic substrate Ecoflex (Figure 4.16c), the sensor becomes insensitive to strain and temperature. This feature effectively mitigates the impact of strain and temperature variations during real-world monitoring.

Body temperature is a vital sign of health, with abnormal increases often associated with conditions such as inflammation and fever. Elevated body temperature is also a significant clinical indicator of COVID-19 infection, making continuous

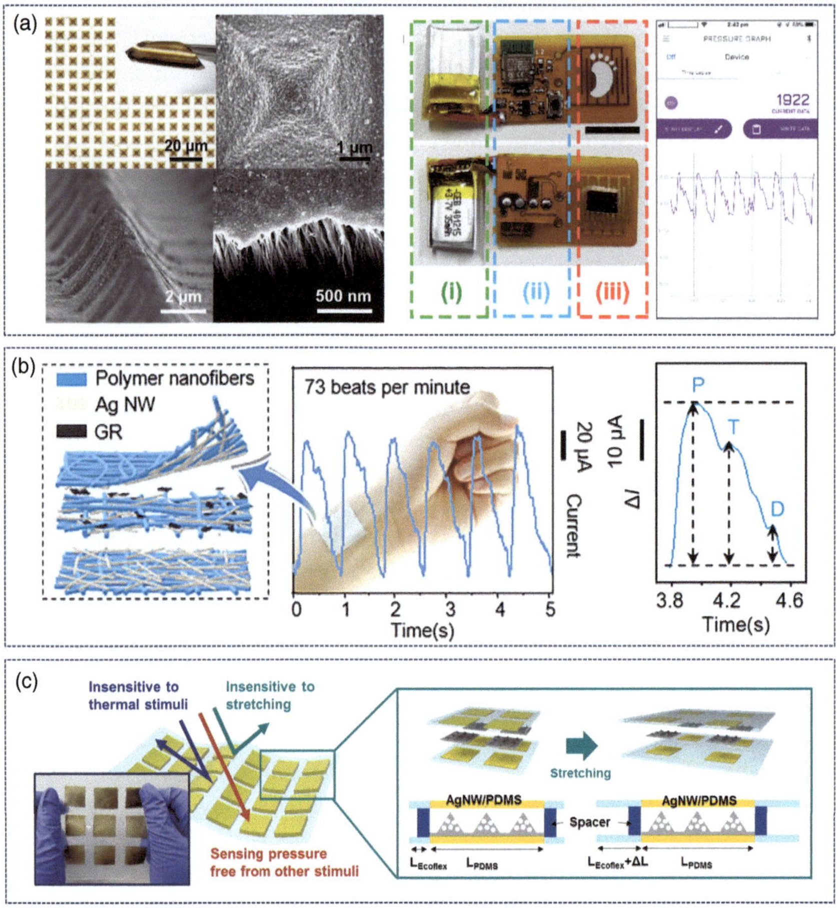

Figure 4.16 Flexible physical signal biosensors. (a) Vertical AuNWs pressure sensor with hierarchical pyramid microarrays. *Source:* Ref. [32] / American Chemical Society. (b) Flexible piezoresistive sensor with layered Ag NW, GR, and PANF structure. *Source:* Ref. [188] / American Chemical Society. (c) Capacitive sensor with porous pyramid dielectric layer embedded in Ecoflex. *Source:* Ref. [211] / American Chemical Society.

temperature monitoring particularly important. Resistive temperature sensors are widely used due to their simple readout systems, with their sensitivity closely linked to the temperature coefficient of resistance (TCR). A higher TCR indicates greater sensor sensitivity. Metallic nanomaterial films are considered an ideal choice for resistive temperature sensors because the spacing between metal nanoparticles and the mechanical strain on the film can significantly influence electron transport pathways, resulting in changes in the film's resistance.

For example, Oh et al. reported a PDMS-based AgNPs film, where the thermal expansion of PDMS increases the spacing between AgNPs as the temperature rises, leading to an increase in the film's resistance [190]. This film exhibits an exceptionally high TCR of 0.5 K^{-1}, demonstrating ultra-high temperature sensitivity.

Additionally, resistive sensing has been widely applied in human motion detection, playing a crucial role in rehabilitation therapy [26]. This application primarily relies on stretchable strain and pressure sensors. 1D metallic nanowires, such as AgNWs and AuNWs, are particularly promising for manufacturing motion-monitoring biosensors due to their excellent conductivity, stretchability, and flexibility. Currently, motion sensors based on AgNWs and AuNWs are being used to monitor movements in various parts of the body, including facial expressions, throat motion, neck movement, elbow/knee flexion, finger movements, and wrist motion [30].

4.5.3 Soft Wearable Electrochemical Biosensors

Wearable electrochemical sensors represent a significant advancement in personal health monitoring. These sensors are designed to be flexible, lightweight, and comfortable, allowing them to be seamlessly integrated into clothing or directly onto the skin. By combining the principles of electrochemistry with advanced materials science, wearable electrochemical sensors can continuously monitor various physiological and biochemical parameters in real time [212]. Soft electrochemical electrodes are crucial components in wearable electrochemical biosensors, playing a key role in biosensing performance [213, 214]. Gold-based nanomaterials exhibit unique advantages, including high electrical conductivity, a wide potential window, excellent mechanical flexibility and robustness, long-term stability, and high chemical stability [215]. They are also easy to fabricate/modify and demonstrate excellent biocompatibility for skin and in vivo applications, making them the most widely used electrode materials for manufacturing soft electrochemical biosensors [214]. For example, Wu et al. encapsulated lactate oxidase (LOx) in chitosan (CS) hydrogel and fabricated a gold nanoelectrode array on a nanoporous polycarbonate (PC) membrane for simultaneous detection of body temperature and sweat lactate [216]. The flexible gold nanoporous electrodes not only increase the electrode surface area but also provide a nanoconfined space to accelerate the catalytic reaction of LOx and control the substrate concentration on the LOx surface to reduce substrate inhibition (Figure 4.17a). Moreover, the temperature-dependent transmembrane current of the sensor can be used to estimate body temperature, and the influence of temperature on lactate detection can be adjusted using multiple linear regression. Lim et al. reported a simple and scalable synthesis method for whisker-like gold nanosheets (W-AuNS) used in stretchable, conductive, and biocompatible nanocomposites (Figure 4.17b) [218]. By adding platinum-coated W-AuNSs (W-AuNSs@Pt), impedance is significantly reduced and charge storage capacity is enhanced. The electrochemical performance remains stable even under 40% strain.

Cheng group has reported a series of soft electrochemical electrodes based on vertical AuNWs [30, 203, 207, 219, 220]. During the growth of vertical AuNWs, a specially designed shadow mask is used to obtain a patterned three-electrode system [207]. This three-electrode system is intrinsically stretchable and maintains its conductivity under low strain. The vertical AuNWs electrodes modified with Gox and Prussian blue are used as the working electrode, unmodified vertical AuNWs are used as the counter electrode, and Ag/AgCl-modified vertical AuNWs are used as the reference electrode (Figure 4.17c, d). The resulting electrochemical biosensor exhibits a

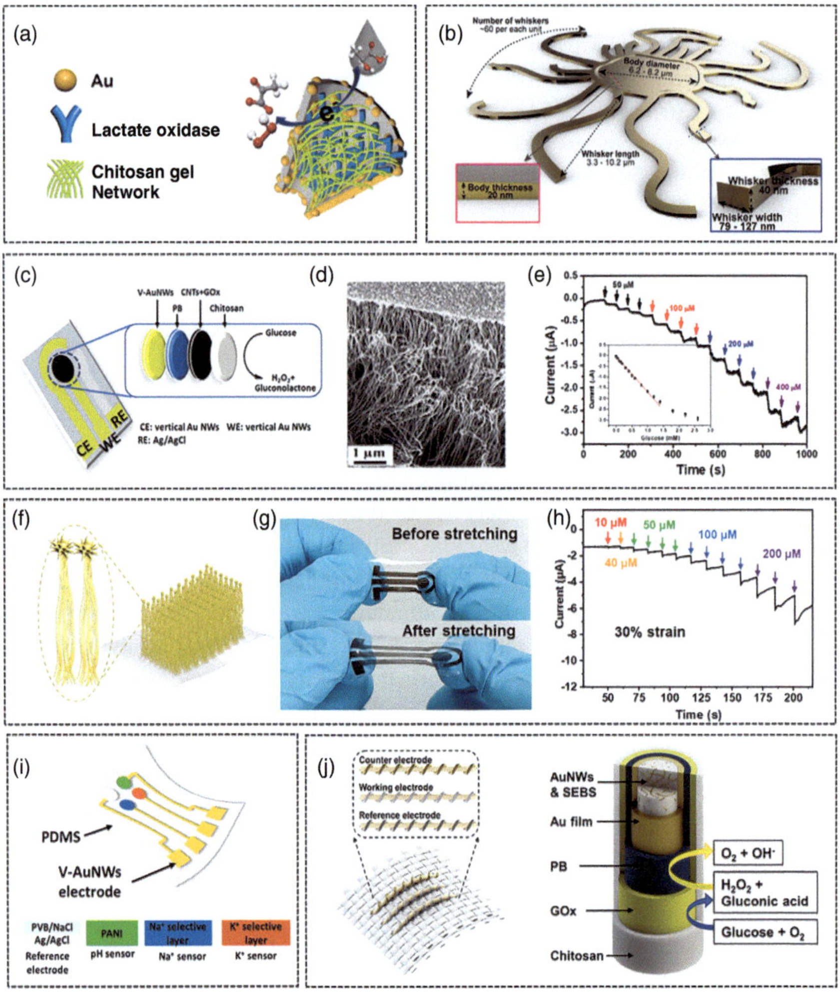

Figure 4.17 Plasmonic-based soft electrochemical electrodes. (a) Schematic illustration of the stretchable LOx@CS PC sensor. *Source:* Ref. [216] / with permission of American Chemical Society. (b) Schematic illustrations of the morphology of a W-AuNS. *Source:* Ref. [216] / with permission of American Chemical Society. (c) Modification process of the vertical AuNW-based stretchable glucose biosensor. *Source:* Ref. [207] / with permission of American Chemical Society. (d) SEM image of the vertical Au NWs. *Source:* Ref. [207] / American Chemical Society. (e) Chronoamperometric response of the fabricated stretchable biosensor upon successive addition of glucose [207]. (f) Schematic illustration of the jellyfish-like AuNWs array. *Source:* Ref. [30] / with permission of Elsevier. (g) Optical images of the three-electrode system before and after stretching. *Source:* Ref. [30] / with permission of Elsevier. (h) Glucose response at 30% strain. *Source:* Ref. [30] / with permission of Elsevier. (i) Schematic illustration of the wearable ion-to-electron transducers. *Source:* Ref. [203] / with permission of American Chemical Society. (j) Stretchable fiber-based glucose sensor. *Source:* Ref. [217] / with permission of American Chemical Society.

good catalytic response to glucose (Figure 4.17e). Similarly, Zhang et al. constructed jellyfish-like AuNWs (Figure 4.17f) on a PDMS elastomeric substrate, resulting in a stretchable three-electrode system [30]. This three-electrode system did not exhibit any visible cracks before and after stretching (Figure 4.17g), and it was able to sensitively detect glucose even under 30% strain (Figure 4.17h). Additionally, vertical AuNWs have been shown to serve as flexible and wearable ion-to-electron transducers for multiplexed in-situ potentiometric analysis of pH, Na^+, and K^+ in sweat [203]. By modifying the vertical AuNWs electrodes with polyaniline, sodium ionophore X, and valinomycin-based selective membranes, these electrodes can specifically detect pH, Na^+, and K^+ with high selectivity, reproducibility, and stability (Figure 4.17i). Notably, their electrochemical performance remains stable even under 30% strain and during repeated stretch-release cycles, eliminating the need for extra structural design. In addition to being used for human sweat detection, the excellent biocompatibility of vertical gold nanowires allows live cells to grow directly on their surface [219]. By further stimulating the cells with chemicals to induce H_2O_2 production, the vertical gold nanowires electrochemical electrode platform can be used to detect the natural and stretched states of cells in real time and in situ.

The Cheng group also developed ultrathin gold wires, which they combined with the highly elastic block copolymer poly(styrene-ethylene-butylene-styrene) (SEBS) to form stretchable AuNWs/SEBS composites [217, 221]. To further enhance the stretchability of AuNWs/SEBS, the fibers were prestretched to 300%, and then a gold film was chemically plated onto their surface. The resulting gold fibers could be easily woven into commercially available textiles. Subsequently, glucose-sensing electrodes were prepared by depositing Prussian Blue (PB) and immobilizing Gox on the stretchable gold fibers (Figure 4.17j). These electrodes exhibited a sensitive response to glucose even under 200% strain [217].

4.5.4 Plasmonics-enhanced Electrochemical Sensors

Electrochemistry is a powerful biosensing technology capable of converting biological interactions into measurable electrical signals [222]. Consequently, electrochemical technology plays a pivotal role in the development of the biosensing field, characterized by unmatched sensitivity, specificity, and low detection limits. Its strengths include rapid response, portability, and ease of use, making electrochemical biosensors widely applicable tools [214]. Plasmonic materials, owing to their excellent electrical and thermal conductivity, high specific surface area, unique SPR optical properties, and SERS effects, are extensively used in the electrochemical sensing domain [223, 224]. Moreover, plasmonic materials can be easily functionalized with various biological ligands, and numerous plasmonic nanocomposites have been developed for electrochemical applications [225]. These composite nanomaterials successfully integrate the advantages of plasmonics with electrochemistry, enabling researchers to develop novel electrochemical sensing strategies with enhanced sensitivity, selectivity, and stability.

Plasmonic-enhanced electrochemistry is an advanced field that leverages the unique properties of plasmonic materials to enhance the performance of electrochemical sensors. By integrating plasmonic materials with electrochemical systems, researchers can achieve significant improvements in sensitivity, selectivity, and detection limits. When plasmonic metal nanostructures (such as gold, silver, or copper) are coupled with other substrates, such as semiconductors (including TiO_2, CdS, WO_3, graphene, MoS_2, and Fe_2O_3) [226–228], the plasmon-excited hot electrons can overcome the Schottky barrier and inject from the metal nanoparticles into the conduction band of the semiconductor.

Chen and colleagues reported a plasmon-activated hydrogen evolution reaction (HER) electrochemical system comprising hybrids of Au nanorods and MoS_2 nanosheets (Au–MoS_2) [229]. This system consists of two essential components: (i) Au nanomaterials that act as light absorbers and excite electron–hole pairs during the SPR process and (ii) MoS_2 nanosheets that serve as active sites and electron acceptors to facilitate HER. Through the SPR excitation of Au nanostructures, hot electrons are injected from Au into ce-MoS_2. This phenomenon increases the charge density of ce-MoS_2 and, consequently, modulates the energy level of the catalyst to be more suitable for HER (Figure 4.18a, b). As a result, enhanced electrocatalytic activity for HER is achieved through plasmonic effects. Similarly, plasmonic metals can function as light absorbers and catalytic sites, efficiently harvesting light for chemical transformations. Using glucose electrocatalysis as a model reaction system, Wang et al. observed the direct plasmon-accelerated electrochemical reaction (PAER) on gold nanoparticles [230]. The wavelength- and solution pH-dependent electrochemical oxidation rate, along with dark-field scattering spectroscopy results, confirm that the hot charge carriers generated during plasmon decay are responsible for the enhanced electrocatalysis performance (Figure 4.18c, d). Based on the proposed PAER mechanism, they construct a plasmon-improved glucose electrochemical sensor, showcasing the enhanced performance of the non-enzyme sensor upon LSPR excitation. Furthermore, the nonradiative decay of plasmons generates hot carriers (HCs), which are electrons significantly above the Fermi level of the metal and holes significantly below the Fermi level of the metal (HC effect), and induces local heating (thermal effect). Ou et al. provided a novel strategy to distinguish HCs from the photothermal effect based on the different characteristics of heat dissipation (long distance) and HC transport (short distance) [233]. By investigating the rapid and slow response currents in plasmon-mediated electrochemistry on nanostructured silver electrodes, they quantitatively reveal the dominant role of the photothermal effect and its potential-dependent behavior. Additionally, it is found that the plasmonic surface potential contributes to the rapid response current.

Zhao et al. investigated the electrochemical oxidation activity of Pt NPs and AuPt NPs under both illuminated and nonilluminated conditions [231]. Their study revealed that AuPt NPs exhibit an electrochemical oxidation peak around the potential of −0.7 V, which is significantly amplified under illumination (Figure 4.18e). Based on this finding, they developed a ratiometric electrochemical aptamer sensor to achieve sensitive and reliable monitoring of the epithelial cell adhesion molecule (EpCAM) (Figure 4.18f). Bai et al. developed a self-powered biosensor based on a photofuel cell (PFC-SPB), which consists of a plasmonic AuNPs/organic

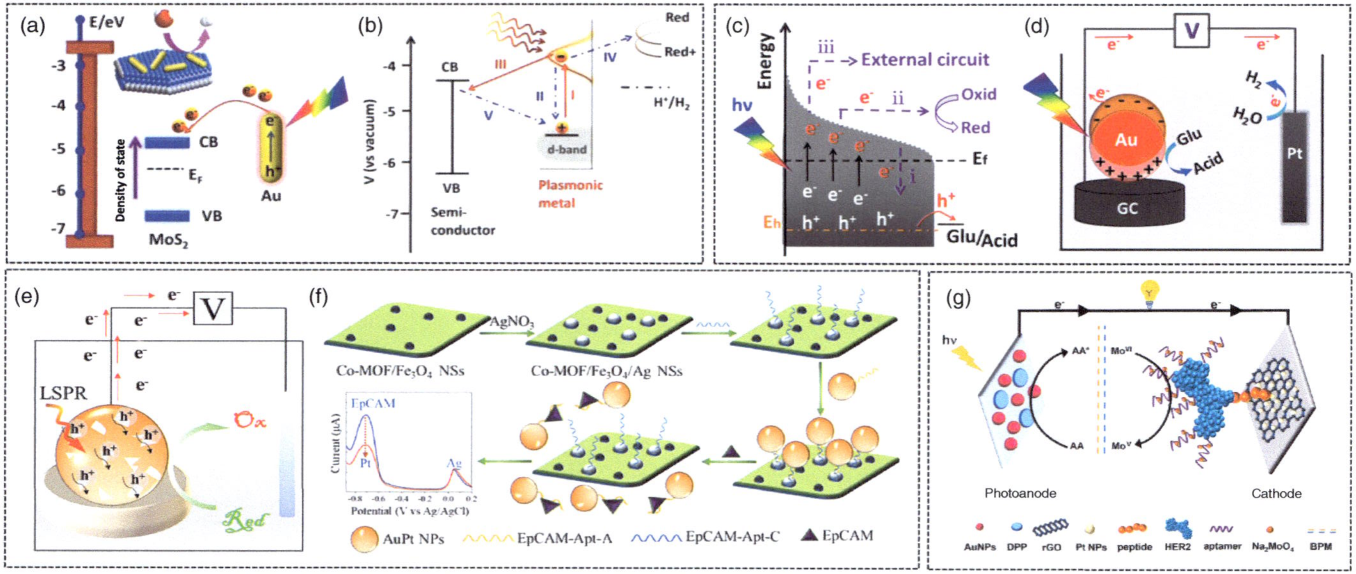

Figure 4.18 Plasmonic-enhanced electrochemistry. (a) Schematic Au–MoS_2 and energy level diagram illuminating hot electrons injection and change of MoS_2 Fermi level. *Source:* Ref. [229] / with permission of American Chemical Society. (b) Different plausible hot electrons transfer pathways likely to occur during SPR. *Source:* Ref. [229] / with permission of American Chemical Society. (c) Possible transfer channels of LSPR-excited hot charge carriers. *Source:* Ref. [230] / with permission of American Chemical Society. The dashed yellow line indicates the energy level (Eh) of the hot holes. (d) Illustration of the electrochemical setup and the principle of PAER. *Source:* Ref. [230] / with permission of American Chemical Society. (e) The mechanism of plasmon-driven electrochemical oxidation enhancement for AuPt NPs. *Source:* Ref. [231] / with permission of Elsevier. (f) Schematic illustration of a constructed ratiometric electrochemical aptasensor for specific detection of EpCAM. *Source:* Ref. [231] / with permission of Elsevier. (g) Schematic illustration of HER2 detection based on PFC-SPB. *Source:* Ref. [232] / with permission of Elsevier.

semiconductor hybrid photoanode and an electrochemical sandwich structure biosensing strategy cathode [232]. The local surface plasmon resonance effect of the AuNPs significantly enhances the separation efficiency of photogenerated electron–hole pairs, contributing to the sensitivity and stability of the PFC-SPB. The PFC-SPB is capable of detecting epidermal growth factor receptor 2 (HER2) within the range of 0.1 to 500 pg/mL, with a low detection limit of 0.02 pg/mL (Figure 4.18g). Li et al. employed positively charged Ag@Au core-shell nanoparticles ((+)Ag@Au CSNPs) as electrochemical signal enhancers and used DNA hybridization chain reaction to prepare an electrochemical biosensor for detecting Hg^{2+}. Additionally, the development of plasmonic metal-based electrochemical immunosensors has progressed [234–236]. For instance, Zhang et al. utilized Au@Ag nanoparticles loaded by polydopamine functionalized phenolic resin microporous carbon spheres (Au@Ag/PDA-PR-MCS) for signal amplification and designed a sandwich-type electrochemical immunosensor for the detection of α-fetoprotein (AFP) [234]. Huang et al. used Ag/AuNPs coated on graphene as signal amplifiers to develop a sandwich-type electrochemical immunosensor for detecting carcinoembryonic antigen [235]. Wang et al. constructed a novel immunosensor for carbohydrate antigen 199 (CA199) by enhancing the catalytic current of the oxygen reduction reaction (ORR) using bimetallic alloyed AuAg hollow nanocrystals (AuAg HNCs) [236].

4.5.5 Plasmonic-FET Biosensors

Field-effect transistors (FETs) [237] are integral to modern microelectronics technology due to a host of benefits they offer, such as high input resistance, minimal noise, low power usage, wide dynamic range, smooth integration, lack of secondary breakdown phenomenon, and a broad safe operating area. Biosensors that employ FETs have gained considerable attention because they can detect a variety of target analytes with exceptional sensitivity, eliminating the need for labels. This method is further improved by its compatibility with large-scale manufacturing processes.

The integration of plasmonic building blocks, especially Au NPs, further enhances the performance of FET biosensors. Au NPs offer a large surface area, robust stability, excellent sensitivity, and abundant functionalization sites. These properties make them ideal for surface functionalization, thereby expanding the range of detection methods available and improving overall biosensor performance.

A FET is a semiconductor device that consists of channel materials, source (S), drain (D), and gate (G) electrodes (Figure 4.19a) [238]. The FET operates by modulating the conductance between the insulating layer and the semiconductor channel layer through the electric field at the gate. A current flow between the source and drain electrodes (Ids) when gate voltages (Vds) are applied (Figure 4.19b). For selective biosensing, FET devices must be functionalized with biorecognition elements such as aptamers, antibodies, DNA, or enzymes (Figure 4.19c). During biosensing, the interaction between the biorecognition element and target molecules leads to the consumption or accumulation of charge in the channel materials, generating a measurable current signal (Figure 4.19d). AuNPs are commonly used as an important

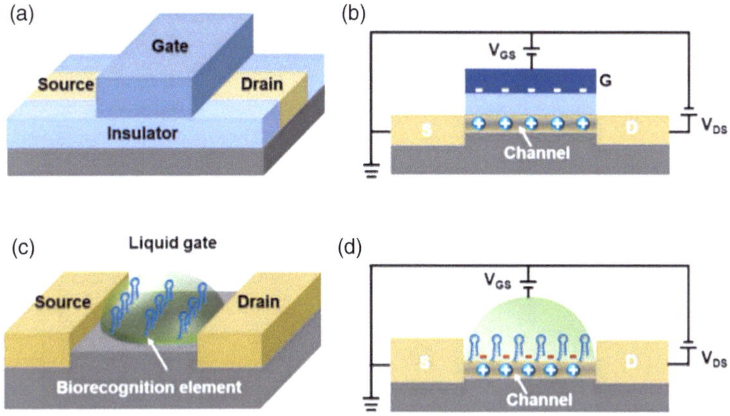

Figure 4.19 Operating principles of FET biosensors. *Source:* Ref. [238] / John Wiley & Sons / CC BY 4.0. (a) Diagram of a typical FET device. (b) Operating principle of a p-channel type FET. (c) Diagram of a liquid-gate FET biosensor. (d) Sensing mechanism of a p-channel type FET biosensor.

molecular linkage medium in plasmonic-FET sensors due to their stable properties and their ability to form strong Au-S covalent bonds with thiol functional groups.

To date, AuNPs have been extensively utilized in the preparation of a variety of FET biosensors, showcasing their versatility and efficacy. These biosensors have been employed in detecting a wide range of biological targets, including nucleic acids such as DNA or RNA [239–241], proteins [242–247] involved in various physiological processes, and small molecules [10–12] that are crucial in numerous biochemical pathways. The incorporation of AuNPs into FET biosensors has significantly enhanced their sensitivity and specificity, making them powerful tools in medical diagnostics, environmental monitoring, and biochemical research.

Building on this foundation, recent studies have demonstrated innovative applications of AuNP-functionalized FETs in various biosensing contexts. For instance, Shao's research [11] leverages FETs based on single-walled carbon nanotubes (SWCNTs) adorned with AuNPs to discern fentanyl from similar compounds (Figure 4.20a). The sensor's efficacy stems from the specific interactions between fentanyl and AuNPs, observable through shifts in the FET device's transfer characteristics (Figure 4.20b, c). This interaction is quantifiable, with a linear detection range from 2 to 50 μg/mL, which is significantly broadened to 10 fg/mL to 100 pg/mL through the incorporation of fentanyl antibodies, thus enhancing the sensor's sensitivity and selectivity.

Similarly, Li et al. have harnessed the unique properties of graphene and AuNPs to develop a graphene FET (G-FET) sensor for COVID-19 detection [248]. This sensor utilizes the high surface-to-volume ratio of AuNPs and the specificity of PMO probes to detect the SARS-CoV-2 RdRp, demonstrating anti-interference capabilities and clinical applicability. It differentiates COVID-19 patients from non-infected individuals within two minutes, providing a faster alternative to RT-PCR (Figure 4.20d).

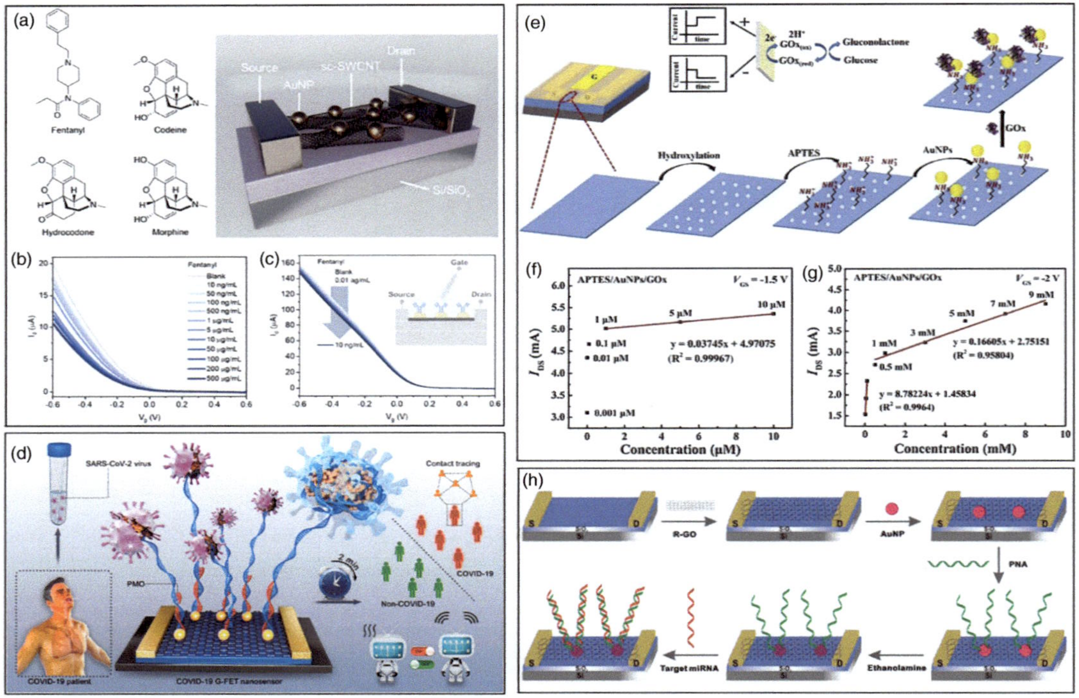

Figure 4.20 Application of plasmonic-FET biosensors. (a) Detection of opioid drugs using an AuNP-decorated sc-SWCNT-based FET device. *Source:* Ref. [11] / John Wiley & Sons / CC BY 4.0. (b) FET transfer characteristics of the Au-SWCNT FET device when increasing concentrations of fentanyl from 10 ng/mL to 500 μg/mL are added. *Source:* Ref. [11] / John Wiley & Sons / CC BY 4.0. (c) FET transfer characteristics of the fentanyl-ab@Au-SWCNT FET device when increasing concentrations of fentanyl from 1 ag/mL to 10 ng/mL are added. *Source:* Ref. [11] / John Wiley & Sons / CC BY 4.0. (d) Schematic diagram of PCR-free rapid direct identification of COVID-19 using the PMO-functionalized G-FET nanosensor. *Source:* Ref. [248] / with permission of Elsevier. (e) Diagram illustrating the process of functionalization. *Source:* Ref. [249] / with permission of Elsevier. (f) Calibration plot showing the steady-state current against glucose concentration at a gate voltage of −1.5 V and (g) −2 V. *Source:* Ref. [249] / with permission of Elsevier. (h) Diagram depicting the AuNPs-decorated graphene FET biosensor for miRNA detection. *Source:* Ref. [241] / with permission of Elsevier.

Furthering the application of FETs in biosensing, Liu et al. have created a highly sensitive glucose sensor using an AlGaN/GaN-based high electron mobility transistor (HEMT) [249]. The sensor's gate surface was functionalized with APTES and AuNPs, with GOx immobilized through electrostatic attraction (Figure 4.20e). This configuration facilitated efficient charge transfer through the uniformly distributed AuNPs, resulting in a sensor with a broad linear response and high sensitivity, suitable for glucose concentration ranges from 0.001 to 9 mM (Figure 4.20f-g).

Additionally, Cai et al. have developed a graphene FET biosensor decorated with AuNPs for miRNA detection [241]. The biosensor, which involved immobilizing a peptide nucleic acid (PNA) probe on the AuNPs (Figure 4.1h), exhibited a detection limit as low as 10 fM and was capable of detecting miRNA in serum samples, indicating its potential for diagnosing gene-related diseases.

Recent findings further emphasize the pivotal role of AuNPs. Baraban and colleagues [250] introduced a multiplexed extended gate FET biosensor with gold nanoantennae as signal amplifiers. It achieves sensitivities comparable to custom nanofabricated sensors, detecting assays down to 0.2 fM with horseradish peroxidase-linked secondary antibodies, surpassing standard microplate methods (0.4 nM). AuNPs bioconjugation enhances potentiometric response approximately fivefold to a detection limit of 13.3 fM, crucially enhancing signal via nanoparticle-mediated disruption of the diffusion barrier layer.

These studies collectively demonstrate the synergy between FET technology and AuNP functionalization, highlighting the potential for sensitive, selective, and rapid detection of various biological targets. The integration of specific recognition elements, such as antibodies or PMO probes, into the FET framework not only enhances the performance of these sensors but also opens new avenues for their application in medical diagnostics and environmental monitoring. The logical progression from the foundational benefits of FETs to the specific applications in opioid detection, COVID-19 identification, glucose monitoring, and miRNA analysis underscores the transformative impact of this technology in the biosensing field.

4.6 Challenges and Future Perspectives

Metallic nanomaterials-based plasmonic biosensors and flexible bioelectronics have demonstrated considerable promise in biomedical applications, particularly for disease diagnostics. However, significant technical challenges must be overcome to enable their widespread clinical implementation. Addressing these challenges will require interdisciplinary approaches encompassing advances in nanotechnology, materials science, and computational methods.

Sensitivity and selectivity remain key areas requiring improvement, particularly in complex biological matrices such as blood or saliva. Nonspecific adsorption of biomolecules continues to compromise sensor accuracy, limiting their clinical utility. Future efforts should focus on engineering advanced nanostructures and refining surface chemistries to enhance target specificity while mitigating nonspecific interactions. Additionally, computational approaches, including machine learning, may prove critical in extracting meaningful diagnostic signals from the inherent noise in biological samples.

Stability is another limiting factor, particularly for long-term, real-world applications. Plasmonic nanostructures are susceptible to gradual degradation, impacting sensor reliability over time. Developing more robust fabrication techniques, alongside durable surface coatings, could provide greater operational stability across variable environmental conditions. Furthermore, achieving reproducibility at scale remains a persistent challenge, with current manufacturing processes yielding inconsistent sensor performance across batches. Standardization of fabrication methods will be essential to ensure uniformity and reliability in future clinical devices.

The cost and scalability of these technologies present additional barriers to their translation. Current fabrication protocols are often labor-intensive and reliant on expensive materials, making large-scale production commercially impractical. Developing streamlined, high-throughput manufacturing processes that maintain analytical performance while reducing costs will be critical for broader adoption. Additionally, miniaturization of devices could pave the way for portable, point-of-care diagnostic platforms, which are likely to be pivotal for widespread clinical use.

The growing complexity of data interpretation in multiplexed sensor formats poses another challenge. As the number of detectable targets increases, so too does the difficulty in resolving specific signals from complex biological samples. Advances in data science, particularly the integration of machine learning algorithms, offer promising solutions for identifying patterns within complex datasets. Rigorous comparative studies using orthogonal validation methods will be necessary to ensure the robustness of these computational models in clinical applications.

The clinical translation of plasmonic biosensors and flexible bioelectronics will require rigorous demonstration of safety, accuracy, and usability in real-world settings. Close collaboration with clinicians and end-users will be essential to ensure that these technologies are developed in line with practical healthcare needs. In parallel, establishing regulatory frameworks and quality control systems will be critical for the approval and standardization of these devices in clinical practice.

In conclusion, realizing the full potential of metallic nanomaterials-based plasmonic biosensors and flexible bioelectronics for diagnostics and therapeutics will necessitate continued innovation across multiple technical and regulatory fronts. Collaborative efforts between experts in engineering, chemistry, data science, and clinical medicine will be essential in overcoming current limitations and expediting the translation of these technologies into clinical settings.

References

1 Shrivastav, A.M., Cvelbar, U., and Abdulhalim, I. (2021). A comprehensive review on plasmonic-based biosensors used in viral diagnostics. *Communications Biology* 4 (1): 12.

2 Xu, W., Wang, L., Zhang, R. et al. (2020). Diagnosis and prognosis of myocardial infarction on a plasmonic chip. *Nature Communications* 11 (1): 9.

3 Sun, W., Huang, X., and Wang, X. (2023). CRISPR-based molecular diagnostics: a review. *Shengwu Gongcheng Xuebao/Chin. J. Biotechnol.* 39 (1): 60–73.

4 Fleming, K.A., Horton, S., Wilson, M.L. et al. (2021). The Lancet Commission on diagnostics: transforming access to diagnostics. *Lancet* 398 (10315): 1997–2050.

5 Amirjani, A., Salehi, K., and Sadrnezhaad, S.K. (2022). Simple SPR-based colorimetric sensor to differentiate Mg^{2+} and Ca^{2+} in aqueous solutions. *Spectrochimica Acta. Part A, Molecular and Biomolecular Spectroscopy* 268: 120692.

6 Amirjani, A., Kamani, P., Hosseini, H.R.M., and Sadrnezhaad, S.K. (2022). SPR-based assay kit for rapid determination of Pb^{2+}. *Analytica Chimica Acta* 1220: 340030.

7 Lerdsri, J., Chananchana, W., Upan, J. et al. (2020). Label-free colorimetric aptasensor for rapid detection of aflatoxin B1 by utilizing cationic perylene probe and localized surface plasmon resonance of gold nanoparticles. *Sensors and Actuators B: Chemical* 320: 128356.

8 Chen, Y., Bai, Y., Wang, X. et al. (2022). Plasmonic/magnetic nanoarchitectures: from controllable design to biosensing and bioelectronic interfaces. *Biosensors & Bioelectronics* 114744.

9 Song, L., Chen, J., Xu, B.B., and Huang, Y. (2021). Flexible plasmonic biosensors for healthcare monitoring: progress and prospects. *ACS Nano* 15 (12): 18822–18847.

10 Kim, D., Park, C., Choi, W. et al. (2020). Improved long-term responses of Au-decorated Si nanowire FET sensor for NH_3 detection. *IEEE Sensors Journal* 20 (5): 2270–2277.

11 Shao, W.T., Sorescu, D.C., Liu, Z.R., and Star, A. (2024). Machine learning discrimination and ultrasensitive detection of fentanyl using gold nanoparticle-decorated carbon nanotube-based field-effect transistor sensors. *Small* 20 (35): 2311835.

12 Tian, M., Wang, J.H., Li, C.H. et al. (2022). Qualitative and quantitative detection of microcystin-LR based on SERS-FET dual-mode biosensor. *Biosensors & Bioelectronics* 212: 114434.

13 Gong, S., Lu, Y., Yin, J.L. et al. (2024). Materials-driven soft wearable bioelectronics for connected healthcare. *Chemical Reviews* 124 (2): 455–553.

14 Ling, Y.Z., An, T.C., Yap, L.W. et al. (2020). Disruptive, soft, wearable sensors. *Advanced Materials* 32 (18): 13.

15 Willets, K.A. and Van Duyne, R.P. (2007). Localized surface plasmon resonance spectroscopy and sensing. *Annual Review of Physical Chemistry* 58: 267–297.

16 Manivannan, S., Park, S., Jeong, J., and Kim, K. (2020). Aggregation-free optical and colorimetric detection of Hg(II) with M13 bacteriophage-templated Au nanowires. *Biosensors & Bioelectronics* 161: 112237.

17 Xiang-ling, T., Xin, L., and Yanwei, Z. (2020). Estimation of future extreme precipitation changes in Xinjiang based on RegCM4.4 simulations. *Natural Hazards* 102 (1): 201–218.

18 Wang, F., Zhang, Y., Lu, M. et al. (2021). Near-infrared band Gold nanoparticles-Au film "hot spot" model based label-free ultratrace lead (II) ions detection via fiber SPR DNAzyme biosensor. *Sensors and Actuators B: Chemical* 337: 129816.

19 Wang, Y., Haick, H., Guo, S.Y. et al. (2022). Skin bioelectronics towards long-term, continuous health monitoring. *Chemical Society Reviews* 51 (9): 3759–3793.

20 Zhang, Z.M., Zhu, Z.T., Zhou, P.C. et al. (2023). Soft bioelectronics for therapeutics. *ACS Nano* 17 (18): 17634–17667.

21 Zhang, P.K., Zhu, B.C., Du, P., and Travas-Sejdic, J. (2023). Electrochemical and electrical biosensors for wearable and implantable electronics based on conducting polymers and carbon-based materials. *Chemical Reviews* 124 (3): 722–767.

22 Li, P., Lee, G.H., Kim, S.Y. et al. (2021). From diagnosis to treatment: recent advances in patient-friendly biosensors and implantable devices. *ACS Nano* 15 (2): 1960–2004.

23 Zhang, M.L., Tang, Z.J., Liu, X.L., and van der Spiegel, J. (2020). Electronic neural interfaces. *Nature Electronics* 3 (4): 191–200.

24 Zheng, Q., Tang, Q.Z., Wang, Z.L., and Li, Z. (2021). Self-powered cardiovascular electronic devices and systems. *Nature Reviews. Cardiology* 18 (1): 7–21.

25 Wu, L., Luo, R., Xue, Z. et al. (2022). Self-powered technology and implantable cardiac electronic medical devices. *Science and Technology Review* 40 (12): 53–65.

26 Wang, K., Yap, L.W., Gong, S. et al. (2021). Nanowire-based soft wearable human–machine interfaces for future virtual and augmented reality applications. *Advanced Functional Materials* 31 (39): 2008347.

27 Zhang, Y.J., Liu, Y., Lu, Y.R. et al. (2024). Tailor-made gold nanomaterials for applications in soft bioelectronics and optoelectronics. *Advanced Materials* 38.

28 Fu, R., Warnakula, T., Shi, Q. et al. (2020). Plasmene nanosheets as optical skin strain sensors. *Nanoscale Horizons* 5 (11): 1515–1523.

29 Fu, R., Lu, Y., and Cheng, W. (2022). Soft plasmonics: design, fabrication, characterization, and applications. *Advanced Optical Materials* 10 (1).

30 Zhang, H., Lin, F., Chen, G.S. et al. (2024). Intrinsically stretchable jellyfish-like gold nanowires film as multifunctional wearable chemical and physical sensors. *Chemical Engineering Journal* 490: 11, 151798.

31 Gong, S., Yap, L.W., Zhang, Y. et al. (2022). A gold nanowire-integrated soft wearable system for dynamic continuous non-invasive cardiac monitoring. *Biosensors & Bioelectronics* 205: 114072.

32 Zhu, B., Ling, Y., Yap, L.W. et al. (2019). Hierarchically structured vertical gold nanowire array-based wearable pressure sensors for wireless health monitoring. *ACS Applied Materials & Interfaces* 11 (32): 29014–29021.

33 Faraday, M. (1857). X. The Bakerian Lecture. — Experimental relations of gold (and other metals) to light. *Philosophical Transactions of the Royal Society of London* 147: 145–181.

34 Mie, G. (1908). Articles on the optical characteristics of turbid tubes, especially colloidal metal solutions. *Annalen der Physik-Berlin* 25 (3): 377–445.

35 de Aberasturi, D.J., Serrano-Montes, A.B., and Liz-Marzan, L.M. (2015). Modern applications of plasmonic nanoparticles: from energy to health. *Advanced Optical Materials* 3 (5): 602–617.

36 Jang, Y.H., Jang, Y.J., Kim, S. et al. (2016). Plasmonic solar cells: from rational design to mechanism overview. *Chemical Reviews* 116 (24): 14982–15034.

37 Kazuma, E. and Kim, Y. (2019). Mechanistic studies of plasmon chemistry on metal catalysts. *Angewandte Chemie (International Ed. in English)* 58 (15): 4800–4808.

38 Kim, J.M., Lee, C., Lee, Y. et al. (2021). Synthesis, assembly, optical properties, and sensing applications of plasmonic gap nanostructures. *Advanced Materials* 33 (46): e2006966.

39 Li, S., Miao, P., Zhang, Y. et al. (2021). Recent advances in plasmonic nanostructures for enhanced photocatalysis and electrocatalysis. *Advanced Materials* 33 (6): e2000086.

40 Lim, W.Q. and Gao, Z. (2016). Plasmonic nanoparticles in biomedicine. *Nano Today* 11 (2): 168–188.

41 Kermanshahian, K., Yadegar, A., and Ghourchian, H. (2021). Gold nanorods etching as a powerful signaling process for plasmonic multicolorimetric chemo-/biosensors: strategies and applications. *Coordination Chemistry Reviews* 442: 213934.

42 Langer, J., de Aberasturi, D.J., Aizpurua, J. et al. (2020). Present and future of surface-enhanced Raman scattering. *ACS Nano* 14 (1): 28–117.

43 Zhang, X.M., Chen, Y.L., Liu, R.S., and Tsai, D.P. (2013). Plasmonic photocatalysis. *Reports on Progress in Physics* 76 (4): 41.

44 Kelly, K.L., Coronado, E., Zhao, L.L., and Schatz, G.C. (2003). The optical properties of metal nanoparticles: the influence of size, shape, and dielectric environment. *The Journal of Physical Chemistry. B* 107 (3): 668–677.

45 Haiss, W., Thanh, N.T.K., Aveyard, J., and Fernig, D.G. (2007). Determination of size and concentration of gold nanoparticles from UV-Vis spectra. *Analytical Chemistry* 79 (11): 4215–4221.

46 Tan, S., Campolongo, M., Luo, D., and Cheng, W. (2011). Building plasmonic nanostructures with DNA. *Nature Nanotechnology* 6: 268–276.

47 Wang, P., Krasavin, A., Liu, L. et al. (2022). Molecular plasmonics with metamaterials. *Chemical Reviews* 122 (19): 15031–15081.

48 Nath, N. and Chilkoti, A. (2002). A colorimetric gold nanoparticle sensor to interrogate biomolecular interactions in real time on a surface. *Analytical Chemistry* 74 (3): 504–509.

49 Link, S., Mohamed, M.B., and El-Sayed, M.A. (1999). Simulation of the optical absorption spectra of gold nanorods as a function of their aspect ratio and the effect of the medium dielectric constant. *The Journal of Physical Chemistry. B* 103 (16): 3073–3077.

50 Sajanlal, P.R., Sreeprasad, T.S., Samal, A.K., and Pradeep, T. (2011). Anisotropic nanomaterials: structure, growth, assembly, and functions. *Nanotechnology Reviews* 2 (1): 5883.

51 Chen, Y., Liu, H., Yin, H.J. et al. (2020). Hierarchical fabrication of plasmonic superlattice membrane by aspect-ratio controllable nanobricks for label-free protein detection. *Frontiers in Chemistry* 8: 307.

52 Park, J.E., Kim, M., Hwang, J.H., and Nam, J.M. (2017). Golden opportunities: plasmonic gold nanostructures for biomedical applications based on the second near-infrared window. *Small Methods* 1 (3): 6.

53 Kim, M., Lee, J.H., and Nam, J.M. (2019). Plasmonic photothermal nanoparticles for biomedical applications. *Advanced Science (Weinh)* 6 (17): 1900471.

54 Ortiz-Castillo, J.E., Gallo-Villanueva, R.C., Madou, M.J., and Perez-Gonzalez, V.H. (2020). Anisotropic gold nanoparticles: a survey of recent synthetic methodologies. *Coordination Chemistry Reviews* 425: 213489.

55 Niu, W.X., Chua, Y.A.A., Zhang, W.Q. et al. (2015). Highly symmetric gold nanostars: crystallographic control and surface-enhanced Raman scattering property. *Journal of the American Chemical Society* 137 (33): 10460–10463.

56 Kumar, A., Kumar, S., Rhim, W.K. et al. (2014). Oxidative nanopeeling chemistry-based synthesis and photodynamic and photothermal therapeutic applications of plasmonic core-petal nanostructures. *Journal of the American Chemical Society* 136 (46): 16317–16325.

57 Park, J.E., Lee, Y., and Nam, J.M. (2018). Precisely shaped, uniformly formed gold nanocubes with ultrahigh reproducibility in single-particle scattering and surface-enhanced Raman scattering. *Nano Letters* 18 (10): 6475–6482.

58 Khanal, B.P. and Zubarev, E.R. (2019). Chemical transformation of nanorods to nanowires: reversible growth and dissolution of anisotropic gold nanostructures. *ACS Nano* 13 (2): 2370–2378.

59 Wang, Y., Gong, S., Wang, S.J. et al. (2018). Standing Enokitake-like nanowire films for highly stretchable elastronics. *ACS Nano* 12 (10): 9742–9749.

60 Chen, Y., Milenkovic, S., and Hassel, A.W. (2008). Arrays of Iso-oriented gold nanobelts. *Nano Letters* 8 (2): 737–742.

61 Liebig, F., Sarhan, R.M., Sander, M. et al. (2017). Deposition of gold nanotriangles in large scale close-packed monolayers for X-ray-based temperature calibration and SERS monitoring of plasmon-driven catalytic reactions. *ACS Applied Materials & Interfaces* 9 (23): 20247–20253.

62 Yoo, Y., Kim, S.I., Han, S. et al. (2019). Epitaxially aligned submillimeter-scale silver nanoplates grown by simple vapor transport. *Nanoscale* 11 (37): 17436–17443.

63 Golze, S.D., Hughes, R.A., Rouvimov, S. et al. (2019). Plasmon-mediated synthesis of periodic arrays of gold nanoplates using substrate-immobilized seeds lined with planar defects. *Nano Letters* 19 (8): 5653–5660.
64 Hubert, C., Chomette, C., Desert, A. et al. (2021). Versatile template-directed synthesis of gold nanocages with a predefined number of windows. *Nanoscale Horizons* 6 (4): 311–318.
65 Zhang, H., Wang, R., Sikdar, D. et al. (2022). Plasmonic superlattice membranes based on bimetallic Nano-Sea Urchins as high-performance label-free surface-enhanced Raman spectroscopy platforms. *ACS Sensors* 7 (2): 622–631.
66 Bridges, C.R., DiCarmine, P.M., Fokina, A. et al. (2013). Synthesis of gold nanotubes with variable wall thicknesses. *Journal of Materials Chemistry A* 1 (4): 1127–1133.
67 Armelles, G., Cebollada, A., Garcia-Martin, A., and Gonzalez, M.U. (2013). Magnetoplasmonics: combining magnetic and plasmonic functionalities. *Advanced Optical Materials* 1 (1): 10–35.
68 Aslam, U., Chavez, S., and Linic, S. (2017). Controlling energy flow in multimetallic nanostructures for plasmonic catalysis. *Nature Nanotechnology* 12 (10): 1000–1005.
69 Phatangare, A.B., Dhole, S.D., Dahiwale, S.S., and Bhoraskar, V.N. (2018). Ultra-high sensitive substrates for surface enhanced Raman scattering, made of 3 nm gold nanoparticles embedded on SiO_2 nanospheres. *Applied Surface Science* 441: 744–753.
70 Hwang, C.S.H., Ahn, M.S., Lee, Y. et al. (2019). Ag/Au alloyed nanoislands for wafer-level plasmonic color filter arrays. *Scientific Reports* 9 (1): 9082.
71 Paldi, R.L., Wang, X., Sun, X. et al. (2020). Vertically aligned AgxAu1-x alloyed nanopillars embedded in ZnO as nanoengineered low-loss hybrid plasmonic metamaterials. *Nano Letters* 20 (5): 3778–3785.
72 Park, M., Hwang, C.S.H., and Jeong, K.H. (2018). Nanoplasmonic alloy of Au/Ag nanocomposites on paper substrate for biosensing applications. *ACS Applied Materials & Interfaces* 10 (1): 290–295.
73 Wang, D., Hermes, M., Kotni, R. et al. (2018). Interplay between spherical confinement and particle shape on the self-assembly of rounded cubes. *Nature Communications* 9 (1): 2228.
74 Xu, Z., Xie, H., Ye, W. et al. (2021). 4.9% Au stabilizes Ag in an atomically homogenous bimetallic alloy for anisotropic nanocrystals with enhanced stability under light irradiation. *Nanoscale* 13 (23): 10335–10341.
75 Pujari, A. and Thomas, T. (2021). Aluminium nanoparticles alloyed with other earth-abundant plasmonic metals for light trapping in thin-film a-Si solar cells. *Sustainable Materials and Technologies* 28: 12.
76 Jawad, M., Ali, S., Waseem, A. et al. (2019). Plasmonic effects and size relation of gold-platinum alloy nanoparticles. *Advances in Nano Research* 7 (3): 167–178.
77 Hao, L., Leng, Y., Zeng, L. et al. (2020). Core-shell-heterostructured magnetic-plasmonic nanoassemblies with highly retained magnetic-plasmonic activities for ultrasensitive bioanalysis in complex matrix. *Adv. Sci. (Weinh)* 7 (2): 1902433.

78 Lee, S., Hwang, H., Lee, W. et al. (2020). Core–shell bimetallic nanoparticle trimers for efficient light-to-chemical energy conversion. *ACS Energy Letters* 5 (12): 3881–3890.

79 Lopez-Ortega, A., Takahashi, M., Maenosono, S., and Vavassori, P. (2018). Plasmon induced magneto-optical enhancement in metallic Ag/FeCo core/shell nanoparticles synthesized by colloidal chemistry. *Nanoscale* 10 (39): 18672–18679.

80 Nguyen, L., Dass, M., Ober, M.F. et al. (2020). Chiral assembly of gold-silver core-shell plasmonic nanorods on DNA origami with strong optical activity. *ACS Nano* 14 (6): 7454–7461.

81 Susarrey-Arce, A., Czajkowski, K.M., Darmadi, I. et al. (2019). A nanofabricated plasmonic core-shell-nanoparticle library. *Nanoscale* 11 (44): 21207–21217.

82 Tymoczko, A., Kamp, M., Rehbock, C. et al. (2019). One-step synthesis of Fe–Au core–shell magnetic-plasmonic nanoparticles driven by interface energy minimization. *Nanoscale Horizons* 4 (6): 1326–1332.

83 Wang, T., Wang, S., Cheng, Z. et al. (2021). Emerging core–shell nanostructures for surface-enhanced Raman scattering (SERS) detection of pesticide residues. *Chemical Engineering Journal* 424: 130323.

84 Zhang, Y.J., Radjenovic, P.M., Zhou, X.S. et al. (2021). Plasmonic core-shell nanomaterials and their applications in spectroscopies. *Advanced Materials* 33 (50): e2005900.

85 Wang, J., Wu, X., Ma, W., and Xu, C. (2020). Chiral AuCuAu heterogeneous nanorods with tailored optical activity. *Advanced Functional Materials* 30 (17): 2000670.

86 Yang, X.Q., Lu, Y., Liu, Y. et al. (2021). Heterostructures built through site-selective deposition on anisotropic plasmonic metal nanocrystals and their applications. *Small Structures* 2 (12): 2100101.

87 Zheng, G., Mourdikoudis, S., and Zhang, Z. (2020). Plasmonic metallic heteromeric nanostructures. *Small* 16 (38): e2002588.

88 Zhu, X., Yip, H.K., Zhuo, X. et al. (2017). Realization of Red Plasmon shifts up to approximately 900 nm by AgPd-tipping elongated Au nanocrystals. *Journal of the American Chemical Society* 139 (39): 13837–13846.

89 Shan, B.B., Li, L.H., Zhao, Y.W. et al. (2021). Near-infrared II plasmonic Au@Au-Ag dot-in-cubic nanoframes for in vivo surface-enhanced raman spectroscopic detection and photoacoustic imaging. *Advanced Functional Materials* 31 (29): 12.

90 Kim, D., Resasco, J., Yu, Y. et al. (2014). Synergistic geometric and electronic effects for electrochemical reduction of carbon dioxide using gold-copper bimetallic nanoparticles. *Nature Communications* 5 (1): 4948.

91 Su, T., Xiao, H.L., Shen, W. et al. (2018). Nonlinear size-dependent melting of silica-encapsulated Ag-Cu alloy nanoparticles. *Journal of Physical Chemistry C* 122 (48): 27761–27768.

92 Huang, J., Phuah, X.L., McClintock, L.M. et al. (2021). Core-shell metallic alloy nanopillars-in-dielectric hybrid metamaterials with magneto-plasmonic coupling. *Materials Today* 51: 39–47.

93 Feng, J. and Yin, Y. (2019). Self-templating approaches to hollow nanostructures. *Advanced Materials* 31 (38): e1802349.

94 Bai, T., Tan, Y., Zou, J. et al. (2015). AuBr2–engaged galvanic replacement for citrate-capped Au–Ag Alloy nanostructures and their solution-based surface-enhanced raman scattering activity. *Journal of Physical Chemistry C* 119 (51): 28597–28604.

95 González, E., Arbiol, J., and Puntes, V.F. (2011). Carving at the nanoscale: sequential galvanic exchange and Kirkendall growth at room temperature. *Science* 334 (6061): 1377–1380.

96 Huang, Z.L., Meng, G.W., Hu, X.Y. et al. (2019). Plasmon-tunable Au@Ag core-shell spiky nanoparticles for surface-enhanced Raman scattering. *Nano Research* 12 (2): 449–455.

97 Sunwoo, S.H., Han, S.I., Kang, H. et al. (2019). Stretchable low-impedance nanocomposite comprised of Ag–Au core–shell nanowires and Pt black for epicardial recording and stimulation. *Advanced Materials Technologies* 5 (3): 1900768.

98 Wang, W.W., Hao, C.L., Sun, M.Z. et al. (2018). Spiky Fe_3O_4@Au supraparticles for multimodal in vivo imaging. *Advanced Functional Materials* 28 (22): 1800310.

99 Jiang, J.L., Zou, S.M., Ma, L.W. et al. (2018). Surface-enhanced raman scattering detection of pesticide residues using transparent adhesive tapes and coated silver nanorods. *ACS Applied Materials & Interfaces* 10 (10): 9129–9135.

100 Muhammad, M., Yao, G.H., Zhong, J. et al. (2020). A facile and label-free SERS approach for inspection of fipronil in chicken eggs using SiO_2@Au core/shell nanoparticles. *Talanta* 207: 120324.

101 Chen, D., Shi, J., and Shen, H.Y. (2020). High-dispersed catalysts of core-shell structured Au@SiO_2 for formaldehyde catalytic oxidation. *Chemical Engineering Journal* 385: 123887.

102 Ouyang, L., Yao, L., Zhou, T.H., and Zhu, L.H. (2018). Accurate SERS detection of malachite green in aquatic products on basis of graphene wrapped flexible sensor. *Analytica Chimica Acta* 1027: 83–91.

103 Lu, M., Qu, A., Li, S. et al. (2020). Mitochondria-targeting plasmonic spiky nanorods increase the elimination of aging cells in vivo. *Angewandte Chemie (International Ed. in English)* 59 (22): 8698–8705.

104 Zeng, J., Zheng, Y.Q., Rycenga, M. et al. (2010). Controlling the shapes of silver nanocrystals with different capping agents. *Journal of the American Chemical Society* 132 (25): 8552–8553.

105 Liu, X., Ma, J., Jiang, P. et al. (2020). Large-scale flexible surface-enhanced Raman scattering (SERS) sensors with high stability and signal homogeneity. *ACS Applied Materials & Interfaces* 12 (40): 45332–45341.

106 Si, K.J., Sikdar, D., Yap, L.W. et al. (2015). Dual-coded plasmene nanosheets as next-generation anticounterfeit security labels. *Advanced Optical Materials* 3 (12): 1710–1717.

107 Zhang, M., Pacheco-Pena, V., Yu, Y. et al. (2018). Nanoimprinted chiral plasmonic substrates with three-dimensional nanostructures. *Nano Letters* 18 (11): 7389–7394.

108 Boles, M.A., Engel, M., and Talapin, D.V. (2016). Self-assembly of colloidal nanocrystals: from intricate structures to functional materials. *Chemical Reviews* 116 (18): 11220–11289.

109 Dong, D.S., Fu, R.F., Shi, Q.Q., and Cheng, W.L. (2019). Self-assembly and characterization of 2D plasmene nanosheets. *Nature Protocols* 14 (9): 2691–2706.

110 Si, K.J., Sikdar, D., Chen, Y. et al. (2014). Giant plasmene nanosheets, nanoribbons, and origami. *ACS Nano* 8 (11): 11086–11093.

111 Shi, Q.Q., Si, K.J., Sikdar, D. et al. (2016). Two-dimensional bipyramid plasmonic nanoparticle liquid crystalline superstructure with four distinct orientational packing orders. *ACS Nano* 10 (1): 967–976.

112 Bigioni, T.P., Lin, X.M., Nguyen, T.T. et al. (2006). Kinetically driven self assembly of highly ordered nanoparticle monolayers. *Nature Materials* 5 (4): 265–270.

113 Wei, W., Bai, F., and Fan, H. (2019). Oriented gold nanorod arrays: self-assembly and optoelectronic applications. *Angewandte Chemie (International Ed. in English)* 58 (35): 11956–11966.

114 Shi, Q., Di, W., Dong, D. et al. (2019). A general approach to free-standing nanoassemblies via acoustic levitation self-assembly. *ACS Nano* 13 (5): 5243–5250.

115 Si, S., Liang, W., Sun, Y. et al. (2016). Facile fabrication of high-density sub-1-nm gaps from Au nanoparticle monolayers as reproducible SERS substrates. *Advanced Functional Materials* 26 (44): 8137–8145.

116 Ding, W.K., Xia, Y., Song, H.Y. et al. (2024). Macroscopic superlattice membranes self-assembled from gold nanobipyramids with precisely tunable tip arrangements for SERS. *Angewandte Chemie, International Edition* 136 (21): e202401945.

117 Lu, B., Vegso, K., Micky, S. et al. (2023). Tunable subnanometer gaps in self-assembled monolayer gold nanoparticle superlattices enabling strong plasmonic field confinement. *ACS Nano* 17 (13): 12774–12787.

118 Moretti, C., Goldmann, C., Imperor-Clerc, M., and Abécassis, B. (2023). Gold nanoparticle superlattices: conditions for long-range order, Moire patterns, and binary phase from a single population. *Chemistry of Materials* 35 (17): 6637–6650.

119 Song, L., Xu, B.B., Cheng, Q. et al. (2021). Instant interfacial self-assembly for homogeneous nanoparticle monolayer enabled conformal "lift-on" thin film technology. *Science Advances* 7 (52): eabk2852.

120 Schulz, F., Pavelka, O., Lehmkühler, F. et al. (2020). Structural order in plasmonic superlattices. *Nature Communications* 11 (1): 3821.

121 Li, C. and Jin, Y. (2021). Shell-isolated plasmonic nanostructures for biosensing, catalysis, and advanced nanoelectronics. *Advanced Functional Materials* 31 (7): 2008031.

122 Li, J.-F., Zhang, Y.-J., Ding, S.-Y. et al. (2017). Core–shell nanoparticle-enhanced Raman spectroscopy. *Chemical Reviews* 117 (7): 5002–5069.

123 Lee, D., Lee, H., Jeong, Y. et al. (2016). Highly sensitive, transparent, and durable pressure sensors based on Sea-Urchin shaped metal nanoparticles. *Advanced Materials* 28 (42): 9364–9369.

124 Liu, Z., Wang, H., Huang, P. et al. (2019). Highly stable and stretchable conductive films through thermal-radiation-assisted metal encapsulation. *Advanced Materials* 31 (35): 1901360.

125 Fleischmann, M., Hendra, P.J., and McQuillan, A.J. (1974). Raman spectra of pyridine adsorbed at a silver electrode. *Chemical Physics Letters* 26 (2): 163–166.

126 Jeanmaire, D.L. and Van Duyne, R.P. (1977). Surface Raman spectroelectrochemistry: Part I. Heterocyclic, aromatic, and aliphatic amines adsorbed on the anodized silver electrode. *Journal of Electroanalytical Chemistry and Interfacial Electrochemistry* 84 (1): 1–20.

127 Albrecht, M.G. and Creighton, J.A. (1977). Anomalously intense Raman spectra of pyridine at a silver electrode[J]. *Journal of the American Chemical Society* 99 (15): 5215–5217.

128 Moskovits, M. (1978). Surface roughness and the enhanced intensity of Raman scattering by molecules adsorbed on metals[J]. *The Journal of Chemical Physics* 69 (9): 4159–4161.

129 Goel, R., Chakraborty, S., Awasthi, V. et al. (2024). Exploring the various aspects of surface enhanced Raman spectroscopy (SERS) with focus on the recent progress: SERS-active substrate, SERS-instrumentation, SERS-application. *Sensors and Actuators A: Physical* 376: 115555.

130 Li, L. and Chin, W.S. (2020). Rapid fabrication of a flexible and transparent Ag nanocubes@PDMS film as a SERS substrate with high performance. *ACS Applied Materials & Interfaces* 12 (33): 37538–37548.

131 Jin, X., Zhu, Q., Feng, L. et al. (2021). Light-trapping SERS substrate with regular bioinspired arrays for detecting trace dyes. *ACS Applied Materials & Interfaces* 13 (9): 11535–11542.

132 Zhang, C., Chen, S., Jiang, Z. et al. (2021). Highly sensitive and reproducible SERS substrates based on ordered micropyramid array and silver nanoparticles. *ACS Applied Materials & Interfaces* 13 (24): 29222–29229.

133 Shafi, M., Liu, R., Zha, Z. et al. (2021). Highly efficient SERS substrates with different Ag interparticle nanogaps based on hyperbolic metamaterials. *Applied Surface Science* 555: 149729.

134 Wang, W., Chen, Y., Xiao, C. et al. (2023). Flexible SERS wearable sensor based on nanocomposite hydrogel for detection of metabolites and pH in sweat. *Chemical Engineering Journal* 474: 145953.

135 Chen, Q.-Q., Hou, R.-N., Zhu, Y.-Z. et al. (2021). Au@ZIF-8 core–shell nanoparticles as a SERS substrate for volatile organic compound gas detection. *Analytical Chemistry* 93 (19): 7188–7195.

136 Kim, S., Kim, T.G., Lee, S.H. et al. (2020). Label-free surface-enhanced raman spectroscopy biosensor for on-site breast cancer detection using human tears. *ACS Applied Materials & Interfaces* 12 (7): 7897–7904.

137 Kim, D., Kim, J., Henzie, J. et al. (2021). Mesoporous Au films assembled on flexible cellulose nanopaper as high-performance SERS substrates. *Chemical Engineering Journal* 419: 129445.

138 Lu, X., Huang, Y., Liu, B. et al. (2018). Light-controlled shrinkage of large-area gold nanoparticle monolayer film for tunable SERS activity. *Chemistry of Materials* 30 (6): 1989–1997.

139 Al-Saadi, A.A., Haroon, M., Popoola, S.A., and Saleh, T.A. (2020). Sensitive SERS detection and characterization of procaine in aqueous media by reduced gold nanoparticles. *Sensors and Actuators B: Chemical* 304: 127057.

140 Krajczewski, J., Michałowska, A., and Kudelski, A. (2020). Star-shaped plasmonic nanostructures: new, simply synthetized materials for Raman analysis of surfaces. *Spectrochimica Acta. Part A, Molecular and Biomolecular Spectroscopy* 225: 117469.

141 Pardehkhorram, R., Alshawawreh, F.A., Gonçales, V.R. et al. (2021). Functionalized gold nanorod probes: a sophisticated design of SERS immunoassay for biodetection in complex media. *Analytical Chemistry* 93 (38): 12954–12965.

142 Chen, Z., Tan, R., Zeng, M. et al. (2024). SERS detection of triazole pesticide residues on vegetables and fruits using Au decahedral nanoparticles. *Food Chemistry* 439: 138110.

143 Kołątaj, K., Krajczewski, J., and Kudelski, A. (2018). Dipyramidal-Au@SiO_2 nanostructures: new efficient electromagnetic nanoresonators for Raman spectroscopy analysis of surfaces. *Applied Surface Science* 456: 932–940.

144 Palermo, G., Rippa, M., Conti, Y. et al. (2021). Plasmonic metasurfaces based on pyramidal nanoholes for high-efficiency SERS biosensing. *ACS Applied Materials & Interfaces* 13 (36): 43715–43725.

145 Yang, J., Hahm, D., Kim, K. et al. (2020). High-resolution patterning of colloidal quantum dots via non-destructive, light-driven ligand crosslinking. *Nature Communications* 11 (1): 2874.

146 Liu, Y., Wu, S.-H., Du, X.-Y., and Sun, J.-J. (2021). Plasmonic Ag nanocube enhanced SERS biosensor for sensitive detection of oral cancer DNA based on nicking endonuclease signal amplification and heated electrode. *Sensors and Actuators B: Chemical* 338: 129854.

147 Plou, J., Valera, P.S., García, I. et al. (2022). Prospects of surface-enhanced raman spectroscopy for biomarker monitoring toward precision medicine. *ACS Photonics* 9 (2): 333–350.

148 Li, Y., Guo, Y., Chen, H. et al. (2024). Flexible wearable plasmonic paper-based microfluidics with expandable channel and adjustable flow rate for portable surface-enhanced raman scattering sweat sensing. *ACS Photonics* 11 (2): 613–625.

149 He, X., Fan, C., Luo, Y. et al. (2022). Flexible microfluidic nanoplasmonic sensors for refreshable and portable recognition of sweat biochemical fingerprint. *npj Flexible Electronics* 6 (1): 60.

150 Lee, W.-C., Koh, E.H., Kim, D.-H. et al. (2021). Plasmonic contact lens materials for glucose sensing in human tears. *Sensors and Actuators B: Chemical* 344: 130297.

151 Wang, D., Xu, G., Zhang, X. et al. (2022). Dual-functional ultrathin wearable 3D particle-in-cavity SF-AAO-Au SERS sensors for effective sweat glucose and lab-on-glove pesticide detection. *Sensors and Actuators B: Chemical* 359: 131512.

152 Chowdhury, A.D., Nasrin, F., Gangopadhyay, R. et al. (2020). Controlling distance, size and concentration of nanoconjugates for optimized LSPR based biosensors. *Biosensors & Bioelectronics* 170: 112657.

153 Zeng, J., Zhang, Y., Zeng, T. et al. (2020). Anisotropic plasmonic nanostructures for colorimetric sensing. *Nano Today* 32: 100855.

154 Wang, H., Xu, C.-H., Zhao, W. et al. (2021). Alkaline phosphatase-triggered etching of Au@FeOOH nanoparticles for enzyme level assay under dark-field microscopy. *Analytical Chemistry* 93 (30): 10727–10734.

155 Jiao, F., Li, F., Shen, J. et al. (2021). Wafer-scale flexible plasmonic metasurface with passivated aluminum nanopillars for high-sensitivity immunosensors. *Sensors and Actuators B: Chemical* 344: 130170.

156 Ni, B., Mychinko, M., Gómez-Graña, S. et al. (2023). Chiral seeded growth of gold nanorods into fourfold twisted nanoparticles with plasmonic optical activity. *Advanced Materials* 35 (1): 2208299.

157 Guo, Y., Wu, J., Li, J., and Ju, H. (2016). A plasmonic colorimetric strategy for biosensing through enzyme guided growth of silver nanoparticles on gold nanostars. *Biosensors & Bioelectronics* 78: 267–273.

158 Lee, S., Sim, K., Moon, S.Y. et al. (2021). Controlled assembly of plasmonic nanoparticles: from static to dynamic nanostructures. *Advanced Materials* 33 (46): 2007668.

159 Zhang, X., Chen, Z., and Zuo, X. (2020). Chloroauric Acid/Silver Nanoparticle Colorimetric Sensors for Antioxidant Discrimination Based on a Honeycomb Ag-Au Nanostructure. *ACS Sustainable Chemistry & Engineering* 8 (9): 3922–3928.

160 Yang, Y.Z., Tang, W.J., Wang, J.Y. et al. (2022). Stress-deconcentrated ultrasensitive strain sensor with hydrogen-bonding-tuned fracture resilience for robust biomechanical monitoring. *Science China Materials* 65 (8): 2289–2297.

161 Li, H.F., Han, F., Wang, L.L. et al. (2023). A hybrid strategy-based ultra-narrow stretchable microelectrodes with cell-level resolution. *Advanced Functional Materials* 33 (29): 2300859.

162 Lee, Y., Chung, J.W., Lee, G.H. et al. (2021). Standalone real-time health monitoring patch based on a stretchable organic optoelectronic system. *Science Advances* 7 (23): eabg9180.

163 Kim, D.W., Kwon, J., Kim, H.S., and Jeong, U. (2021). Printed stretchable single-nanofiber interconnections for individually-addressable highly-integrated transparent stretchable field effect transistor array. *Nano Letters* 21 (13): 5819–5827.

164 Zhao, Q.A., Gribkova, E., Shen, Y.Y. et al. (2024). Highly stretchable and customizable microneedle electrode arrays for intramuscular electromyography. *Science Advances* 10 (18): eadn7202.

165 Fan, Q., Miao, J.L., Liu, X.H. et al. (2022). Biomimetic hierarchically silver nanowire Interwoven MXene Mesh for flexible transparent electrodes and invisible Camouflage electronics. *Nano Letters* 22 (2): 740–750.

166 Sunwoo, S.H., Han, S.I., Jung, D. et al. (2023). Stretchable low-impedance conductor with Ag-Au-Pt core-shell-shell nanowires and in situ formed

Pt nanoparticles for wearable and implantable device. *ACS Nano* 17 (8): 7550–7561.

167 Jeong, W., Lee, S., Choi, H. et al. (2022). Washable, stretchable, and reusable core-shell metal nanowire network-based electronics on a breathable polymer nanomesh substrate. *Materials Today* 61: 30–39.

168 Loh Joel, Y.Y., Zeineddine, A., Shayegannia, M. et al. (2022). A one-step, tunable method of selective reactive sputter deposition as a wrinkling approach for silver/polydimethylsiloxane for electrically conductive pliable surfaces. *Microsystems & Nanoengineering* 8 (1): 89.

169 Fan, Q.C., Fan, H.W., Li, K.R. et al. (2023). Stretchable, electrochemically-stable electrochromic devices based on semi-embedded Ag@Au nanowire network. *Small* 19 (22): 2208234.

170 Drack, M., Graz, I., Sekitani, T. et al. (2015). An imperceptible plastic electronic wrap. *Advanced Materials* 27 (1): 34–40.

171 Xu, Y.D., Ye, Z.L., Zhao, G.G. et al. (2024). Phase-separated porous nanocomposite with ultralow percolation threshold for wireless bioelectronics. *Nature Nanotechnology* 19: 1158–1167.

172 Zhang, H.X., Shao, Y., Xia, R. et al. (2024). Stretchable electrodes with interfacial percolation network. *Advanced Materials* 36 (26): 2401550.

173 Choi, J., Min, J., Kim, D. et al. (2023). Hierarchical 3D percolation network of Ag-Au core-shell nanowire-hydrogel composite for efficient biohybride electrodes. *ACS Nano* 17 (18): 17966–17978.

174 Kwon, C., Seong, D., Ha, J. et al. (2020). Self-bondable and stretchable conductive composite fibers with spatially controlled percolated Ag nanoparticle networks: novel integration strategy for wearable electronics. *Advanced Functional Materials* 30 (49): 2005447.

175 Woo, J., Lee, H., Yi, C. et al. (2020). Ultrastretchable helical conductive fibers using percolated Ag nanoparticle networks encapsulated by elastic polymers with high durability in omnidirectional deformations for wearable electronics. *Advanced Functional Materials* 30 (29): 1910026.

176 Jiang, Y., Ji, S.B., Sun, J. et al. (2023). A universal interface for plug-and-play assembly of stretchable devices. *Nature* 614 (7948): 456–462.

177 Jeong, S., Heo, S., Kang, M., and Kim, H.J. (2020). Mechanical durability enhancement of gold-nanosheet stretchable electrodes for wearable human bio-signal detection. *Materials and Design* 196: 109178.

178 Jung, D., Lim, C., Shim, H.J. et al. (2021). Highly conductive and elastic nanomembrane for skin electronics. *Science* 373 (6558): 1022–1026.

179 Zhao, Y.C., Wang, B., Tan, J.W. et al. (2022). Soft strain-insensitive bioelectronics featuring brittle materials. *Science* 378 (6625): 1222–1227.

180 Wang, Y., Gong, S., Gomez, D. et al. (2018). Unconventional Janus properties of Enokitake-like gold nanowire films. *ACS Nano* 12 (8): 8717–8722.

181 Ling, Y.Z., Gong, S., Zhai, Q.F. et al. (2019). Embedding pinhole vertical gold nanowire electronic skins for Braille recognition. *Small* 15 (13): 1804853.

182 Gong, S., Yap, L.W., Zhu, Y. et al. (2020). A soft resistive acoustic sensor based on suspended standing nanowire membranes with point crack design. *Advanced Functional Materials* 30 (25): 1910717.

183 Yong, Z.J., Gong, S., Chesman, A.S.R. et al. (2023). Conformal coating of CdS onto flexible enokitake-like standing gold nanowire arrays for omnidirectional low-light-intensity photocatalysis. *Nano Energy* 108: 108227.

184 Lu, Y., Yong, Z.J., Gong, S. et al. (2023). Pd-conformally coated, one-end-embedded gold nanowire percolation network for intrinsically stretchable, epidermal tattoo fuel cell. *Biosensors & Bioelectronics* 221: 114924.

185 Gong, S., Yap, L.W., Zhu, B. et al. (2019). Local crack-Programmed gold nanowire electronic skin tattoos for in-plane multisensor integration. *Advanced Materials* 31 (41): e1903789.

186 Gong, S., Zhang, X., Nguyen, X.A. et al. (2023). Hierarchically resistive skins as specific and multimetric on-throat wearable biosensors. *Nature Nanotechnology* 18 (8): 889–897.

187 Zhang, H., Lin, F.G., Cheng, W.L. et al. (2023). Hierarchically oriented Jellyfish-like gold nanowires film for elastronics. *Advanced Functional Materials* 33 (2): 2209760.

188 Li, X., Fan, Y.J., Li, H.Y. et al. (2020). Ultracomfortable hierarchical nanonetwork for highly sensitive pressure sensor. *ACS Nano* 14 (8): 9605–9612.

189 Chen, D., Gao, J., He, D. et al. (2023). Plasmonic bridge sensor enabled by carbon nanotubes and Au–Ag nano-rambutan for multifunctional detection of biomechanics and bio/chemical molecules. *ACS Applied Materials & Interfaces* 15 (6): 8783–8793.

190 Bang, J.; Lee, W. S.; Park, B.; Joh, H.; Woo, H. K.; Jeon, S.; Ahn, J.; Jeong, C.; Kim, T.-i.; Oh, S. J. Highly sensitive temperature sensor: ligand-treated Ag nanocrystal thin films on PDMS with thermal expansion strategy. *Advanced Functional Materials* 2019, 29 (32), 1903047.

191 Zhang, W., Huang, X., Liu, W. et al. (2023). Semiconductor plasmon enhanced upconversion toward a flexible temperature sensor. *ACS Applied Materials & Interfaces* 15 (3): 4469–4476.

192 Hoon Hong, S.; Yeon Kim, T.; Cheong, S.; Bae, H.; Hyun Yu, K.; Kwang Hahn, S. Multifunctional breathable electronic devices for smart wearable healthcare applications. *Chemical Engineering Journal* 2023, 476, 146559.

193 Tang, Q., Zou, M., Chang, L., and Guo, W. (2022). A super-flexible and transparent wood film/silver nanowire electrode for optical and capacitive dual-mode sensing wood-based electronic skin. *Chemical Engineering Journal* 430: 132152.

194 Zhang, H., Zhang, D., Zhang, B. et al. (2022). Wearable pressure sensor array with layer-by-layer assembled MXene nanosheets/Ag nanoflowers for motion monitoring and Human–Machine interfaces. *ACS Applied Materials & Interfaces* 14 (43): 48907–48916.

195 Yang, L., Wang, H., Yuan, W. et al. (2021). Wearable pressure sensors based on MXene/tissue papers for wireless human health monitoring. *ACS Applied Materials & Interfaces* 13 (50): 60531–60543.

196 Zhao, Z., Li, Q., Dong, Y. et al. (2022). Washable patches with gold nanowires/textiles in wearable sensors for health monitoring. *ACS Applied Materials & Interfaces* 14 (16): 18884–18900.

197 Yu, P., Li, X., Li, H. et al. (2021). All-fabric ultrathin capacitive sensor with high pressure sensitivity and broad detection range for electronic skin. *ACS Applied Materials & Interfaces* 13 (20): 24062–24069.

198 Guan, F., Xie, Y., Wu, H. et al. (2020). Silver nanowire–bacterial cellulose composite fiber-based sensor for highly sensitive detection of pressure and proximity. *ACS Nano* 14 (11): 15428–15439.

199 Xiao, J., Wang, J., Luo, Y. et al. (2023). Wearable plasmonic sweat biosensor for acetaminophen drug monitoring. *ACS Sensors* 8 (4): 1766–1773.

200 Mogera, U., Guo, H., Namkoong, M. et al. (2022). Wearable plasmonic paper–based microfluidics for continuous sweat analysis. *Science Advances* 8 (12), eabn1736.

201 Yuan, Q., Fang, H., Wu, X. et al. (2023). Self-adhesive, biocompatible, wearable microfluidics with erasable liquid metal plasmonic hotspots for glucose detection in sweat. *ACS Applied Materials & Interfaces* .

202 Chung, M., Skinner, W.H., Robert, C. et al. (2021). Fabrication of a wearable flexible sweat ph sensor based on SERS-active Au/TPU electrospun nanofibers. *ACS Applied Materials & Interfaces* 13 (43): 51504–51518.

203 Zhai, Q., Yap, L.W., Wang, R. et al. (2020). Vertically aligned gold nanowires as stretchable and wearable epidermal ion-selective electrode for noninvasive multiplexed sweat analysis. *Analytical Chemistry* 92 (6): 4647–4655.

204 Teymourian, H., Barfidokht, A., and Wang, J. (2020). Electrochemical glucose sensors in diabetes management: an updated review (2010–2020). *Chemical Society Reviews* 49 (21): 7671–7709.

205 Lee, H., Hong, Y.J., Baik, S. et al. (2018). Enzyme-based glucose sensor: from invasive to wearable device. *Advanced Healthcare Materials* 7 (8): 1701150.

206 Imamura, A., Zakashansky, J., Cho, K. et al. (2020). Stretchable sensors for nanomolar glucose detection. *Advanced Materials Technologies* 5 (4): 1900843.

207 Zhai, Q., Gong, S., Wang, Y. et al. (2019). Enokitake mushroom-like standing gold nanowires toward wearable noninvasive bimodal glucose and strain sensing. *ACS Applied Materials & Interfaces* 11 (10): 9724–9729.

208 Sehit, E. and Altintas, Z. (2020). Significance of nanomaterials in electrochemical glucose sensors: an updated review (2016-2020). *Biosensors & Bioelectronics* 159: 112165.

209 Dervisevic, E., Dervisevic, M., Wang, Y. et al. (2020). Highly selective nanostructured electrochemical sensor utilizing densely packed Ultrathin gold nanowires film. *Electroanalysis* 32 (8): 1850–1858.

210 Meng, K., Xiao, X., Wei, W. et al. (2022). Wearable pressure sensors for pulse wave monitoring. *Advanced Materials* 34 (21): 2109357.

211 Yang, J.C., Kim, J.O., Oh, J. et al. (2019). Microstructured porous pyramid-based ultrahigh sensitive pressure sensor insensitive to strain and temperature. *ACS Applied Materials & Interfaces* 11 (21): 19472–19480.

212 Gao, F.P., Liu, C.X., Zhang, L.C. et al. (2023). Wearable and flexible electrochemical sensors for sweat analysis: a review. *Microsystems & Nanoengineering* 9 (1): 1–21.

213 Lyu, Q., Gong, S., Dyson, J.M., and Cheng, W.L. (2022). Soft, disruptive and wearable electrochemical biosensors. *Current Analytical Chemistry* 18 (6): 689–704.

214 Zhai, Q. and Cheng, W. (2019). Soft and stretchable electrochemical biosensors. *Materials Today Nano* 7: 100041.

215 Zhu, B.W., Gong, S., and Cheng, W.L. (2019). Softening gold for elastronics. *Chemical Society Reviews* 48 (6): 1668–1711.

216 Wu, Z.Q., Cao, X.Q., Hua, Y., and Yu, C.M. (2024). A bifunctional wearable sensor based on a nanoporous membrane for simultaneous detection of sweat lactate and temperature. *Analytical Chemistry* 96 (7): 3087–3095.

217 Zhao, Y.M., Zhai, Q.F., Dong, D.S. et al. (2019). Highly stretchable and strain-insensitive fiber-based wearable electrochemical biosensor to monitor glucose in the sweat. *Analytical Chemistry* 91 (10): 6569–6576.

218 Lim, C., Park, C., Sunwoo, S.H. et al. (2022). *Facile and scalable synthesis of whiskered gold nanosheets for stretchable, conductive, and biocompatible nanocomposites. ACS Nano* 16 (7): 10431–10442.

219 Lyu, Q., Zhai, Q.F., Dyson, J. et al. (2019). Real-time and in-situ monitoring of H_2O_2 release from living cells by a stretchable electrochemical biosensor based on vertically aligned gold nanowires. *Analytical Chemistry* 91 (21): 13521–13527.

220 Zhai, Q., Wang, Y., Gong, S. et al. (2018). Vertical gold nanowires stretchable electrochemical electrodes. *Analytical Chemistry* 90 (22): 13498–13505.

221 Wang, R., Zhai, Q.F., An, T.C. et al. (2021). Stretchable gold fiber-based wearable textile electrochemical biosensor for lactate monitoring in sweat. *Talanta* 222: 8.

222 Wu, J., Liu, H., Chen, W. et al. (2023). Device integration of electrochemical biosensors. *Nature Reviews Bioengineering* 1 (5): 346–360.

223 Hemmerová, E. and Homola, J. (2024). Combining plasmonic and electrochemical biosensing methods. *Biosensors & Bioelectronics* 251: 116098.

224 Zhao, J., Xue, S., Ji, R.R. et al. (2021). Localized surface plasmon resonance for enhanced electrocatalysis. *Chemical Society Reviews* 50 (21): 12070–12097.

225 Xiao, T., Huang, J.S., Wang, D.W. et al. (2020). Au and Au-based nanomaterials: synthesis and recent progress in electrochemical sensor applications. *Talanta* 206: 120210.

226 Kong, B., Tang, J., Selomulya, C. et al. (2014). Oriented mesoporous nanopyramids as versatile plasmon-enhanced interfaces. *Journal of the American Chemical Society* 136 (19): 6822–6825.

227 Wu, K.F., Rodríguez-Córdoba, W.E., Yang, Y., and Lian, T.Q. (2013). Plasmon-induced hot electron transfer from the Au tip to CdS rod in CdS-Au nanoheterostructures. *Nano Letters* 13 (11): 5255–5263.

228 Tanaka, A., Hashimoto, K., and Kominami, H. (2014). Visible-light-induced hydrogen and oxygen formation over Pt/Au/WO_3 photocatalyst utilizing two types of photoabsorption due to surface plasmon resonance and band-gap excitation. *Journal of the American Chemical Society* 136 (2): 586–589.

229 Shi, Y., Wang, J., Wang, C. et al. (2015). Hot electron of Au nanorods activates the electrocatalysis of hydrogen evolution on MoS_2 nanosheets. *Journal of the American Chemical Society* 137 (23): 7365–7370.

230 Wang, C., Nie, X.G., Shi, Y. et al. (2017). Direct plasmon-accelerated electrochemical reaction on gold nanoparticles. *ACS Nano* 11 (6): 5897–5905.

231 Zhao, Y., Yao, J., Wu, Z.L. et al. (2022). AuPt NPs with enhanced electrochemical oxidization activity for ratiometric electrochemical aptasensor. *Biosensors & Bioelectronics* 196: 113733.

232 Bai, L.P., Gu, C.C., Liu, J.H. et al. (2023). Photofuel cell-based self-powered biosensor for HER2 detection by integration of plasmonic-metal/conjugated molecule hybrids and electrochemical sandwich structure. *Biosensors & Bioelectronics* 220: 114850.

233 Ou, W.H., Zhou, B.B., Shen, J.D. et al. (2020). Thermal and nonthermal effects in plasmon-mediated electrochemistry at nanostructured Ag electrodes. *Angewandte Chemie, International Edition* 59 (17): 6790–6793.

234 Zhang, X.B., Li, Y.Y., Lv, H. et al. (2018). Sandwich-type electrochemical immunosensor based on Au@Ag supported on functionalized phenolic resin microporous carbon spheres for ultrasensitive analysis of α-fetoprotein. *Biosensors & Bioelectronics* 106: 142–148.

235 Huang, J.L., Tian, J.N., Zhao, Y.C., and Zhao, S.L. (2015). Ag/Au nanoparticles coated graphene electrochemical sensor for ultrasensitive analysis of carcinoembryonic antigen in clinical immunoassay. *Sensors and Actuators B: Chemical* 206: 570–576.

236 Wang, R., Feng, J.J., Liu, W.D. et al. (2017). A novel label-free electrochemical immunosensor based on the enhanced catalytic currents of oxygen reduction by AuAg hollow nanocrystals for detecting carbohydrate antigen 199. *Biosensors & Bioelectronics* 96: 152–158.

237 Sengupta, J. and Hussain, C.M. (2024). Molybdenum disulfide-based field effect transistor biosensors for medical diagnostics: exploring a decade of advancements (2014-2024). *Trac-Trends in Analytical Chemistry* 176: 117742.

238 Wang, J., Chen, D., Huang, W. et al. (2023). Aptamer-functionalized field-effect transistor biosensors for disease diagnosis and environmental monitoring. *Exploration (Beijing, China)* 3 (3): 20210027.

239 Danielson, E., Sontakke, V.A., Porkovich, A.J. et al. (2020). Graphene based field-effect transistor biosensors functionalized using gas-phase synthesized gold nanoparticles. *Sensors and Actuators B: Chemical* 320: 128432.

240 Xing, Y.Q., Hu, J.Q., Wu, Z. et al. (2024). Ultrasensitive detection of miRNA based on single-walled carbon nanotube@MoS2 field-effect transistor regulated by atomic layer deposition. *Carbon* 223: 119022.

241 Cai, B.J., Huang, L., Zhang, H. et al. (2015). Gold nanoparticles-decorated graphene field-effect transistor biosensor for femtomolar MicroRNA detection. *Biosensors & Bioelectronics* 74: 329–334.

242 Dalila, N.R., Arshad, M.K.M., Gopinath, S.C.B. et al. (2020). Molybdenum disulfide-gold nanoparticle nanocomposite in field-effect transistor back-gate for enhanced C-reactive protein detection. *Microchimica Acta* 187 (11): 1–15.

243 Wang, S.Y., Hossain, M.Z., Han, T. et al. (2020). Avidin-biotin technology in gold nanoparticle-decorated graphene field effect transistors for detection of biotinylated macromolecules with ultrahigh sensitivity and specificity. *ACS Omega* 5 (46): 30037–30046.

244 Surya, S.G., Majhi, S.M., Agarwal, D.K. et al. (2020). A label-free aptasensor FET based on Au nanoparticle decorated Co_3O_4 nanorods and a SWCNT layer for detection of cardiac troponin T protein. *Journal of Materials Chemistry B* 8 (1): 18–26.

245 Presnova, G., Presnov, D., Krupenin, V. et al. (2017). Biosensor based on a silicon nanowire field-effect transistor functionalized by gold nanoparticles for the highly sensitive determination of prostate specific antigen. *Biosensors & Bioelectronics* 88: 283–289.

246 Chen, Y.T., Ren, R., Pu, H.H. et al. (2017). Field-effect transistor biosensors with two-dimensional black phosphorus nanosheets. *Biosensors & Bioelectronics* 89: 505–510.

247 Choi, J.H., Kim, H., Choi, J.W., and Oh, B.K. (2013). Signal enhancement of silicon nanowire-based biosensor for detection of matrix metalloproteinase-2 using DNA-Au nanoparticle complexes. *ACS Applied Materials & Interfaces* 5 (22): 12023–12028.

248 Li, J.H., Wu, D., Yu, Y. et al. (2021). Rapid and unamplified identification of COVID-19 with morpholino-modified graphene field-effect transistor nanosensor. *Biosensors & Bioelectronics* 183: 113206.

249 Liu, J., Zhang, H.Q., Xia, X.C. et al. (2020). High sensitivity detection of glucose with negatively charged gold nanoparticles functionalized the gate of AlGaN/GaN High Electron Mobility Transistor. *Sensors and Actuators, A: Physical* 312: 112128.

250 Janicijevic, Z., Nguyen-Le, T.A., Alsadig, A. et al. (2024). Multiplexed extended gate field-effect transistor biosensor with gold nanoantennae as signal amplifiers (vol 241, 115701, 2023). *Biosensors & Bioelectronics* 243: 115715.

5

Deciphering Pathogens

The Power of Genomic Sequencing in Public Health

Hai-Yan Wang

Southeast University, Jiangsu Key Laboratory for Design and Manufacturing of Precise Medicine Equipment, Nanjing 211189, China
Southeast University, School of Mechanical Engineering, Nanjing 211189, China

CHAPTER MENU

5.1 Introduction

The incidence of infectious diseases worldwide has increased with pathogens, which shows a trend of diversification and complication. New infectious diseases such as severe acute respiratory syndrome (SARS), novel coronavirus infection (COVID-19), new variant Creutzfeldt-Jakob disease, and H7N9 avian influenza continue to emerge. Meanwhile, pathogens of classic infectious diseases such as HIV (human immunodeficiency virus), multidrug-resistant or extensively drug-resistant Mycobacterium tuberculosis, and Mycoplasma trachomatis have re-emerged or developed new pathogenic characteristics. A variety of emerging and re-emerging infectious diseases, difficult-to-detect multiple infections, and fevers with unknown etiology pose significant threats to human health. Therefore, there is an increased demand for accuracy and timeliness in the diagnosis of infectious diseases across clinics. Infectious diseases are often caused by pathogens such as bacteria, viruses, fungi, and parasites, as well as their products, and lead to localized or systemic inflammation or organ dysfunction, which result in relatively high mortality rates and pose unpredictable risks to human health. The etiological factors of infectious diseases are complex. Their clinical manifestations are diverse as well. To rapidly, comprehensively, and accurately identify pathogens is a basic prerequisite for clinical doctors to ensure timely therapy for specific diseases.

Diagnostic Techniques in Public Health: Analytical and Chemometric Approaches in Disease Detection and Management, First Edition. Edited by Jinghong Li and Yan Du.

Traditional laboratory methods for detecting pathogenic microorganisms include morphological examination, culture and isolation, biochemical testing, and immunological testing. These methods have disadvantages such as long testing cycles, low positivity rates (insufficient sensitivity and inadequate detection coverage), and high technical skill requirements for operators. Molecular biology detection methods such as quantitative fluorescent PCR (qPCR) and isothermal amplification techniques have partially addressed these issues. By detecting specific sequences of nucleic acid, they can quickly identify the type of pathogen within two to four hours, as well as help in accurate typing and drug-resistance gene detection. Currently, PCR-based molecular detection methods are suitable for detecting common clinical pathogens. However, traditional detection methods, even such as quantitative fluorescent PCR, may be insufficient for mixed-type or rare pathogens. With advances in genomics technology, high-throughput sequencing technology, or next-generation sequencing (NGS), has revolutionized the field by circumventing the conventional microbial culture approach. The automated Sanger method is considered a "first-generation" sequencing technology. This method allows for the direct, high-throughput sequencing of nucleic acids extracted from clinical samples. Subsequent comparison of these sequences with existing databases facilitates a range of applications, including source tracking, identification, classification, and assessment of drug resistance in infectious diseases. This approach has garnered increasing attention from clinical and scientific researchers. There are two main methods for identifying microorganisms based on NGS: metagenomic NGS (mNGS) and targeted NGS (tNGS).

The current clinical application of metagenomic sequencing primarily refers to shotgun metagenomics [1], which involves high-throughput sequencing of total nucleic acids (DNA or RNA) from human samples or specific environmental specimens. By aligning these sequences with databases, relevant information on infectious pathogens can be obtained. Given its ability to detect emerging and rare pathogens, mNGS has played an irreplaceable role in diagnosing critical infections, particularly those caused by difficult-to-diagnose and rare diseases. In 2016, the FDA (U.S. Food and Drug Administration) approved NGS for the diagnosis of pathogenic microorganisms and the analysis of drug resistance and virulence [2]. The incorporation of metagenomic sequencing in the diagnosis and treatment of hospital-acquired pneumonia and ventilator-associated pneumonia in adults in 2019 has contributed to guidelines (Guidance Document: Considerations for Design, Development, and Analytical Validation of Next Generation Sequencing [NGS]—Based In Vitro Diagnostics [IVDs] Intended to Aid in the Diagnosis of Suspected Germline Diseases) [3]. Furthermore, the Chinese Journal of Emergency Medicine published the "Expert Consensus on the Application of Metagenomic Diagnostic Technology in Acute and Critical Infections" in 2019, and the Chinese Critical Care Medicine Journal released the "Expert Consensus on the Clinical Application of Metagenomic Sequencing Technology in Moderate to Severe Infections (First Edition)" in 2020 [4]. These consensuses affirm the value of mNGS's rapid and comprehensive pathogen identification in saving time for targeted therapy and reducing the use of other empirical medications for critically ill patients, particularly those in intensive care units (ICUs). However, due to its relatively high cost and demanding experimental requirements, mNGS has not yet been incorporated into routine clinical testing [5].

The mNGS detection process can be divided into several steps: sample processing, nucleic acid extraction, library preparation, template preparation and sequencing, data analysis, and report issuance (Figure 5.1). During the process, it is essential to set quality-control points and appropriate controls at critical steps to minimize the occurrence of false-positive and false-negative results, placing high demands on companies offering mNGS detection services. As the clinical application of mNGS increases, given the long period and uncontrollable process of external outsourcing solutions, establishing standard procedures and in-house testing within hospitals has become a matter of consideration.

The mNGS detection method also faces numerous challenges. The sensitivity of detection results is influenced by host and other microbial background nucleic acids, particularly showing significant deficiencies in detecting antibiotic resistance genes (ARGs). In tissue samples, human DNA content often exceeds 99% [5], posing challenges for removing human background and requiring adequate sequencing throughput. Additionally, contamination and colonization bacteria resulting in insufficient database completeness and accuracy. Clinical interpretation of positive and negative results also requires continuous consideration and improvement. Currently, most research and applications focus on qualitative analysis of pathogens, necessitating the expansion of database sizes and conversion of qualitative information into quantitative information to ensure the accuracy of clinical diagnostic results. Whether an assembly-free or assembly-based method is used, when evaluating the metabolic potential of microbial communities, the main limitation is that most microbial species lack annotations for accessory genes (except for selected model organisms; Box 5.1). Examples of the limitations and opportunities in metagenomics, as well as opportunities presented by shotgun metagenomic studies are detailed in Box 1.

Box 5.1 Limitations and Opportunities in Metagenomics

Shotgun metagenomic studies face several challenges and limitations such as:

1) ***Accessibility and Cost:***
Access to sequencing and computational resources remains costly, particularly for large-scale metagenome projects. Advances in sequencing platforms and cloud computing are expected to reduce these initial costs.
2) ***Genome Catalog Comprehensiveness:***
The existing catalog of over 50,000 microbial genomes is biased toward model organisms, pathogens, and easily culturable bacteria. Metagenomic tools, which depend on these genomes, inherit the biases found in the reference sequences.
3) ***Functional Profiling Biases:***
The functional profiling of metagenomes is impeded by the lack of validated annotations for most genes, a problem that can only be addressed

(Continued)

Box 5.1 (Continued)

through low-throughput and expensive functional studies focused on specific genes. Additionally, microbiome characteristics, such as average genome size, significantly influence quantitative profiling.

4) ***Microbial Dark Matter:***
Many microbiome members may remain uncharacterized by culture-based methods or metagenomics. Assembly-based approaches can recover some of the "microbial dark matter," yet some reads may stay unassigned after assembly. This is due to community structure, complexity, sequencing noise, contaminant DNA, and the vague taxonomic status of certain microbes and plasmids.
5) ***Live vs. Dead Dilemma:***
DNA persists in the environment after cell death, potentially leading to sequencing results that do not accurately represent the active microbial population. Techniques like metatranscriptomics or the use of compounds that bind free DNA, such as propidium monazide, can be employed to study active microbes.
6) ***Curse of Compositionality:***
Metagenomic features are reported as relative fractions, devoid of absolute concentration references, leading to potential false correlations due to variations in organism concentration across samples.
7) ***Mucosa-Associated Microbiome Sequencing:***
Sequencing the mucosal microbiome is especially tough due to the high proportion of human DNA and the low level of microbial biomass.

Opportunities presented by shotgun metagenomic studies include:

1) ***Integrative Meta-Omics:***
While shotgun metatranscriptomics, metaproteomics based on mass-spectrometry, and metabolomics are complementary to DNA sequencing, integrating and analyzing multi-omic data within a unified framework remains a challenge.
2) ***Virome Shotgun Sequencing:***
Shotgun metagenomics can detect viral organisms, but virome enrichment techniques are often necessary for broader viral detection. Virome analysis is computationally challenging due to limited viral genome availability and sparse inter-family phylogenetic signals.
3) ***Strain-Level Profiling:***
Single isolate sequencing offers higher genomic resolution for individual microbial genomes compared to metagenomic approaches. Improving profiling resolution to the strain level is vital for advancing population genomics and microbial epidemiology.
4) ***Longitudinal Study Design:***
Most metagenomic research is cross-sectional in nature, restricting the evaluation of variability between and within subjects as well as the dynamic changes

in the microbiome over time. Although there are tools available for longitudinal studies, more methods and data are required to delve into the temporal aspect.

5) ***Disentangling Cause from Effect:***
   Metagenomic hypotheses should be validated through experimental work to confirm correlations and associations. Longitudinal and prospective study designs can offer insights into the causative dynamics of conditions of interest.
6) ***Validation of Microbiome Biomarkers:***
   Microbiome biomarkers are often study-specific, necessitating cross-technology and cohort validation to improve reproducibility and minimize batch effects.
7) ***Data Sharing and Analysis Reproducibility:***
   Raw data deposition is often required before publication. Metagenomics lacks the standardization from other established high-throughput techniques, which makes open-source software highly desirable for analysis reproducibility.

Compared to mNGS, targeted next-generation sequencing (tNGS) technology employs ultra-multiplex PCR to detect pathogens and related resistance genes in samples [6]. tNGS technology focuses on predefined pathogens, specifically amplifying target genes for examination, thereby eliminating interference from human nucleic acids and background microbial flora. According to the design of specific primers, it can amplify hundreds to thousands of amplicons in a single run. This enables single-tube library construction for virus, bacteria, and drug-resistance detection without establishing RNA and DNA libraries. This approach offers lower sequencing costs than mNGS and exhibits higher sensitivity in detecting low-abundance viruses, resistance genes, and mutations. Additionally, the tNGS data analysis process is more straightforward, and the entire workflow is more readily automated [7, 8], which makes it suitable for routine clinical pathogen detection. It is important to note that during the design phase of reagent kits, clinical and laboratory departments can collaborate to screen for common pathogen types and resistance genes. It will be helpful to customers if a detection panel corresponds with the hospitals' needs, thereby facilitating precise treatment of infectious diseases.

In addition to its applications in identifying infectious pathogens and detecting drug resistance, NGS has emerged as a new technology in infectious disease research for exploring the interactions between pathogens and their hosts. With the application of NGS in RNA metagenomic sequencing of samples, it is possible to not only identify RNA viruses, but also to capture information on the host transcriptome, enabling further investigation into hosts' gene expression regulation. Furthermore, the response of the human immune system following viral infections and vaccination is another current research direction in infectious diseases. High-throughput sequencing of immune repertoires, such as T-cell receptor (TCR) and B-cell receptor (BCR) repertoires, provides insights into how the body combats pathogen infections and helps assess vaccine efficacy.

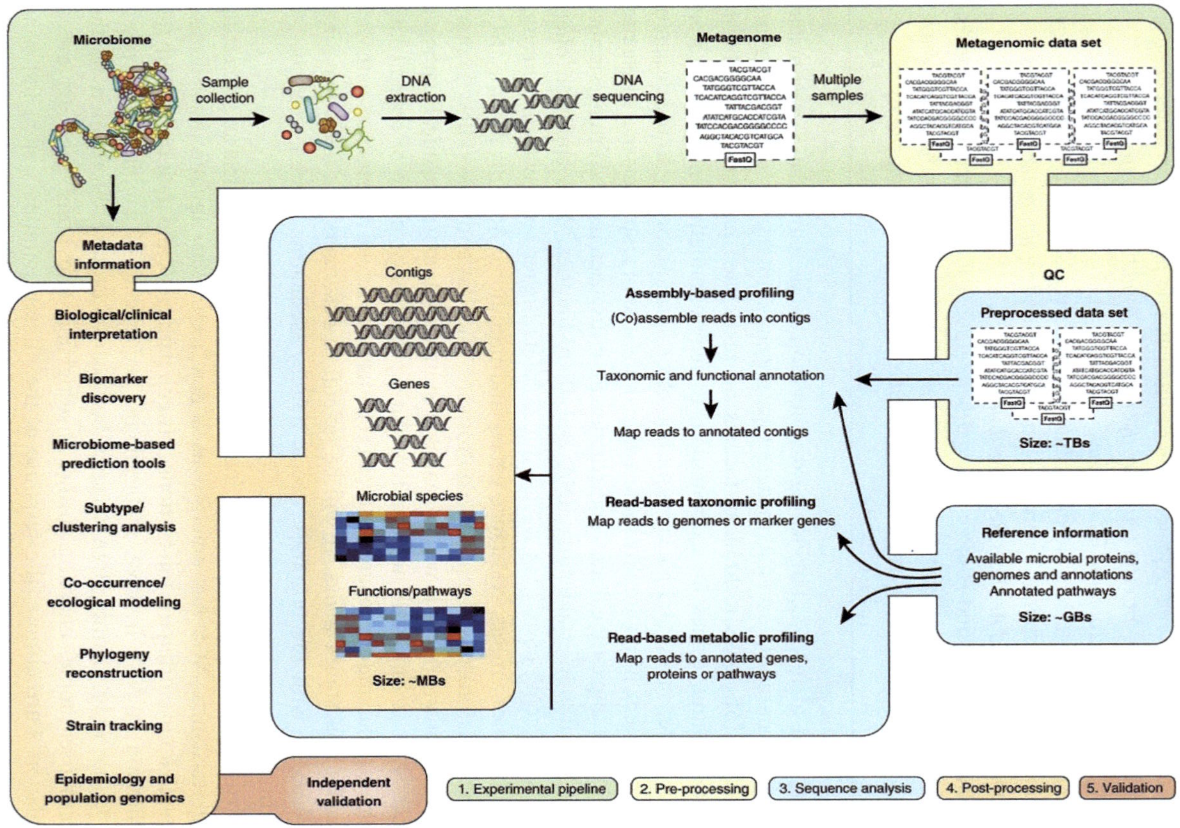
Microbiome
Sample collection
DNA extraction
DNA sequencing
Metagenome
Multiple samples
Metagenomic data set
QC
Preprocessed data set
Size: ~TBs
Reference information
Available microbial proteins, genomes and annotations
Annotated pathways
Size: ~GBs
Assembly-based profiling
(Co)assemble reads into contigs
Taxonomic and functional annotation
Map reads to annotated contigs
Read-based taxonomic profiling
Map reads to genomes or marker genes
Read-based metabolic profiling
Map reads to annotated genes, proteins or pathways
Contigs
Genes
Microbial species
Functions/pathways
Size: ~MBs
Metadata information
Biological/clinical interpretation
Biomarker discovery
Microbiome-based prediction tools
Subtype/ clustering analysis
Co-occurrence/ ecological modeling
Phylogeny reconstruction
Strain tracking
Epidemiology and population genomics
Independent validation
1. Experimental pipeline
2. Pre-processing
3. Sequence analysis
4. Post-processing
5. Validation

Figure 5.1 The outline of a typical metagenomics workflow is as follows. Firstly, study design and the establishment of experimental protocols, a step that is frequently underemphasized/overlooked in metagenomic studies. Next comes computational pre-processing. Computational quality-control procedures are essential here, as they reduce basic sequence biases and artifacts. This includes removing sequencing adaptors, carrying out quality trimming, and eliminating sequencing duplicates, which can be done using tools such as FastQC, Trimmomatic (a flexible trimmer for Illumina sequence data), or Picard. Additionally, foreign or non-target DNA sequences are filtered out. If comparisons of taxonomic or functional diversity are to be made, samples are also subsampled to equalize read numbers. Then there is sequence analysis, which should involve a mix of "read-based" and "assembly-based" methods, depending on the specific goals of the experiment. Following this is post-processing, where various multivariate statistical techniques can be applied to interpret the data. The final step is validation. Given that the conclusions drawn from high-dimensional biological data are prone to study-related biases, follow-up analyses are critical. *Source:* Reproduced from Ref. [1] / with permission of Springer Nature.

The application of high-throughput sequencing technologies in clinical pathogen detection demonstrates immense potential. However, compared to its use in non-invasive prenatal testing (NIPT) and companion diagnostics for targeted cancer therapies, the clinical utilization of high-throughput sequencing for pathogen identification still faces great challenges such as standardization and maturity [9]. In the detection of infectious diseases, being able to flexibly select testing protocols (balancing factors such as time, cost, and clinical guidance information) is crucial to achieving precise medication and maximizing patient benefits. While the molecular detection and sequencing solutions based on nucleic acid are effectively complementary to traditional detection methods, they cannot yet fully replace phenotype-based tests such as traditional drug sensitivity assays in many aspects. The clinical demand for "speed, accuracy, and stability" is driving innovation among equipment and reagent providers across multiple fronts. At present, the second-generation sequencer manufactured by Illumina is the sequencer most commonly used in clinical setting, followed by products manufactured by the Beijing Genomics Institute (BGI).

Illumina platforms perform sequencing according to the synthesis and reversible terminator chemistry methods [10, 11], coupled with the bridge amplification technique. This process begins with the attachment of individual DNA molecules to a flow cell, followed by local amplification to form a clonal cluster. Sequencing by synthesis then takes place, where complementary DNA synthesis occurs with the addition of a single nucleotide per cycle, and the fluorescently labeled nucleotides are optically read out to identify them as A, G, T, or C [10]. Illumina provides a range of platforms, such as HiSeq, MiniSeq, MiSeq, NextSeq, and NovaSeq [12–15]. They are also designed for various sequencing-scale needs and applications, from ultra-large-scale to more specific requirements. The MiSeq instrument was a pioneer in clinical pathogen detection with its successful diagnosis of Leptospira infection in an immunocompromised boy within 48 hours, marking a new epoch for the clinical use of mNGS [14, 16]. The HiSeq series is typically used for clinical pathogen detection with large-scale samples and is known for its high throughput and long read lengths. Its runtime can take as long as 3.5 days, which makes it less suitable for rapid single-sample detection [17–19]. In contrast, the NextSeq500 and NextSeq550 series provides moderate throughput and shorter runtimes (12–30 hours), making it ideal for clinical pathogen detection [20–23].

BGI sequencing utilizes combinatorial probe anchor ligation (cPAL) and deoxyribonucleic acid nanoballs (DNB) as the principal sequencing techniques. Single-stranded circular DNA undergoes rolling circle amplification (RCA) by 2–3 orders of magnitude to generate DNBs. These DNBs are fixed to a silicon chip, and fluorescent probes are polymerized on the DNBs. Optical signals from the high-resolution imaging system are digitally processed to determine the target sequence [24]. Due to its cost-effectiveness, rapid sequencing time, and other benefits, the BGI sequencing platform is gaining popularity in clinical pathogen mNGS detection. BGISEQ sequencers are frequently used to detect various pathogens, including those causing focal infections and Mycobacterium tuberculosis (MTB) [23, 25–27].

The advent of second-generation sequencing technologies has significantly advanced biological and medical research and marked a new era in clinical pathogen detection. The mainstream methods yield short read lengths of approximately 50–300 bp during fragment assembly with different platforms and technologies, which can introduce errors. Additionally, PCR amplification during library preparation can lead to the underrepresentation of sequences with low initial abundance, causing the loss of certain information. Furthermore, the runtime for second-generation sequencing requires over 12 hours and prevents real-time data acquisition. Thus, there is still scope for improvement in these sequencing technologies. The widely used Illumina sequencing platform utilizes a sequencing-by-synthesis method based on reversible dye terminators, which limits the accuracy of sequencing. Another upcoming technology, SOLiD sequencing (Sequencing by Oligonucleotide Ligation and Detection), employs a ligation-based approach using reversible terminators to determine the DNA sequence. These second-generation sequencing technologies have significantly increased the throughput and speed of DNA sequencing, enabling a wide range of applications in genomics research and clinical diagnostics. These platforms have enabled whole-genome sequencing, transcriptome analysis, and targeted sequencing, leading to breakthroughs in genetic variation, disease research, and personalized medicine. Many developments in the second generation of sequencing methods have been achieved over the years. But with rapid developments in single-molecule level studies, researchers have found that novel technologies provide long-read sequencing capabilities, enabling the sequencing of much larger DNA fragments compared to earlier methods. Examples include PacBio Sequencing, which uses a single-molecule real-time (SMRT) approach with fluorescently labeled nucleotides, enabling long-read sequencing of DNA fragments up to tens of kilobases in length. [5] Another example is Oxford nanopore technology (ONT) based on nanopore sequencing, where a single-stranded DNA molecule passes through a nanopore, and changes in electrical current are measured to determine the DNA sequence. ONT sequencing provides long read lengths, portability, low costs, and real-time analysis. In principle, a nanopore is a tiny hole with a nano-size diameter, embedded in a membrane that separates two chambers filled with electrolyte solutions. After applying voltage across the hole, the potential in the supporting electrolyte steadily drives the DNA molecule through the nanopore along with a signatural ionic current. Specialized software then determines its sequence by analyzing how much of the short sequences of individual nucleotides impede the flow of ions, along with the minute changes in electrical current. Both DNA and ribonucleic acid (RNA) contain adenosine (A), cytosine (C), and guanine (G) nucleotides, but they differ by one nucleotide. DNA contains thymine (T), while RNA contains uracil (U). Nanopore sequencing has become one of the most promising third-generation sequencing methods for decoding genetic variations at the molecular level. Unlike previous sequencing technologies, nanopore sequencing does not require initial amplification of the sample for sequencing. Moreover, it can sequence an entire nucleic acid strand over a million bases in length. There is no longer a need to use DNA sequences that must be computationally assembled from fragments just a few hundred bases long. Additionally, it can directly read a sequence from a biological sample in real time [28].

In summary, the choice between second-generation and third-generation sequencing technologies depends on the specific research or clinical application, with each offering distinct advantages and challenges. Second-generation sequencing technology provides rapid, high-throughput short-read sequencing, while third-generation sequencing offers long reads and real-time analysis capabilities. The latter is particularly valuable for resolving complex genomic regions and structural variations. It is worth mentioning that nanopore sequencing makes it possible to sequence nucleic acids (including DNA and RNA) directly from biological samples in real time. Table 5.1 provides a comparison of the three generations of sequencing technologies across different aspects, including the underlying technology, resolution, accuracy, read length, throughput, cost, RNA sequencing methods, time efficiency, sample preparation complexity, data analysis difficulty, and types of results. Researchers can choose the sequencing technology that best fits their needs based on these characteristics.

The advent of highly portable sequencing devices, capable of being connected directly to laptops for data upload to cloud storage, has revolutionized the capability to perform genomic sequencing in remote locales. These areas, often characterized by limited access to laboratory facilities, unreliable internet connections, and inconsistent electrical power, can now engage in sequencing endeavors, thus expanding the frontiers of genetic research to regions previously constrained by logistical challenges. Nanopore sequencing technology also eliminates the need for specialized personnel, rendering nanopore sequencing an indispensable asset in the swift identification and surveillance of pathogens behind emerging disease outbreaks and in tracing the transmission routes. For instance, pocket-sized nanopore sequencers have facilitated the rapid acquisition of viral genome sequences from blood samples during the Ebola virus outbreaks in the remote areas of West Africa and for the Zika virus in the inaccessible forested areas of Brazil. Nanopore sequencing has been instrumental in sequencing approximately 25% of all SARS-CoV-2 virus genomes globally [29]. Recently, this technology was also

Table 5.1 The comparison of three generations of sequencing.

Feature	First-generation sequencing	Second-generation sequencing	Third-generation sequencing
Underlying technology	Fragment sizing through SBS or degradation to produce specifically terminated DNA fragments	Elution-scanning SBS	SBS through degradation or direct physical inspection of DNA molecules
Resolution	Average across multiple DNA molecules	Average across multiple DNA molecules	Single-molecule resolution
Raw read accuracy	High	High	Moderate
Read Length	Moderate (800–1,000 bp)	Short, usually much shorter than Sanger sequencing	Long, 1,000 bp and longer in commercial systems
Throughput	Low	High	Moderate
Cost per Base	High cost per base	Low cost per base	Low-to-moderate cost per base
RNA sequencing methods	cDNA sequencing	cDNA sequencing	Direct RNA sequencing and cDNA sequencing
Time from sequencing reaction to results	Hours	Days	Hours
Sample preparation complexity	Moderate complexity, no PCR amplification required	Complex, requires PCR amplification	Varies from complex to very simple, depending on the sequence technology
Data analysis complexity	Routine	Complex, due to large data volumes and short reads complicating assembly and alignment algorithms	Complex, due to large data volumes and the technology generating new types of information and new signal processing challenges
Main results	Base calls with quality values	Base calls with quality values	Base calls with quality values, possibly other base information such as kinetics

employed in China for the sequencing and identification of SARS-CoV-2, the virus responsible for the COVID-19 pandemic, and its variants, highlighting its critical role in epidemiological studies and public health responses.

5.2 DNA Sequencing Technology

5.2.1 History of DNA Sequencing Technology

Fred Sanger devoted his scientific life to unravel the primary structure of DNA. He held a firm conviction that to truly comprehend the biological world at a deeper level, it was essential to have knowledge of the precise chemical structures of biological macromolecules [30]. Ironically, given the state of the sequencing technologies available for various biopolymers, proteins and RNA were the first to be sequenced. Sanger's groundbreaking work in the early 1950s led to the determination of the first protein sequence, that of insulin [31]. He achieved this by fragmenting its two chains, decoding each fragment individually, and then overlapping them to construct the complete sequence. His research provided conclusive evidence that proteins possess definite patterns of amino acid residues. In his Nobel lecture titled "The Chemistry of Insulin," Sanger delved deeper into this discovery. Subsequently, the development of the Edman degradation method, which involves the stepwise removal of N-terminal residues from peptide chains, greatly simplified the process of protein sequencing [32]. Despite the laborious nature of these early methods, by the late 1960s, numerous proteins had been sequenced. This body of work revealed that protein sequences exhibit variation not only across different species but also among individuals of the same species. Turning our attention to RNA sequencing, the 1960s saw scientists applying a similar general strategy. The process began with the fragmentation of an RNA species using RNases. The fragments were then separated employing chromatography and electrophoresis techniques. Following this, individual fragments were sequenced through sequential exonuclease digestion. Finally, the overall sequence was deduced by identifying overlaps between the fragments. The first RNA sequence to be determined was that of alanine tRNA. This monumental task required five researchers working full-time for three years, utilizing one gram of pure material isolated from 140 kg of yeast, to decipher its 76-nucleotide sequence [33]. The introduction of "fingerprinting" techniques brought significant advancements to the field. These techniques involved the separation of radioactively labelled RNA fragments and their visualization in two dimensions. The positions of these fragments, as seen in the visualizations, served as diagnostic indicators of their size and sequence [34]. Here, this chapter will overview the generations of DNA sequencing based on their technical-development timeline.

5.2.1.1 The Invention of First-Generation DNA Sequencing Technology

Watson and Crick famously solved the three-dimensional structure of DNA in 1953. In 1975, Sanger and Coulson [30] demonstrated a dideoxy chemistry-based process for sequencing DNA, which is referred to as first-generation sequencing. Several years later, Gilbert and Maxam reported the sequence of 24 base pairs using a method that made use of nucleobase-specific chemical modification with radioactive labels and cleavage of the phosphate backbone, and the base-specific fragments could be discriminated using gel electrophoresis [35]. These early methods required breaking the genome into many tiny pieces, creating many copies of them, sequencing each strand, and determining the sequence of a gene with computers. The latter is akin to solving a puzzle from many small interlocking pieces, the complexity increases with the number of bases to sequence. The advent of Sanger sequencing included the first genome (phiX174; 5.4 kb), essentially assembled by hand [36]. Sanger and Gilbert received the Nobel Prize in Chemistry in 1980 for their groundbreaking methods.

Over the next several decades, automated technical advances dramatically sped up, further refining the process of Sanger sequencing. Simultaneously, with the advancement of large-scale dideoxy sequencing endeavors, another technique emerged, paving the way for the initial wave of second-generation DNA sequencers. Second-generation DNA sequencing was developed by applying the Sanger method in a massively parallel manner. This method stood out significantly from existing ones because it did not rely on radio-labeled or fluorescently labeled dNTPs or oligonucleotides for nucleotide identification prior to electrophoresis visualization. This method employs a purified DNA polymerase enzyme to create numerous copies of DNA strands with various lengths. In the Sanger process, four parallel sequencing reactions are needed to sequence a single sample. Each reaction contains a single-stranded template; a specific primer to initiate the reaction, the four standard deoxynucleotides (dATP, dGTP, dCTP, and dTTP), and DNA polymerase are also needed. The polymerase adds bases one by one to a DNA strand that pairs with the single-stranded sample template. In each reaction, one of the four dideoxynucleotides (ddATP, ddGTP, ddCTP, or ddTTP) is added at a lower concentration than the standard deoxynucleotides. Since dideoxynucleotides lack the 3′ OH group, the DNA chain stops growing when they are incorporated by DNA polymerase. Four different ddNTPs are used to ensure that the chain doesn't always end at the same nucleotide (A, G, C, or T), producing DNA strands of various lengths for analysis. Afterward, the samples are run through four gel columns based on which dideoxynucleotide was added. The fragments are arranged by size in each column, which indicates the base by the end of each fragment. This makes the DNA sequence straightforward to read.

5.2.1.2 The Second Generation of DNA Sequencing Technology

To achieve the goal of rapid and lower-cost sequencing, modern DNA sequencing methods based on single-molecule technology were developed [37]. There is considerable discussion over defining the different generations of DNA sequencing technology, particularly in distinguishing the second from the third generation [38–41]. Sequencing technologies involve various methods, including template preparation, sequencing and imaging, and data analysis. The unique combination of certain protocols differentiates one technology from another and establishes the kind of data each platform generates [40]. While every manufacturer offers quality scores and accuracy estimates, there's no universal agreement that a "quality base" from one platform is the same as one from another. Two methods are used to prepare templates for NGS reactions: one involves clonally amplified templates that come from single DNA molecules, and the other uses single-DNA-molecule templates. The term sequencing by synthesis, which is used to describe numerous DNA polymerase-dependent methods in some literatures, was proven to have failed to delineate the different mechanisms involved in sequencing [42, 43]. The methods in question fall into categories such as cyclic reversible termination (CRT), single-nucleotide addition (SNA), and real-time sequencing. Sequencing by ligation (SBL) stands out, as it uses DNA ligase rather than DNA polymerase. Imaging approaches linked to these sequencing methods vary greatly, from monitoring bioluminescent signals to four-color imaging of single-molecule events. The massive data output of NGS platforms significantly elevates the demands on information technology in terms of data storage, tracking, and quality control [44]. Two decades later, sequencing has evolved. Instead of measuring fragment lengths, sequencing involves cycles of biochemical processes, such as the polymerase-mediated incorporation of fluorescently labeled nucleotides. This process, also termed synthesis-based sequencing (SBS), allows direct "read out" *via* gel capillary. Here, we have listed the main companies on the market that use second-generation sequencing in Table 5.2. Each unique platform and methodology has contributed to the field of genomics and molecular biology.

5.2.1.3 The Third Generation of DNA Sequencing

As discussed above, it is hard to distinguish the second generation from the third generation of DNA sequencing technology. Some argue that the defining characteristics of third-generation sequencing should include single-molecule sequencing (SMS), real-time sequencing, and a fundamental divergence from previous technologies [39]. Single-molecule sequencing technology, often referred to as third-generation sequencing, offers significant advantages over NGS [45]. It has the capacity to detect nucleotides at a rate of 10 bp/sec, significantly reducing sequencing time. Additionally, it can produce full-length mRNA sequences from ultra-long reads and sequence original DNA/RNA samples directly, without the need for PCR amplification [46]. It is also capable of detecting methylation information directly, without bias toward CG nucleotides. Despite these benefits, the high error rates and costs associated with third-generation sequencing have limited its widespread adoption in clinical settings [47]. Herein, we select two main sequencing technologies with different principles to discuss the third generation of sequencing.

SMRT Sequencing Single-molecule real-time (SMRT) sequencing is a flagship method of third-generation sequencing that employs a synthesis-based sequencing approach. The SMRT cell features nanoscale zero-mode waveguide (ZMW) wells, with each well containing a DNA polymerase molecule and a DNA template strand. The DNA strands are modified with hairpin adapters at both ends, creating a circular DNA molecule with constant ssDNA regions at the ends and a dsDNA template

Table 5.2 Main commercial companies using second-generation DNA sequencing.

Company	Platform name	Method of sequencing	Method of detection	Approx. read length (bases)	Advantages	Relative Limitations
Illumina	HiSeq, MiSeq	Reversible Terminator Sequencing by Synthesis	Fluorescence/ optical	2×150	Very high throughput	Expensive instrument; significant cost of data reduction and analysis
Roche	454 GS	Pyrosequencing	Optical	230–400	Longest read lengths among second-generation technologies; higher throughput compared to first-generation sequencing	Challenging sample prep; difficulty reading through repetitive/homopolymer regions; expensive instrument
Thermo Fisher Scientific (Formerly Life Technologies)	SOLiD, Ion Torrent	Sequencing by Ligation	Fluorescence/ Optical	25–35	Very high throughput, lowest reagent cost needed to reassemble a human genome among the widely accepted second-generation platforms	Short reads increase cost and difficulty of data analysis and genome assembly; high instrument cost

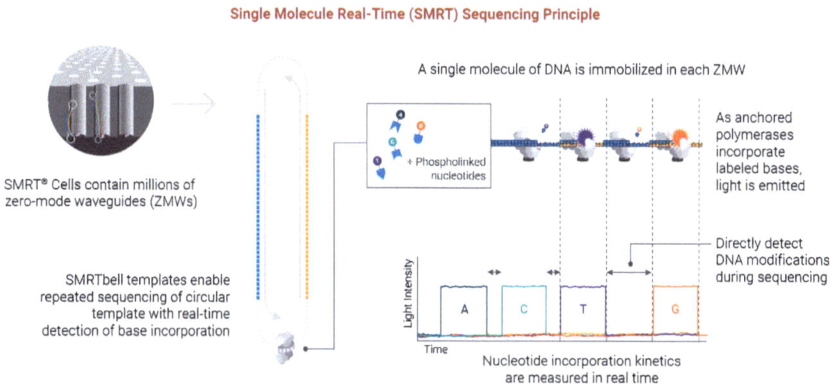

Figure 5.2 Principle of PacBio Sequencing. *Source:* With permission from Pacific Biosciences of California, Inc.

in the middle. This setup enables single-molecule sequencing and real-time fluorescence signal detection from the nucleotides. (Figure 5.2) The incorporation of nucleotides into the growing chain leads to the release of labeling groups, reducing steric hindrance and facilitating continuous DNA synthesis, thus extending read lengths [48, 49]. PacBio's SMRT Sequel series, a recent addition, offers average read lengths of 10–20 kb and can achieve sequencing outputs of 160 GB [50, 51]. However, despite these capabilities, the large footprint and high costs of this system have made it less attractive for clinical pathogen diagnosis. Yet, during the COVID-19 pandemic, PacBio has been instrumental in whole-genome sequencing and assembly of SARS-CoV-2 [52–55].

Nanopore Sequencing is another typical technique developed for third-generation sequencing [56]. It utilizes nanoscale pores made of transmembrane proteins, marking a departure from traditional sequencing methods [57, 58]. By comparison, perhaps the most eagerly awaited advancement in third-generation DNA sequencing is the promise of nanopore sequencing, which is a branch of a broader field that utilizes nanopores for detecting and quantifying a wide range of biological and chemical molecules, as well as its real-time data acquisition and analysis capabilities. The nanopore sequencing of nucleic acids was developed to be the most potential strategy, offering rapid readout, high accuracy, low cost, high throughput, and portability. Improvements in single-molecule DNA sequencing technology will hopefully decrease sequencing costs to less than $1,000/person, taking less than 24 hours per genome [59]. In 2012, Oxford Nanopore Technologies simultaneously introduced the GridION and MinION platforms [60]. The MinION is a portable, 4-inch long, single-molecule nanopore sequencing device that connects to a laptop via USB 3.0, offering a convenient solution for various sequencing needs. It was released in 2014 and is significantly faster in diagnosing infectious diseases compared to other methods. Despite the error rates of 5-40%, the output of a single flow cell in MinlON is 10–25 Gb. The sequencing

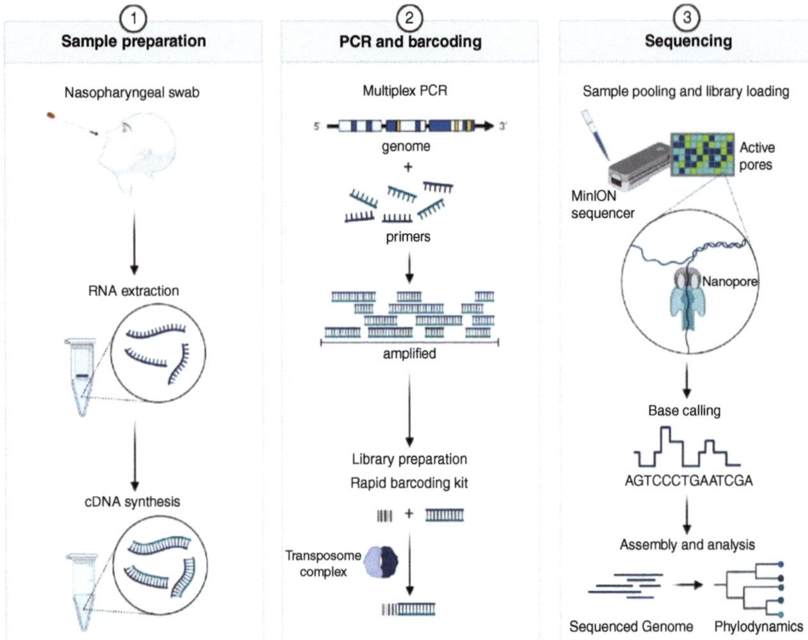

Figure 5.3 SARS-CoV-2 Genome Sequencing using Oxford Nanopore Technologies. *Source:* Sciencevivid/https://sciencevivid.com/nanopore-sequencing/.

times range from 0 to 48 hours, with read lengths of up to 800 kb [61–65]. Due to these advantages, nanopore sequencing has been extensively used in epidemic investigations to identify infectious pathogens and antimicrobial resistance [66–71] (Figure 5.3).

5.3 Nanopore Sequencing Technology, Bioinformatics and Applications in Genomic Sequencing

5.3.1 The Principle and History of Nanopore Sequencing

The idea of nanopore sequencing was first conceived in the 1980s and became a reality, thanks to progress in nanopore tech and related motor proteins [72–75]. Staphylococcus aureus's α-Hemolysin (αHL), a membrane channel protein with a diameter of vestibule entrance~2.4 nm and an exit diameter of ~1.4 nm, was the first biological nanopore inserted into a lipid bilayer shown to identify ionic current blockades from both RNA and DNA homopolymers [76]. Due to its suitable inner diameter and self-assembly ability, αHL is often used to spot biomolecules. Initially, the main way to use nanopores for DNA sequencing was to change the nanopore's size. A key step in achieving single-nucleotide-resolution nanopore sequencing was engineering wild-type αHL protein nanopore to differentiate the four oligonucleotide bases, even though these studies didn't look at complex sequences [77, 78].

The blockade current of ionic flux through the nanopore is partially blocked by the translocating molecule as shown in Figure 5.4. Each translocation molecule generates a transit reduction of the ionic current, which relates to the flux flow through the open and unblocked pore. As we know, even the smallest voltage bias that drives a single-strand DNA molecule through αHL nanopore will move each base through the lumen in less than ~1–10 μs. Identifying nucleobases through small picoampere current fluctuations detected in a nanopore necessitates a signal-to-noise ratio that exceeds the level attainable within a timeframe of less than 10 μs [79]. To achieve higher signal-to-noise ratio, each base has to stay in the sensing area of nanopore's lumen for a time longer than 100 μs. To get such slow velocity to read out each base, it is key to make ssDNA transport through the pore like a ratchet. This is the ideal prototype to make nanopore sequencing possible. Significant decreases in the translocation rate can be achieved by processing DNA enzymes that bind to the DNA strand, thereby restricting its speed within the narrow pore to a pace that does not exceed the enzyme's processing rate [80–82]. Lately, it has been demonstrated that nanopores possess remarkable single-molecule sensitivity, and this has facilitated numerous successful applications [80, 83, 84].

Inspired by nature, a large variety of protein pores exhibit diverse quaternary structures, and there are numerous successful cases of modified protein nanopores. The varied architectures of protein pores have spurred technological progress in label-free biosensing and DNA sequencing [85–87]. Among them, the mutant porin MspA of Mycobacterium smegmatis has been developed for its shorter (≈ 0.6-nm-long) and narrower channel constriction (≈ 1.2-nm-wide) than that of αHL. The engineered mutant MspA nanopore was able to improve the characterization of short segments of a ssDNA molecule by replacing negative charges in the constriction. This provides promising potential for DNA sequencing using shorter and narrow protein nanopore [87], since none of nature protein nanopores with length is shorter than ~5 nm, which is equal to the extend length of 10–15 nucleotides of ssDNA. All these nucleotides together contribute to the ionic current blockade, which is far from reaching single-nucleotide resolution.

The disadvantages of protein nanopore include less stability and limited portability; also, the length is too long to reach the resolution of one nucleotide discrimination. This implies that incorporating electrical measurements directly into commercial sequencing systems, without relying on purified fluorescent reagents or the use of unamplified genomic DNA steps, poses significant challenges in terms of both conceptualization and development. In 2001, thin and insulating SiN solid-state membranes were successfully drilled with a pore size of 3 nm [88, 89]. The size of the SiN nanopore was fabricated to mimic biological nanopores. The integration of sophisticated electronics, optical components and a variety of detection devices into solid-state nanopores was lately developed [90–93]. In general, solid-state nanopores have potential to be realized in an array format to make novel devices that are stable and robust. However, the cost of solid-state nanopore is far higher than next-generation DNA sequencing, and there are other bottlenecks faced in direct DNA sequencing [94, 95]. Solid-state nanopores were only utilized for DNA sensing for their lower resolution and sensitivity during this

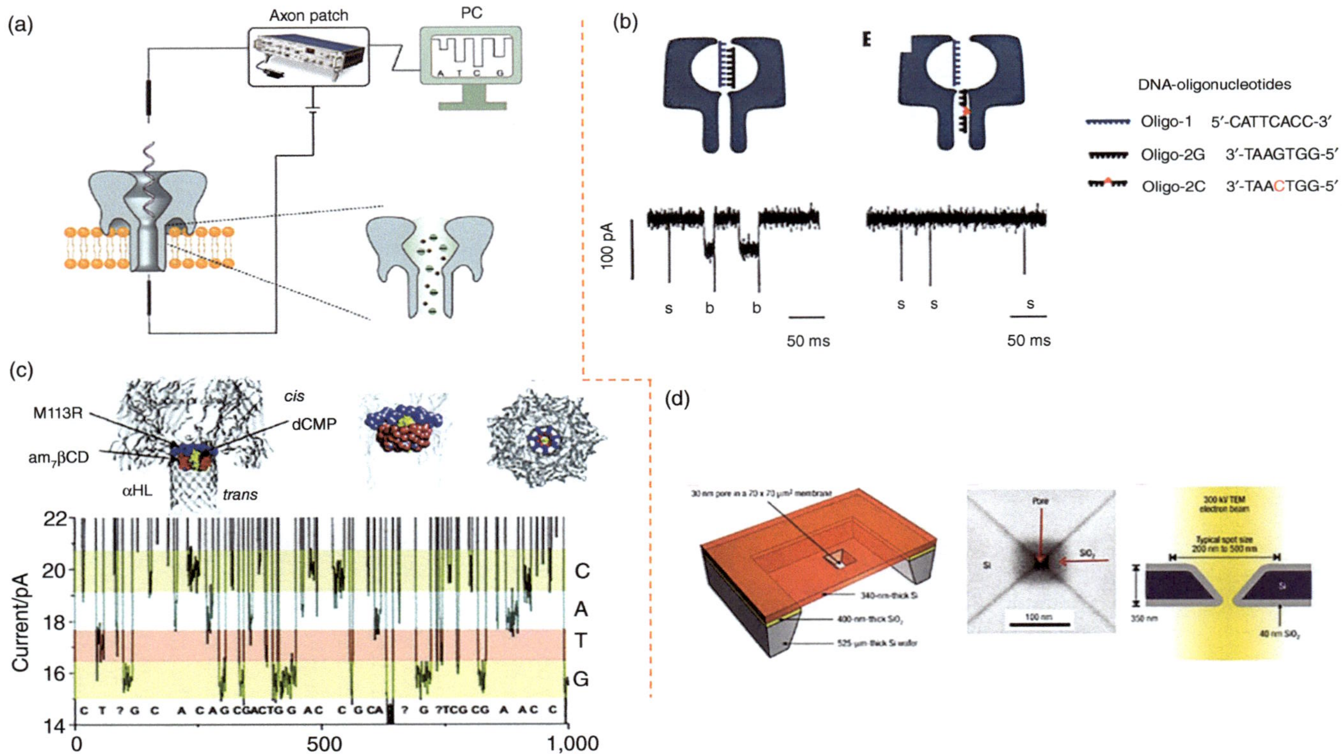

Figure 5.4 (a) Typical biological nanopore setup and working instruments. The nanoscale protein pore is inserted into the bilayer membrane to examine the physical differences in electronic signatures of individual DNA bases. (b) DNA tag is "tethered" to the entrance of the αHL nanopore to increase the resolution of nucleotide sequencing upon the match of paired nucleotide. (c) The αHL-amino cyclodextrin pore increased the reading resolution of the single nucleotide. (d) Solid-state nanopores were fabricated in 2001 and then widely developed. *Source:* Reproduced with permission from Ref. [28] / Nature Publishing Group. Reproduced with permission from Ref. [100] Copyright 2006, American Chemical Society.

period [89, 93, 96–98]. However, none of the fabricated nanopore structures had the appropriate geometry required to detect one nucleotide base at a time during DNA strand translocation through the pore.

Narrowing the constriction size of nanopores was one of the strategies that was developed prior. Two cyclodextrins formed an adapter that lodged simultaneously within the lumen of an engineered αHL pore [99]. The space between the adapters covalently attached on the lumen, forming a nanocavity within which a deoxyribonucleoside monophosphate (dNMP) can be trapped with high resolution in ionic fluctuation [96, 100, 101]. Upon the capture and translocation of a dNMP through the αHL-amino cyclodextrin pore embedded in a lipid bilayer, the ionic current passing through the pore diminishes to four distinct levels, with each level indicative of the passage of one of the four nucleotide bases (A, T, G, or C). Additionally, the clear distinction between the four ionic current blockage levels and the current flowing through the open unblocked pore allows for precise quantification of the total number of each nucleotide base that has traversed the αHL-amino cyclodextrin pore, as evidenced by the current traces [99, 100] (Figure 5.4c). This approach utilizes the nanopore to identify the released nucleotide bases. Lately, DNA hairpins have been designed to reach single-base resolution by constituting precise molecular gauges in nanopore prototype instruments with scalable computational methods that help distinguish between ionic current signatures in real time [102] (Figure 5.4b). But even with these and other ways to slow DNA's journey through the pore, if there's no true nucleobase-by-nucleobase ratcheting and DNA passage sequencing, they fall short. Early studies found that when the Klenow fragment [80] or the *Escherichia coli* exonuclease I [82] binds to the αHL nanopore, the DNA strands' translocation speed slows by orders of magnitude.

The fundamental question around the capability of the nanopore to recognize DNA sequence with single-nucleotide base resolution was initially answered by Bayley's and Ghadiri's groups together in 2005. Upon their investigation, they identified that a solitary adenine nucleotide substitution was occurring at a definite site within a strand of polyd(C). This is distinguishable from all cytosines, owing to its unique influence on the ion conductance of α-HL nanopore [103]. Subsequent research demonstrated that nanopores can distinguish all four DNA nucleobases, whether in homopolymeric or heteropolymeric DNA strands [104]. This capability to identify individual nucleobases in heteropolymeric DNA marked a significant milestone in nanopore sequencing. These experiments relied on fixing DNA strands within the α-HL pore. [105]. A vital enhancement in boosting the signal-to-noise ratio involved using processive enzymes to slow down the DNA during its passing through the nanopore [80–82]. Specifically, motor proteins proved especially effective in advancing DNA through the nanopore in a ratcheting manner [106, 107]. Indeed, this motor protein served as the crucial final piece of the puzzle in the accomplishment of DNA sequencing through the utilization of a heteropolymeric strand of DNA. Around 2012, two groups successfully demonstrated the processive recording of ionic currents generated by single-stranded DNA molecules using αHL [108] and MspA [29], respectively (Figure 5.5). By combining phi29 DNA polymerase with a nanopore, they were able to distinguish signals from individual

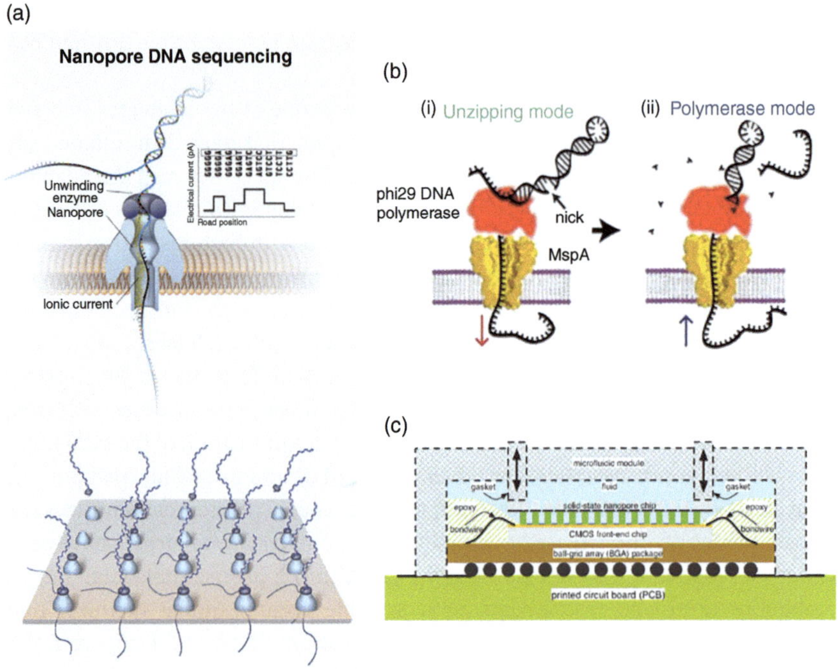

Figure 5.5 The formation of nanopore from protein to solid-state nanopore, as well as nanopore array and chip device. (a) The αHL nanopore is incorporated with motor protein to control the translocation speed of DNA molecules, which achieves a resolution that allows for the identification of individual nucleobases. *Source:* https://www.genome.gov/genetics-glossary/Nanopore-DNA-Sequencing. (b) Similar results were achieved using engineered MspA nanopore, but it can achieve the current resolution to nucleotide, which is specific for its shorter sensing region. *Source:* Ref. [109] / The Biophysical Society of Japan. (c) Cross-section of a solid-state nanopore array construct. *Source:* Adapted from Ref. [110].

nucleotides within these molecules. Employing a similar strategy, the speed of DNA strands translocating through α-HL nanopore was slowed to 2.5–40 nt/s [107]. Similar results were achieved using engineered MspA nanopore, but it could achieve the current resolution to nucleotide-specific for its shorter sensing region [111]. Otherwise, the DNA polymerases of *E. coli* (Klenow fragment) [80, 112–114] and of the bacteriophages, T7 [112] and phi29 [107, 108], have been utilized as molecular motors to control the step movement for DNA in single-nucleotide through α-HL nanopore, respectively. Besides, the utilization of a terminal hairpin or a biotin-streptavidin complex prevented the DNA strand from fully traversing the constriction of nanopore *via* electrophoretic translocation [104, 115]. The stable immobilization of DNA strands within the sensing region of the pore effectively retains the strands and enhances the signal-to-noise ratio of the ionic current, enabling the achievement of a resolution that allows for the identification of individual nucleobases. The results highlighted the importance of ratcheting the DNA strand during its pass through the pore, ensuring sufficient time for the accurate identification of the subsequent nucleobase in sequencing. Otherwise, a diverse range of protein pores are characterized by their unique quaternary structures in nature. Several modified protein nanopores, including Areolysin [85], ClyA [116], FhuA [117], membrane-adapted phi29 motor protein [118] and SP1 [119] and DNA origami nanopore have proven to be successful and have significantly contributed to technological advancements in label-free DNA sequencing. These nanopores are widely utilized in DNA sensing approaches based on nanopore technology [120–122].

Because of the extensive applications of nanopores and the advancements in microfabrication technologies, it is imperative to address the challenges posed by varying nanopore dimensions, to ensure their stability under complex environmental conditions. Overcoming these challenges is essential for the effective utilization of nanopores in various applications. Solid-state nanopores can be fabricated through a process that involves deploying, sculpting with ion beams, and fabrication by electron beams [88, 123–125]. This intricate procedure also encompasses electrochemical reactions, controlled breakdown techniques, laser etching, and laser-assisted controlled breakdown. In contrast to biological nanopores, solid-state nanopores exhibit a broader operational range in terms of temperature and voltage, allowing for diameter adjustments with sub-nanometer precision. They demonstrate superior compatibility with extreme solvent conditions. Notably, SiO_2 and Si_3N_4 nanopores are widely utilized. Their production matches the processes of complementary metal oxide semiconductor integrated circuits. Also, they can be combined with microfluidic chips and CMOS components for fabrication. (Figure 5.5c) Solid-state nanopores are limited by several factors, including a low temporal resolution caused by ionic current noise emanating from membrane materials and their substrate. Additionally, they have a restricted bandwidth due to the presence of high-frequency ionic current noise, hindering their performance and application potential. The speed of DNA translocation through solid-state nanopores reaches up to 0.01–1 μs per base, which is too fast to meet the requirements of DNA sequencing (the ideal DNA translocation velocity is 1–100 base per microseconds in a nanopore). To enhance temporal resolution, various studies have endeavored to decelerate the velocity and prolong the translocation time of small biological molecules through solid-state nanopores [126–130]. By employing substrate modification and biomolecule manipulation strategies, the movement of these molecules during translocation is impeded via a physical mesh or manipulation [131]. in addition to modifying nanopores, experimental conditions were also investigated. Researchers decreased the DNA translocation speed by regulating the electrolyte temperature and viscosity in buffer [132, 133], unsymmetric salt concentration [127], and electrical bias voltage across the nanopores. Feng et al. presented a viscosity gradient system based on room-temperature ionic liquids to manage the speed of DNA translocation through MoS_2 nanopores [134]. Compared to potassium ions in solution, the smaller size of lithium ions allows them to bind with DNA, resulting in at least a tenfold improvement in resolution for nanopore applications [128]. This improvement is due to the partial neutralization of DNA by the transient binding of counterions. In addition to the charges of translocating biomolecules, the surface charges within the nanopore wall also influence the speed of molecular translocation and corresponding ionic currents [135].

To date, tons of research studies have identified the ionic current blockades resulting from DNA translocation. These blockades are used for characterizing molecules and determining DNA sequence configuration in response to transmembrane electric fields. In the case of short ssDNA sequences (under 100 nt), relatively shorter molecules may experience information loss. This is because the translocation time may be briefer than the data acquisition interval of the measurement systems or equivalent to the cutoff frequency of the digital lowpass filters. Along with the rapid development of nanofabrication, computer science, and intelligent composite advanced materials, several other routes have been explored to sequence DNA *via* nanopore technology, leading to the establishment of several related commercial enterprises. As mentioned above, solid-state nanopores are principally produced or inserted in a thin film of materials such as graphene, SiN, phosphorene, Al_2O_3, nanotube, as well as 2D materials such as MoS_2 monolayer [136–139], DNA origami nanopore [140, 141], and so on [142]. Along with prospective interfacial nanopores, there's great potential to boost solid-state nanopore technology's spatial resolution. These advanced materials include atom-layer thin membranes composed of graphene [143], MoS_2 [134], and related 2D materials [144], which match the molecular dimension of nucleotides and allow for better ionic current discrimination between different nucleotides when compared to thicker membranes composed of SiN [145]. The thickness of a 2D single-layer material is usually within the range of 3.0–11.0 Å. This is akin to the space between two adjacent nucleotides in single-stranded DNA, roughly 3.5–5.2 Å. By slowing the biomolecules via solid-state nanopores and improving the signal-noise ratio, we can control the translocation speed. Thus, ultrathin nanopores can manipulate biomolecule translocation with high precision. Also, solid-state nanopore DNA sequencing applications need a low-noise/high-bandwidth amplifier for single-molecule recognition in the measurement system. All these advances are boosting solid-state nanopore DNA sequencing, making it a widely used, cost-effective and versatile alternative to biological nanopores.

5.3.2 Nanopore Technology Used in Genomic Sequencing

In the first generation of ONT products, the indicated error rates ranged from 5% to 20% [146], which is higher than the error rate of NGS. This was deemed as the first challenge to the use of nanopore technology in genomic sequencing. Nanopore sequencing error rates can affect the accuracy of library preparation and the type of molecules being sequenced [147–149]. Error rates are notably higher in G/C mononucleotide repeats [146]. In recent years, ONT has enhanced nanopore sequencing read accuracy. For example, the R10.1.4 flow cell and the latest 1D chemistry flow cells have an average sequencing error rate of 3.9%. New sequencing kits such as Q23 have been tested, reportedly increasing accuracy to 99.5%. (https://nanoporetech.com/platform/accuracy, accessed on 25 November 2024)

The PromethION platform can achieve a sequencing accuracy of up to 98.3% [150]. Additionally, new basecalling algorithms such as Bonito have significantly improved nanopore sequencer accuracy by aligning probability profiles of two neural networks, achieving a median sequencing error reduction of over 50% and an accuracy rate of 98.3% [151]. It is especially remarkable that nanopore sequencing has higher accuracy for long or ultra-long molecules compared to short-read NGS sequencing.

The second challenge in nanopore genomic sequencing is the significant amount of input DNA and RNA demanded, potentially requiring several micrograms of DNA and hundreds of nanograms of RNA. Specifically, for direct RNA sequencing, a substantial initial RNA quantity (>10 μg) is necessary to ensure an adequate yield of mRNA (500 ng) following the efficient removal of rRNA. Under certain circumstances, the direct amplification of viral genetic material for long-read sequencing is discouraged due to the potential for significant PCR bias, which can lead to an underrepresentation of the viral sequences in the sequencing output. Nonetheless, in the context of viral diagnostics or identification, it is not mandatory to include PCR amplification before nanopore sequencing. Alternative methods such as sequence-independent single-primer amplification (SISPA) can be employed in this case [152]. For the detection of viral genomes, it is recommended to initially perform experiments aimed at enriching the viral genetic material. This enrichment can be accomplished by employing techniques such as filtration and centrifugation of the samples, which are then subjected to DNase and RNase treatments to eliminate any eukaryotic and bacterial contaminants before the extraction of nucleic acids. Another approach is to use ribosomal depletion which specifically targets both eukaryotic and bacterial rRNAs. Generally, the quantity of input material required for these procedures is contingent upon the particular focus of the study at hand.

The third challenge is the analysis of subsequent data generated during sequencing, which can be shared with other high-throughput sequencing platforms. Although various analysis tools for nanopore sequencing data have been developed, many of these tools are Linux-based and require proficiency in command-line operations and programming, which can be a significant hurdle for users without a strong technical background. However, ONT addressed this issue with the launch of EPI2ME in 2020 (https://labs.epi2me.io, accessed on 25 November 2024). It is a comprehensive bioinformatics platform designed to simplify the analysis of nanopore sequencing data. EPI2ME provides a suite of analytical features (including alignment, metagenomic species identification, and the detection of structural variation). More importantly, EPI2ME is crafted to be accessible to researchers, even those without extensive computational skills, enabling them to analyze nanopore data with minimal bioinformatics expertise.

RNA sequencing is another significant component in genomic sequencing. Most common issues include the absence of 5′ ends during direct RNA sequencing. This mostly happens because motor proteins release mRNA molecules too early, causing a loss of control in guiding the RNA through the nanopore [153]. To address this, combining data from various sequencing platforms can facilitate the prediction of putative 5′ ends, thus enhancing the reliability and precision of sequencing outcomes. Additionally, the nanopore community is actively working on improving the detection of 5′ ends in direct RNA sequencing with 5′-dependent adapter ligation techniques [154]. Another common issue in library preparation is template switching, where reverse transcriptase (RT) during reverse transcription can jump templates. This leads to chimeric cDNAs containing false introns that are commonly misidentified as splice isoforms, negatively impacting downstream transcript analysis. Direct RNA sequencing can circumvent the issues associated with template switching, making it essential to choose the most appropriate library construction method based on the specific experimental requirements. While there are certain obstacles to overcome, nanopore sequencing presents numerous advantages and holds great potential to enhance research on viruses in both animals and humans. This method can produce exceedingly long reads and offers insights into a range of biological polymers, such as DNA, RNA, and proteins. The distinctive capabilities of nanopore sequencing will significantly enhance the continuity and completeness of viral genome assemblies. This improvement will, in turn, promote a more profound and holistic comprehension of the origins and evolutionary trajectories of viruses.

5.3.3 Nanopore Technology Used in Pathogen Diagnosis

Bloodstream infections (BSIs) and sepsis represent major global health burdens, contributing significantly to morbidity and mortality worldwide. Current epidemiological models, primarily derived from high-income regions [155], estimate approximately 31.5 million sepsis episodes and 5.3 million associated deaths annually [156]. This projection likely underrepresents the true impact across resource-limited settings. Over recent decades, studies have indicated rising BSI/sepsis incidence [157], disproportionately affecting immunocompromised, elderly, and multimorbid populations—trends exacerbated by escalating antimicrobial resistance (AMR) undermining empirical treatments [158]. Rapid pathogen identification and targeted therapy initiation are critical for improving outcomes, with international bodies advocating ≤2-hour diagnostic windows [159]. While blood culture remains the diagnostic reference, its one to three days turnaround and susceptibility to variables (bacteremia levels, prior antibiotics) necessitate faster alternatives. Consequently, molecular approaches enabling direct pathogen/AMR detection from blood specimens are gaining traction. Target-restricted methods (e.g., hybridization, PCR, microarrays) contrast with whole-genome sequencing (WGS), which delivers exhaustive genomic insights—spanning AMR determinants, strain typing, and virulence profiling. Oxford Nanopore Technology's real-time sequencing platform exemplifies progress here, facilitating both 16S amplicon and metagenomic WGS workflows for culture-free diagnosis [160, 161]. This technology's versatility extends beyond outbreak virology [146] to bacterial/viral genomics and human genome analysis [57], enabled by key innovations including: (i) Direct DNA/RNA sequencing without amplification [162]. (ii) Scalability to large genomes (demonstrated by 2018's human genome milestone [163]). (iii) Dynamic AMR surveillance capabilities.

The evolution of nanopore systems toward true point-of-care deployment could revolutionize sepsis management. Future integration of AI-driven real-time basecalling with automated AMR interpretation may help achieve the two-hours target as required by the World Health Organization (WHO), transforming precision medicine in critical care. However, cost reduction and streamlined bioinformatics pipelines remain essential for global accessibility—particularly in regions bearing the highest AMR burdens. (https://www.whatisbiotechnology.org/index.php/exhibitions/antimicrobial) With the capacity for swift microbial identification, nanopore sequencing is poised to replace the laborious traditional methods of bacterial culture for identification and antibiotic susceptibility testing. (https://www.whatisbiotechnology.org/index.php/exhibitions/antimicrobial/index/susceptibility). This advancement allows physicians to transition from employing broad-spectrum antibiotics to more targeted treatment strategies. Consequently, nanopore sequencers may become pivotal in the responsible management of antibiotics, helping curb the overprescription and ineffective use of these medications. Furthermore, nanopore sequencing enhances patient diagnosis and treatment by enabling the tracking of the evolutionary trajectories of various pathogenic microbes. It also assists in monitoring their transmission pathways, which is crucial for the surveillance of emerging infectious diseases and for controlling outbreaks. Beyond these applications, the use of nanopore sequencing extends to fundamental biological research, ancestry analysis, and forensic science. [159] The technology is currently being utilized to explore microbes in some of the most extreme and remote environments on our planet, such as the Arctic and Antarctic regions [164, 165], and a deep mine in South Wales.

Nanopore sequencing allows for the rapid identification of pathogens, including the diagnosis of bacterial meningitis [166], bacterial lower respiratory tract infection [47], infective endocarditis [167], pneumonia [168], and infection in prosthetic joints [169]. It has been used for the rapid detection of pathogens, antibiotic-resistance genes, and plasmids in blood cultures as well. This is crucial for reducing morbidity and mortality caused by BSI, where rapid diagnosis and appropriate antimicrobial therapy are essential. Unlike PCR, which is limited to detecting a predefined set of genetic targets, nanopore sequencing offers an unbiased approach to diagnostics, with the potential to identify any pathogen and antimicrobial resistance-encoding gene [170]. This technology is especially crucial in scenarios where the pathogen remains unidentified or in the context of emerging infectious diseases. The real-time sequencing capabilities of nanopore technology facilitate immediate data analysis for pathogen identification and the detection of plasmids and genes associated with antimicrobial resistance (AMR). This feature significantly outperforms PCR, which lacks real-time output and necessitates further post-amplification processing and analysis. Moreover, nanopore sequencing delivers comprehensive genomic insights through whole-genome sequencing (WGS), potentially uncovering all AMR-encoding genes within a bacterial genome. Nanopore sequences individual DNA or RNA molecules without PCR amplification or chemical sample labeling, thus avoiding biases and enhancing the detection of a broader range of RNA species. It also holds the potential for cost-effective genotyping earlier than traditional diagnostics, offering a particular advantage for immunocompromised patients who are prone to infections and may exhibit non-typical symptoms [171]. MinION has proven to be faster (six hours) and more sensitive than traditional culture-based gold standard methods (over two days) in diagnosing bacterial lower respiratory tract infections [47]. Concurrently, iterative refinement of basecalling algorithms has markedly improved the precision of nanopore sequencing platforms [29]. For example, algorithms such as Bonito improve accuracy by aligning the probability profiles of two neural networks to reach a consensus, reducing the median sequencing error rate by over 50% and achieving an accuracy of 98.3% [151]. Importantly, when comparing the accuracy of nanopore sequencing on long or ultra-long molecules with the error rates of short-read NGS sequencing, the former shows a remarkable advantage. The portability of nanopore devices enables on-site sequencing and expedites the turnaround time, which is invaluable during outbreaks or in resource-limited settings where swift diagnostics are essential.

Nanopore sequencing technology, mainly represented by ONT sequencing, has shown significant advantages in the field of pathogen detection. It can not only accurately identify pathogens but also rapidly profile antibiotic and antimicrobial resistance in bacteria and other microorganisms. For example, the MinION platform has been used to directly identify 51 acquired resistance genes from clinical urine samples without cultivation. In comparison, Illumina sequencing identified 55 such genes from cultured bacteria [172]. Furthermore, in a recent large-scale study on colistin resistance involving 12,053 Salmonella strains, a combined approach using data from ONT, PacBio, and Illumina was employed [173]. ONT sequencing is particularly effective in detecting specific species and strains with virulent characteristics from complex microbiome samples. Its longer reads provide clearer mapping ability, offering more precise estimates of microbiome composition compared to traditional 16S rRNA and DNA amplicon-based methods [174, 175].

In summary, nanopore sequencing technology has made remarkable progress in pathogen diagnosis. It provides rapid and comprehensive genomic information, and holds great potential for detecting all AMR-encoding genes that are present in bacterial genomes.

5.4 Conclusion and Perspective

The advent of NGS technologies has revolutionized our approach to scientific research, spanning from fundamental to applied and clinical domains. The ability to sequence complete genomes across a multitude of related species has facilitated in-depth comparative and evolutionary studies that were previously deemed unattainable. Perhaps the most profound application of NGS is the resequencing of human genomes, which significantly deepens our comprehension of the impact of genetic variations on health and disease. Given the diverse capabilities of NGS, it is anticipated that multiple platforms will continue to coexist, each offering distinct advantages tailored to specific applications.

Overall, nanopore sequencing is a relatively inexpensive sequencing technology compared to other commercial forms of DNA sequencing technology. It is a technology that can potentially decrease sequencing costs to less than $1,000/person, taking less than 24 hours per genome. Despite its advantages, nanopore sequencing does necessitate the use of disposable flow cells that need periodic replacement. As a relatively recent innovation, nanopore sequencing is still in the process of gaining broader acceptance. A part of this challenge stems from its disruption to the established norms of sequencing practices. Many researchers, preferring familiarity, have been reluctant to adopt the technology, favoring PCR methods instead. The concept of performing nanopore sequencing outside of centralized, large-scale laboratories also contradicts the current paradigm. The strength of this resistance was evident during the early stages of the COVID-19 pandemic when ONT offered to provide decentralized testing options. The UK government, however, chose to invest in large laboratories equipped with PCR machines, adhering to the established and tested model. Developing a rapid, easy-to-use, portable test for ONT to diagnose COVID-19 or other diseases at the point of care also faces challenges due to the prevalent demand for clear yes/no results, which lateral-flow-assay tests currently provide more conveniently and at a lower cost. Although multiple firms are developing nanopore sequencing platforms, none has yet produced a competitive alternative to ONT. This situation underscores the complexities and high costs associated with academic research.

In brief, the applications of nanopore sequencing in global public health are very broad, but particularly focused on the following areas:

1) Infectious Disease Surveillance and Control. Nanopore sequencing can rapidly identify and monitor infectious disease pathogens, including emerging viruses and drug-resistant bacteria, which is crucial for the prevention and control of infectious diseases worldwide.
2) Rapid Response to Epidemics. In public health emergencies, such as the COVID-19 pandemic, nanopore sequencing provides real-time genomic data, helping researchers quickly understand virus mutations and providing a scientific basis for epidemic control.
3) Antimicrobial Resistance Monitoring. Nanopore sequencing can detect antibiotic resistance genes directly from clinical samples, aiding in the establishment of global antimicrobial resistance monitoring networks and the tracking of resistance patterns.
4) Personalized Medicine and Precision Treatment. The real-time genomic profiling capability of nanopore sequencing is advancing precision infectious disease management, where therapies are dynamically optimized against evolving pathogen genotypes.
5) Primary Healthcare and Telemedicine. Portable nanopore sequencing devices enable on-site sequencing in primary and remote areas, enhancing the regions' infectious disease diagnostic and response capabilities.
6) Public Health Decision Support. The large datasets generated by nanopore sequencing can support public health decision-making by helping analyze

the spread patterns and mutation trends of pathogens, thereby providing a scientific basis for policy formulation.

7) Global Health Cooperation. Nanopore sequencing technology can strengthen global health cooperation through the sharing of data and resources, helping low-income countries improve their disease surveillance and control capabilities.
8) Biosecurity and Biodefense. In the field of biosecurity and biodefense, nanopore sequencing can be used to identify and monitor potential biological threats, enhancing the prevention and response to bioterrorism attacks.
9) Environmental and Ecological Monitoring. Nanopore sequencing can be used for the analysis of microbial communities in environmental samples, helping monitor the impact of environmental changes on microbial populations and the spread of pathogens in the environment.
10) Ethical and Privacy Issues. With the widespread application of nanopore sequencing in public health, ethical and privacy issues have become increasingly prominent, requiring the development of corresponding policies and regulations to protect personal data.

In summary, nanopore sequencing technology has tremendous potential in the field of global public health, but it also needs to overcome challenges in technology, economy, ethics, and other areas to achieve widespread application globally.

References

1 Quince, C., Walker, A.W., Simpson, J.T. et al. (2017). Shotgun metagenomics, from sampling to analysis. *Nature Biotechnology* 35 (9): 833–844. https://doi.org/10.1038/nbt.3935.

2 Jackson, S., Sichtig, H., Goldberg, B. et al. (2015). *Standards for Pathogen Detection via Next Generation Sequencing*. Gaithersburg, MD: NIST.

3 Chiu, C.Y. and Miller, S.A. (2019). Clinical metagenomics. *Nature Reviews Genetics* 20 (6): 341–355. https://doi.org/10.1038/s41576-019-0113-7.

4 Consensus Group of Experts on Application of Metagenomic Next Generation Sequencing in the Pathogen Diagnosis in Clinical Moderate and Severe Infections Professional Committee of Sepsis and Shock, Chinese Research Hospital Association Professional Committee of Microbial Toxins, Chinese Society for Microbiology Professional Committee of Critical Care Medicine, Shenzhen Medical Association, Li, Y., Ma, J., Yao, Y., Feng, Y., Expert consensus for the application of metagenomic next generation sequencing in the pathogen diagnosis in clinical moderate and severe infections. *Zhonghua Wei Zhong Bing Ji Jiu Yi Xue 中华危重病急救医学* (2020) 32 (5): 531–536. https://doi.org/10.3760/cma.j.cn121430-20200228-00095.

5 Tang, W.J., Zhang, Y., Luo, C. et al. (2021). Clinical application of metagenomic next-generation sequencing for suspected infections in patients with primary immunodeficiency disease. *Frontiers in Immunology* 12. ARTN 696403 10.3389/fimmu.2021.696403.

6 Ye, J., Huang, K., Xu, Y. et al. (2024). Clinical application of nanopore-targeted sequencing technology in bronchoalveolar lavage fluid from patients with pulmonary infections. *Microbiology Spectrum* 12 (6): e0002624. https://doi.org/10.1128/spectrum.00026-24.

7 Nekrutenko, A. and Taylor, J. (2012). Next-generation sequencing data interpretation: enhancing reproducibility and accessibility. *Nature Reviews Genetics* 13 (9): 667–U693. https://doi.org/10.1038/nrg3305.

8 Kathiresan, N., Temanni, R., Almabrazi, H. et al. (2017). Accelerating next generation sequencing data analysis with system level optimizations. *Scientific Reports-Uk* 7. ARTN 9058 10.1038/s41598-017-09089-1.

9 Li, N., Cai, Q.Q., Miao, Q. et al. (2021). High-throughput metagenomics for identification of pathogens in the clinical settings. *Small Methods* 5 (1). ARTN 2000792 10.1002/smtd.202000792.

10 Bentley, D.R., Balasubramanian, S., Swerdlow, H.P. et al. (2008). Accurate whole human genome sequencing using reversible terminator chemistry. *Nature* 456 (7218): 53–59. https://doi.org/10.1038/nature07517.

11 Hu, Z., Cheng, L., and Wang, H. (2015). The Illumina-Solexa sequencing protocol for bacterial genomes. In: *Bacterial Pangenomics: Methods and Protocols* (ed. A. Mengoni, M. Galardini, and M. Fondi), 91–97. New York: Springer.

12 Lang, J.D., Zhu, R.R., Sun, X. et al. (2021). Evaluation of the MGISEQ-2000 sequencing platform for Illumina target capture sequencing libraries. *Frontiers in Genetics* 12. ARTN 730519 10.3389/fgene.2021.730519.

13 Wang, S.W., Tan, Y., Li, S.J., and Zhu, T.H. (2022). Structural and dynamic analysis of leaf-associated fungal community of walnut leaves infected by leaf spot disease based illumina high-throughput sequencing technology. *Polish Journal of Microbiology* 71 (3): 429–441. https://doi.org/10.33073/pjm-2022-038.

14 Lin, L.J., Du, F.M., Zeng, J. et al. (2020). Deep insights into fungal diversity in traditional Chinese sour soup by Illumina MiSeq sequencing. *Food Research International* 137. ARTN 109439 10.1016/j.foodres.2020.109439.

15 Kim, Y., Cheng, W.Q., Cho, C.S. et al. (2024). Seq-Scope: repurposing Illumina sequencing flow cells for high-resolution spatial transcriptomics. *Nature Protocols*. https://doi.org/10.1038/s41596-024-01065-0.

16 Chen, M.J., Qin, Y.Y., Deng, F.M. et al. (2021). Illumina MiSeq sequencing reveals microbial community succession in salted peppers with different salinity during preservation. *Food Research International* 143. DOI: ARTN 110234 10.1016/j.foodres.2021.110234.

17 Langelier, C., Kalantar, K.L., Moazed, F. et al. (2018). Integrating host response and unbiased microbe detection for lower respiratory tract infection diagnosis in critically ill adults. *P Natl Acad Sci USA* 115 (52): E12353–E12362. https://doi.org/10.1073/pnas.1809700115.

18 Wilson, M.R., Sample, H.A., Zorn, K.C. et al. (2019). Clinical metagenomic sequencing for diagnosis of meningitis and encephalitis. *The New England Journal of Medicine* 380 (24): 2327–2340. https://doi.org/10.1056/NEJMoa1803396.

19 Wilson, M.R., Sample, H.A., Zorn, K.C. et al. (2018). Clinical metagenomic next-generation sequencing for diagnosis of infectious meningitis and encephalitis. *Annals of Neurology* 84: S110–S110.

20 Ganesan Prakash, A.K., Sheoran, N., Aggarwal, R. et al. (2019). First draft genome sequence of a Pearl Millet Blast pathogen, Magnaporthe grisea strain PMg_Dl, obtained using PacBio single-molecule real-time and Illumina NextSeq 500 sequencing. *Microbiology Resoure Announcements* 8 (20): e01499–e01418. https://doi.org/10.1128/MRA.01499-18.

21 Chen, L.L., Xu, Y., Liu, C.F. et al. (2020). Clinical features of aseptic meningitis with varicella zoster virus infection diagnosed by next-generation sequencing: case reports. *BMC Infectious Diseases* 20 (1): https://doi.org/10.1186/s12879-020-05155-8.

22 van Boheemen, S., van Rijn, A.L., Pappas, N. et al. (2020). Retrospective Validation of a Metagenomic Sequencing Protocol for Combined Detection of RNA and DNA Viruses Using Respiratory Samples from Pediatric Patients. *Journal of Molecular Diagnostics* 22 (2): 196–207. https://doi.org/10.1016/j.jmoldx.2019.10.007.

23 Zhou, X., Wu, H.L., Ruan, Q.L. et al. (2019). Clinical Evaluation of Diagnosis Efficacy of Active Complex Infection via Metagenomic Next-Generation Sequencing of Direct Clinical Samples. *Frontiers in Cellular and Infection Microbiology* 9: ARTN 351 10.3389/fcimb.2019.00351.

24 Zhu, F.Y., Chen, M.X., Ye, N.H. et al. (2018). Comparative performance of the BGISEQ-500 and Illumina HiSeq4000 sequencing platforms for transcriptome analysis in plants. *Plant Methods* 14. https://doi.org/10.1186/s13007-018-0337-0.

25 Drmanac, R., Sparks, A.B., Callow, M.J. et al. (2010). Human genome sequencing using unchained base reads on self-assembling DNA nanoarrays. *Science* 327 (5961): 78–81. https://doi.org/10.1126/science.1181498.

26 Wang, S.N., Chen, Y.L., Wang, D.M. et al. (2019). The feasibility of metagenomic next-generation sequencing to identify pathogens causing Tuberculous Meningitis in cerebrospinal fluid. *Frontiers in Microbiology* 10: DOI: ARTN 1993 10.3389/fmicb.2019.01993.

27 Zhang, H.C., Ai, J.W., Cui, P. et al. (2019). Incremental value of metagenomic next generation sequencing for the diagnosis of suspected focal infection in adults. *The Journal of Infection* 79 (5): 419–425. https://doi.org/10.1016/j.jinf.2019.08.012.

28 Howorka, S., Cheley, S., and Bayley, H. (2001). Sequence-specific detection of individual DNA strands using engineered nanopores. *Nature Biotechnology* 19 (7): 636–639. https://doi.org/10.1038/90236.

29 Ji, C.M., Feng, X.Y., Huang, Y.W., and Chen, R.A. (2024). The applications of nanopore sequencing technology in animal and human virus research. *Viruses-Basel* 16 (5): DOI: ARTN 798 10.3390/v16050798.

30 Sanger, F. and Coulson, A.R. (1975). A rapid method for determining sequences in DNA by primed synthesis with DNA polymerase. *Journal of Molecular Biology* 94 (3): 441–448. https://doi.org/10.1016/0022-2836(75)90213-2.

31 https://www.nobelprize.org/nobel_prizes/chemistry/laureates/1958/sanger-lecture.html.

32 Edman, P. (1950). Method for determination of the amino acid sequence in peptides. *Acta Chemica Scandinavica* 4 (2): 283–293. https://doi.org/10.3891/acta.chem.scand.04-0283.

33 Holley, R.W., Apgar, J., Everett, G.A. et al. (1965). Structure of a ribonucleic acid. *Science* 147 (3664): 1462. https://doi.org/10.1126/science.147.3664.1462.

34 Sanger, F. (1981). Citation classic –two-dimensional fractionation procedure for radioactive nucleotides. *Cc/Life Sci* 3: 16–16.

35 Gilbert, W. and Maxam, A. (1973). The nucleotide sequence of the lac operator. *Proceedings of the National Academy of Sciences of the United States of America* 70 (12): 3581–3584. https://doi.org/10.1073/pnas.70.12.3581.

36 Air, G.M., Sanger, F., Barrell, B.G. et al. (1977). Nucleotide-sequence of DNA of bacteriophage-Phix174. *Proceedings of the Australian Biochemical Society* 10: 60–60.

37 Kasianowicz, J.J., Brandin, E., Branton, D., and Deamer, D.W. (1996). Characterization of individual polynucleotide molecules using a membrane channel. *Proceedings of the National Academy of Sciences of the USA* 93 (24): 13770–13773. https://doi.org/10.1073/pnas.93.24.13770.

38 Gut, I.G. (2013). New sequencing technologies. *Clinical & Translational Oncology* 15 (11): 879–881. https://doi.org/10.1007/s12094-013-1073-6.

39 Schadt, E.E., Turner, S., and Kasarskis, A. (2010). A window into third-generation sequencing. *Human Molecular Genetics* 19 (R2): R227–R240. https://doi.org/10.1093/hmg/ddq416.

40 Niedringhaus, T.P., Milanova, D., Kerby, M.B. et al. (2011). Landscape of next-generation sequencing technologies. *Analytical Chemistry* 83 (12): 4327–4341. https://doi.org/10.1021/ac2010857.

41 Pareek, C.S., Smoczynski, R., and Tretyn, A. (2011). Sequencing technologies and genome sequencing. *Journal of Applied Genetics* 52 (4): 413–435. https://doi.org/10.1007/s13353-011-0057-x.

42 Metzker, M.L. (2005). Emerging technologies in DNA sequencing. *Genome Research* 15 (12): 1767–1776. https://doi.org/10.1101/gr.3770505.

43 Fan, J.B., Chee, M.S., and Gunderson, K.L. (2006). Highly parallel genomic assays. *Nature Reviews Genetics* 7 (8): 632–644. https://doi.org/10.1038/nrg1901.

44 Pop, M. and Salzberg, S.L. (2008). Bioinformatics challenges of new sequencing technology. *Trends in Genetics* 24 (3): 142–149. https://doi.org/10.1016/j.tig.2007.12.006.

45 Petersen, L.M., Martin, I.W., Moschetti, W.E. et al. (2020). Third-generation sequencing in the clinical laboratory: exploring the advantages and challenges of nanopore sequencing. *Journal of Clinical Microbiology* 58 (1): DOI: ARTN e01315-19 10.1128/JCM.01315-19.

46 Akeson, M., Branton, D., Kasianowicz, J.J. et al. (1999). Microsecond time-scale discrimination among polycytidylic acid, polyadenylic acid, and polyuridylic acid as homopolymers or as segments within single RNA molecules. *Biophysical Journal* 77 (6): 3227–3233. https://doi.org/10.1016/S0006-3495(99)77153-5.

47 Charalampous, T., Kay, G.L., Richardson, H. et al. (2019). Nanopore metagenomics enables rapid clinical diagnosis of bacterial lower respiratory infection. *Nature Biotechnology* 37 (7): 783–+. https://doi.org/10.1038/s41587-019-0156-5.

48 Goodwin, S., McPherson, J.D., and McCombie, W.R. (2016). Coming of age: ten years of next-generation sequencing technologies. *Nature Reviews Genetics* 17 (6): 333–351. https://doi.org/10.1038/nrg.2016.49.

49 Eid, J., Fehr, A., Gray, J. et al. (2009). Real-time DNA sequencing from single polymerase molecules. *Science* 323 (5910): 133–138. https://doi.org/10.1126/science.1162986.

50 Ardui, S., Ameur, A., Vermeesch, J.R., and Hestand, M.S. (2018). Single molecule real-time (SMRT) sequencing comes of age: applications and utilities for medical diagnostics. *Nucleic Acids Research* 46 (5): 2159–2168. https://doi.org/10.1093/nar/gky066.

51 Jiao, N.B., Xu, J.R., Wang, Y. et al. (2024). Genome-wide characterization of post-transcriptional processes related to wood formation in. *BMC Genomics* 25 (1): DOI: ARTN 372 10.1186/s12864-024-10300-7.

52 Kim, H.S., Jeon, S., Kim, C. et al. (2019). Chromosome-scale assembly comparison of the Korean Reference Genome KOREF from PromethION and PacBio with Hi-C mapping information. *GigaScience* 8 (12): DOI: ARTN giz125 10.1093/gigascience/giz125.

53 Yu, X.L., Jiang, W.Q., Shi, Y. et al. (2019). Applications of sequencing technology in clinical microbial infection. *Journal of Cellular and Molecular Medicine* 23 (11): 7143–7150. https://doi.org/10.1111/jcmm.14624.

54 Gonzalez-Reiche, A.S., Hernandez, M.M., Sullivan, M.J. et al. (2020). Introductions and early spread of SARS-CoV-2 in the New York City area. *Science* 369 (6501): 297–+. https://doi.org/10.1126/science.abc1917.

55 Maurano, M.T., Ramaswami, S., Zappile, P. et al. (2020). Sequencing identifies multiple early introductions of SARS-CoV-2 to the New York City region. *Genome Research* 30 (12): https://doi.org/10.1101/gr.266676.120.

56 Feng, Y.X., Zhang, Y.C., Ying, C.F. et al. (2015). Nanopore-based fourth-generation DNA Sequencing Technology. *Genomics, Proteomics & Bioinformatics* 13 (1): 4–16. https://doi.org/10.1016/j.gpb.2015.01.009.

57 Bowden, R., Davies, R.W., Heger, A. et al. (2019). Sequencing of human genomes with nanopore technology. *Nature Communications* 10: DOI: ARTN 1869 10.1038/s41467-019-09637-5.

58 Leggett, R.M. and Clark, M.D. (2017). A world of opportunities with nanopore sequencing. *Journal of Experimental Botany* 68 (20): 5419–5429. https://doi.org/10.1093/jxb/erx289.

59 Schloss, J.A. (2008). How to get genomes at one ten-thousandth the cost. *Nature Biotechnology* 26 (10): 1113–1115. https://doi.org/10.1038/nbt1008-1113.

60 Eisenstein, M. (2012). Oxford Nanopore announcement sets sequencing sector abuzz. *Nature Biotechnology* 30 (4): 295–296. https://doi.org/10.1038/nbt0412-295.

61 Goodwin, S., Gurtowski, J., Ethe-Sayers, S. et al. (2015). Oxford Nanopore sequencing, hybrid error correction, and de novo assembly of a eukaryotic genome. *Genome Research* 25 (11): 1750–1756. https://doi.org/10.1101/gr.191395.115.

62 Mikheyev, A.S. and Tin, M.M. (2014). A first look at the Oxford Nanopore MinION sequencer. *Molecular Ecology Resources* 14 (6): 1097–1102. https://doi.org/10.1111/1755-0998.12324 From NLM Medline.

63 Herzog, K.S., Wu, R., Hawdon, J.M. et al. (2024). Assessing de novo parasite genomes assembled using only Oxford Nanopore Technologies MinION data. *iScience* 27 (9): 110614. https://doi.org/10.1016/j.isci.2024.110614.

64 Esnault, G., Earley, B., Cormican, P. et al. (2022). Assessment of rapid MinION Nanopore DNA virus meta-genomics using calves experimentally infected with Bovine Herpes Virus-1. *Viruses-Basel* 14 (9): DOI: ARTN 1859 10.3390/v14091859.

65 Rang, F.J., Kloosterman, W.P., and de Ridder, J. (2018). From squiggle to basepair: computational approaches for improving nanopore sequencing read accuracy. *Genome Biology* 19. DOI: ARTN 90 10.1186/s13059-018-1462-9.

66 Prazsák, I., Moldován, N., Balázs, Z. et al. (2018). Long-read sequencing uncovers a complex transcriptome topology in varicella zoster virus. *BMC Genomics* 19: DOI: ARTN 873 10.1186/s12864-018-5267-8.

67 De Jesus, J.G., Giovanetti, M., Faria, N.R., and Alcantara, L.C. (2019). Acute Vector-Borne Viral Infection:Zika and MinION Surveillance. *Microbiology Spectrum* 7 (4): DOI: ARTN AME-0008-2019 10.1128/microbiolspec.AME-0008-2019.

68 Deng, X.D., Achari, A., Federman, S. et al. (2020). Metagenomic sequencing with spiked primer enrichment for viral diagnostics and genomic surveillance. *Nature Microbiology* 5 (3): 443–454. https://doi.org/10.1038/s41564-019-0637-9.

69 Quick, J., Grubaugh, N.D., Pullan, S.T. et al. (2017). Multiplex PCR method for MinION and Illumina sequencing of Zika and other virus genomes directly from clinical samples. *Nature Protocols* 12 (6): 1261–1276. https://doi.org/10.1038/nprot.2017.066.

70 Quick, J., Loman, N.J., Duraffour, S. et al. (2016). Real-time, portable genome sequencing for Ebola surveillance. *Nature* 530 (7589): 228–+. https://doi.org/10.1038/nature16996.

71 Runtuwene, L.R., Tuda, J.S.B., Mongan, A.E. et al. (2018). Nanopore sequencing of drug-resistance-associated genes in malaria parasites. *Sci Rep-Uk* 8. DOI: ARTN 8286 10.1038/s41598-018-26334-3.

72 Bezrukov, S.M. and Kasianowicz, J.J. (1993). Current noise reveals protonation kinetics and number of ionizable sites in an open protein ion channel. *Physical Review Letters* 70 (15): 2352–2355. https://doi.org/10.1103/PhysRevLett.70.2352.

73 Krasilnikov, O.V., Sabirov, R.Z., Ternovsky, V.I. et al. (1988). The structure of Staphylococcus aureus alpha-toxin-induced ionic channel. *General Physiology and Biophysics* 7 (5): 467–473. From NLM.

74 Krasilnikov, O.V. and Sabirov, R.Z. (1989). Ion transport through channels formed in lipid bilayers by Staphylococcus aureus alpha-toxin. *General Physiology and Biophysics* 8 (3): 213–222. From NLM.

75 Bezrukov, S.M. and Vodyanoy, I. (1995). Noise-induced enhancement of signal transduction across voltage-dependent ion channels. *Nature* 378 (6555): 362–364. https://doi.org/10.1038/378362a0 From NLM.

76 Song, L.Z., Hobaugh, M.R., Shustak, C. et al. (1996). Structure of staphylococcal alpha-hemolysin, a heptameric transmembrane pore. *Science* 274 (5294): 1859–1866. https://doi.org/10.1126/science.274.5294.1859.

77 Kasianowicz, J.J., Henrickson, S.E., Misakian, M. et al. (2001). Applications for DNA transport in a single nanopore. *Biophysical Journal* 80 (1): 339a–339a.

78 Stoddart, D., Maglia, G., Mikhailova, E. et al. (2010). Multiple base-recognition sites in a biological nanopore: two heads are better than one. *Angewandte Chemie International Edition* 49: 556–559. https://doi.org/10.1002/anie.200905483.

79 Deamer, D., Akeson, M., and Branton, D. (2016). Three decades of nanopore sequencing. *Nature Biotechnology* 34 (5): 518–524. https://doi.org/10.1038/nbt.3423.

80 Benner, S., Chen, R.J.A., Wilson, N.A. et al. (2007). Sequence-specific detection of individual DNA polymerase complexes in real time using a nanopore. *Nature Nanotechnology* 2 (11): 718–724. https://doi.org/10.1038/nnano.2007.344.

81 Cockroft, S.L., Chu, J., Amorin, M., and Ghadiri, M.R. (2008). A single-molecule nanopore device detects DNA polymerase activity with single-nucleotide resolution. *Journal of the American Chemical Society* 130 (3): 818–+. https://doi.org/10.1021/ja077082c.

82 Hornblower, B., Coombs, A., Whitaker, R.D. et al. (2007). Single-molecule analysis of DNA-protein complexes using nanopores. *Nature Methods* 4 (4): 315–317. https://doi.org/10.1038/Nmeth1021.

83 Meller, A. and Branton, D. (2002). Single molecule measurements of DNA transport through a nanopore. *Electrophoresis* 23 (16): 2583–2591. https://doi.org/10.1002/1522-2683(200208)23:16<2583::Aid-Elps2583>3.0.Co;2-H.

84 Deamer, D.W. and Branton, D. (2002). characterization of nucleic acids by nanopore analysis. *Accounts of Chemical Research* 35 (10): 817–825.

85 Cao, C., Liao, D.F., Ying, Y.L., and Long, Y.T. (2016). Detection of single oligonucleotide by an Aerolysin Nanopore. *Acta Chimica Sinica* 74 (9): 734–737. https://doi.org/10.6023/A16070352.

86 Zhou, W.Q., Qiu, H., Guo, Y.F., and Guo, W.L. (2020). Molecular Insights into distinct detection properties of alpha-Hemolysin, MspA, CsgG, and Aerolysin Nanopore Sensors. *The Journal of Physical Chemistry. B* 124 (9): 1611–1618. https://doi.org/10.1021/acs.jpcb.9b10702.

87 Butler, T.Z., Pavlenok, M., Derrington, I.M. et al. (2008). Single-molecule DNA detection with an engineered MspA protein nanopore. *P Natl Acad Sci USA* 105 (52): 20647–20652. https://doi.org/10.1073/pnas.0807514106.

88 Li, J., Stein, D., McMullan, C. et al. (2001). Ion-beam sculpting at nanometre length scales. *Nature* 412 (6843): 166–169. https://doi.org/10.1038/35084037.

89 Li, J.L., Gershow, M., Stein, D. et al. (2003). DNA molecules and configurations in a solid-state nanopore microscope. *Nature Materials* 2 (9): 611–615. https://doi.org/10.1038/nmat965.

90 Soni, G.V. and Meller, A. (2007). Progress toward ultrafast DNA sequencing using solid-state nanopores. *Clinical Chemistry* 53 (11): 1996–2001. https://doi.org/10.1373/clinchem.2007.091231.

91 Meng, S., Maragakis, P., Papaloukas, C., and Kaxiras, E. (2007). DNA nucleoside interaction and identification with carbon nanotubes. *Nano Letters* 7 (1): 45–50. https://doi.org/10.1021/nl0619103.

92 Storm, A.J., Storm, C., Chen, J.H. et al. (2005). Fast DNA translocation through a solid-state nanopore. *Nano Letters* 5 (7): 1193–1197. https://doi.org/10.1021/nl048030d.

93 Gershow, M. and Golovchenko, J.A. (2007). Recapturing and trapping single molecules with a solid-state nanopore. *Nature Nanotechnology* 2 (12): 775–779. https://doi.org/10.1038/nnano.2007.381.

94 Iqbal, S.M., Akin, D., and Bashir, R. (2007). Solid-state nanopore channels with DNA selectivity. *Nature Nanotechnology* 2 (4): 243–248. https://doi.org/10.1038/nnano.2007.78.

95 Liang, F., Zeng, Y., and Wang, L. (2014). DNA sequencing by recognition and its potential application with Nanopore sequencing. *Current Organic Chemistry* 18 (15): 1948–1956. https://doi.org/10.2174/1385272819666140514004615.

96 Yan, H. and Xu, B.Q. (2006). Towards rapid DNA sequencing: detecting single-stranded DNA with a solid-state nanopore. *Small* 2 (3): 310–312. https://doi.org/10.1002/smll.200500464.

97 Fologea, D., Gershow, M., Ledden, B. et al. (2005). Detecting single stranded DNA with a solid state nanopore. *Nano Letters* 5 (10): 1905–1909. https://doi.org/10.1021/nl051199m.

98 Jenkins, J., Sengupta, D., and Sundaram, S. (2005). Characterization of a solid state DNA nanopore sequencer using multi-scale (nano-to-device) modeling. *Lecture Notes on Computer Science* 3516: 309–316.

99 Gu, L.Q., Cheley, S., and Bayley, H. (2001). Capture of a single molecule in a nanocavity. *Science* 291 (5504): 636–640. https://doi.org/10.1126/science.291.5504.636.

100 Astier, Y., Braha, O., and Bayley, H. (2006). Toward single molecule DNA sequencing: direct identification of ribonucleoside and deoxyribonucleoside 5′-monophosphates by using an engineered protein nanopore equipped with a molecular adapter. *Journal of the American Chemical Society* 128 (5): 1705–1710. https://doi.org/10.1021/ja057123+.

101 Wu, H.C., Astier, Y., Maglia, G. et al. (2007). Protein nanopores with covalently attached molecular adapters. *Journal of the American Chemical Society* 129 (51): 16142–16148. https://doi.org/10.1021/ja0761840.

102 Vercoutere, W., Winters-Hilt, S., Olsen, H. et al. (2001). Rapid discrimination among individual DNA hairpin molecules at single-nucleotide resolution using an ion channel. *Nature Biotechnology* 19 (3): 248–252. https://doi.org/10.1038/85696.

103 Ashkenasy, N., Sánchez-Quesada, J., Bayley, H., and Ghadiri, M.R. (2005). Recognizing a single base in an individual DNA strand: a step toward DNA sequencing in nanopores. *Angewandte Chemie. International Edition* 44 (9): 1401–1404. https://doi.org/10.1002/anie.200462114.

104 Stoddart, D., Heron, A.J., Mikhailova, E. et al. (2009). Single-nucleotide discrimination in immobilized DNA oligonucleotides with a biological nanopore. *P Natl Acad Sci USA* 106 (19): 7702–7707. https://doi.org/10.1073/pnas.0901054106.

105 Stoddart, D., Heron, A.J., Klingelhoefer, J. et al. (2010). Nucleobase recognition in ssDNA at the Central constriction of the α-hemolysin pore. *Nano Letters* 10 (9): 3633–3637. https://doi.org/10.1021/nl101955a.

106 Cherf, G., Lieberman, K., Rashid, H. et al. (2012). Automated forward and reverse ratcheting of DNA in a nanopore at 5-Å precision. *Nat. Biotechnol.* 30: 344–348. https://doi.org/10.1038/nbt.2147.

107 Lieberman, K.R., Cherf, G.M., Doody, M.J. et al. (2010). Processive replication of single DNA molecules in a nanopore catalyzed by phi29 DNA polymerase. *Journal of the American Chemical Society* 132 (50): 17961–17972. https://doi.org/10.1021/ja1087612.

108 Cherf, G.M., Lieberman, K.R., Rashid, H. et al. (2012). Automated forward and reverse ratcheting of DNA in a nanopore at 5-angstrom precision. *Nature Biotechnology* 30 (4): 344–348. https://doi.org/10.1038/nbt.2147.

109 Iizuka, R., Yamazaki, H., and Uemura, S. (2022). Zero-mode103 waveguides and nanopore-based sequencing technologies accelerate single-molecule studies. *Biophysics and Physicobiology* 19: DOI: ARTN e190032 10.2142/biophysico.bppb-v19.0032.

110 Magierowski, S., Huang, Y., Wang, C., and Ghafar-Zadeh, E. (2016). Nanopore-CMOS interfaces for DNA sequencing. *Biosensors* 6 (3): 42.

111 Manrao, E.A., Derrington, I.M., Laszlo, A.H. et al. (2012). Reading DNA at single-nucleotide resolution with a mutant MspA nanopore and phi29 DNA polymerase. *Nature Biotechnology* 30 (4): 349–U174. https://doi.org/10.1038/nbt.2171.

112 Gyarfas, B., Olasagasti, F., Benner, S. et al. (2009). Mapping the position of DNA polymerase-bound DNA templates in a nanopore at 5 angstrom resolution. *ACS Nano* 3 (6): 1457–1466. https://doi.org/10.1021/nn900303g.

113 Hurt, N., Wang, H.Y., Akeson, M., and Lieberman, K.R. (2009). Specific nucleotide binding and rebinding to individual DNA polymerase complexes captured on a nanopore. *Journal of the American Chemical Society* 131 (10): 3772–3778. https://doi.org/10.1021/ja809663f.

114 Wilson, N.A., Abu-Shumays, R., Gyarfas, B. et al. (2009). Electronic control of DNA polymerase binding and unbinding to single DNA molecules. *ACS Nano* 3 (4): 995–1003. https://doi.org/10.1021/nn9000897.

115 Purnell, R.F. and Schmidt, J.J. (2009). Discrimination of single base substitutions in a DNA strand immobilized in a biological nanopore. *ACS Nano* 3 (9): 2533–2538. https://doi.org/10.1021/nn900441x.

116 Nomidis, S.K., Hooyberghs, J., Maglia, G., and Carlon, E. (2018). DNA capture into the ClyA nanopore: diffusion-limited versus reaction-limited processes. *Journal of Physics. Condensed Matter* 30 (30): DOI: ARTN 304001 10.1088/1361-648X/aacc01.

117 Niedzwiecki, D.J., Mohammad, M.M., and Movileanu, L. (2012). Inspection of the engineered FhuA ΔC/Δ4L protein nanopore by polymer exclusion. *Biophysical Journal* 103 (10): 2115–2124. https://doi.org/10.1016/j.bpj.2012.10.008.

118 Wendell, D., Jing, P., Geng, J. et al. (2009). Translocation of double-stranded DNA through membrane-adapted phi29 motor protein nanopores. *Nature Nanotechnology* 4 (11): 765–772. https://doi.org/10.1038/Nnano.2009.259.

119 Wang, H.Y., Li, Y., Qin, L.X. et al. (2013). Single-molecule DNA detection using a novel SP1 protein nanopore. *Chemical Communications* 49 (17): 1741–1743. https://doi.org/10.1039/c3cc38939a.

120 Thomsen, R.P., Malle, M.G., Okholm, A.H. et al. (2019). A large size-selective DNA nanopore with sensing applications. *Nature Communications* 10. DOI: ARTN 5655 10.1038/s41467-019-13284-1.

121 Burns, J.R., Al-Juffali, N., Janes, S.M., and Howorka, S. (2014). Membrane-spanning DNA nanopores with cytotoxic effect. *Angewandte Chemie. International Edition* 53 (46): 12466–12470. https://doi.org/10.1002/anie.201405719.

122 Howorka, S. (2016). Changing of the guard. *Science* 352 (6288): 890–891. https://doi.org/10.1126/science.aaf5154.

123 Jou, I.A., Melnikov, D.V., McKinney, C.R., and Gracheva, M.E. (2012). DNA translocation through a nanopore in a single-layered doped semiconductor membrane. *Physical Review E* 86 (6): DOI: ARTN 061906 10.1103/PhysRevE.86.061906.

124 Haque, F., Li, J.H., Wu, H.C. et al. (2013). Solid-state and biological nanopore for real-time sensing of single chemical and sequencing of DNA. *Nano Today* 8 (1): 56–74. https://doi.org/10.1016/j.nantod.2012.12.008.

125 Shekar, S., Niedzwiecki, D.J., Chien, C.C. et al. (2016). Measurement of DNA translocation dynamics in a solid-state nanopore at 100 ns temporal resolution. *Nano Letters* 16 (7): 4483–4489. https://doi.org/10.1021/acs.nanolett.6b01661.

126 Fologea, D., Uplinger, J., Thomas, B. et al. (2005). Slowing DNA translocation in a solid-state nanopore. *Nano Letters* 5 (9): 1734–1737. https://doi.org/10.1021/nl051063o.

127 Ghosal, S. (2007). Effect of salt concentration on the electrophoretic speed of a polyelectrolyte through a nanopore. *Physical Review Letters* 98 (23): DOI: ARTN 238104 10.1103/PhysRevLett.98.238104.

128 Kowalczyk, S.W., Wells, D.B., Aksimentiev, A., and Dekker, C. (2012). Slowing down DNA translocation through a nanopore in lithium chloride. *Nano Letters* 12 (2): 1038–1044. https://doi.org/10.1021/nl204273h.

129 Plesa, C., van Loo, N., and Dekker, C. (2015). DNA nanopore translocation in glutamate solutions. *Nanoscale* 7 (32): 13605–13609. https://doi.org/10.1039/c5nr02793d.

130 Goto, Y., Yanagi, I., Matsui, K. et al. (2016). Integrated solid-state nanopore platform for nanopore fabrication via dielectric breakdown, DNA-speed deceleration and noise reduction. *Sci Rep-Uk* 6: DOI: ARTN 31324 10.1038/srep31324.

131 Soni, N., Verma, N.C., Talor, N., and Meller, A. (2023). Over 30-fold enhancement in DNA translocation dynamics through nanoscale pores coated with an anionic surfactant. *Nano Letters* 23 (10): 4609–4616. https://doi.org/10.1021/acs.nanolett.3c01096.

132 Zhang, Y., Wu, G.S., Ma, J. et al. (2015). Temperature effect on translocation speed and capture rate of nanopore-based DNA detection. *Science China Technological Sciences* 58 (3): 519–525. https://doi.org/10.1007/s11431-014-5674-2.

133 Plesa, C., van Loo, N., Ketterer, P. et al. (2015). Velocity of DNA during Trans location through a Solid-State Nanopore. *Nano Letters* 15 (1): 732–737. https://doi.org/10.1021/nl504375c.

134 Feng, J.D., Liu, K., Bulushev, R.D. et al. (2015). Identification of single nucleotides in MoS nanopores. *Nature Nanotechnology* 10 (12): 1070–+. https://doi.org/10.1038/Nnano.2015.219.

135 He, Y.H., Tsutsui, M., Fan, C. et al. (2011). Controlling DNA translocation through gate modulation of nanopore wall surface charges. *ACS Nano* 5 (7): 5509–5518. https://doi.org/10.1021/nn201883b.

136 Graf, M., Lihter, M., Altus, D. et al. (2019). Transverse detection of DNA using a MoS_2 nanopore. *Nano Letters* 19 (12): 9075–9083. https://doi.org/10.1021/acs.nanolett.9b04180.

137 Thakur, M., Macha, M., Chernev, A. et al. (2020). Wafer-scale fabrication of nanopore devices for single-molecule DNA biosensing using MoS_2. *Small Methods* 4 (11): DOI: ARTN 2000072 10.1002/smtd.202000072.

138 Chakraborty, R., Xiong, M.Y., Athreya, N. et al. (2023). Solid-state MoS_2 nanopore membranes for discriminating among the lengths of RNA tails on a double-stranded DNA: a new simulation-based differentiating algorithm. *ACS Applied Nano Materials* 6 (6): 4651–4660. https://doi.org/10.1021/acsanm.3c00129.

139 Wang, F., Zhao, C., Zhao, P. et al. (2023). MoS_2 nanopore identifies single amino acids with sub-1 Dalton resolution. *Nature Communications* 14 (1): https://doi.org/10.1038/s41467-023-38627-x.

140 Diederichs, T., Pugh, G., Dorey, A. et al. (2019). Synthetic protein-conductive membrane nanopores built with DNA. *Nature Communications* 10 (1): https://doi.org/10.1038/s41467-019-12639-y.

141 Xing, Y., Dorey, A., Jayasinghe, L., and Howorka, S. (2022). Highly shape- and size-tunable membrane nanopores made with DNA. *Nature Nanotechnology* 17 (7): 708–713. https://doi.org/10.1038/s41565-022-01116-1.

142 Yadav, P., Cao, Z.L., and Farimani, A.B. (2021). DNA detection with single-layer Ti3C2 MXene nanopore. *ACS Nano* 15 (3): 4861–4869.

143 Traversi, F., Raillon, C., Benameur, S.M. et al. (2013). Detecting the translocation of DNA through a nanopore using graphene nanoribbons. *Nature Nanotechnology* 8 (12): 939–945. https://doi.org/10.1038/Nnano.2013.240.

144 Mojtabavi, M., VahidMohammadi, A., Liang, W.T. et al. (2019). Single-molecule sensing using nanopores in two-dimensional transition metal carbide (MXene) membranes. *ACS Nano* 13 (3): 3042–3053. https://doi.org/10.1021/acsnano.8b08017.

145 Schneider, G.F., Xu, Q., Hage, S. et al. (2013). Tailoring the hydrophobicity of graphene for its use as nanopores for DNA translocation. *Nature Communications* 4: DOI: ARTN 2619 10.1038/ncomms3619.

146 Ashton, P.M., Nair, S., Dallman, T. et al. (2015). MinION nanopore sequencing identifies the position and structure of a bacterial antibiotic resistance island. *Nature Biotechnology* 33 (3): 296–300. https://doi.org/10.1038/nbt.3103 From NLM Medline.

147 Diensthuber, G., Pryszcz, L.P., Llovera, L. et al. (2024). Enhanced detection of RNA modifications and read mapping with high-accuracy nanopore RNA basecalling models. *Genome Research*. https://doi.org/10.1101/gr.278849.123.

148 Liu-Wei, W., van der Toorn, W., Bohn, P. et al. (2024). Sequencing accuracy and systematic errors of nanopore direct RNA sequencing. *BMC Genomics* 25 (1): 528. https://doi.org/10.1186/s12864-024-10440-w.

149 Gao, R.X., Liu, X.X., Xiong, Z.Q. et al. (2024). Research progress on detection of foodborne pathogens: the more rapid and accurate answer to food safety. *Food Research International* 193: DOI: ARTN 114767 10.1016/j.foodres.2024.114767.

150 Wang, Y., Zhao, Y., Bollas, A. et al. (2021). Nanopore sequencing technology, bioinformatics and applications. *Nature Biotechnology* 39 (11): 1348–1365. https://doi.org/10.1038/s41587-021-01108-x From NLM Medline.

151 Silvestre-Ryan, J. and Holmes, I. (2021). Pair consensus decoding improves accuracy of neural network basecallers for nanopore sequencing. *Genome Biology* 22 (1): ARTN 38 10.1186/s13059-020-02255-1.

152 Lewandowski, K., Xu, Y.F., Pullan, S.T. et al. (2020). Metagenomic nanopore sequencing of influenza virus direct from clinical respiratory samples. *Journal of Clinical Microbiology* 58 (1): DOI: ARTN e00963-19 10.,1128/JCM.00963-19.

153 Moldován, N., Tombácz, D., Szücs, A. et al. (2018). Multi-platform sequencing approach reveals a novel transcriptome profile in pseudorabies virus. *Frontiers in Microbiology* 8: DOI: ARTN 2708 10.3389/fmicb.2017.02708.

154 Grünberger, F., Ferreira-Cerca, S., and Grohmann, D. (2022). Nanopore sequencing of RNA and cDNA molecules in. *RNA* 28 (3): 400–417. https://doi.org/10.1261/rna.078937.121.

155 Fleischmann, C., Scherag, A., Adhikari, N.K.J. et al. (2015). Assessment of global incidence and mortality of hospital-treated sepsis. current estimates and limitations. *American Journal of Respiratory and Critical Care Medicine* 193 (3): https://doi.org/10.1164/rccm.201504-0781OC.

156 Reinhart, K., Daniels, R., Kissoon, N. et al. (2017). Recognizing sepsis as a Global Health Priority – a WHO resolution. *New Engl J Med* 377 (5): 414–417. https://doi.org/10.1056/NEJMp1707170.

157 Goto, M. and Al-Hasan, M.N. (2013). Overall burden of bloodstream infection and nosocomial bloodstream infection in North America and Europe. *Clinical Microbiology and Infection* 19 (6): 501–509. https://doi.org/10.1111/1469-0691.12195.

158 Cassini, A., Högberg, L.D., Plachouras, D. et al. (2019). Attributable deaths and disability-adjusted life-years caused by infections with antibiotic-resistant bacteria in the EU and the European Economic Area in 2015: a population-level modelling analysis. *The Lancet Infectious Diseases* 19 (1): 56–66. https://doi.org/10.1016/S1473-3099(18)30605-4.

159 Buehler, S.S., Madison, B., Snyder, S.R. et al. (2016). Effectiveness of practices to increase timeliness of providing targeted therapy for inpatients with bloodstream infections: a laboratory medicine best practices systematic review and meta-analysis. *Clinical Microbiology Reviews* 29 (1): 59–103. https://doi.org/10.1128/cmr.00053-14.

160 Ashikawa, S., Tarumoto, N., Imai, K. et al. (2018). Rapid identification of pathogens from positive blood culture bottles with the MinION nanopore sequencer. *Journal of Medical Microbiology* 67 (11): 1589–1595. https://doi.org/10.1099/jmm.0.000855.

161 Sakai, J., Tarumoto, N., Kodana, M. et al. (2019). An identification protocol for ESBL-producing Gram-negative bacteria bloodstream infections using a MinION nanopore sequencer. *Journal of Medical Microbiology* 68 (8): 1219–1226. https://doi.org/10.1099/jmm.0.001024.

162 Garalde, D.R., Snell, E.A., Jachimowicz, D. et al. (2018). Highly parallel direct RNA sequencing on an array of nanopores. *Nature Methods* 15 (3): 201–+. https://doi.org/10.1038/Nmeth.4577.

163 Jain, M., Koren, S., Miga, K.H. et al. (2018). Nanopore sequencing and assembly of a human genome with ultra-long reads. *Nature Biotechnology* 36 (4): 338–+. https://doi.org/10.1038/nbt.4060.

164 Edwards, A., Soares, A., Rassner, S.M.E. et al. (2017). Deep Sequencing: Intra-terrestrial metagenomics illustrates the potential of off-grid Nanopore DNA sequencing. *bioRxiv* 133413. https://doi.org/10.1101/133413.

165 Johnson, S.S., Zaikova, E., Goerlitz, D.S. et al. (2017). Real-time DNA sequencing in the Antarctic Dry Valleys using the Oxford Nanopore Sequencer. *Journal of Biomolecular Techniques* 28 (1): 2–7. https://doi.org/10.7171/jbt.17-2801-009 From NLM.

166 Moon, J., Kim, N., Kim, T.J. et al. (2019). Rapid diagnosis of bacterial meningitis by nanopore 16S amplicon sequencing: a pilot study. *International Journal of Medical Microbiology* 309 (6): DOI: ARTN 151338 10.1016/j.ijmm.2019.151338.

167 Cheng, J., Hu, H., Kang, Y. et al. (2018). Identification of pathogens in culture-negative infective endocarditis cases by metagenomic analysis. *Annals of Clinical Microbiology and Antimicrobials* 17. DOI: ARTN 43 10.1186/s12941-018-0294-5.

168 Gorrie, C.L., Mirceta, M., Wick, R.R. et al. (2018). Antimicrobial-resistant carriage and infection in specialized geriatric care wards linked to acquisition in the referring hospital. *Clinical Infectious Diseases* 67 (2): 161–170. https://doi.org/10.1093/cid/ciy027.

169 Sanderson, N.D., Street, T.L., Foster, D. et al. (2018). Real-time analysis of nanopore-based metagenomic sequencing from infected orthopaedic devices. *BMC Genomics* 19: DOI: ARTN 714 10.1186/s12864-018-5094-y.

170 Taxt, A.M., Avershina, E., Frye, S.A. et al. (2020). Rapid identification of pathogens, antibiotic resistance genes and plasmids in blood cultures by nanopore sequencing. *Sci Rep-Uk* 10 (1): DOI: ARTN 7622 10.1038/s41598-020-64616-x.

171 Deng, Q.M., Cao, Y.Q., Wan, X.F. et al. (2022). Nanopore-based metagenomic sequencing for the rapid and precise detection of pathogens among immunocompromised cancer patients with suspected infections. *Front Cell Infect Mi* 12. DOI: ARTN 943859 10.3389/fcimb.2022.943859.

172 Schmidt, K., Mwaigwisya, S., Crossman, L.C. et al. (2017). Identification of bacterial pathogens and antimicrobial resistance directly from clinical urines by nanopore-based metagenomic sequencing. *The Journal of Antimicrobial Chemotherapy* 72 (1): 104–114. https://doi.org/10.1093/jac/dkw397.

173 Lu, X., Zeng, M., Xu, J.L. et al. (2019). Epidemiologic and genomic insights on-harbouring from diarrhoeal outpatients in Shanghai, China, 2006-2016. *eBioMedicine* 42: 133–144. https://doi.org/10.1016/j.ebiom.2019.03.006.

174 Hu, Y., Fang, L., Nicholson, C., and Wang, K. (2020). Implications of error-prone long-read whole-genome shotgun sequencing on characterizing reference microbiomes. *Iscience* 23 (6): DOI: ARTN 101223 10.1016/j.isci.2020.101223.

175 Nicholls, S.M., Quick, J.C., Tang, S.Q., and Loman, N.J. (2019). Ultra-deep, long-read nanopore sequencing of mock microbial community standards. *GigaScience* 8 (5): DOI: ARTN giz043 10.1093/gigascience/giz043.

6

The Art of Molecular Diagnosis

Techniques and Implications for Public Health

Lu Gao[1,#], Zhiyin Wang[1,#], Pengkun Yin[1,#], Sitong Lv[1,#], Ye Peng[1,#], Yao Liu[1,#], and Feng Li[1,2,3]

[1] *Sichuan University, College of Chemistry, Key Laboratory of Green Chemistry and Technology of Ministry of Education, Chengdu, Sichuan, 610064, China*
[2] *Brock University, Centre for Biotechnology, Department of Chemistry, St. Catharines, Ontario L2S 3A1, Canada*
[3] *Sichuan University, West China Hospital, Med+X Center for Manufacturing, Department of Laboratory Medicine, Chengdu, Sichuan 610041, China*

CHAPTER MENU

6.1 Introduction

Public health relies on timely and accurate diagnostic information to manage health risks, prevent disease outbreaks, and implement evidence-based interventions. With its remarkable sensitivity and specificity, as well as rapid turnaround time, molecular diagnostics has evolved as one of the most widely used tests in public health, playing critical roles in disease prevention, epidemiological surveillance, genetic screening, and personalized medicine. The COVID-19 pandemic, caused by SARS-CoV-2, highlighted the critical role of molecular diagnostics in rapidly identifying and containing public health threats [1, 2]. This same technological foundation has also been extended to other areas of public safety, such as prenatal screening [3]. Noninvasive molecular diagnostics, such as cell-free fetal DNA testing [4], enable early detection of genetic abnormalities, reducing the risk of congenital diseases and promoting safer prenatal care through timely intervention and

# These authors contributed equally to this work.

Diagnostic Techniques in Public Health: Analytical and Chemometric Approaches in Disease Detection and Management, First Edition. Edited by Jinghong Li and Yan Du.

informed decision-making. Facing the ongoing need for test accuracy, accessibility, and affordability, analytical and chemometric approaches have played increasing roles in equipping conventional molecular diagnostics with better nucleic acid amplification routes [5–9], signal readout strategies [10–12], microfluidic devices [13–15], and miniaturized detection systems [16–19].

Molecular diagnostics harnesses the principle of molecular biology to identify and detect primarily nucleic acid biomarkers in varying clinical and biological samples (Figure 6.1). As all living organisms can be precisely defined by their genetic code, molecular diagnostics is one of the most confirmatory techniques for the identification of infectious pathogens. It also allows the accurate profiling of small changes in genetic materials, such as single-nucleotide polymorphisms (SNPs), insertions, deletions, and gene fusions, which are directly related to many devastating diseases, such as genetic disorders, drug-resistant infections, and cancer. Quantitative nucleic acid detection techniques have also been employed in molecular diagnostics for the quantification of levels of disease-related gene expression (Figure 6.1). As early as 1987 [20], the advent of polymerase chain

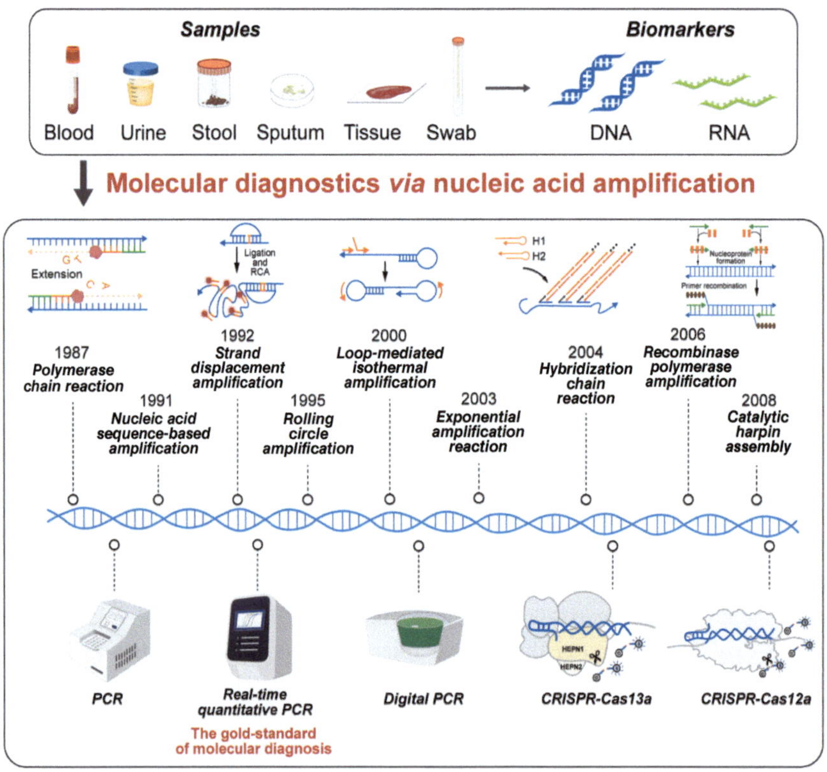

Figure 6.1 Schematic representation of the development of molecular diagnostics via nucleic acid amplification. Clinical translation-driven nucleic acid amplification methods for nucleic acid biomarker detection in clinical samples are summarized. *Source:* Figure created by the authors.

reaction (PCR) enabled exponential amplification of DNA in vitro, revolutionizing molecular diagnostics for the detection of low-abundance and specific biomarkers. As the field progressed, quantitative PCR (qPCR) emerged in 1993 [21], combining real-time quantification with amplification to facilitate semiquantitative analysis of viral load and gene expression levels, thus promoting the development of molecular diagnostics from qualitative to quantitative. This transition has established qPCR as the gold standard in most molecular diagnostic applications. To achieve absolute quantification, digital PCR (dPCR) leverages partitioning of the reaction into thousands of discrete droplets, allowing nucleic acid quantification based on the principles of Poisson distribution [22]. This approach significantly enhances sensitivity, making dPCR particularly valuable for detecting rare mutations and low-abundance biomarkers in oncology and prenatal screening. However, the dependence of PCR on precision thermal cycler temperature control instruments highlights the need for simpler and faster methods, which has promoted the development of isothermal amplification technologies. As the first isothermal amplification technique, nucleic acid sequence-based amplification (NASBA) was developed in 1991, providing a continuous, isothermal process for the amplification of single-stranded RNA [9]. NASBA can achieve amplification efficiencies comparable to PCR at 41 °C (10^9-fold amplification in 1.5–2 hours). Following NASBA, significant advancements were made in the development of isothermal amplification methods. In 1992, strand displacement amplification (SDA) was developed by Walker, making it possible to amplify double-stranded DNA at 37 °C [8, 23]. Subsequently, rolling circle amplification (RCA) was developed in 1995 by utilizing a circular template (called a padlock probe) to synthesize long, repetitive single-stranded products, and has been widely used in molecular diagnostics, nanobiomedicine, nanotechnology, and other fields [5]. To improve sequence specificity and simplify the amplification system, loop-mediated isothermal amplification (LAMP) was first proposed by Notomi and co-workers in 2000 [6]. LAMP employs a set of four to six primers that specifically target different regions of the nucleic acid sequence, resulting in a high level of specificity and amplification efficiency in a single-step reaction. The high specificity and efficiency of LAMP has led to exceptional utility in pathogen detection and SNP identification. For example, the LMAP-based Lucira COVID-19 All-In-One Test Kit has become the first Food and Drug Administration (FDA)-approved test kit for home testing of COVID-19. Another critical advancement in isothermal technology came with the development of recombinase polymerase amplification (RPA) [7]. RPA utilizes recombinase to initiate strand invasion and primer binding at a constant temperature (37–42 °C), offering rapid amplification within 10–20 minutes. Its robustness across diverse environmental conditions makes it ideal for point-of-care diagnostics. The method's speed, flexibility, and compatibility with various detection platforms establish RPA as a valuable tool in molecular diagnostics, particularly in resource-limited settings. These isothermal amplification technologies play a key role in signal amplification for molecular diagnostics. Moreover, emerging approaches, including DNA nanotechnology [24] and CRISPR/Cas systems [25], are also essential prerequisites for improving the research of molecular diagnosis with high sensitivity, specificity, and selectivity.

In this chapter, we will focus on the development of molecular diagnostic techniques, where the advancement of analytical approaches serves as one of the main drivers. By reviewing these advancements, we aim to provide a comprehensive understanding of the key milestones and future directions in the field of molecular diagnosis. This chapter also delves into the development of molecular diagnostics and their applications in public safety, with a particular focus on how analytical and chemometric approaches enhance the accuracy and robustness of these methods. By incorporating metrological standards, the chapter aligns with the overarching theme of the book, demonstrating the critical role of analytical techniques in improving disease detection and management. It underscores the significance of metrology in optimizing diagnostic tools and ensuring their reliability across various clinical and public health contexts, as explored through the analytical and chemometric frameworks presented in this book.

6.2 Criteria for Developing Molecular Diagnostic Technologies

Molecular diagnosis typically requires an extended period for verification, spanning from the initial development of methodologies to clinical sample testing and, ultimately, clinical translation. As such, the development of high-performance molecular diagnostic methods is critical for ensuring public safety. To address the clinical challenges posed by public health emergencies, an ideal molecular diagnostic approach should adhere to the following key principles:

6.2.1 *Robustness and Reliability*

The analytical performance of diagnostic methods is often influenced by clinical sample matrix interference and variability in the detection scenarios. Consequently, it is imperative that these methods demonstrate robust reliability across a range of conditions, including different operators, instrumentation, and sample matrices. To achieve this, stringent validation and standardization protocols are essential, ensuring the reproducibility and consistency of results across diverse clinical and laboratory settings.

6.2.2 *Specificity*

High specificity is essential to minimize false positives, particularly in infectious disease diagnostics. Improving probe-target recognition can significantly reduce misdiagnosis rates. Therefore, molecular diagnostic methods must optimize probe specificity to ensure accurate and reliable results.

6.2.3 *Sensitivity*

Highly sensitive molecular diagnostic technologies are capable of rapidly detecting low-abundance disease markers at early stages, facilitating timely clinical decision-making. Achieving this requires highly efficient amplification techniques that can rapidly and accurately amplify target nucleic acids from complex biological samples within a short time frame.

6.2.4 *Speed*

Rapid turnaround times are vital for effective clinical decision-making. Molecular diagnostics should be designed to deliver results quickly, ideally within a few hours, to facilitate timely interventions. Innovations in sample preparation and automated workflows can contribute to reduced processing times.

The ideal molecular diagnostic technology must balance these performance criteria to achieve optimal diagnostic accuracy, efficiency, and usability. Continuous advancements in technology and methodology will further enhance these standards, ultimately improving patient outcomes in clinical practice.

6.3 Amplifying the Invisible: PCR's Pivotal Role in Genetic Detection

The detection and analysis of nucleic acids within biological systems are often hindered by the naturally low abundance of these molecules in clinical samples, such as blood, urine, tissue, swabs, and so on. Whether diagnosing infectious diseases, identifying genetic mutations, or monitoring gene expression, the limited quantity of nucleic acid available for analysis presents a significant challenge. To overcome this, amplification of these nucleic acids is not only necessary but critical to achieving accurate and reliable clinical results. The concept of amplifying genetic material is inspired by natural biological processes, where nucleic acids are replicated and transcribed with high fidelity within living organisms. This natural phenomenon provided the basis for the development of the PCR, a technique that mimics the in vivo DNA replication process, enabling the exponential amplification of specific DNA sequences in vitro. Since its introduction by Kary Mullis in 1983, PCR has become a milestone in the history of molecular biology, with applications ranging from biological and medical sciences to clinical diagnostics. The ability to specifically amplify trace amounts of nucleic acids has transformed various fields, particularly in public health.

6.3.1 Overview of PCR

The classic PCR technique involves three cyclic steps that result in the exponential amplification of the target DNA (Figure 6.2a). **Denaturation**: The double-stranded

DNA has to be heated to around 95 °C, causing the hydrogen bonds to break and resulting in the separation of the complementary DNA (cDNA) into two single strands. **Annealing**: The temperature is then lowered to 50–65 °C, allowing the primers to bind (anneal) to their complementary sequences on the single-stranded DNA. This step is critical for the specificity of the PCR, as the primers define the region of DNA to be amplified. **Extension**: The temperature is raised to approximately 72 °C, the optimal temperature for DNA polymerase activity. During this step, the DNA polymerase synthesizes a new strand of DNA by adding nucleotides to the primer in a sequence-specific manner, extending the DNA strand complementary to the template. Typically, the time for this extension step depends on the DNA polymerase and the length of the DNA fragment to be amplified, generally 1 kb per minute. Through the cyclic process of the above three key steps, PCR can exponentially amplify specific DNA. Each cycle effectively doubles the amount of target DNA, resulting in a geometric increase in the number of amplicons. Therefore, even starting from a trace amount of target, PCR can produce millions to billions of copies of the target DNA, which is also an important basis for the widespread application of PCR in the field of molecular diagnosis.

6.3.2 Types of PCR

However, with the expansion of diagnostic targets and scenarios, the limitations of the original technique became apparent, particularly when dealing with complex samples or requiring high sensitivity and specificity. With the continuous demand for more sensitive, accurate, and rapid diagnostic methods, numerous PCR-based variant techniques have emerged, each tailored to address specific needs.

6.3.2.1 *Reverse Transcription PCR*

Reverse transcription PCR (RT-PCR) is another powerful technique that combines reverse transcription of RNA into cDNA with subsequent amplification using PCR (Figure 6.2c) [26]. Traditional PCR is limited to amplifying DNA, which excludes a significant portion of the genetic material in cells, particularly in cases where the focus is on gene expression or RNA viruses. Driven by the need to analyze and quantify RNA, the RT-PCR begins with the reverse transcription of RNA by reverse transcriptase, producing cDNA that serves as a template for PCR amplification. This dual capability makes RT-PCR especially crucial in diagnosing viral infections caused by RNA viruses, such as HIV, hepatitis C, and influenza, studying gene expression patterns, and understanding the role of messenger RNA in protein synthesis.

6.3.2.2 *Asymmetric PCR*

Asymmetric PCR is a variant of the traditional PCR technique designed to preferentially amplify one strand of the DNA target, generating single-stranded DNA (ssDNA) rather than the typical double-stranded product (Figure 6.2b) [27]. By using unequal molar amounts of the two primers, where one primer is in excess, excess single-stranded selected strands are generated for direct sequencing or identification of hybridization probes. In fact, exclusively dsDNA will be produced during the

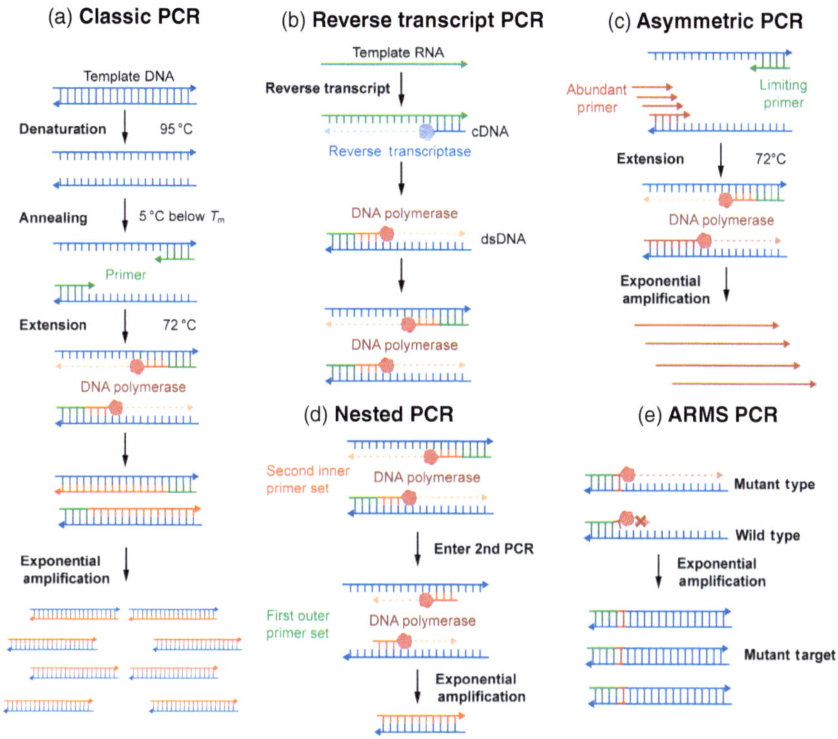

Figure 6.2 Overview of representative PCR reactions. (a) Standard PCR reaction showing the basic amplification cycle of denaturation, annealing, and extension. (b) Reverse transcription PCR (RT-PCR reaction, where template RNA is first reverse-transcribed to cDNA before amplification. (c) Asymmetric PCR reaction. By using an unequal concentration of forward and reverse primers, the reaction produces single-stranded DNA. (d) Nested PCR reaction, using two rounds of amplification to improve specificity. (e) ARMS PCR reaction, designed for detecting specific mutations or alleles using allele-specific primers. *Source:* Figure created by the authors.

first 10–15 cycles. However, when the limiting number of added primers has been used up, an excess of ssDNA amplicon will be produced in each cycle. Theoretically, the amount of dsDNA should increase exponentially, while the production of ssDNA should only follow a linear increase. Therefore, it is important to allow dsDNA production to reach a certain level before starting ssDNA production.

In disease diagnostics, asymmetric PCR is highly valued for its ability to create ssDNA probes for hybridization assays, which are crucial for detecting specific mutations or pathogens. For instance, it has been effectively employed in the detection of mutations associated with genetic disorders, such as cystic fibrosis, where allele-specific oligonucleotide (ASO) probes generated by asymmetric PCR enhance the precision of mutation identification [28]. Additionally, asymmetric PCR is used in viral diagnostics to create probes that can sensitively and specifically identify viral RNA sequences, contributing to early and accurate disease detection.

6.3.2.3 *Nested PCR*

Nested PCR is an advanced variation of the traditional PCR technique, developed to enhance specificity and sensitivity in detecting target DNA sequences [29]. The process begins with the standard PCR steps: denaturation, annealing, and extension. However, the challenge of nonspecific amplification—especially in complex samples—prompted the need for a more refined approach. Nested PCR addresses this by introducing a second round of amplification using a new set of primers that target a region within the first PCR product (Figure 6.2d). This two-step process significantly reduces the likelihood of amplifying nonspecific sequences, as the second primer set further narrows down the target region. The result is a highly specific amplification, even from low-abundance or degraded DNA samples. The enhanced specificity of Nested PCR makes it a critical tool in clinical diagnostics, particularly in challenging diagnostic scenarios. For instance, Bhagya developed a modified nested PCR (Mo-STNPCR) for detecting *Leishmania donovani*, achieving 100% sensitivity and specificity by optimizing primer combinations. This method enables timely diagnosis and control of leishmaniasis. Additionally, Mo-STNPCR has been successfully applied to various pathogens, including *Schistosoma mansoni*, *Plasmodium falciparum*, *Yersinia pestis*, dengue virus serotypes, *Vibrio cholerae* O1, *Leishmania chagasi*, Porcine Circovirus type 2, and *Mycobacterium tuberculosis*, demonstrating its broad diagnostic utility across infectious diseases [30].

6.3.2.4 *ARMS PCR*

Amplification Refractory Mutation System PCR (ARMS PCR) is a specialized variant of traditional PCR designed to detect specific mutations or single-nucleotide polymorphisms (SNPs) within a DNA sequence [31]. This technique enhances mutation detection by using allele-specific primers, where the 3′ end of the primer is specifically designed to hybridize with the mutant sequence. This selective hybridization during the annealing step ensures that amplification occurs only when the mutant allele is present, allowing for precise detection of genetic variations (Figure 6.2e). This selectivity enables the differentiation between closely related DNA sequences, such as a single-nucleotide change. Therefore, ARMS PCR is particularly valuable for genotyping and mutation analysis in genetic diseases, oncology, and pharmacogenetics [32, 33]. It is widely applied to detect mutations in the KRAS gene in colorectal cancer, where specific mutations inform targeted therapy decisions [34]. Additionally, ARMS PCR plays a critical role in identifying mutations linked to cystic fibrosis and other genetic disorders. Its high sensitivity and specificity for detecting genetic variants make it a powerful tool in personalized medicine, enabling accurate diagnoses and facilitating tailored treatment strategies.

6.3.3 Quantitative PCR

qPCR, also known as real-time PCR, marks a significant advancement in molecular biology by enabling the simultaneous amplification and real-time quantification of DNA. Developed in the early 1990s, qPCR emerged to meet the growing

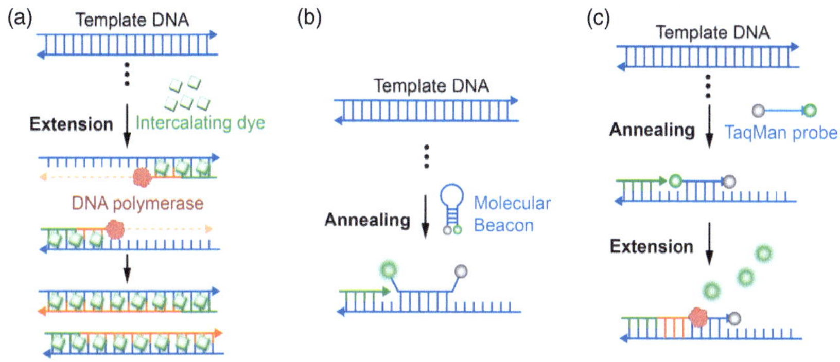

Figure 6.3 Overview of representative qPCR reactions. (a) Dye-based qPCR, using intercalating dyes such as SYBR Green I to detect template DNA for real-time monitoring. (b) Molecular beacon-based qPCR. In the absence of template DNA, the beacon forms a hairpin structure, quenching the fluorescence. Upon hybridization to the target sequence during qPCR, the beacon opens, separating the fluorophore and quencher, thus emitting fluorescence for real-time detection. (c) TaqMan probe-based qPCR relies on the exonuclease activity of DNA polymerase to cleave the TaqMan probe during the extension process to release a fluorescent signal. *Source:* Figure created by the authors.

demand for precise and dynamic nucleic acid measurements in clinical diagnostics [35]. Unlike traditional PCR, which quantifies DNA only at the end of amplification, qPCR allows for the real-time tracking of DNA levels at each cycle, providing continuous insight into the cumulative product formation throughout the amplification process. This capability offers greater sensitivity and accuracy in detecting target sequences. This breakthrough in qPCR technology is attributed to the use of fluorescent intercalating dyes, such as SYBR Green, and sequence-specific probes, including molecular beacons and TaqMan probes (Figure 6.3) [36]. These reagents enable real-time monitoring of DNA amplification by releasing fluorescent signals during each cycle, directly proportional to the amount of DNA being amplified. By tracking these signals, the initial DNA template concentration can be accurately quantified based on the cycle threshold (Ct) [37], which is defined as the cycle number at which fluorescence exceeds 10 times the background standard deviation.

6.3.3.1 Dye-Based qPCR

In detail, Higuchi developed the first demonstration of qPCR by including a common fluorescence dye called ethidium bromide (EtBr) in the extension step under ultraviolet light (Figure 6.3a) [21]. Subsequently, the study of dye–DNA interactions has led to the utilization of various DNA-specific dyes in qPCR reactions, including SYBR Green 1, BEBO, YOYO-1, TOTO-1, SYBR Gold, and EvaGreen. Notably, SYBR Green 1 is the predominant intercalating dye utilized in commercial dye-based qPCR. Upon binding to the minor groove of double-stranded DNA, SYBR Green 1 exhibits a fluorescent signal 1,000 times more intense than in its free state [38]. Despite its widespread use in public health applications such as infectious disease screening and early diagnosis, its nonspecific binding to nucleic acids limits its suitability for target-specific detection and multiplex diagnostics. This limitation has driven the development of more advanced, specific hybridization probe-based technologies, which offer greater specificity for accurate molecular diagnostics.

6.3.3.2 *Molecular Beacon-Based qPCR*

In probe-based chemistries, various short fluorescent probes have been successively developed as internal probes that specifically hybridize with the amplified DNA region, generating enhanced fluorescence signals, which provide superior specificity to SYBR Green 1. As a classic hairpin probe, the molecular beacon was first reported by Tyagi and Kramer (Figure 6.3b) and is usually composed of a hairpin-shaped oligonucleotide [39]. The structure consists of a loop region, typically 18–30 base pairs, complementary to the target sequence, and a stem region of 5–7 base pairs. The 5′ end of the probe is labeled with a fluorophore, while the 3′ end carries a quencher. In the absence of the target, the probe remains in a closed conformation, with the quencher suppressing fluorescence. Upon hybridization with the target sequence, the hairpin structure unfolds, separating the fluorophore and quencher, resulting in the emission of fluorescence. Due to its specific hybridization capability, molecular beacons are widely applied in various molecular diagnostic contexts, including single or multiplex pathogen detection, viral and bacterial load quantification, genotyping, and mutation detection [40, 41]. To improve the single-base resolution of molecular beacon probes, Guo et al. recently developed the entropy-compensated probe (Sprobe), which addresses the limitations of narrow temperature recognition ranges caused by the high entropy penalty in traditional molecular beacons [42]. By incorporating intrinsic disorder into the hybridization probe design, Sprobe expanded the temperature detection range (ΔT) from less than 10 °C to over 30 °C. This innovation has been successfully applied in detecting EGFR L858R mutations in clinical tumor samples from lung cancer patients, demonstrating 100% clinical sensitivity and specificity.

6.3.3.3 TaqMan Probe-Based qPCR

Another widely used hybridization probe is the TaqMan probe, first reported by Holland in 1991 [43]. It consists of a short oligonucleotide (typically 20–30 base pairs) labeled with a fluorophore at the 5′ end and a quencher at the 3′ end (Figure 6.3c). In the intact probe, the fluorophore and quencher are in proximity, preventing fluorescence. The reaction leverages the 5′–3′ exonuclease activity of Taq DNA polymerase. During the extension phase of PCR, the polymerase hydrolyzes the probe bound to the target sequence, separating the fluorophore and quencher, and releasing a fluorescent signal. This process occurs in every cycle, enabling real-time monitoring of target DNA amplification. To further improve the binding specificity and sensitivity of target DNA, subsequent research teams incorporated minor groove binding (MGB) ligands at the 3′ or 5′ ends of TaqMan probes, leading to the development of MGB-TaqMan probes [44]. Examples of such ligands include dihydrocyclopyrroloindole tripeptide (DIP) and 1,2-dihydro-(3H)-pyrrolo[3.2-e]indole-7-carboxylate (CDPI). By increasing the melting temperature (T_m) of the probe-target binding, these modifications helped reduce nonspecific amplification. Additionally, background fluorescence was minimized through the introduction of nonfluorescent quenchers (NFQ) or dark quenchers.

In addition to internal probes, numerous primer-based hybridization probes are also employed in qPCR reactions. For example, the Scorpion primer probe, developed by Whitcombe in 1999 [45], blocks polymerase extension of the primer by connecting the 3′ end of a hairpin structure—containing a reporter group at the 5′ end and a quencher group at the 3′ end—to the 5′ end of the primer via a hexaethylene glycol (HEG) blocking agent. The hairpin structure opens and releases a fluorescent signal only in the presence of the target sequence. This primer-specific binding mechanism effectively prevents the formation of primer dimers and nonspecific amplification, making it particularly useful in mutation detection assays. In 2017, Wu et al. also developed a temperature-stable PCR method called blocker displacement amplification (BDA) [46]. Through the rare allele enrichment method, it allows accurate detection and quantification of single nucleotide variants (SNVs) and indices as low as 0.01% variant allele frequency (VAF) in a highly multiplexed environment and has been successfully used for the detection of rare driver mutations in cfDNA in the plasma of lung cancer patients with a clinical sensitivity of 95%.

6.3.4 Digital PCR

Unlike qPCR, which requires standard curves for quantification, dPCR is an absolute quantification method. The principle of dPCR can be divided into three parts (Figure 6.4): **Partitioning**—where the sample containing the nucleic acid of interest is mixed with various reagents needed for amplification and then divided into many individual reaction units using different partitioning techniques. These techniques include microfluidic chips, droplet generation, and microplate technologies, with each reaction unit containing only a few or no target sequences. **Amplification**—where the reaction units are incubated at the required temperature to amplify the nucleic acids independently within each unit. Each reaction unit acts as an independent PCR reaction vessel, and because they are separate, this method avoids cross-contamination and competitive effects between samples, enhancing the specificity and sensitivity of detection. **Detection**—where fluorescence signals are used to identify which reaction units contain amplification products. By statistically analyzing the fluorescence signals from all reaction units

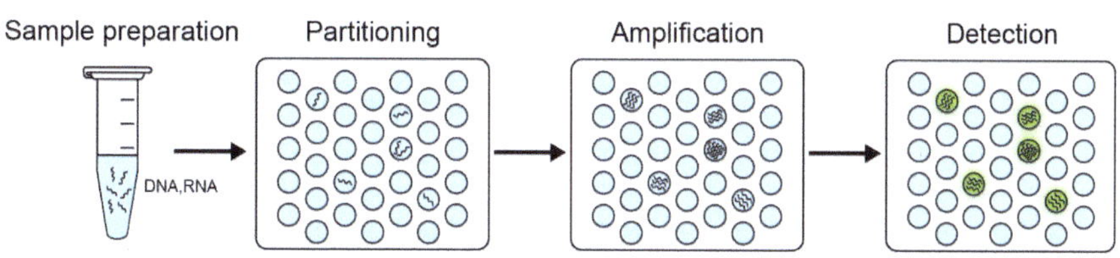

Figure 6.4 Overview of the Digital Droplet PCR (ddPCR) workflow. The workflow involves partitioning a DNA sample into thousands of droplets for individual PCR amplification, followed by fluorescence analysis to quantify target DNA based on the number of positive droplets. *Source:* Figure created by the authors.

and using Poisson statistics to calculate the absolute concentration of the target sequence, the number of positive reaction units is directly proportional to the initial concentration of the target nucleic acid in the sample, achieving absolute quantification.

Despite the concept of dPCR being proposed in the 1990s, its application did not see widespread development due to many technical limitations [47]. In early dPCR experiments, samples were primarily diluted continuously to achieve reactions at the single-molecule level, a process also known as limiting dilution, making it more akin to single-molecule amplification reactions than true dPCR. The term "digital PCR" was first used by Kinzler and Vogelstein in their research by partitioning samples into 384-well microplates for PCR amplification, thereby achieving quantitative analysis of ras mutations [22]. However, these early technologies and experimental methods struggled to partition samples into a large number of independent reaction units and to conduct target nucleic acid amplification within each individual unit, while simultaneously employing Poisson distribution and other mathematical models for absolute quantification. It was not until the development of microfluidic chips and emulsion droplet techniques that dPCR became a practical technology. In the following content, we will primarily summarize existing dPCR methods, focusing on two different microfluidic droplet partitioning techniques: droplet-based and chip-based methods.

6.3.4.1 Droplet-Based Microfluidic Technology and Chip-Based Microfluidic Technology

Droplet-based microfluidic technology generates a large number of independent droplets as reaction units, achieving highly efficient and precise target molecule detection. In 1998, Burns et al. introduced an integrated nanoliter DNA analysis device with microfabricated fluid channels, heaters, sensors, and detectors to process DNA samples (Figure 6.5a) [48]. In the early stages of microfluidic technology, chip-based devices were easier to produce with more mature manufacturing processes. As a result, chip-based microfluidic dPCR was developed earlier and had more advanced partitioning techniques compared to droplet-based systems. This section will briefly introduce the development of chip-based dPCR and notable research achievements.

In 2006, Stephen R. Quake and colleagues published two pioneering papers on chip-based microfluidic dPCR. They first used digital RT-PCR to analyze transcription factor expression in individual hematopoietic progenitor cells, providing high-resolution insights into gene expression and early hematopoietic differentiation (Figure 6.5b) [13, 57]. They then developed a dPCR method to detect chromosomal aneuploidy in 2007. Stephen R. Quake's pioneering work in microfluidics significantly influenced the development of Fluidigm's first commercial chip-based dPCR product. This system, launched in 2006, utilized integrated fluidic circuits (IFCs) for high-throughput qPCR and dPCR, allowing precise and efficient analysis of DNA, RNA, and microRNA (miRNA). This innovation marked a significant milestone in dPCR technology, facilitating its application

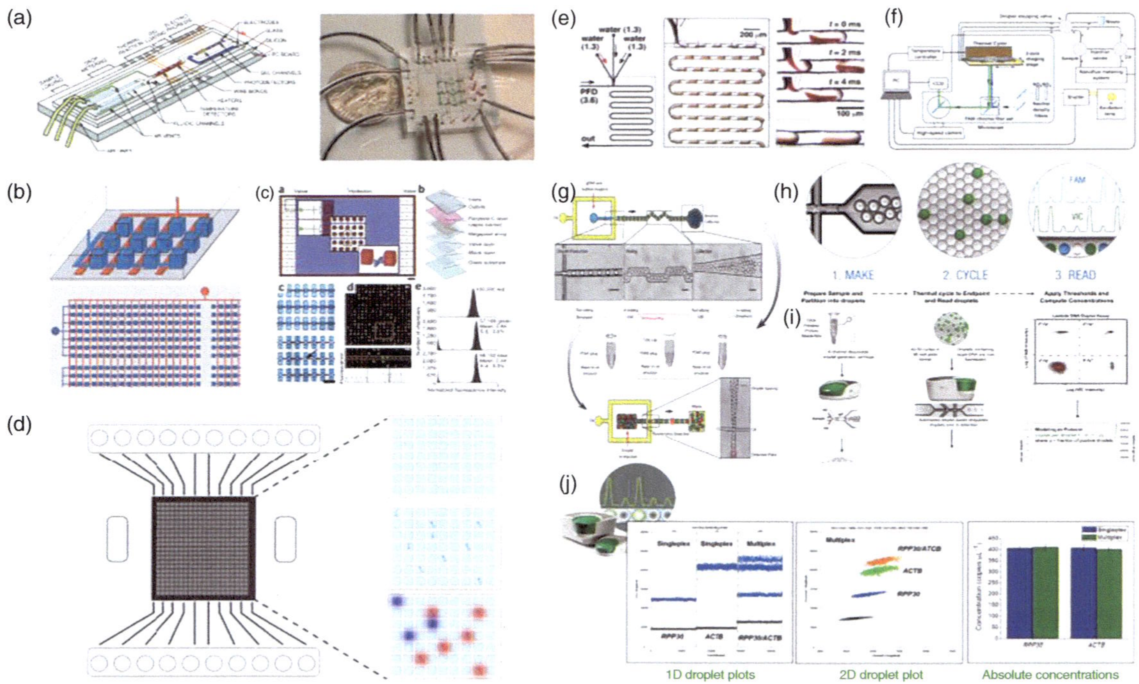

Figure 6.5 Droplet-based microfluidic technology and chip-based microfluidic technology. (a) Early nanoliter-scale DNA analysis device integrating microfluidic channels, heaters, and sensors for DNA processing (*Source:* Reprinted with permission from Ref. [48]. Copyright [1998] The American Association for the Advancement of Science). (b) Chip-based digital RT-PCR for analyzing gene expression in single cells, providing high-resolution genetic insights (*Source:* Reprinted with permission from Ref. [13]. Copyright [2006] The American Association for the Advancement of Science)
(c) Megapixel dPCR device designed for high-throughput and precise absolute DNA quantification with over a million reactions (*Source:* Reprinted with permission from Ref. [49]. Copyright (2011) Nature Publishing Group). (d) Chip-based digital PCR technology for noninvasive prenatal diagnostics and accurate quantification of fetal DNA in maternal plasma (*Source:* Reprinted with permission from Ref. [50]. Copyright (2008) Oxford University Press).
(e) Microfluidic system offering precise millisecond-scale control for high-throughput chemical reactions (*Source:* Reprinted with permission from Ref. [51]. Copyright [2003] John Wiley & Sons). (f) Picoliter droplet-based real-time PCR system enabling single-molecule RNA detection and amplification (*Source:* Reprinted with permission from Ref. [52]. Copyright (2008) American Chemical Society). (g) Droplet-based platform for detecting rare mutations with high sensitivity (*Source:* Reprinted with permission from Ref. [53]. Copyright (2011) Royal Society of Chemistry). (h) ddPCR system with 2 million PCR reactions, improving sensitivity and throughput (*Source:* Reprinted with permission from Ref. [54]. Copyright (2011) American Chemical Society). (i) ddPCR used for precise DNA copy number quantification (*Source:* Reprinted with permission from Ref. [55]. Copyright (2012) American Chemical Society).
(j) Enhanced ddPCR system combining TaqMan probes with DNA-binding dyes for multiplex detection (*Source:* Reprinted with permission from Ref. [56]. Copyright [2013] American Chemical Society).

in various fields, including clinical diagnostics, environmental monitoring, and genetic research [58].

Following this, research based on Fluidigm's microfluidics dPCR technology or other similar chip-based microfluidic dPCR technologies also flourished. Qin et al. explored the application of Fluidigm Corporation's digital array technology for the study of copy number variations (CNVs) in the human genome [59]. This innovative nanofluidic platform enabled precise quantification of gene copy numbers in DNA samples, demonstrating significant advantages over traditional methods. Tadmor et al. used microfluidic dPCR to detect viral infections in single bacterial cells [60], while Sanders et al. evaluated its precision and sensitivity for absolute DNA quantification [61], Heyries et al. introduced a "megapixel" dPCR device capable of over one million reactions (Figure 6.5c) [49], and Baker reviewed these advancements and the growing adoption of dPCR in research and diagnostics. Lun et al. demonstrated that Fluidigm's microfluidics dPCR technology is a powerful and precise tool for measuring fetal DNA in maternal plasma, offering significant advancements for noninvasive prenatal diagnostics and other medical applications requiring accurate nucleic acid quantification (Figure 6.5d) [50].

The miniaturized, low-power device is ideal for portable use, marking the first full integration of DNA analysis with a microdevice and paving the way for future microfluidic devices and dPCR integration. After the research by Burns et al., two studies laid the foundation for microdroplet generation, providing technical support for conducting PCR reactions within droplets. In 2001, Thorsen et al. introduced a microfluidic device that generates reverse micelles, forming complex, ordered patterns away from thermodynamic equilibrium [62]. In 2003, Song et al. describe a microfluidic system that controls chemical reactions on a millisecond timescale, offering a compact platform for high-throughput chemical analysis and synthesis with precise control, useful for complex reactions, screening, and diagnostics (Figure 6.5e) [51]. However, the successful implementation of PCR reactions within droplets can be traced back to the consecutive works published by Beer et al. They developed the first lab-on-chip system that generates picoliter droplets for real-time PCR with fluorescence detection. And in 2008, Beer et al. expanded this system to perform real-time RT-PCR in picoliter droplets, isolating and amplifying single RNA molecules (Figure 6.5f) [52, 63].

Afterward, research and exploration based on droplet microfluidics and PCR flourished. Kiss et al. developed a high-throughput microfluidic chip designed to perform qPCR in picoliter droplets [64]. Mazutis et al. developed a droplet-based microfluidic system designed for high-throughput isothermal amplification and analysis of single DNA molecules [65]. This system utilizes droplet microfluidics to compartmentalize DNA molecules in picoliter droplets, allowing for efficient and accurate digital quantification of DNA. Hatch et al. introduced a droplet microfluidic platform for high-throughput dPCR, combining rapid droplet generation, on-chip thermocycling, and wide-field fluorescence imaging [66]. This system processes and quantifies over 1 million picoliter droplets, greatly improving the dynamic range and throughput of dPCR. Pekin et al. developed a highly sensitive and quantitative method for detecting rare mutations using droplet-based microfluidics (Figure 6.5g) [53]. Research has greatly advanced the integration of droplet-based microfluidic technology with PCR, showcasing its precision and sensitivity in clinical diagnostics and molecular research.

Following these advancements, droplet dPCR technology achieved a landmark breakthrough with Bio-Rad's release of the first droplet digital PCR (ddPCR) instrument, the QX100, in 2011. Led by Bio-Rad Laboratories, Hindson et al. developed a high-throughput ddPCR system that handled around 2 million PCR reactions using standard TaqMan reagents and a 96-well plate, significantly improving the accuracy and sensitivity of qPCR for genomic research and diagnostics (Figure 6.5h) [54]. Pinheiro et al. evaluated ddPCR for DNA copy number quantification, demonstrating its high precision and reliability for nucleic acid measurement (Figure 6.5i) [55]. McDermott et al. further enhanced ddPCR by combining TaqMan probes with DNA-binding dyes, enabling multiplex target detection in a single reaction, improving sensitivity and precision for genetic analysis and diagnostics (Figure 6.5j) [56].

Ye et al. introduced OsciDrop, a chip-free system that generates size-tunable droplets on demand by oscillating a pipette tip, creating monodisperse droplets from 200 pL to 2 μL (Figure 6.6a) [67]. Li et al. (2024) advanced this with the all-in-one OsciDrop dPCR system, automating droplet generation, thermal cycling, and fluorescence detection for high-throughput nucleic acid quantification [18]. Chen et al. developed MiCA, a centrifugal microchannel array technology for producing uniform water-in-oil droplets using a standard centrifuge, minimizing contamination and material loss (Figure 6.6b) [68].

Xu et al. developed cross-interface emulsification (XiE) to produce monodisperse droplets from picoliters to nanoliters using a vibrating capillary at the air–oil interface, with droplet size controlled by flow rate and vibration frequency (Figure 6.6c) [69]. Dangla et al. introduced a droplet microfluidics method driven by confinement gradients, simplifying monodisperse droplet production and improving process stability and flexibility. This approach decouples droplet size from production rate and physical forces, allowing for efficient, loss-free emulsion production, ideal for high-throughput biological and chemical applications (Figure 6.6d) [70]. Schuler et al. introduced a centrifugal step emulsification method for producing monodisperse droplets with diameters of 120–170 μm and variation coefficients of 2–4% [73]. Chen et al. developed the Spinning Micropipette Liquid Emulsion (SiMPLE) generator, an off-chip method for producing stable, monodisperse water-in-oil emulsions (Figure 6.6e) [71]. This technique uses a moving micropipette in an oil-filled tube to create picoliter droplets without dead volume. Droplet size, ranging from 25 to 230 μm with a dispersity under 5%, is controlled by adjusting the flow rate and micropipette tip speed.

In 2010, Shen et al. introduced a groundbreaking dPCR method using the SlipChip, a cost-effective platform with two plates that creates 1,280 reaction compartments (2.6 nL each) through "slipping" movement [74]. This method demonstrated high sensitivity in detecting *Staphylococcus aureus* DNA, enabling single DNA molecule amplification without cross-contamination. They later developed the rotational SlipChip for multiplexed nucleic acid quantification using multivolume digital RT-PCR, with wells from 0.2 to 625 nL. This platform showed

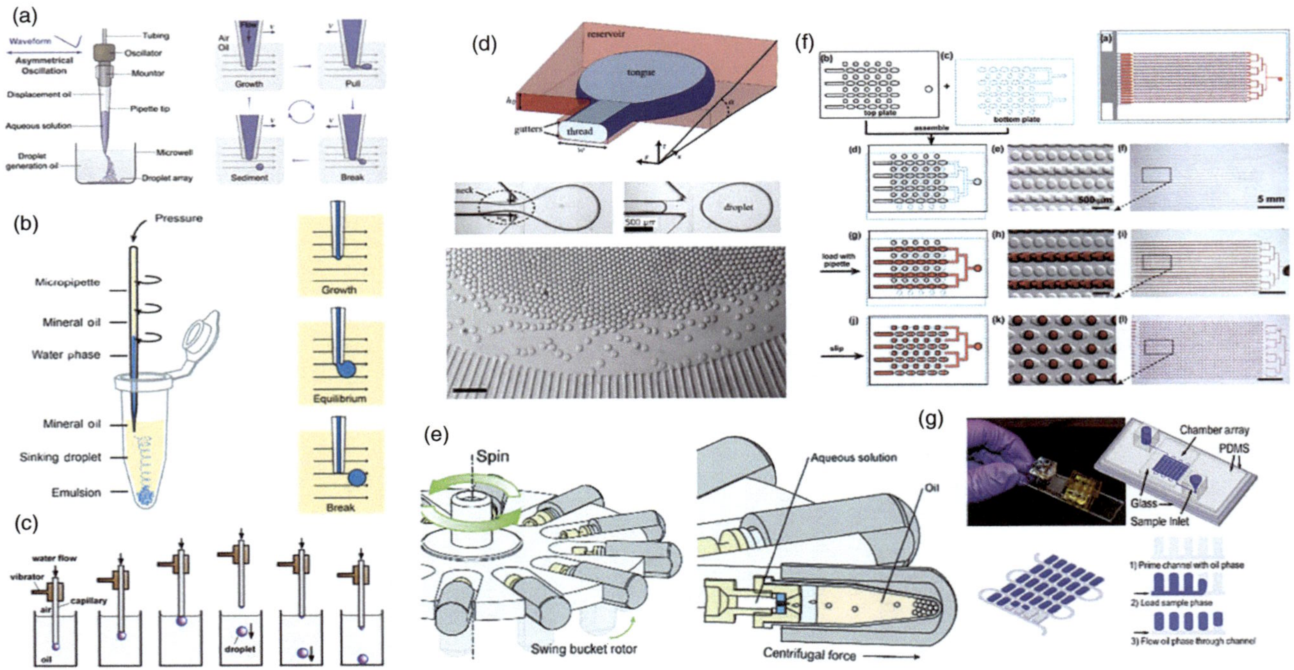

Figure 6.6 The latest microfluidic digital PCR technology. (a) OsciDrop system generates on-demand, size-tunable droplets by oscillating a pipette tip, with volumes ranging from 200 pL to 2 μL (*Source:* Reprinted with permission from Ref. [67]. Copyright (2022) American Chemical Society). (b) MiCA technology uses a centrifugal microchannel array to produce uniform droplets with a standard centrifuge, minimizing contamination (*Source:* Reprinted with permission from Ref. [68]. Copyright [2016] Royal Society of Chemistry). (c) Cross-interface emulsification (XiE) creates monodisperse droplets using a vibrating capillary at the air/oil interface (*Source:* Reprinted with permission from Ref. [69]. Copyright [2016] American Chemical Society). (d) Confinement gradient method decouples droplet size from production rate, allowing precise control over droplet size and efficient, loss-free emulsion production (*Source:* Reprinted with permission from Ref. [70]. Copyright (2013) NAS). (e) SiMPLE generator produces stable water-in-oil emulsions using a moving micropipette, controlling droplet size by flow rate (*Source:* Reprinted with permission from Ref. [71]. Copyright [2017] Royal Society of Chemistry). (f) SlipChip platform with 1,280 compartments enables cost-effective, high-sensitivity digital PCR through slipping movement (*Source:* Reprinted with permission from Ref. [71]. Copyright (2011) Royal Society of Chemistry). (g) Sample self-digitization (SD) chip automates droplet formation for digital LAMP, allowing accurate DNA quantification with a minimal sample volume (*Source:* Reprinted with permission from Ref. [72]. Copyright [2014] American Chemical Society).

high accuracy in detecting viral RNA, including HIV and HCV, making it suitable for point-of-care diagnostics in resource-limited settings (Figure 6.6f).

Sundberg et al. introduced a spinning disk platform for dPCR, using a low-cost plastic disk to compartmentalize samples into 1,000 wells (33 nL each) via centrifugation, with rapid air thermocycling enabling 33-second PCR cycles and 94% efficiency [75]. Cohen et al. developed a simple microfluidic device that digitizes samples by self-dividing them into chambers with an immiscible phase, achieving 100% digitization through fluidic forces and channel geometry without sample loss [76].

After that, in 2012, Men et al. developed a compact microfluidic device for high dynamic-range dPCR using an array of 36-femtoliter microreactors [77]. Made from PDMS, the device loads and seals samples in over 20,000 microreactors per mm^2, ensuring efficient reactions with minimal reagent and sample use. Thompson et al. introduced a sample self-digitization (SD) chip for digital loop-mediated DNA amplification (dLAMP), providing a simple, cost-effective method for DNA amplification [72]. The SD chip automates droplet formation, enabling accurate quantification of DNA with minimal sample volume, making it ideal for point-of-care diagnostics. Zhu et al. developed a self-priming compartmentalization (SPC) method for digital LAMP, enabling point-of-care HIV detection without external pumps or valves [78].

Gansen et al. introduced an SD microfluidic chip for absolute mRNA quantification in single cells. The chip, which has no valves or moving parts, enables accurate digital RT-PCR without preamplification, aligning well with traditional qPCR and single-molecule FISH [79]. This chip is ideal for studying cellular heterogeneity and low-abundance gene expression in single cells (Figure 6.6g).

6.3.5 Portable PCR for Point-of-Care Testing

To fully harness the potential of PCR technology in molecular diagnostics, the development of PCR equipment has been relentless. Initially, PCR reactions were achieved by manually transferring samples between water baths at different temperatures to cycle through the required thermal profiles. Subsequently, the advent of automated thermal cyclers made PCR technology more user-friendly. PCR microfluidic chips, or lab-on-a-chip technologies, have enabled PCR reactions to be performed at faster cycling rates, with higher throughput, and with increased portability, thereby further expanding the applications of PCR (Figure 6.7).

The heating method in PCR significantly influences reaction efficiency. PCR heating methods can be broadly categorized into two types: contact and noncontact heating. Among these, contact heating is the most widely applied, with common examples including metal heating blocks, Peltier-effect-based heating blocks, and thin-film heating elements. However, contact heating makes it difficult to achieve rapid thermal cycling because the contact between the PCR solution chamber and the heat source results in an increased overall thermal mass. Researchers have also attempted to develop various noncontact thermal cycling methods. In the early stages, hot air, infrared radiation, and microwave radiation were all utilized for rapid noncontact heating [80].

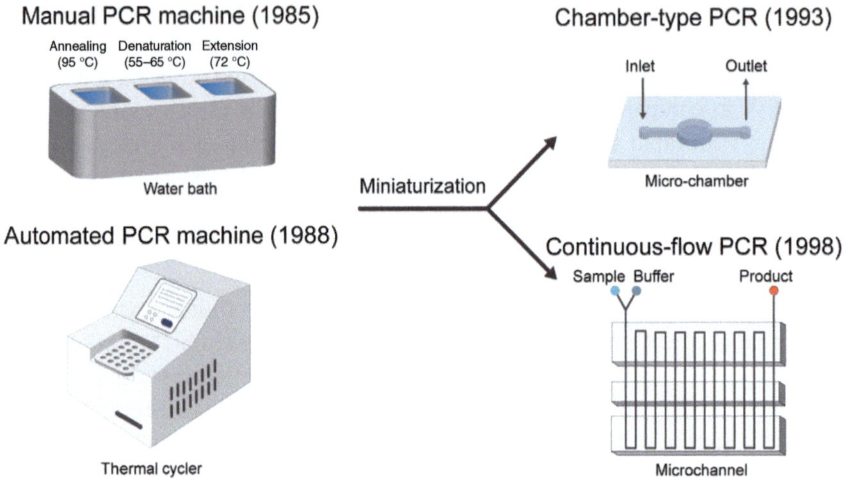

Figure 6.7 The development process of PCR instruments. *Source:* Figure created by the authors.

Recently, the photothermal effects provided a unique method for the necessary temperature control of PCR. Plasmonic photothermal PCR utilizes the photothermal effect of metallic nanomaterials (MNs) to heat the PCR solution under incident light. Compared with the traditional PCR, plasmonic photothermal PCR can complete noncontact heating of the reaction solution in a very short time. Moreover, plasmonic photothermal PCR is compatible with various light wavelengths by adjusting the shape, size, and dielectric environment of MNs. Roche et al. first demonstrated a PCR reaction driven by nanoparticle plasmonic heaters in direct contact with DNA and the polymerase [81]. Laser-induced plasmonic heating from a continuous wave laser (532 nm) and gold nanoparticles (60 nm) obtained a fast heating rate, which demonstrates a very attractive PCR thermal cycling method (Figure 6.8a). Son et al. presented an ultrafast photonic PCR through plasmonic photothermal heating of thin Au films driven by light-emitting diodes (LEDs) (Figure 6.8b) [82]. This simple, robust, and low-cost approach could be generally integrated into a variety of devices or procedures, including on-chip thermal lysis and heating for isothermal amplifications. Cheong et al. further expand the potential of photonic PCR for point-of-care (POC) detection (Figure 6.8c) [83]. They used magneto-plasmonic nanoparticles (MPNs) that have both plasmonic heating and magnetic separation capabilities. However, the properties of the current designs preclude the possibility of real-time analysis, the overlap of the absorption spectra of these nanostructures with those of widely used fluorescent dyes. Lee et al. demonstrated that it is possible to achieve real-time fluorescence detection during photonic PCR by using gold bipyramid nanoreactors (AuBPs) (Figure 6.8d) [84]. The spacer of AuBPs with silica coating and the near-infrared (NIR) resonant wavelength (LSPR ~846 nm) all contributed to preventing quenching of fluorescent dyes. Blumenfeld et al. used gold nanorods (AuNRs, LSPR ~850 nm) to achieve the photothermal effect. Moreover, the AuNRs would not interfere with fluorescence

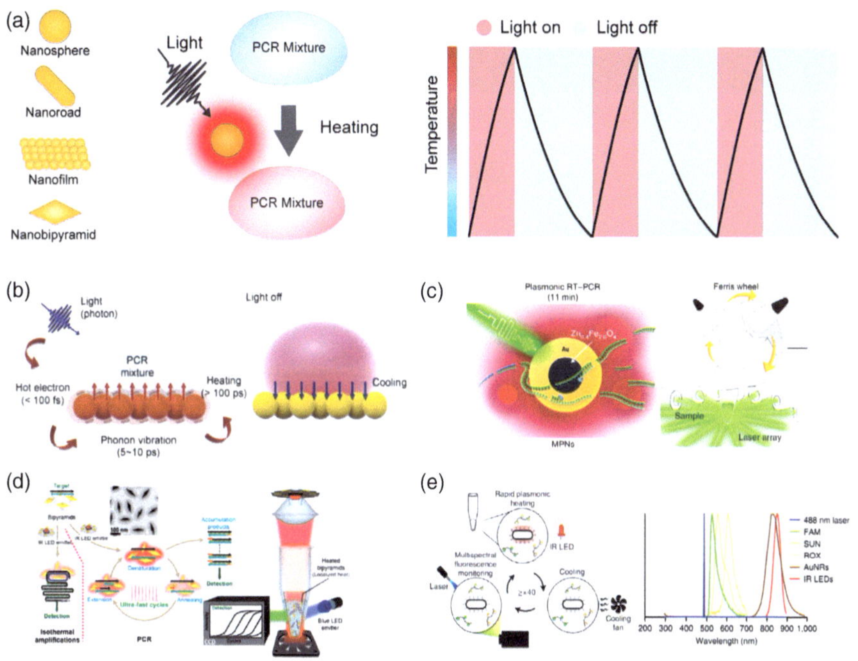

Figure 6.8 (a) Schematic of the plasmonic photothermal PCR. (b) Thin gold film-based ultrafast photonic PCR (*Source:* Reprinted with permission from Ref. [82]. Copyright (2015) Nature Publishing Group). (c) Magneto-plasmonic nanoparticles-based photonic PCR (*Source:* Reprinted with permission from Ref. [83]. Copyright [2020] Nature Publishing Group). (d) Gold bipyramid nanoreactors for ultrafast real-time photonic PCR (*Source:* Reprinted with permission from Ref. [84]. Copyright [2017] ACS Publishing Group). (e) Multiplexed RT-qPCR using plasmonic nanorods (*Source:* Reprinted with permission from Ref. [85]. Copyright (2022) Nature Publishing Group).

measurements and allowed the four real-time fluorescence probes (FAM, SUN, and ROX dyes) without the need to remove AuNRs during detection (Figure 6.8e) [85]. Overall, plasmon-driven heating significantly enhances the heating rate, yet its impact on the cooling rate is less pronounced. Nanomaterials necessitate sophisticated coatings to mitigate the absorption of biomolecules and to preserve their monodispersity at elevated temperatures. Additionally, the plasmon resonance wavelengths might coincide with the emission spectra of fluorescent dyes, potentially restricting the detection capabilities of plasmon-driven ultrafast photon PCR in fluorescence-based qPCR applications.

As semiconductor and nanotechnologies advance, the dimensions of devices have considerably shrunk, allowing for the miniaturization of microprocessors and chemical/biological reactors. This reduction in size boosts the surface-to-volume ratio, which facilitates faster analytical processes due to more efficient heat dissipation. The smaller scale also requires less sample volume and makes it possible to conduct high-throughput parallel analyses, enhancing device portability and reducing energy consumption. In the field of miniaturized PCR, there are two main types: chamber-type PCR and continuous-flow PCR (Figure 6.9).

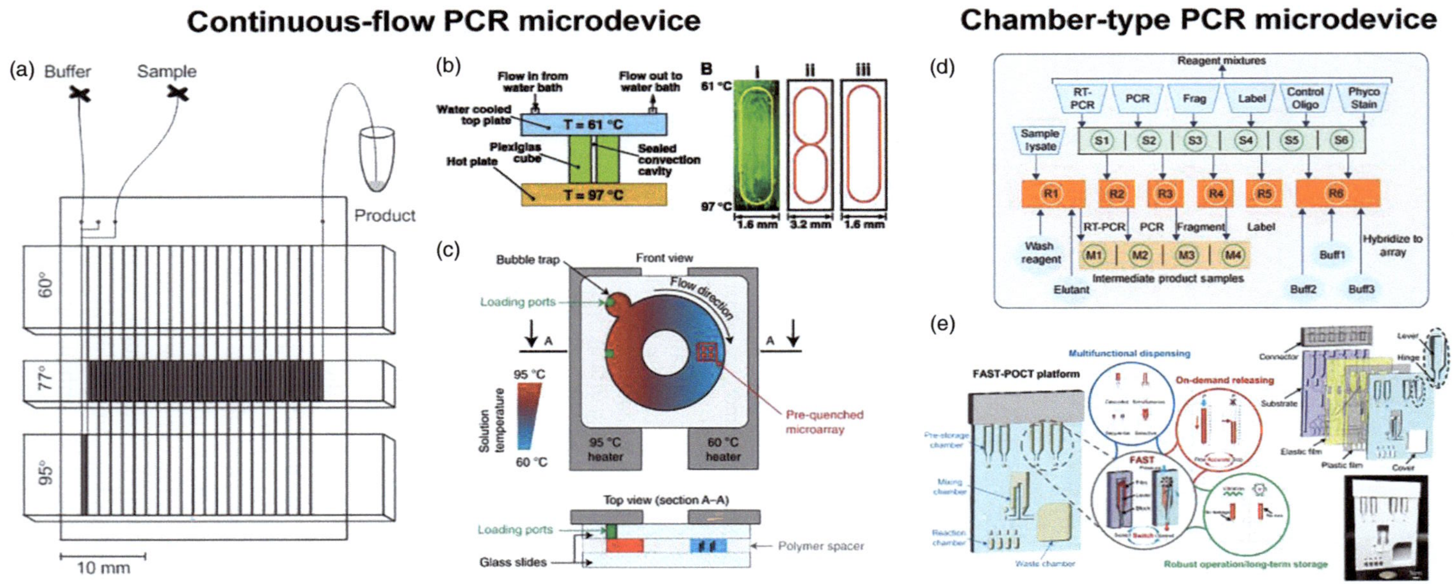

Figure 6.9 The different continuous-flow PCR (CF-PCR) microdevices. (a) Continuous-Flow PCR on a chip (*Source:* Reprinted with permission from Ref. [86], Copyright [1998] AAAS). (b) PCR in a Rayleigh–Benard Convection Cell (*Source:* Reprinted with permission from Ref. [87], Copyright [2002] AAAS). (c) Highly multiplexed rapid DNA detection with single-nucleotide specificity via convective PCR (*Source:* Reprinted with permission from Ref. [15], Copyright [2021] Nature Publishing Group). (d) Polycarbonate cartridge designed for automated multistep genetic assays (*Source:* Reprinted with permission from Ref. [88], Copyright [2000] Oxford University Press). (e) Fully integrated chip for POCT qPCR detection (*Source:* Reprinted with permission from Ref. [89], Copyright (2022) Nature Publishing Group).

CF-PCR relies on the movement of samples through thermostatically controlled temperature zones on a microchip. Unlike traditional PCR, CF-PCR achieves time–space conversion, where constant temperatures are maintained at different locations within the system, while the sample is transported through these zones. This method allows the sample to undergo the necessary temperature cycles for PCR—DNA denaturation, primer annealing, and extension—while in continuous flow. The design of the CF-PCR instrument typically includes two essential temperature zones: a denaturation zone (95 °C) and an annealing zone (50 °C to 65 °C). In addition to directly setting the corresponding thermostatic blocks for the extended zone (72–77 °C), the thermal convection phenomenon is also used to simplify the system design. The CF-PCR chip layout is engineered such that the sample experiences PCR temperature cycling as it traverses through these zones. Nakano et al. first proposed the prototype of the CF-PCR device [90]. A thin Teflon capillary tube was used as a tubular reactor. The reaction mixture of PCR was driven by a pump through different temperature zones. The total reaction time is from 12 to 18 minutes. Kopp and colleagues described a miniaturized CF-PCR device on a glass microchip, which relied on the movement of samples between three constant temperature zones [86]. At different flow rates, 20 cycles of PCR amplification can be completed in 90 seconds to 18.7 minutes. CF-PCR in a closed system eliminates the need for an external driving mechanism and instead takes advantage of the temperature gradient present (Figure 6.9a). Krishnan et al. introduced a novel method for CF-PCR using Rayleigh–Bénard convection [87]. The device is designed as a cylindrical chamber that achieves convection-driven temperature cycling by maintaining different temperatures at the top and bottom of the chamber (61 and 97 °C, respectively) (Figure 6.9b). Wheeler et al. used lower-cost reactors and improved PCR speed by adjusting the shape of the reaction chamber and side heating [91]. Zhang and coworkers presented a toroidal chamber-based PCR system for scalable large-scale multiplexing, rapid turnaround, single-nucleotide discrimination, and precise quantification [15]. The chip design utilized the Rayleigh–Bénard thermal convection principle. By setting heaters at 95 and 60 °C on both sides of the chip, the fluid can complete the thermal cycle (Figure 6.9c).

Chamber PCR uses different processing technologies to create reaction chambers on substrates such as silicon wafers, glass, and various polymers. Chamber PCR takes up relatively little space and has less nonspecific surface adsorption. However, it usually requires complex valves to control fluids and fix the reaction solution in the chamber. High-precision micronanostructures can be fabricated for miniaturized and parallelized PCR reactions by photolithographic micromachining of silicon. Burns et al. described the construction of thermal-cycling chambers, gel electrophoresis channels, and radiolabeled DNA detectors that are compatible with the thermocapillary pump channels [92]. The thermocapillary pump utilized surface tension and local heating to move liquid through enclosed channels for the entire process of PCR. Anderson presented a highly integrated, monolithic miniaturized polycarbonate cartridge designed for automated multistep genetic assays [88]. The device was tested for detecting mutations in a 1.6-kb region of the HIV genome from serum samples containing as few as 500 copies of RNA.

It integrates multiple enzymatic reactions, including reverse transcription, PCR, DNase digestion, and terminal transferase labeling (Figure 6.9d). Lee Gwo-Bin and colleagues have developed an innovative system where sample droplets are precisely generated, deftly transported, and efficiently mixed within a digital microfluidic chip (DMC), enabling fully automated PCR processes with a heating rate of 38.0 °C/s and a cooling rate of 7.9 °C/s [93]. Neuzil et al. reported an ultrafast microchamber PCR chip which can complete 40 PCR cycles in five minutes and 40 seconds [94]. The reduced volume of the PCR reaction (100 nl) and the cantilevered design with a disc-shaped end facilitated rapid and uniform heating. Beer et al. described a groundbreaking lab-on-chip system that performs real-time, single-copy PCR in picoliter droplets [63]. Picoliter water-in-oil droplets are isolated from the microfluidic channel walls and each other by an oil-phase carrier. An off-chip valving system stored droplets on-chip for subsequent thermal cycling with heating and cooling rates set at 1.8 °C/s and 1.4 °C/s, respectively. Jiang and coworkers highlighted a lab-on-a-chip system that overcomes challenges in multifunctionality and robustness, featuring a film-lever actuated switch technology for precise liquid handling and integrated PCR capabilities (Figure 6.9e) [89].

6.4 Isothermal Amplification: Revolutionizing Molecular Diagnostics with Constant Temperature

As molecular diagnostics have progressed, the limitations of PCR, particularly its dependence on thermal cycling and complex instrumentation, have driven the development of isothermal amplification technologies. These methods offer several advantages, including simplified workflows, faster reaction times, and operation under constant temperature conditions. Such features make isothermal amplification particularly valuable in time-sensitive and resource-limited settings, such as public safety incidents, where rapid and accurate molecular diagnostics are essential for detecting and managing pathogens, toxins, or other hazardous agents. Unlike PCR, which requires cycling through various temperature stages, isothermal methods enable amplification at a constant temperature, facilitating faster response times with minimal equipment in the field. The following sections will examine the development of key isothermal amplification technologies, including LAMP, RPA, and NASBA, and explore their specific applications in enhancing molecular diagnostics during public safety emergencies.

6.4.1 Loop-Mediated Isothermal Amplification

LAMP has become a prominent method in molecular diagnostics due to its simplicity, speed, and high sensitivity. In the following sections, we will provide a brief summary of the various detection methods for LAMP amplification products. In their groundbreaking 2000 paper, Notomi et al. introduced the LAMP reaction, which allows for the rapid, efficient, and highly specific amplification of DNA under isothermal conditions [6]. This method employs a DNA polymerase and four specially designed primers that recognize six distinct sequences on the target DNA. Operating at a constant temperature of approximately 65 °C, LAMP can amplify a few copies of DNA to 10^9 copies in less than an hour (Figure 6.10a).

LAMP achieves high specificity by recognizing six sequences in the initial stages and four in later stages of amplification, producing easily detectable stem-loop DNAs and cauliflower-like structures. The reaction requires minimal equipment, such as a water bath or heat block, and is unaffected by nontarget DNA, making it ideal for clinical diagnostics. LAMP can also detect both DNA and RNA (with reverse transcription), making it versatile for molecular diagnostics, including pathogen detection and genetic disorder screening [100].

Based on LAMP's high efficiency and specificity, its detection methods are more diverse compared to PCR. Below is a summary of key detection methods for LAMP amplification products: **Gel electrophoresis** is widely used for detecting LAMP products by visually confirming the presence and size of amplified DNA fragments [101]. It validates amplification success through characteristic bands, assesses specificity with distinct patterns, and quantifies the product by comparing band intensity to standards. **Turbidity detection** monitors LAMP reactions in real time by measuring the formation of magnesium pyrophosphate, a byproduct of amplification. This method is simple, rapid, and highly sensitive, capable of detecting DNA concentrations as low as a few copies [102, 103]. **Calcein-based detection** provides a clear visual indicator of LAMP amplification. During the reaction, pyrophosphate ions released bind with metal ions, freeing calcein to bind with magnesium, causing a color change that can be observed under UV light or daylight [95] (Figure 6.10b). **Colorimetric detection** uses dyes, like hydroxy naphthol blue (HNB), which change color when DNA is amplified. This method simplifies LAMP detection, providing rapid visual confirmation without the need for specialized equipment [104]. **pH-sensitive dye detection** detects the pH change during DNA amplification, as hydrogen ions are released during the reaction. This method is simple, requires basic equipment, and offers rapid detection, making it ideal for clinical diagnostics and field use [105].

6.4.1.1 Integration of LAMP with Various Detection Technologies

Although we have summarized various methods for detecting LAMP amplification products, the mentioned techniques are nonspecific. Due to the use of 4–6 primers and the high amplification efficiency, LAMP tends to produce nonspecific amplification. To address this, researchers have developed methods for more specific detection and have integrated LAMP with other technologies, further expanding its application potential.

In 2012, Li et al. developed an enzyme-free DNA circuit known as catalytic hairpin assembly (CHA) to improve the specificity of LAMP product detection [96]. CHA uses hairpin acceptors that bind to single-stranded loops from the LAMP reaction, reducing false positives from nonspecific amplification. The dual-CHA system further enhances specificity by only producing a signal when two specific LAMP loops are present, allowing for sensitive detection of target DNA at concentrations below 10 molecules/μL while avoiding interference from nonspecific

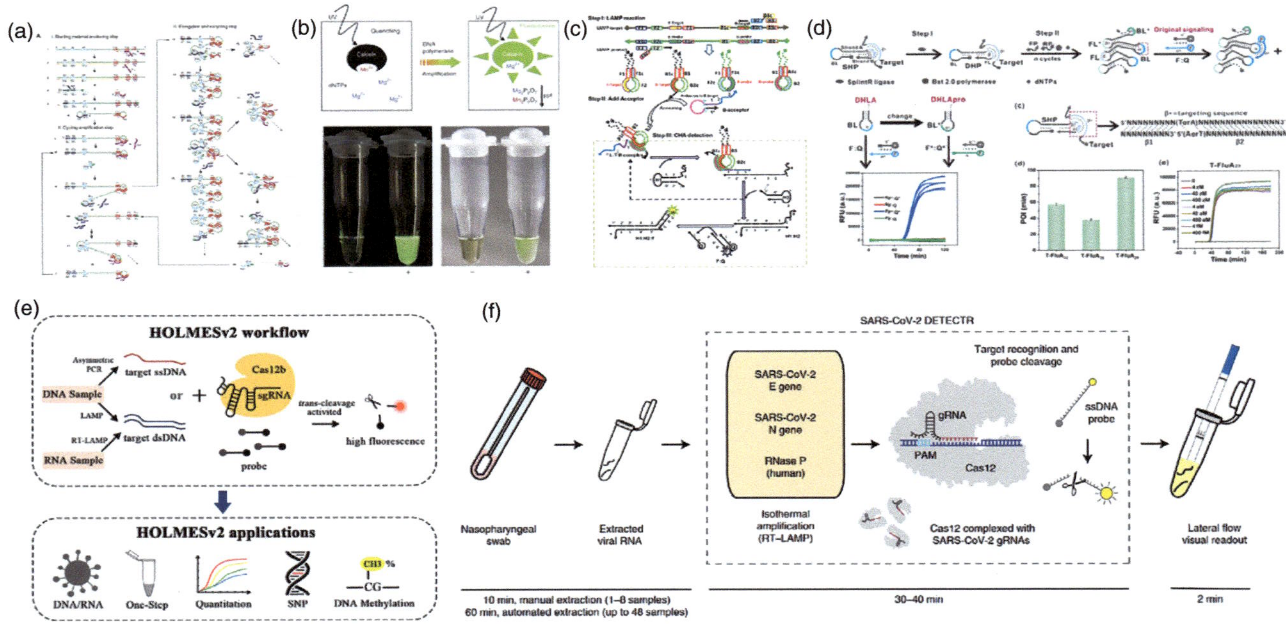

Figure 6.10 Integration of LAMP with various detection technologies. (a) Standard LAMP reaction using multiple primers (*Source:* Reprinted with permission from Ref. [6]. Copyright [2000] Oxford University Press). (b) Calcein-based fluorescence color change for visual detection under UV light (*Source:* Reprinted with permission from Ref. [95], Copyright (2008) Nature Publishing Group). (c) CHA combined with LAMP for enhanced specificity by targeting specific LAMP structures, minimizing nonspecific amplification (*Source:* Reprinted with permission from Ref. [96]. Copyright (2012) American Chemical Society). (d) DHLApro uses dual hairpin probes and ligase to increase sensitivity and specificity, enabling detection of very low DNA concentrations and supporting multiplex detection (*Source:* Reprinted with permission from Ref. [97]. Copyright (2022) John Wiley & Sons). (e) HOLMESv2 integrates LAMP and CRISPR-Cas12b, enabling single-step detection of SNPs and DNA methylation with high precision (*Source:* Reprinted with permission from Ref. [98]. Copyright [2019] American Chemical Society). (f) SARS-CoV-2 DETECTR combines RT-LAMP and CRISPR-Cas12a for rapid virus detection, providing visual results on a lateral flow strip within 30–40 minutes (*Source:* Reprinted with permission from Ref. [99]. Copyright [2020] Nature Publishing Group).

templates. This method significantly improves the accuracy of LAMP product detection for molecular diagnostics (Figure 6.10c). Li et al. also introduced dual hairpin ligation-induced isothermal amplification (DHLApro), an ultrasensitive method for detecting pathogens like influenza and coronaviruses, with a detection limit of 0.45 copies/μL [97]. Using dual hairpin probes and SplintR ligase, DHLApro is adaptable, requiring minimal changes for new targets. It offers high sensitivity and specificity, supports multiplex assays, and reduces cross-interference, making it efficient and cost-effective. DHLApro is ideal for pathogen screening, variant detection, and high-throughput screening, representing a significant advance in isothermal amplification for clinical diagnostics and genetic research (Figure 6.10d).

The LAMP amplification method can also be combined with CRISPR protein detection. The HOLMESv2 platform, developed by Li et al., integrates the thermophilic CRISPR-Cas12b system for highly sensitive and specific nucleic acid detection and DNA methylation quantification [98]. It is capable of detecting SNPs, viral RNA, and human mRNA, using a one-step process that combines CRISPR-Cas12b with LAMP. HOLMESv2 offers a portable, cost-effective solution for clinical diagnostics and epigenetic studies, with applications in pathogen detection and DNA methylation analysis (Figure 6.10e).

Pang and coworkers developed a single-tube assay combining RT-LAMP with CRISPR-Cas12a for the detection of SARS-CoV-2, enhancing specificity by leveraging the sequence recognition of CRISPR-Cas12a and the amplification efficiency of RT-LAMP [99]. The assay runs RT-LAMP at 62 °C for 30 minutes, followed by CRISPR detection at room temperature, achieving 94% sensitivity and eliminating false positives in clinical tests. This integrated single-tube system minimizes contamination risks and simplifies the workflow, making it ideal for point-of-care diagnostics, demonstrating improved sensitivity and specificity for molecular testing. Broughton et al. developed SARS-CoV-2 DETECTR, a CRISPR-Cas12-based lateral flow assay for rapid and accurate detection of SARS-CoV-2. It combines RT-LAMP with Cas12 detection, targeting the E and N genes of the virus, providing visual results on a lateral flow strip within 30–40 minutes. The assay showed 95% positive and 100% negative predictive agreement, offering a reliable alternative to traditional qRT-PCR methods. Validated with clinical samples, this method is ideal for rapid point-of-care diagnostics, especially in resource-limited settings (Figure 6.10f).

These combinations effectively address LAMP's issue of nonspecific amplification, enabling precise, rapid, and reliable pathogen and genetic mutation detection. These methods have shown success in clinical diagnostics, point-of-care testing, and field applications. Looking ahead, continued development and integration of LAMP with emerging technologies promise to further revolutionize molecular diagnostics, offering high sensitivity and specificity in nucleic acid detection, especially in resource-limited settings.

6.4.1.2 Integration of LAMP with Various Detection Equipments

The simplicity and robustness of LAMP make it ideal for integration with other technologies, leading to hybrid diagnostic platforms. LAMP has been combined with digital detection, electrochemical sensors, and high-throughput sequencing, enhancing its diagnostic range and precision. These innovations have addressed limitations in traditional diagnostics, creating more accessible, rapid, and precise detection methods for various settings. Below, we will introduce notable examples of LAMP combined with other technologies.

LAMP can be combined with digital detection technology. The paper by Jin et al. introduces StratoLAMP, a novel label-free, multiplex digital LAMP method [106]. This technique uses the visual stratification of the precipitate produced during LAMP reactions to distinguish between different nucleic acid targets in droplets. By combining specially designed primers with deep learning image analysis, the system accurately detects and quantifies multiple nucleic acids in a single assay without the need for fluorescent labels. This method shows promise for cost-effective and scalable molecular diagnostics, particularly in resource-limited settings (Figure 6.11a).

LAMP can also be combined with microfluidic chip technology. The paper by Dong et al. presents a portable nucleic acid testing platform that integrates photosensitization-based color development, a 3D-printed multipiece chip, and digital color sensing for rapid detection of nucleic acids [107]. The platform combines LAMP amplification with colorimetric readouts and uses a smartphone app for quantification. It was successfully applied to the detection of tuberculosis in clinical samples and food adulteration in commercial salmon products, demonstrating high sensitivity, simplicity, and applicability in resource-limited settings (Figure 6.11b). Chandrasekaran et al. developed the DISCoVER system, an RNA-extraction-free, microfluidic platform that integrates LAMP amplification with Cas13 detection for rapid and sensitive detection of SARS-CoV-2 in saliva samples [109]. The assay achieves attomolar sensitivity within 60 minutes, with a detection limit of 40 copies/μL, and demonstrate 94% sensitivity and 100% specificity when validated against qPCR. This portable system offers a rapid, point-of-care diagnostic solution, crucial for infectious disease surveillance, especially in decentralized and resource-limited settings (Figure 6.11f).

LAMP can also be combined with electrochemical detection platforms. Najjar et al. developed a lab-on-a-chip device combining LAMP technology with electrochemical detection to simultaneously analyze two prostate cancer biomarkers [14]. This platform conducts DNA amplification at a constant temperature, eliminating the need for thermal cycling. The real-time, label-free monitoring through electrochemical detection significantly enhances the test's sensitivity and specificity, making it suitable for point-of-care diagnostics (Figure 6.11d). Similarly, Tayyab et al. introduced a digital assay combining isothermal amplification with electrical impedance detection for the rapid quantification of pathogens, including SARS-CoV-2 and HIV, with high sensitivity and cost-effectiveness [10]. Both studies highlight advancements in molecular diagnostics, especially in resource-limited settings.

LAMP can be combined with droplet manipulation technology. Lin et al. introduced an automated viral testing platform using millimeter-sized "ferrobots" to manipulate magnetic nanoparticle-laden droplets [17]. This system supports high-throughput, precise nucleic acid amplification, specifically LAMP, in a

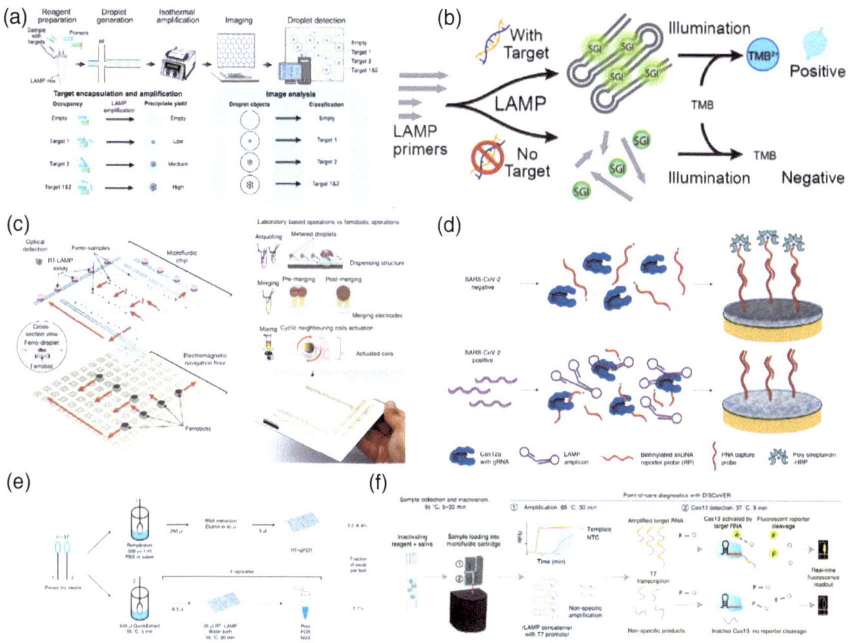

Figure 6.11 Integration of LAMP with various detection equipments. (a) StratoLAMP combines digital detection and deep learning to quantify multiple nucleic acids in droplets using visual stratification, without fluorescent labels (*Source:* Reprinted with permission from Ref. [106]. Copyright (2024) NAS.). (b) Microfluidic chip platform integrates LAMP with colorimetric detection and smartphone-based quantification for portable and sensitive testing (*Source:* Reprinted with permission from Ref. [107]. Copyright (2024) American Chemical Society). (c) Ferrobotic droplet manipulation uses magnetic "ferrobots" to automate LAMP in droplet-based reactions, enabling high-throughput and precise nucleic acid amplification (*Source:* Reprinted with permission from Ref. [17]. Copyright (2022) Nature Publishing Group). (d) Electrochemical Detection is a lab-on-a-chip platform combining LAMP and electrochemical sensors for real-time, label-free detection of cancer biomarkers (*Source:* Reprinted with permission from Ref. [14]. Copyright (2022) Nature Publishing Group). (e) LAMP-Seq combines RT-LAMP with next-generation sequencing for high-throughput, low-cost testing of unpurified samples, ideal for large-scale diagnostics (*Source:* Reprinted with permission from Ref. [108]. Copyright (2021) Nature Publishing Group). (f) DISCoVER system is a microfluidic platform integrating LAMP and Cas13 for rapid SARS-CoV-2 detection in saliva, achieving high sensitivity and specificity (*Source:* Reprinted with permission from Ref. [109]. Copyright (2022) Nature Publishing Group).

compact, programmable format. The ferrobotic method overcomes the limitations of traditional liquid handling by being scalable, low-cost, and suitable for decentralized testing. The platform demonstrated full accuracy in detecting SARS-CoV-2, matching standard off-chip results, and has potential for large-scale screening and various biomedical applications, while significantly reducing reagent and equipment costs (Figure 6.11c).

LAMP and sequencing technology can also be combined. Ludwig et al. introduced LAMP-Seq, a modified RT-LAMP method combined with molecular barcoding for high-throughput COVID-19 testing [108]. This approach allows direct testing from unpurified samples and integrates RT-LAMP with next-generation sequencing (NGS), achieving a detection limit of 2.2 molecules per μL with near-perfect specificity. Clinical validation showed 100% sensitivity for samples with RT-qPCR cycle threshold values up to 33, and 99.7% specificity (Figure 6.11e). LAMP-Seq is low-cost, fast, and scalable, making it ideal for large-scale diagnostics and public health surveillance, with potential for use beyond SARS-CoV-2. The paper by Dao Thi et al. presents a colorimetric RT-LAMP assay and LAMP-sequencing for rapid and accurate detection of SARS-CoV-2 RNA from clinical samples [11]. The study demonstrates the high sensitivity and specificity of the colorimetric RT-LAMP assay, making it a valuable tool for large-scale, point-of-care diagnostics. Additionally, the development of LAMP-sequencing offers a scalable method for validating RT-LAMP results through NGS, enhancing the robustness of molecular testing during the COVID-19 pandemic.

Recent research highlights significant advancements in LAMP applications, showcasing its versatility and integration with various detection technologies. These studies have demonstrated LAMP's robust adaptability in numerous diagnostic scenarios, including the rapid and sensitive detection of pathogens, cancer biomarkers, and genetic material. Innovations include combining LAMP with digital and electrochemical detection, high-throughput sequencing, and automated platforms, which have enhanced the specificity, sensitivity, and throughput of nucleic acid diagnostics. These advancements underline LAMP's potential to significantly advance molecular diagnostics, public health applications, and point-of-care testing, making it an invaluable tool in both clinical and research settings.

6.4.2 Recombinase Polymerase Amplification

First introduced in 2006 by Piepenburg et al. of TwistDx, UK, RPA has attracted much attention [7]. Unlike conventional PCR, which requires an instrument to perform cyclic amplification, RPA operates at 25–42 °C and requires little sample preparation to amplify only 1–10 copies of DNA in 10 minutes, offering the advantages of high sensitivity, selectivity, portability, and speed with the ability to perform multiplex amplification. RPA has been widely used to amplify a variety of different targets, including RNA, miRNA, ssDNA, and dsDNA. Compared to PCR, two key proteins are essential components in RPA, serving as a substitute for the general heat denaturation step: *Escherichia coli* RecA recombinase and single-strand DNA-binding protein (SSB). Starting with the binding of recombinase and primers, the resulting filament scans the double-stranded DNA to locate homologous sequences (Figure 6.12a). The filament then enters the template DNA to form a D-loop structure, while the complementary strand is stabilized by SSBs, and the target strand is hybridized to the primer. Finally, the recombinase dissociates from the complex, allowing DNA polymerase to synthesize from the free 3′-OH at the end of the primer. As the elongation continues, the free recombinase can initiate another cycle of strand displacement. It is through this cyclic process that exponential amplification is achieved.

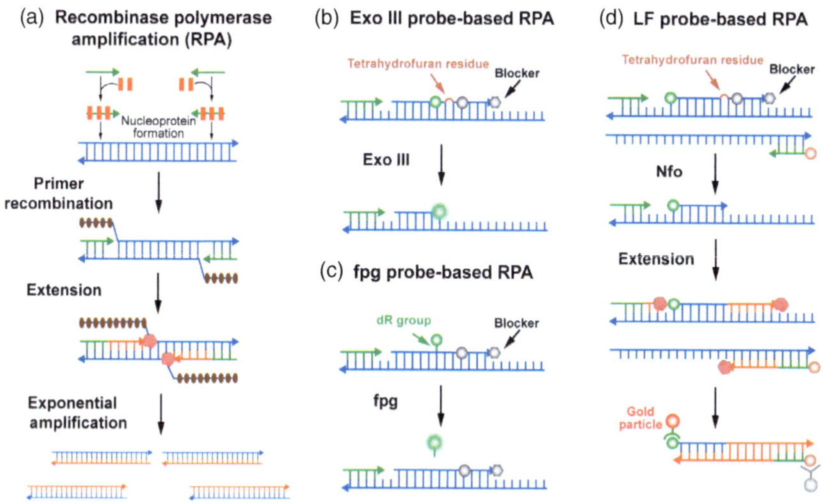

Figure 6.12 RPA reaction mechanism and RPA probes. (a) RPA amplification schematic diagram. (b) TwistAmp exo probe. The probe is cleaved by the *E. coli* exonuclease III at the abasic site to dislodge the fluorophore from the quencher, thereby generating an extensible 3′-OH group for polymerization. (c) TwistAmp™ fpg probe. The probe is cleaved by the glycosylase/lyase *E. coli* Fpg at the dR position (the deoxyribose of the abasic site via a C–O–C linker) to displace the fluorophore from the quencher and generate a 3′-phosphate group, which is nonextensible for polymerization. (d) TwistAmp™ LF probe. The probe is cleaved by the Nfo endonuclease IV at the abasic site to generate an extensible 3′-OH group for polymerization. The DNA polymerases extend and displace the 3′-ends of the primers and cleaved probe, thereby producing the minor amplicons and a displaced strand. The displaced strand then combines with the labeled reverse primer, resulting in the production of a dual-labeled amplicon for subsequent detection. *Source:* Figure created by the authors.

In recent years, many detection strategies have been developed to monitor RPA amplicons. As the founder of RPA, TwistDx™, UK, also exploits several commercially used strategies to detect the RPA amplification of nucleic acid: TwistAmp™ exo, TwistAmp™ fpg, and TwistAmp® nfo probes. Of the aforementioned probes, the TwistAmp exo probe is most commonly employed for real-time monitoring. The exo probe comprises a fluorophore and a quencher (Figure 6.12b). The emission of fluorescence occurs only when the probe binds to the target sequences, resulting in the cleavage of the abasic site, which is situated in the middle of the probe and is a target for exonuclease. Similarly, the formamidopyrimidine DNA glycosylase (FPG) probe is also based on the cleavage of fluorogenic probes at an abasic site, albeit with a change from THF to dR (Figure 6.12c). The nfo probes are employed for lateral flow strip detection. This probe is 5′-labeled, has a 3′ blocker and an internal abasic site (Figure 6.12d). Upon cleavage of the abasic site, an extendable 3′-OH group is generated for polymerization.

The RPA assay demonstrates excellent performance when used in conjunction with a CRISPR diagnostic system. In 2017, with the revealing of significant findings on the dual functionality of CRISPR-Cas12a, DETECTR was developed to detect DNA sensitively and accurately [110]. Combined with RPA, DETECTR is

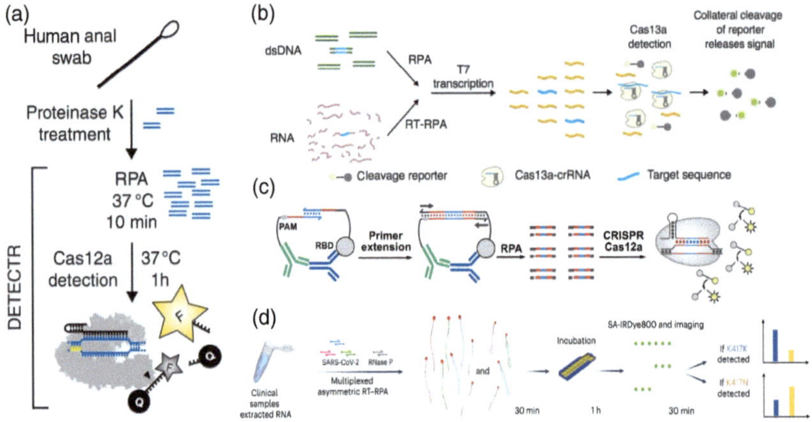

Figure 6.13 Diagnostic methods for diseases leveraging RPA amplification methods. (a) RPA integrated with CRISPR/Cas 12a for rapid detection of HPV16 and HPV18 in human samples (*Source:* Reprinted with permission from Ref. [110]. Copyright [2018] AAAS). (b) RPA and RT-RPA integrated with CRISPR/Cas 12a for single-molecule nucleic acid detection (*Source:* Reprinted with permission from Ref. [111]. Copyright [2017] AAAS). (c) RPA amplified DNA barcode for subsequent CRISPR detection. (d) Multiplexed RPA for SARS-CoV-2 detection and viral lineage identification (*Source:* Reprinted with permission from Ref. [19], Copyright [2023] Nature Publishing Group).

demonstrated to enable rapid and specific detection of human papillomavirus (HPV) in patient samples, showcasing its potential for molecular diagnostics (Figure 6.13a). At the same time, Zhang et al. introduced a novel diagnostic platform, designated SHERLOCK, which employs the CRISPR-Cas13a system for the rapid and sensitive detection of nucleic acids [111]. The SHERLOCK platform integrates RPA with Cas13a to achieve the detection of infectious diseases in the field through the use of paper spotting and lyophilization, as the components of RPA are compatible with freeze-drying (Figure 6.13b).

Ever since the outbreak of coronavirus disease 2019(COVID-19), numerous different tests have been further studied to detect SARS-CoV-2. Li's group reported an ultrasensitive CRISPR-based antibody detection (UCAD) assay that translates the detection of anti-SARS-CoV-2 antibodies into CRISPR-based nucleic acid detection in a homogeneous solution and is 10,000 times more sensitive than the classic immunoassays. (Figure 6.13c) [112]. RPA was used to amplify the double-stranded DNA barcode, along with the subsequent indiscriminate single-stranded DNase (ssDNase) activity of Cas12a that can be unleashed by the antibody.

The urgent requirements of multiplexed point-of-care technologies are driven by the emergence of new SARS-CoV-2 variants and subvariants. The use of simple thermal lysis and asymmetric RT-RPA for direct probe hybridization eliminates the need for complex RNA extraction steps, offering a promising solution for rapid and accurate detection [19]. The assay demonstrated the capacity to accurately discriminate 12 lineages of SARS-CoV-2, with the potential for expansion to include the detection of other viruses and variants (Figure 6.13d).

6.4.3 Strand Displacement Amplification

SDA is also an isothermal amplification method that relies on DNA polymerase. In addition, it necessitates the involvement of another nicking enzyme in order to facilitate the cycling and amplify the signal. SDA was first proposed in 1992 by Walker et al. (Figure 6.14a) [23]. In this piece of work, two probes, which individually contained two parts: a HincII recognition sequence and a complementary sequence to the template, were designed for the target. The mechanism of SDA is based on the continuous nicking and polymerization/displacement process catalyzed by endonuclease and DNA polymerase. In the same year, Walker et al. improved this strategy by using four primers to increase the amplification efficiency and sensitivity (Figure 6.14b) [8]. In this case, SDA can achieve nearly 10^7-fold amplification in two hours at 37 °C. The product can be detected by a variety of methods, including molecular beacons [115] and intercalating dyes [116], the fluorescence intensity of which is enhanced by binding to dsDNA. The SDA technique has been combined with a ligation reaction (2011) to detect SNPs (Figure 6.14c) [113]. The utilization of two consecutive SDAs allowed highly efficient amplification to create a great abundance of peroxidase-mimic signal-reporter sequences. It can achieve the limit of detection as low as 1 fM with a signal-to-noise ratio of 150. A new type of technology named CRISDA (2018) has been developed based on the combination of SDA and CRISPR system (Figure 6.14d) [114]. In this method, Cas9 is utilized to identify and generate substrates for subsequent reaction, which are cleaved and extended in the SDA process to achieve exponential amplification. Although SDA is nominally isothermal, thermal denaturation (95 °C) for target generation is often required when detecting genomic DNA.

6.4.4 Nucleic Acid Sequence-Based Amplification

NASBA is a primer-dependent technique that can be used for continuous amplification of nucleic acids in a single mixture at a single temperature (41 °C) (Figure 6.15a) [9]. The standard NASBA reaction mixture includes avian myeloblastosis virus reverse transcriptase (AMV-RT), T7 RNA polymerase, *E. coli* RNase H, nucleoside triphosphates, two specific primers, and appropriate buffer components. The amplification process of NASBA can be divided into two steps: one step is called the "noncyclic phase," and the other step is called the "cyclic phase." First, in the "noncyclic phase," we use natural RNA as a template and anneal Primer 1 with the RNA target sequence. The AMV reverse transcriptase in the reaction mixture extends the 3′ end of Primer 1 using the same deoxyribonucleoside triphosphates to form a DNA copy of the template and generate an RNA: DNA hybrid. At this point, RNase H can hydrolyze this type of RNA: RNA in DNA hybrids, leaving only one DNA strand. Primer 2 is then annealed with it, and reverse transcriptase synthesizes a second DNA strand, turning the promoter region into double-stranded DNA. The T7 RNA polymerase in the mixture transcribes RNA copies from promoters with transcriptional activity, and each template molecule produces up to 100 RNA copies. The RNA molecules generated above will serve as templates for reverse

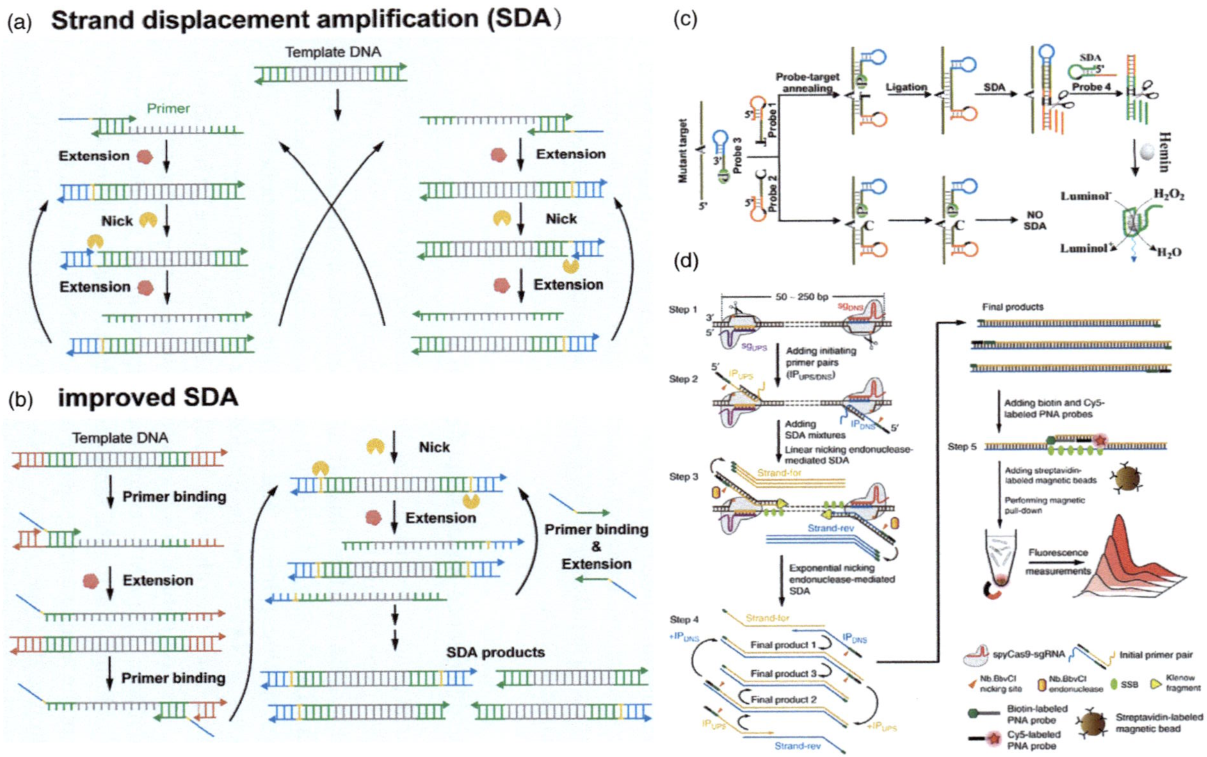

Figure 6.14 Different formats and combinations based on strand displacement amplification (SDA). (a) Scheme of SDA. The first generation of SDA utilizes two primers with a dsDNA target fragment, which constitutes an exponential reaction (*Source:* Figure created by the authors). (b) The improved SDA strategy, using two pairs of primers, is able to improve the reaction efficiency to a large extent. (*Source:* Figure created by the authors). (c) SNV detection based on SDA. A label-free homogeneous SNP genotyping technique is developed on the basis of ligation-mediated SDA with DNAzyme-based chemiluminescence detection (*Source:* Reprinted with permission from Ref. [113]. Copyright (2011) American Chemical Society). (d) Scheme of CRISDA. An isothermal approach for amplifying and detecting a dsDNA target based on a CRISPR–Cas9-triggered nicking endonuclease mediated SDA method (*Source:* Reprinted with permission from Ref. [114]. Copyright (2018) Nature Publishing Group).

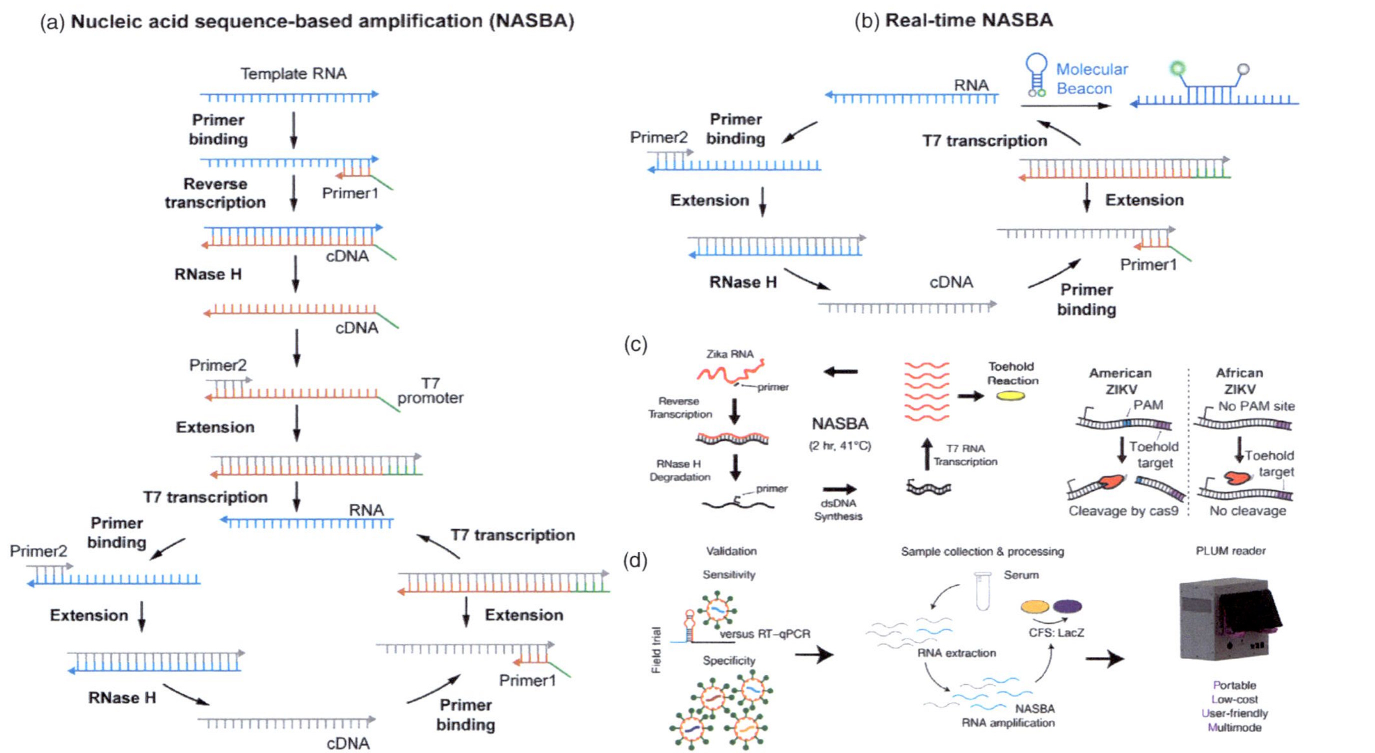

Figure 6.15 Overview of NASBA reaction. (a) The schematic diagram of the standard NASBA reaction includes two stages: "noncyclic process" and "cyclic process," which can achieve exponential amplification at a single temperature (*Source:* Figure created by the authors). (b) Real-time NASBA technology utilizes molecular beacons to detect the combination process of amplicons in real-time (*Source:* Figure created by the authors). (c) The flowchart of a cell-free paper-based sensor combining NASBA with a foothold switch RNA sensor for detecting the RNA genome of the Zika virus (*Source:* Reprinted with permission from ref. [117]. Copyright [2016] Cell Press). (d) Schematic diagram of cell-free protein expression response (PURExpress) combining NASBA and foothold transition reaction for detecting Zika virus and chikungunya virus (*Source:* Reprinted with permission from Ref. [16]. Copyright (2022) Nature Publishing Group).

transcriptase during the “cycling” stage. In the “cycle” stage, Primer 2 first binds to the RNA template, which is extended by reverse transcriptase to produce a similar RNA: DNA hybrid. Then, RNase H hydrolyzes the RNA strand, Primer 1 binds to the generated single-stranded DNA, and reverse transcriptase synthesizes DNA, producing a double-stranded DNA promoter with transcriptional activity again.

The amplification process of NASBA is very stable, and NASBA only needs to be incubated at a single temperature for 1.5 to 2 hours to achieve about 10^9 times amplification. Compared with PCR, fewer “cycles” are required to reach the same amplification amount, resulting in a reduction in the total incubation time and total error frequency of NASBA [118]. NASBA can convert single-stranded RNA, double-stranded nucleic acids (which need to be converted into single-stranded molecules compatible with the process), and nucleic acids with discrete ends. If they are first converted into single strands, they directly enter the “cycle” stage mentioned above.

When used as a research tool, NASBA can simplify the detection and sequencing of mRNA sequences, and there is no thermal denaturation during the NASBA process. Contaminated genomic DNA will not be amplified, and NASBA does not require specialized equipment, making it more suitable for laboratories with limited requirements or budgets for nucleic acid amplification.

In addition to being used as a research tool, as mentioned above, NASBA can also be clinically applied. In terms of in vitro diagnostics, NASBA can be used for screening infectious and genetic diseases, diagnosing and monitoring cancer, and testing disease susceptibility [119, 120].

For example, the NucliSENS EasyQ® HPV v1 test, developed by Jeantet et al. (2009), aims to detect the expression of oncogenic E6/E7 mRNA in the five most common oncogenic HPV types (16, 18, 31, 33, and 45) in cervical smears (Figure 6.15b) [121]. The NucliSENS EasyQ HPV v1 assay is a combination process based on NASBA technology and molecular beacons for the simultaneous detection of amplicons. All primers and probes used in the detection are designed to detect full-length mRNA transcripts of the U1A and E6/E7 regions of the five HPV types 16, 18, 31, 33, and 45 mentioned above. Six different molecular beacons were used in the detection, with one specific beacon used for each HPV subtype-specific amplicon and one specific beacon used for the U1A amplicon. Two fluorescent dyes (6-FAM and 6-ROX) can be used to detect HPV target RNAs in three dual amplification reactions: U1A and HPV 16, 18, 31, 33, and 45. The characteristic of the NASBA process is the early exponential growth of double-stranded intermediate DNA amplicons, followed by a linear increase in RNA levels, which is then converted into a linear increase in fluorescence. This test can be used for the identification and risk stratification of individuals confirmed to be infected with high-risk papillomavirus.

Subsequently, in 2016, Collins and coworkers reported a process for rapid design, assembly, and validation of cell-free paper-based sensors for detecting the RNA genome of the Zika virus [117]. This work combines isothermal RNA amplification (NASBA) with a foothold switch RNA sensor to detect clinically relevant concentrations of Zika virus sequences and demonstrates its specificity for closely related dengue virus sequences (Figure 6.15c). When combined with a novel module based on CRISPR/Cas9, the authors utilized the sequence-specific nuclease activity of CRISPR/Cas9 to distinguish virus lineages and developed NASBA-CRISPR cleavage (NASBACC). NASBACC successfully distinguished between African and American Zika virus strains by utilizing Cas9's ability to selectively cleave DNA only in the presence of NGG prototype spacer adjacent motifs (PAM).

NASBA technology can currently be applied to some on-site detection instruments for clinical diagnosis and screening of diseases [122]. In 2022, Karlikow et al. reported the results of a double-blind study on paper-based diagnostic tests for Zika virus and chikungunya virus in field environments in Latin America [16]. This test used cell-free protein expression response based on NASBA and foothold switching reaction (PURExpress), a dedicated portable reader, and onboard software for computer vision image analysis (Figure 6.15d). NASBA can achieve a limit of detection as low as 3 fM. The PURExpress reaction first transcribes an RNA-based foothold switch from a DNA template. If the target Zika virus sequence is present, the translated reporter protein produces a colorimetric output (yellow → purple) by enzymatic cleavage of chlorophenol red-β-D-galactopyranoside. Then the author also developed a low-cost, computer vision-supported automated tablet reader (PLUM reader), and the final combination of PLUM reader and paper-based Zika sensor showed comparable sensitivity and specificity for analyzing the Zika virus to RT qPCR. At the same time, the authors demonstrated that by changing the molecular components that confer detection specificity, such as NASBA primers and toehold switches, it is possible to detect chikungunya virus.

The above work combines NASBA with new detection technologies that can be applied to a wide range of public health threats. With the continuous advancement of technology, the development of in vitro portable sensors can help the world quickly respond to current and future health crises. The above work can use the engineering principles of synthetic biology to quickly design and implement biomolecular diagnostic methods for emerging pathogens, indicating that this field will play an increasingly important role in supporting and improving human health.

6.4.5 Exponential Amplification Reaction

Several isothermal amplification methods have been mentioned above, such as LAMP, RPA, NASBA, and SDA. In these exponential reactions, double-stranded amplicons are required to generate single-stranded templates for recycling the amplification. For example, the techniques of RPA and NASBA use recombinase and RNaseH in order to separate dsDNA or DNA–RNA hybrids into single-stranded forms. And in SDA, the denaturation process is required. Instead of using additional enzymes, the strategy of EXPAR makes use of the strand displacement activity of DNA polymerase to displace a single strand from dsDNA. It is precisely because the system has relatively few components and simpler steps that it is advantageous for application in real-time detection. For instance, during the COVID-19 pandemic, ID NOW was developed by Abbott Laboratories, based on the principle of EXPAR, for rapid and instant self-testing.

First reported (2003) by Galasand coworkers (Figure 6.16a) [123], EXPAR is currently most used in nucleic acid detection, in addition to small molecules, proteins, and cells. It is a completely isothermal process, and in less than 30 minutes, the amplification efficiency can reach 10^6–10^8. EXPAR employs a functional template instead of the primers used in other isothermal strategies. The template consists of two copies complementary to the target fragment separated by the recognition sequence of NEase. In the original study, Galas and co-workers used SYBR Green I for the real-time monitoring. Duan et al. combined molecular beacons (2013) with EXPAR to achieve ultrasensitive detection of breast cancer marker (Figure 6.16b) [124]. Similar to CRISDA, Xing et al. first developed CAS-EXPAR (2018) technology by combining EXPAR with Cas9 (Figure 6.16c) [125]. This strategy showed excellent specificity in discriminating single-base mismatches. Moreover, the applicability to detect DNA methylation and *Listeria monocytogenes* total RNA was also verified. EXPAR can be initiated in a variety of modes, either in combination with a heteroduplex nicking enzyme [126] (Figure 6.16d) or relying on a hairpin template for self-initiation (Figure 6.16e) [127]. In addition, EXPAR can be employed in a digital assay (Figure 6.16f), monitoring the amplification reaction of multiple enzyme molecules in real time to determine the enzyme activity [128]. Although EXPAR has an extremely wide range of applications, its practical applicability is still constrained due to the high background. However, there have been many studies to improve the high background of EXPAR [129, 130].

6.4.6 Rolling Circle Amplification

RCA is arguably one of the most straightforward of the various DNA amplification methods. It is an efficient enzymatic isothermal reaction that uses a circular probe as a template to generate long tandem single-stranded DNA or RNA products initiated by short primers [5]. In the mid-1990s, the "padlock probe" was identified as a valuable technique for the detection of nucleic acids [131]. The two ends of the linear oligonucleotide probes are brought into close proximity through the process of hybridization with a target sequence (Figure 6.17a). Subsequently, several studies demonstrated that a polymerase can utilize this probe to synthesize a long, repetitive ssDNA product (Figure 6.17b) [132]. Over the past two decades, RCA has emerged as a highly effective amplification method, with numerous subtypes having been developed. The most straightforward and fundamental form of RCA is linear, also referred to as single-primer RCA. The combination of the pre-made circle and target nucleic acid will result in the generation of a long product by DNA or RNA polymerase.

The introduction of multiple primers designed in advance facilitates the achievement of exponential amplification in RCA. The exponential amplification of DNA sequences has been successfully achieved using a variety of RCA techniques, including multiply-primed RCA, hyperbranched RCA (HRCA), circle-to-circle RCA, and primer generation (PG) RCA. The introduction of multiple primers for hybridization with a single circle results in a multiplication of events (Figure 6.17c) [133]. HRCA, which is a variant of linear RCA, incorporates a second novel primer into the amplification system, utilizing the RCA product as a template for subsequent

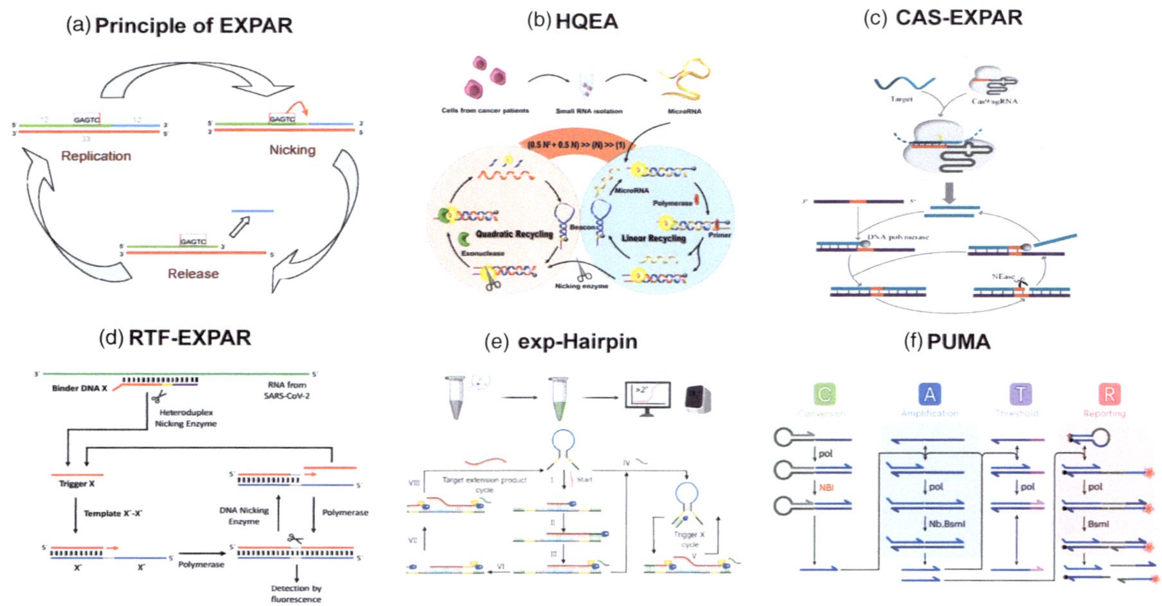

Figure 6.16 Schematic representation of EXPAR and applications of EXPAR based on target recognition. (a) Principle of classical EXPAR. One functional template is utilized in EXPAR to amplify the target nucleic acids as primers. EXPAR has an extremely fast amplification speed and is suitable for detecting short single-stranded targets (*Source:* Reprinted with permission from Ref. [123]. Copyright (2003) NAS). (b) Hairpin-mediated quadratic enzymatic amplification (HQEA). An ultrasensitive one-pot miRNA detection scheme uses a molecular beacon with detection limits of 10 fM at 37 °C (*Source:* Reprinted with permission from Ref. [124]. Copyright (2013) American Chemical Society). (c) CRISPR/Cas9-triggered isothermal exponential amplification reaction (CAS-EXPAR) strategy. It is based on CRISPR/Cas9 cleavage and nicking endonuclease-mediated nucleic acid amplification that was developed for rapid and site-specific nucleic acid detection (*Source:* Reprinted with permission from Ref. [125]. Copyright (2018) American Chemical Society). (d) Reverse transcription-free EXPAR (RTF-EXPAR). This method involves the formation of an RNA/DNA heteroduplex, whose selective cleavage generates a short DNA trigger strand, which is then rapidly amplified using EXPAR (*Source:* Reprinted with permission from Ref. [126]. Copyright (2021) NAS). (e) A hairpin template tailored for the EXPAR system (exp-Hairpin). Hairpin templates are constructed by adding identical trigger sequences and enzyme cleavage sites on two arms of the hairpin, achieving theoretically more than $2n$ amplification efficiency and minimal background (*Source:* Reprinted with permission from Ref. [127]. Copyright (2024) American Chemical Society). (f) Programmable ultrasensitive molecular amplifier (PUMA). Combining the sensitivity of an exponential molecular amplifier with the modularity of DNA–enzyme circuits and droplet readout makes it possible to specifically detect, at the single-molecule level, virtually any DNA/RNA-related enzymatic activity (*Source:* Reprinted with permission from Ref. [128]. Copyright (2024) Springer Nature).

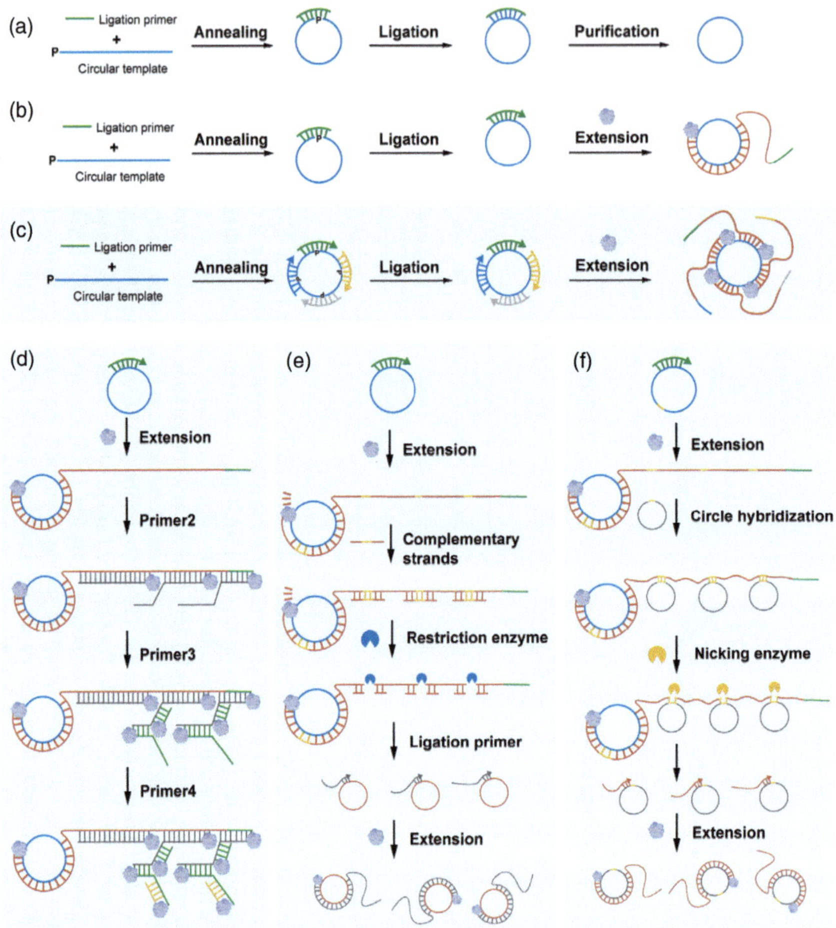

Figure 6.17 An overview of the fundamental principles of RCA. (a) The construction of a DNA circle is achieved through the process of template-mediated enzymatic ligation. (b) Linear RCA reaction is employed to generate long ssDNA with a premade circle and target nucleic acids. (c) Multiply primed is employed to generate multiple copies of the RCA product from a single circle. (d) Hyperbranched RCA uses the template for further amplification. (e) Exponential DNA amplification by generating multiple circles from the RCA products. (f) Circle-to-circle RCA by primer generation using nicking enzymes. *Source:* Figure created by the authors.

amplification (Figure 6.17e) [134]. With regard to circle-to-circle RCA (C2CA), the products are subjected to digestion in order to be converted into monomers, which are then circularized. The circles then function as templates for the subsequent RCA (Figure 6.17f) [135]. Also referred to as "nicking RCA," "PG RCA" employs nicking enzymes to cut multiple circular probe hybrids, thereby enabling the realization of multiple reaction cycles (Figure 6.17g) [136].

The RCA template cyclization step can be designed to achieve highly specific target recognition, and thus RCA-based biosensors are applied to a variety of

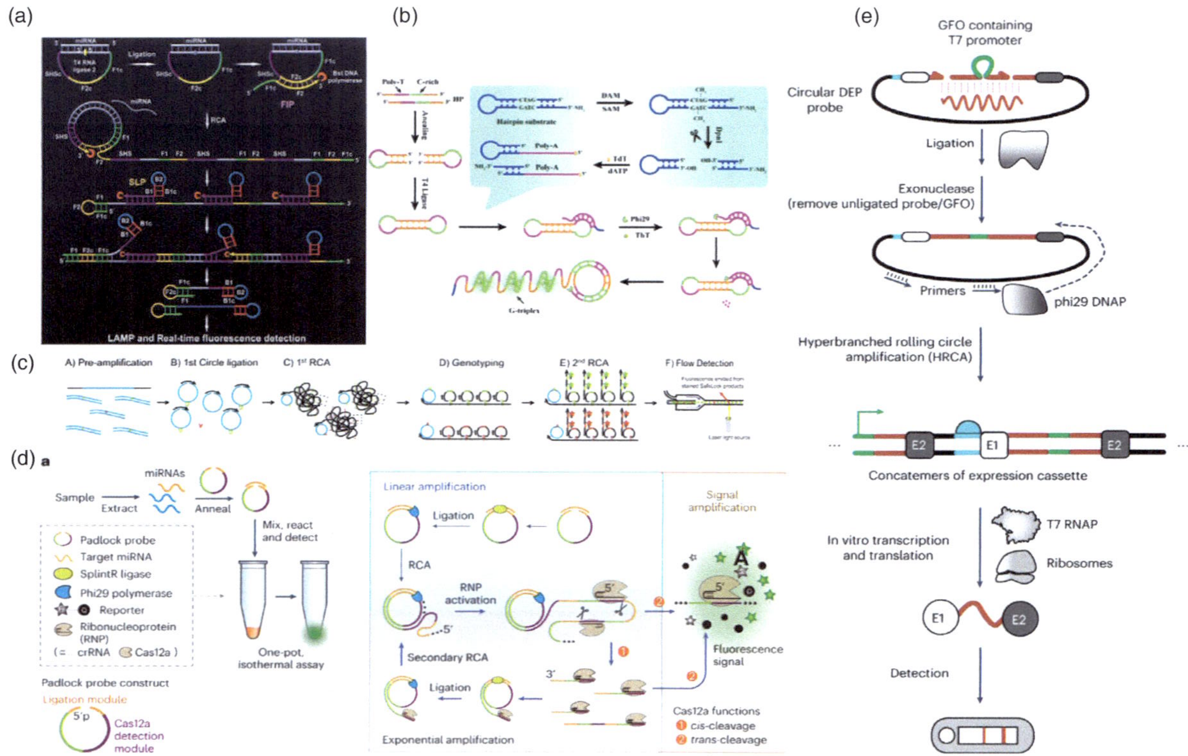

Figure 6.18 Applications of RCA-based assays in point-of-care settings. (a) The integration of loop-mediated isothermal amplification (LAMP) and rolling circle amplification (RCA) to detect miRNAs (*Source:* Reprinted with permission from Ref. [137], Copyright [2019]Elsevier). (b) Terminal deoxynucleotidyl transferase (TdT)-induced RCA to analyze DNA MTase activity (*Source:* Reprinted with permission from Ref. [138], Copyright [2019] Elsevier). (c) Using a standard flow cytometer to visualize RCA products achieves ultrasensitive detection of DNA (*Source:* Reprinted with permission from Ref. [139], Copyright (2022) Nature Publishing Group). (d) A one-step and one-pot isothermal assay that leverages RCA and the endonuclease Cas12a for the accurate detection of specific miRNAs (*Source:* Reprinted with permission from Ref. [140], Copyright [2023] Nature Publishing Group). (e) HRCA combined with internal splint-pairing expression-cassette translation reaction enables sensitive detection of target (*Source:* Reprinted with permission from Ref. [141], Copyright [2023] Nature Publishing Group).

requirements in combination with other techniques. In a recent study, Li et al. developed a novel approach for the detection of miRNAs through the integration of LAMP and RCA [137]. The fundamental premise of this concept is to transform the output of RCA into the template for LAMP. The method exhibits a detection limit of 10 aM (Figure 6.18a).

Due to the numerous RCA products, they can be programmed by manipulating the circular template with label-free fluorescent dyes. Yan and coworkers developed a transferase-induced RCA strategy that is highly sensitive and can be used for fluorescence-based biosensing of DNA MTase activity (Figure 6.18b) [138]. The dumbbell-like RCA template and a palindromic sequence serve to initiate the RCA reaction, with the generation of a fluorescence signal achieved through the combination of ThT and G-triplexes. In conclusion, the MTase can be detected at a concentration as low as 0.058 U/mL.

Furthermore, flow cytometry represents a powerful and creative tool that can be employed in the construction of an RCA-based biosensor. In a study conducted by Landegren and colleagues, a strategy referred to as the superRCA assay was developed. This approach employs a conventional flow cytometer to quantify the production of large superRCA products, each of which represents a single initial DNA circle (Figure 6.18c) [139].

Given that RCA is an optimal pre-amplification method for sensitive CRISPR miRNA assays, Yan et al. have successfully developed one-step, one-pot CRISPR assays for miRNA sensing (Figure 6.18d) [140]. The authors incorporate multiple reactions, including those for target-mediated ligation, RCA, Cas12a binding, and nucleolytic cleavage, into a single reaction network. The analysis is dependent on the targets and samples, with a completion time of between 20 minutes and three hours.

The primary enzymes of RCA are active at temperatures close to ambient, with the ability to operate below 30°C. In their INSPECTR technology for the detection of viral RNA, such as SARS-CoV-2, Phillips et al. selected HRCA as the amplification method [141]. RCA was instrumental in reducing background signal, thereby facilitating the sensitive detection of targets (Figure 6.18e).

6.5 Isothermal and Enzyme-Free Nucleic Acid Amplification

Nucleic acid amplification techniques are generally categorized into two types, namely thermal cycling amplification and isothermal amplification. In turn, isothermal amplification methods can be subdivided into enzyme-mediated and enzyme-independent reactions. In particular, toehold-mediated strand displacement (TMSD) amplification, displaying high amplification efficiency and controllable kinetics, is one of the most popular enzyme-free amplification strategies. In the TMSD system, hybridization chain reaction (HCR) is a concise and efficacious isothermal amplification process proposed by Dirks and Pierce [142] in 2004 (Figure 6.19a). In the initial HCR, the initiator triggers cascade

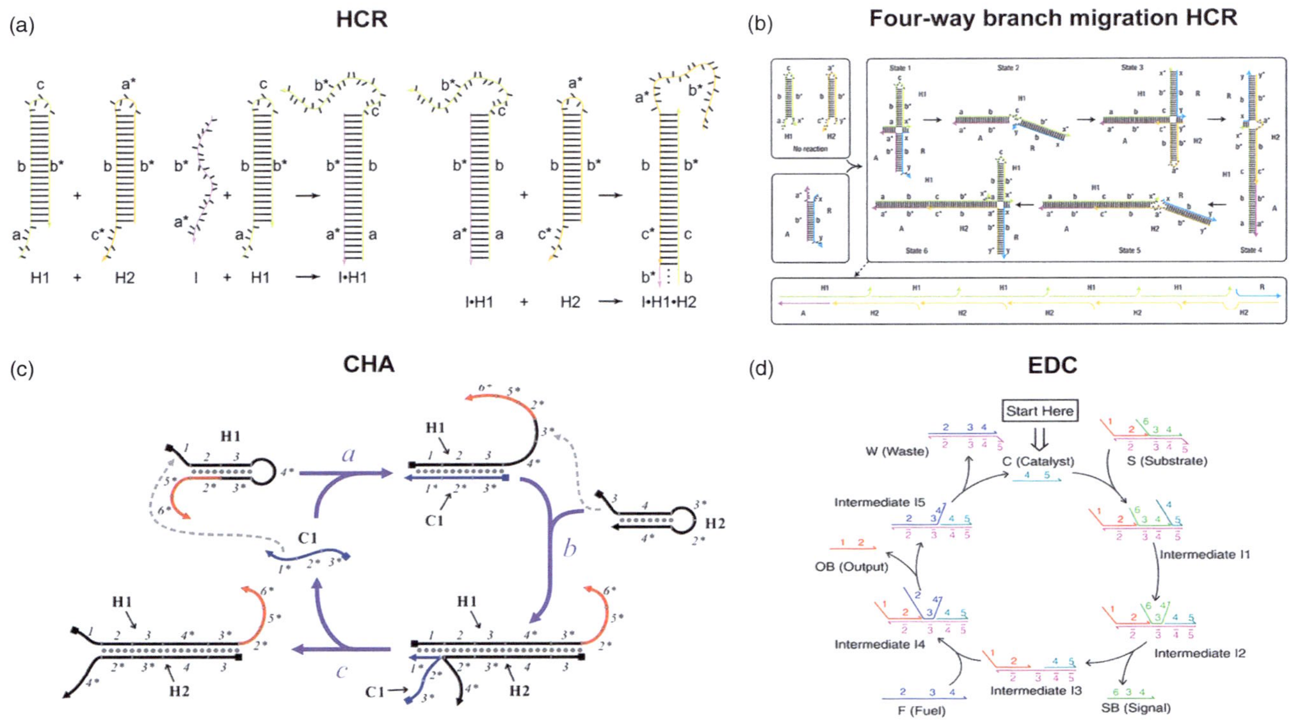

Figure 6.19 Enzyme-free amplification strategies. (a) Scheme of the hybridization chain reaction (HCR) system. Two stable species of DNA hairpins coexist in solution until the introduction of initiator strands triggers a cascade of hybridization events (*Source:* Reprinted with permission from Ref. [142]. Copyright [2004] NAS). (b) Diagram of four-way branch migration HCR. A formal improvement has been made over the initial version (*Source:* Reprinted with permission from Ref. [143]. Copyright [2007] Springer Nature). (c) Principle of catalyzed hairpin assembly (CHA). Signal amplification in CHA is achieved by modular circuits of DNA hybridization (self-assembling and disassembling) rather than a chain reaction (*Source:* Reprinted with permission from Ref. [144]. Copyright (2011) PMC). (d) Illustration of entropy-driven circuit (EDC). This strategy, driven by entropy, allows a specified input oligonucleotide to catalyze the release of a specified output oligonucleotide, which in turn can serve as a catalyst for other reactions, with the simple, fast, modular, composable, and robust performance (*Source:* Reprinted with permission from Ref. [145]. Copyright [2007] The American Association for the Advancement of Science).

hybridization between two DNA hairpins based on the toehold, forming a double helix structure with hundreds of repeating units, which continues until the hairpin probes are depleted. In 2007, the same research group proposed an autonomous polymerization motor in which the target triggers a four-way branch migration to perform HCR (Figure 6.19b) [143]. Differing from the initial linear form, in recent years, researchers have devised more complex template forms, such as branching [146] and dendritic [147]. The mild reaction condition is a significant advantage of HCR, which makes it an ideal means for signal amplification. HCR-based amplified biosensing methods are widely used for nucleic acids, proteins, enzyme activities, tumor cells, and small molecules. It is worth mentioning that, in recent years, HCR-based FISH (HCR-FISH) techniques have been developed to generate sufficient signal-to-background ratios. Using HCR-FISH, Pierce's group conducted a series of innovative studies [148–150].

CHA is another enzyme-free amplification technique, which is driven by enthalpy. CHA was originally introduced by Yin et al. [151] and engineered for multiple analytical applications by Li et al. [144] (Figure 6.19c). Comparable to HCR, CHA uses two metastable hairpins to store potential energy as well. The two hairpins do not initially interact with each other but can catalytically form a duplex in the presence of an ssDNA input. In subsequent research, the same research group developed the layer-CHA method [152], which increased amplification efficiency by 7,000 times. This technique can be paired with a variety of signal outputs, such as fluorescence, electrochemical, and colorimetric signal outputs. CHA is widely used for nucleic acid detection and live cell imaging. miRNA sensors based on CHA account for the maximum portion of CHA-based biosensors. With the use of CHA, Li et al. detected three miRNAs and successfully applied this technology to clinical samples [153].

Apart from enthalpy-driven reactions, the entropy-driven circuit (EDC) is another enzyme-free, isothermal, DNA signal-amplification reaction that is propagated by the net gain of entropy within the system. The entropy-driven catalytic gate is substantially simpler than hairpin hybridization on the level of design; moreover, it is faster and more modular. Zhang et al. pioneered the concept of EDC and designed an autocatalyst system (Figure 6.19d) [145]. A typical EDC system consists of a three-stranded substrate complex (S), signal strands (SB), and a fuel strand (F). In the absence of the catalyst strand (C), the reaction between S and F is very slow because the toehold for the reaction with F has been caged. In the presence of the catalyst strand, the TMSD reactions are carried out sequentially, releasing the signal strand and output strand. The catalytic DNA is recycled, generating multiple outputs of DNA. The total number of base pairs is not changed and is thus driven by a gain in entropy from producing liberated molecules. Multilayered EDC reactions and autocatalytic cascades have been constructed for exponential amplification [145]. EDC can also be combined with CHA for increased efficiency of amplification [154]. Zhang et al. employed nanoparticles to deliver an EDC system intracellularly to detect miRNA let-7a [155].

Compared with PCR and enzyme-associated isothermal amplification technology, the enzyme-free amplification systems have significant advantages because it is not limited by enzyme activity and it is better in reagent cost, storage, and transportation. Although the principles are well understood and the processes do not require the involvement of enzymes, they are also subject to several challenges. In the first instance, its performance relies on DNA-assembly, so that the rate is governed by spatial structure and kinetic factors. Second, the stability and robustness of the whole nucleic acid system are insufficient, making it difficult to achieve stable results in real clinical testing. Third, to improve the sensitivity, it is necessary to increase the number of probes, but this carries the risk of high background due to the crosstalk between DNA probes. Despite many challenges, enzyme-free amplification strategies continue to attract the attention of researchers. There is a consensus that through technological integration and innovation, these types of enzyme-free strategies will make more practical breakthroughs.

6.6 Summary and Outlook

Advancements in molecular diagnostics, particularly PCR-based amplification techniques and isothermal amplification methods, have significantly bolstered the field of public health. By integrating advanced analytical and chemometric approaches, modern molecular diagnostic techniques have demonstrated remarkably enhanced sensitivity, specificity, speed, and accessibility for the detection of various disease biomarkers, providing timely diagnostic data that support clinical decision-making during infectious disease outbreaks, bioterrorism monitoring, and early disease screening. Innovations in PCR formats, including qPCR and digital droplet PCR, alongside the rise of isothermal methods such as LAMP and RPA, have transformed the molecular diagnostic landscape. This chapter systematically examines the evolution of molecular diagnostic technology, with a focus on the development and clinical application of key nucleic acid amplification techniques.

Despite the notable progress in molecular diagnostics, several challenges persist in meeting the evolving clinical environment and diagnostic needs. While PCR-based methods are highly accurate, they require complex thermal cycling equipment, limiting their use in resource-constrained settings. In contrast, isothermal amplification methods offer simplicity and speed, yet require further optimization—particularly regarding probe design—and greater clinical accessibility before they can be widely adopted. Additionally, issues such as nonspecific amplification and limited multiplexing capabilities in isothermal technologies must be addressed to fully unlock their potential.

Looking ahead, the integration of molecular diagnostics with emerging fields presents exciting opportunities. Portable, point-of-care devices powered by microfluidics and nanotechnology are facilitating the broader adoption of isothermal amplification technologies, enabling rapid and accurate testing in areas with limited medical resources. The incorporation of CRISPR-based detection systems is poised to enhance diagnostic precision, offering highly specific and customizable tools for genetic material detection. Furthermore, advances in bioinformatics and machine learning are likely to play pivotal roles in managing and interpreting the vast amounts of data generated by next-generation diagnostics. As these technologies continue to evolve, there is substantial potential to develop more efficient, accessible, and robust molecular diagnostic platforms, further strengthening global public health surveillance and response systems. In conclusion, ongoing innovation in PCR and isothermal amplification technologies, combined with the integration of emerging tools and digital platforms, will continue to drive the future of molecular diagnostics in public health. This progress will improve our ability to respond to current and emerging threats with greater speed and accuracy. At the same time, the integration of metrology into molecular diagnostics not only enhances the precision and reproducibility of these technologies but also strengthens their utility in public safety applications. Looking ahead, the continuous refinement of measurement standards will be crucial for advancing diagnostic accuracy and ensuring consistent performance across diverse healthcare settings. This chapter's focus on metrological and chemometric approaches reflects the broader themes of this book, underscoring the indispensable role of rigorous measurement techniques in the ongoing development of effective disease detection and management strategies.

References

1 Wang, C., Horby, P.W., Hayden, F.G., and Gao, G.F. (2020). A novel coronavirus outbreak of global health concern. *The Lancet.* 395 (10223): 470–473.

2 Wölfel, R., Corman, V.M., Guggemos, W. et al. (2020). Virological assessment of hospitalized patients with COVID-2019. *Nature.* 581 (7809): 465–469.

3 Vermeesch, J.R., Voet, T., and Devriendt, K. (2016). Prenatal and pre-implantation genetic diagnosis. *Nature Reviews Genetics.* 17 (10): 643–656.

4 Pettit, K., Hull, A., Korty, L. et al. (2014). The utilization of circulating cell-free fetal DNA testing and decrease in invasive diagnostic procedures: an institutional experience. *Journal of Perinatology.* 34 (10): 750–753.

5 Fire, A. and Xu, S.-Q. (1995). Rolling replication of short DNA circles. *Proceedings of the National Academy of Sciences.* 92 (10): 4641–4645.

6 Notomi, T., Okayama, H., Masubuchi, H. et al. (2000). Loop-mediated isothermal amplification of DNA. *Nucleic acids research.* 28 (12): e63-e.

7 Piepenburg, O., Williams, C.H., Stemple, D.L., and Armes, N.A. (2006). DNA detection using recombination proteins. *PLoS Biology.* 4 (7): e204.

8 Walker, G.T., Fraiser, M.S., Schram, J.L. et al. (1992). Strand displacement amplification—an isothermal, in vitro DNA amplification technique. *Nucleic Acids Research.* 20 (7): 1691–1696.

9 Compton, J. (1991). Nucleic acid sequence-based amplification. *Nature.* 350 (6313): 91–92.

10 Tayyab, M., Barrett, D., van Riel, G. et al. (2023). Digital assay for rapid electronic quantification of clinical pathogens using DNA nanoballs. *Science Advances.* 9(36):eadi4997.

11 Dao Thi, V.L., Herbst, K., Boerner, K. et al. (2020). A colorimetric RT-LAMP assay and LAMP-sequencing for detecting SARS-CoV-2 RNA in clinical samples. *Science Translational Medicine.* 12 (556): eabc7075.
12 Gupta, R., Gupta, P., Wang, S. et al. (2023). Ultrasensitive lateral-flow assays via plasmonically active antibody-conjugated fluorescent nanoparticles. *Nature Biomedical Engineering.* 7 (12): 1556–1570.
13 Ottesen, E.A., Hong, J.W., Quake, S.R., and Leadbetter, J.R. (2006). Microfluidic digital PCR enables multigene analysis of individual environmental bacteria. *Science.* 314 (5804): 1464–1467.
14 Najjar, D., Rainbow, J., Sharma Timilsina, S. et al. (2022). A lab-on-a-chip for the concurrent electrochemical detection of SARS-CoV-2 RNA and anti-SARS-CoV-2 antibodies in saliva and plasma. *Nature Biomedical Engineering.* 6 (8): 968–978.
15 Khodakov, D., Li, J., Zhang, J.X., and Zhang, D.Y. (2021). Highly multiplexed rapid DNA detection with single-nucleotide specificity via convective PCR in a portable device. *Nature Biomedical Engineering.* 5 (7): 702–712.
16 Karlikow, M., da Silva, S.J.R., Guo, Y. et al. (2022). Field validation of the performance of paper-based tests for the detection of the Zika and chikungunya viruses in serum samples. *Nature Biomedical Engineering.* 6 (3): 246–256.
17 Lin, H., Yu, W., Sabet, K.A. et al. (2022). Ferrobotic swarms enable accessible and adaptable automated viral testing. *Nature.* 611 (7936): 570–577.
18 Li, C., Kang, N., Ye, S. et al. (2024). All-in-one OsciDrop digital PCR system for automated and highly multiplexed molecular diagnostics. *Advanced Science.* 2309557.
19 Liu, Y., Yang, Y., Wang, G. et al. (2023). Multiplexed discrimination of SARS-CoV-2 variants via plasmonic-enhanced fluorescence in a portable and automated device. *Nature Biomedical Engineering.* 7 (12): 1636–1648.
20 Mullis, K., Faloona, F., Scharf, S. et al. (ed.) (1986). Specific enzymatic amplification of DNA in vitro: the polymerase chain reaction. In: *Cold Spring Harbor Symposia on Quantitative Biology*. Cold Spring Harbor Laboratory Press.
21 Higuchi, R., Fockler, C., Dollinger, G., and Watson, R. (1993). Kinetic PCR analysis: real-time monitoring of DNA amplification reactions. *Bio/technology.* 11 (9): 1026–1030.
22 Vogelstein, B. and Kinzler, K.W. (1999). Digital PCR. *Proceedings of the National Academy of Sciences.* 96 (16): 9236–9241.
23 Walker, G.T., Little, M.C., Nadeau, J.G., and Shank, D.D. (1992). Isothermal in vitro amplification of DNA by a restriction enzyme/DNA polymerase system. *Proceedings of the National Academy of Sciences.* 89 (1): 392–396.
24 Zhang, D.Y. and Seelig, G. (2011). Dynamic DNA nanotechnology using strand-displacement reactions. *Nature Chemistry.* 3 (2): 103–113.
25 Tang, Y., Gao, L., Feng, W. et al. (2021). The CRISPR–Cas toolbox for analytical and diagnostic assay development. *Chemical Society Reviews.* 50 (21): 11844–11869.
26 Foley, K.P., Leonard, M.W., and Engel, J.D. (1993). Quantitation of RNA using the polymerase chain reaction. *Trends in Genetics.* 9 (11): 380–385.

27 Sanchez, J.A., Pierce, K.E., Rice, J.E., and Wangh, L.J. (2004). Linear-after-the-exponential (LATE)–PCR: an advanced method of asymmetric PCR and its uses in quantitative real-time analysis. *Proceedings of the National Academy of Sciences.* 101 (7): 1933–1938.

28 Will, K., Reiss, J., Dean, M. et al. (1993). CFTR transcripts are undetectable in lymphocytes and respiratory epithelial cells of a CF patient homozygous for the nonsense mutation R553X. *Journal of Medical Genetics.* 30 (10): 833–837.

29 Green, M.R. and Sambrook, J. (2019). Nested polymerase chain reaction (PCR). *Cold Spring Harbor Protocols* 2019 (2): pdb prot095182.

30 Deepachandi, B., Weerasinghe, S., Soysa, P. et al. (2019). A highly sensitive modified nested PCR to enhance case detection in leishmaniasis. *BMC Infectious Diseases* 19: 1–10.

31 Newton, C., Graham, A., Heptinstall, L. et al. (1989). Analysis of any point mutation in DNA. The amplification refractory mutation system (ARMS). *Nucleic Acids Research.* 17 (7): 2503–2516.

32 Islam, M.T., Alam, A.R.U., Sakib, N. et al. (2021). A rapid and cost-effective multiplex ARMS-PCR method for the simultaneous genotyping of the circulating SARS-CoV-2 phylogenetic clades. *Journal of Medical Virology.* 93 (5): 2962–2970.

33 Cai, Y., Zhang, S., Zhang, J. et al. (2022). Visual detection of ACE I/D polymorphism using T-ARMS-PCR combined with a lateral flow assay and its clinical application. *Analytical Chemistry.* 94 (11): 4686–4694.

34 Markman, B., Javier Ramos, F., Capdevila, J., and Tabernero, J. (2010). EGFR and KRAS in colorectal cancer. *Advances in Clinical Chemistry.* 51: 72.

35 Guiver, M., Borrow, R., Marsh, J. et al. (2000). Evaluation of the applied biosystems automated Taqman polymerase chain reaction system for the detection of meningococcal DNA. *FEMS Immunology & Medical Microbiology.* 28 (2): 173–179.

36 Heid, C.A., Stevens, J., Livak, K.J., and Williams, P.M. (1996). Real time quantitative PCR. *Genome Research.* 6 (10): 986–994.

37 Livak, K.J. and Schmittgen, T.D. (2001). Analysis of relative gene expression data using real-time quantitative PCR and the 2−ΔΔCT method. *Methods.* 25 (4): 402–408.

38 Grimberg, B.T. (2011). Methodology and application of flow cytometry for investigation of human malaria parasites. *Journal of Immunological Methods.* 367 (1–2): 1–16.

39 Tyagi, S. and Kramer, F.R. (1996). Molecular beacons: probes that fluoresce upon hybridization. *Nature Biotechnology.* 14 (3): 303–308.

40 Tyagi, S., Bratu, D.P., and Kramer, F.R. (1998). Multicolor molecular beacons for allele discrimination. *Nature Biotechnology.* 16 (1): 49–53.

41 Huang, Q., Chen, D., Du, C. et al. (2022). Highly multiplex PCR assays by coupling the 5′-flap endonuclease activity of Taq DNA polymerase and molecular beacon reporters. *Proceedings of the National Academy of Sciences.* 119 (9): e2110672119.

42 Guo, C., Deng, H., Yang, Q. et al. (2023). Coding intrinsic disorder into DNA hybridization probes enables discrimination of single nucleotide variants over wide and tunable temperature ranges. *Angewandte Chemie.* 135 (49): e202314386.

43 Holland, P.M., Abramson, R.D., Watson, R., and Gelfand, D.H. (1991). Detection of specific polymerase chain reaction product by utilizing the 5′----3′exonuclease activity of *Thermus aquaticus* DNA polymerase. *Proceedings of the National Academy of Sciences.* 88 (16): 7276–7280.

44 Kutyavin, I.V., Afonina, I.A., Mills, A. et al. (2000). 3′-minor groove binder-DNA probes increase sequence specificity at PCR extension temperatures. *Nucleic Acids Research.* 28 (2): 655–661.

45 Whitcombe, D., Theaker, J., Guy, S.P. et al. (1999). Detection of PCR products using self-probing amplicons and fluorescence. *Nature Biotechnology.* 17 (8): 804–807.

46 Wu, L.R., Chen, S.X., Wu, Y. et al. (2017). Multiplexed enrichment of rare DNA variants via sequence-selective and temperature-robust amplification. *Nature Biomedical Engineering.* 1 (9): 714–723.

47 Ruano, G., Kidd, K.K., and Stephens, J.C. (1990). Haplotype of multiple polymorphisms resolved by enzymatic amplification of single DNA molecules. *Proceedings of the National Academy of Sciences.* 87 (16): 6296–6300.

48 Burns, M.A., Johnson, B.N., Brahmasandra, S.N. et al. (1998). An integrated nanoliter DNA analysis device. *Science.* 282 (5388): 484–487.

49 Heyries, K.A., Tropini, C., VanInsberghe, M. et al. (2011). Megapixel digital PCR. *Nature Methods.* 8 (8): 649–651.

50 Lun, F.M., Chiu, R.W., Allen Chan, K. et al. (2008). Microfluidics digital PCR reveals a higher than expected fraction of fetal DNA in maternal plasma. *Clinical Chemistry.* 54 (10): 1664–1672.

51 Song, H., Tice, J.D., and Ismagilov, R.F. (2003). A microfluidic system for controlling reaction networks in time. *Angewandte Chemie.* 115 (7): 792–796.

52 Beer, N.R., Wheeler, E.K., Lee-Houghton, L. et al. (2008). On-chip single-copy real-time reverse-transcription PCR in isolated picoliter droplets. *Analytical Chemistry.* 80 (6): 1854–1858.

53 Pekin, D., Skhiri, Y., Baret, J.-C. et al. (2011). Quantitative and sensitive detection of rare mutations using droplet-based microfluidics. *Lab on a Chip.* 11 (13): 2156–2166.

54 Hindson, B.J. et al. (2011). High-throughput droplet digital PCR system for absolute quantitation of DNA copy number. *Analytical Chemistry.* 83 (22): 8604–8610.

55 Pinheiro, L.B., Coleman, V.A., Hindson, C.M. et al. (2012). Evaluation of a droplet digital polymerase chain reaction format for DNA copy number quantification. *Analytical Chemistry.* 84 (2): 1003–1011.

56 McDermott, G.P., Do, D., Litterst, C.M. et al. (2013). Multiplexed target detection using DNA-binding dye chemistry in droplet digital PCR. *Analytical Chemistry.* 85 (23): 11619–11627.

57 Warren, L., Bryder, D., Weissman, I.L., and Quake, S.R. (2006). Transcription factor profiling in individual hematopoietic progenitors by digital RT-PCR. *Proceedings of the National Academy of Sciences.* 103 (47): 17807–17812.
58 Baker, M. (2012). Digital PCR hits its stride. *Nature Methods.* 9 (6): 541–544.
59 Qin, J., Jones, R.C., and Ramakrishnan, R. (2008). Studying copy number variations using a nanofluidic platform. *Nucleic acids Research.* 36 (18): e116-e.
60 Tadmor, A.D., Ottesen, E.A., Leadbetter, J.R., and Phillips, R. (2011). Probing individual environmental bacteria for viruses by using microfluidic digital PCR. *Science.* 333 (6038): 58–62.
61 Sanders, R., Huggett, J.F., Bushell, C.A. et al. (2011). Evaluation of digital PCR for absolute DNA quantification. *Analytical Chemistry.* 83 (17): 6474–6484.
62 Thorsen, T., Roberts, R.W., Arnold, F.H., and Quake, S.R. (2001). Dynamic pattern formation in a vesicle-generating microfluidic device. *Physical Review Letters.* 86 (18): 4163.
63 Beer, N.R., Hindson, B.J., Wheeler, E.K. et al. (2007). On-chip, real-time, single-copy polymerase chain reaction in picoliter droplets. *Analytical Chemistry.* 79 (22): 8471–8475.
64 Kiss, M.M., Ortoleva-Donnelly, L., Beer, N.R. et al. (2008). High-throughput quantitative polymerase chain reaction in picoliter droplets. *Analytical Chemistry.* 80 (23): 8975–8981.
65 Mazutis, L., Araghi, A.F., Miller, O.J. et al. (2009). Droplet-based microfluidic systems for high-throughput single DNA molecule isothermal amplification and analysis. *Analytical Chemistry.* 81 (12): 4813–4821.
66 Hatch, A.C., Fisher, J.S., Tovar, A.R. et al. (2011). 1-million droplet array with wide-field fluorescence imaging for digital PCR. *Lab on a Chip.* 11 (22): 3838–3845.
67 Ye, S., Li, C., Zheng, X. et al. (2022). OsciDrop: a versatile deterministic droplet generator. *Analytical Chemistry.* 94 (6): 2918–2925.
68 Chen, Z., Fu, Y., Zhang, F. et al. (2016). Spinning micropipette liquid emulsion generator for single cell whole genome amplification. *Lab on a Chip.* 16 (23): 4512–4516.
69 Xu, P., Zheng, X., Tao, Y., and Du, W. (2016). Cross-interface emulsification for generating size-tunable droplets. *Analytical Chemistry.* 88 (6): 3171–3177.
70 Dangla, R., Kayi, S.C., and Baroud, C.N. (2013). Droplet microfluidics driven by gradients of confinement. *Proceedings of the National Academy of Sciences.* 110 (3): 853–858.
71 Chen, Z., Liao, P., Zhang, F. et al. (2017). Centrifugal micro-channel array droplet generation for highly parallel digital PCR. *Lab on a Chip.* 17 (2): 235–240.
72 Thompson, A.M., Gansen, A., Paguirigan, A.L. et al. (2014). Self-digitization microfluidic chip for absolute quantification of mRNA in single cells. *Analytical Chemistry.* 86 (24): 12308–12314.
73 Schuler, F., Schwemmer, F., Trotter, M. et al. (2015). Centrifugal step emulsification applied for absolute quantification of nucleic acids by digital droplet RPA. *Lab on a Chip.* 15 (13): 2759–2766.

74 Shen, F., Du, W., Kreutz, J.E. et al. (2010). Digital PCR on a SlipChip. *Lab on a Chip.* 10 (20): 2666–2672.

75 Sundberg, S.O., Wittwer, C.T., Gao, C., and Gale, B.K. (2010). Spinning disk platform for microfluidic digital polymerase chain reaction. *Analytical Chemistry.* 82 (4): 1546–1550.

76 Cohen, D.E., Schneider, T., Wang, M., and Chiu, D.T. (2010). Self-digitization of sample volumes. *Analytical Chemistry.* 82 (13): 5707–5717.

77 Men, Y., Fu, Y., Chen, Z. et al. (2012). Digital polymerase chain reaction in an array of femtoliter polydimethylsiloxane microreactors. *Analytical Chemistry.* 84 (10): 4262–4266.

78 Zhu, Q., Gao, Y., Yu, B. et al. (2012). Self-priming compartmentalization digital LAMP for point-of-care. *Lab on a Chip.* 12 (22): 4755–4763.

79 Gansen, A., Herrick, A.M., Dimov, I.K. et al. (2012). Digital LAMP in a sample self-digitization (SD) chip. *Lab on a Chip.* 12 (12): 2247–2254.

80 Lee, N.Y. (2018). A review on microscale polymerase chain reaction based methods in molecular diagnosis, and future prospects for the fabrication of fully integrated portable biomedical devices. *Microchimica Acta.* 185: 1–22.

81 Roche, P.J., Beitel, L.K., Khan, R. et al. (2012). Demonstration of a plasmonic thermocycler for the amplification of human androgen receptor DNA. *Analyst.* 137 (19): 4475–4481.

82 Son, J.H., Cho, B., Hong, S. et al. (2015). Ultrafast photonic PCR. *Light: Science & Applications.* 4 (7): e280-e.

83 Cheong, J., Yu, H., Lee, C.Y. et al. (2020). Fast detection of SARS-CoV-2 RNA via the integration of plasmonic thermocycling and fluorescence detection in a portable device. *Nature Biomedical Engineering* 4 (12): 1159–1167.

84 Lee, J.H., Cheglakov, Z., Yi, J. et al. (2017). Plasmonic Photothermal gold Bipyramid Nanoreactors for ultrafast real-time bioassays. *Journal of the American Chemical Society* 139 (24): 8054–8057.

85 Blumenfeld, N.R., Bolene, M.A.E., Jaspan, M. et al. (2022). Multiplexed reverse-transcriptase quantitative polymerase chain reaction using plasmonic nanoparticles for point-of-care COVID-19 diagnosis. *Nature Nanotechnology* 17 (9): 984–992.

86 Kopp, M.U., Mello, A.J., and Manz, A. (1998). Chemical amplification: continuous-flow PCR on a chip. *Science.* 280 (5366): 1046–1048.

87 Krishnan, M., Ugaz, V.M., and Burns, M.A. (2002). PCR in a Rayleigh-Benard convection cell. *Science.* 298 (5594): 793.

88 Anderson, R.C., Su, X., Bogdan, G.J., and Fenton, J. (2000). A miniature integrated device for automated multistep genetic assays. *Nucleic Acids Research.* 28 (12): e60-e.

89 Liang, C., Yang, Z., and Jiang, H. (2022). A film-lever actuated switch technology for multifunctional, on-demand, and robust manipulation of liquids. *Nature Communications* 13 (1): 4902.

90 Nakano, H., Matsuda, K., Yohda, M. et al. (1994). High speed polymerase chain reaction in constant flow. *Bioscience, Biotechnology, and Biochemistry.* 58 (2): 349–352.

91 Wheeler, E., Benett, W., Stratton, P. et al. (2004). Convectively driven polymerase chain reaction thermal cycler. *Analytical Chemistry.* 76 (14): 4011–4016.
92 Burns, M.A., Mastrangelo, C.H., Sammarco, T.S. et al. (1996). Microfabricated structures for integrated DNA analysis. *Proceedings of the National Academy of Sciences.* 93 (11): 5556–5561.
93 Chang, Y.H., Lee, G.B., Huang, F.C. et al. (2006). Integrated polymerase chain reaction chips utilizing digital microfluidics. *Biomed Microdevices.* 8 (3): 215–225.
94 Neuzil, P., Zhang, C., Pipper, J. et al. (2006). Ultra fast miniaturized real-time PCR: 40 cycles in less than six minutes. *Nucleic Acids Research* 34 (11): e77.
95 Tomita, N., Mori, Y., Kanda, H., and Notomi, T. (2008). Loop-mediated isothermal amplification (LAMP) of gene sequences and simple visual detection of products. *Nature Protocols.* 3 (5): 877–882.
96 Li, B., Chen, X., and Ellington, A.D. (2012). Adapting enzyme-free DNA circuits to the detection of loop-mediated isothermal amplification reactions. *Analytical Chemistry.* 84 (19): 8371–8377.
97 Li, H., Lu, H., Tang, Y. et al. (2022). A rebuilding-free nucleic acid detection strategy enables ultrasensitive genotyping, N-in-1 logic screening and accurate multiplex assay of dangerous pathogens. *Angewandte Chemie International Edition.* 61 (40): e202209496.
98 Li, L., Li, S., Wu, N. et al. (2019). HOLMESv2: a CRISPR-Cas12b-assisted platform for nucleic acid detection and DNA methylation quantitation. *ACS Synthetic Biology.* 8 (10): 2228–2237.
99 Broughton, J.P., Deng, X., Yu, G. et al. (2020). CRISPR–Cas12-based detection of SARS-CoV-2. *Nature Biotechnology.* 38 (7): 870–874.
100 Notomi, T., Mori, Y., Tomita, N., and Kanda, H. (2015). Loop-mediated isothermal amplification (LAMP): principle, features, and future prospects. *Journal of Microbiology.* 53 (1): 1–5.
101 Parida, M., Sannarangaiah, S., Dash, P.K. et al. (2008). Loop mediated isothermal amplification (LAMP): a new generation of innovative gene amplification technique; perspectives in clinical diagnosis of infectious diseases. *Reviews in Medical Virology* 18 (6): 407–421.
102 Mori, Y., Nagamine, K., Tomita, N., and Notomi, T. (2001). Detection of loop-mediated isothermal amplification reaction by turbidity derived from magnesium pyrophosphate formation. *Biochemical and Biophysical Research Communications.* 289 (1): 150–154.
103 Mori, Y., Kitao, M., Tomita, N., and Notomi, T. (2004). Real-time turbidimetry of LAMP reaction for quantifying template DNA. *Journal of Biochemical and Biophysical Methods.* 59 (2): 145–157.
104 Goto, M., Honda, E., Ogura, A. et al. (2009). Colorimetric detection of loop-mediated isothermal amplification reaction by using hydroxy naphthol blue. *Biotechniques.* 46 (3): 167–172.
105 Tanner, N.A., Zhang, Y., and Evans, T.C. Jr., (2015). Visual detection of isothermal nucleic acid amplification using pH-sensitive dyes. *Biotechniques.* 58 (2): 59–68.

106 Jin, M., Ding, J., Zhou, Y. et al. (2024). StratoLAMP: label-free, multiplex digital loop-mediated isothermal amplification based on visual stratification of precipitate. *Proceedings of the National Academy of Sciences.* 121 (2): e2314030121.

107 Dong, T., Ding, R., Zhou, R. et al. (2024). A portable nucleic acid testing platform with photosensitization, a three-dimensionally printed multipiece Chip, and digital color sensing. *Analytical Chemistry.* 96 (17): 6628–6633.

108 Ludwig, K.U., Schmithausen, R.M., Li, D. et al. (2021). LAMP-Seq enables sensitive, multiplexed COVID-19 diagnostics using molecular barcoding. *Nature Biotechnology.* 39 (12): 1556–1562.

109 Chandrasekaran, S.S., Agrawal, S., Fanton, A. et al. (2022). Rapid detection of SARS-CoV-2 RNA in saliva via Cas13. *Nature Biomedical Engineering.* 6 (8): 944–956.

110 Chen, J.S., Ma, E., Harrington, L.B. et al. (2018). CRISPR-Cas12a target binding unleashes indiscriminate single-stranded DNase activity. *Science.* 360 (6387): 436–439.

111 Gootenberg, J.S., Abudayyeh, O.O., Lee, J.W. et al. (2017). Nucleic acid detection with CRISPR-Cas13a/C2c2. *Science.* 356 (6336): 438–442.

112 Tang, Y., Song, T., Gao, L. et al. (2022). A CRISPR-based ultrasensitive assay detects attomolar concentrations of SARS-CoV-2 antibodies in clinical samples. *Nature Communications.* 13 (1): 4667.

113 Wang, H.-Q., Liu, W.-Y., Wu, Z. et al. (2011). Homogeneous label-free genotyping of single nucleotide polymorphism using ligation-mediated strand displacement amplification with DNAzyme-based chemiluminescence detection. *Analytical Chemistry.* 83 (6): 1883–1889.

114 Zhou, W., Hu, L., Ying, L. et al. (2018). A CRISPR–Cas9-triggered strand displacement amplification method for ultrasensitive DNA detection. *Nature Communications.* 9 (1): 5012.

115 Ding, C., Li, X., Ge, Y., and Zhang, S. (2010). Fluorescence detection of telomerase activity in cancer cells based on isothermal circular strand-displacement polymerization reaction. *Analytical Chemistry.* 82 (7): 2850–2855.

116 Xu, H., Wu, B., Wang, J. et al. (2020). Label-free detection of cancer related gene based on target recycling and palindrome-mediated strand displacement amplification. *Talanta.* 215: 120897.

117 Pardee, K., Green, A.A., Takahashi, M.K. et al. (2016). Rapid, low-cost detection of Zika virus using programmable biomolecular components. *Cell.* 165 (5): 1255–1266.

118 Deiman, B., van Aarle, P., and Sillekens, P. (2002). Characteristics and applications of nucleic acid sequence-based amplification (NASBA). *Molecular Biotechnology.* 20 (2): 163–179.

119 Greijer, A.E., Adriaanse, H.M., Dekkers, C.A., and Middeldorp, J.M. (2002). Multiplex real-time NASBA for monitoring expression dynamics of human cytomegalovirus encoded IE1 and pp67 RNA. *Journal of Clinical Virology.* 24 (1–2): 57–66.

120 Loens, K., Ieven, M., Pattyn, S. et al. (2006). Sensitivity of detection of rhinoviruses in spiked clinical samples by nucleic acid sequence-based amplification in the presence of an internal control. *Journal of Microbiological Methods.* 66 (1): 73–78.

121 Jeantet, D., Schwarzmann, F., Tromp, J. et al. (2009). NucliSENS® EasyQ® HPV v1 test–testing for oncogenic activity of human papillomaviruses. *Journal of Clinical Virology.* 45: S29–S37.

122 Casper, E.T., Patterson, S.S., Bhanushali, P. et al. (2007). A handheld NASBA analyzer for the field detection and quantification of Karenia brevis. *Harmful Algae.* 6 (1): 112–118.

123 Van Ness, J., Van Ness, L.K., and Galas, D.J. (2003). Isothermal reactions for the amplification of oligonucleotides. *Proceedings of the National Academy of Sciences.* 100 (8): 4504–4509.

124 Duan, R., Zuo, X., Wang, S. et al. (2013). Lab in a tube: ultrasensitive detection of microRNAs at the single-cell level and in breast cancer patients using quadratic isothermal amplification. *Journal of the American Chemical Society.* 135 (12): 4604–4607.

125 Huang, M., Zhou, X., Wang, H., and Xing, D. (2018). Clustered regularly interspaced short palindromic repeats/Cas9 triggered isothermal amplification for site-specific nucleic acid detection. *Analytical Chemistry.* 90 (3): 2193–2200.

126 Carter, J.G., Orueta Iturbe, L., Duprey, J.-L.H. et al. (2021). Ultrarapid detection of SARS-CoV-2 RNA using a reverse transcription–free exponential amplification reaction, RTF-EXPAR. *Proceedings of the National Academy of Sciences.* 118 (35): e2100347118.

127 Guo, H., Chen, J., Feng, Y., and Dai, Z. (2024). A simple and robust exponential amplification reaction (EXPAR)-based hairpin template (exp-hairpin) for highly specific, sensitive, and universal MicroRNA detection. *Analytical Chemistry.* 96 (6): 2643–2650.

128 Gines, G., Espada, R., Dramé-Maigné, A. et al. (2024). Functional analysis of single enzymes combining programmable molecular circuits with droplet-based microfluidics. *Nature Nanotechnology.* 1–10.

129 Mok, E., Wee, E., Wang, Y., and Trau, M. (2016). Comprehensive evaluation of molecular enhancers of the isothermal exponential amplification reaction. *Scientific Reports.* 6 (1): 37837.

130 Reid, M.S., Paliwoda, R.E., Zhang, H., and Le, X.C. (2018). Reduction of background generated from template-template hybridizations in the exponential amplification reaction. *Analytical Chemistry.* 90 (18): 11033–11039.

131 Nilsson, M., Malmgren, H., Samiotaki, M. et al. (1994). Padlock probes: circularizing oligonucleotides for localized DNA detection. *Science.* 265 (5181): 2085–2088.

132 Baner, J., Nilsson, M., Mendel-Hartvig, M., and Landegren, U. (1998). Signal amplification of padlock probes by rolling circle replication. *Nucleic acids research.* 26 (22): 5073–5078.

133 Dean, F.B., Nelson, J.R., Giesler, T.L., and Lasken, R.S. (2001). Rapid amplification of plasmid and phage DNA using phi29 DNA polymerase and multiply-primed rolling circle amplification. *Genome Research.* 11 (6): 1095–1099.

134 Lizardi, P.M., Huang, X., Zhu, Z. et al. (1998). Mutation detection and single-molecule counting using isothermal rolling-circle amplification. *Nature Genetics.* 19 (3): 225–232.

135 Dahl, F., Banér, J., Gullberg, M. et al. (2004). Circle-to-circle amplification for precise and sensitive DNA analysis. *Proceedings of the National Academy of Sciences.* 101 (13): 4548–4553.

136 Murakami, T., Sumaoka, J., and Komiyama, M. (2009). Sensitive isothermal detection of nucleic-acid sequence by primer generation–rolling circle amplification. *Nucleic Acids Research.* 37 (3): e19-e.

137 Tian, W., Li, P., He, W. et al. (2019). Rolling circle extension-actuated loop-mediated isothermal amplification (RCA-LAMP) for ultrasensitive detection of microRNAs. *Biosensors and Bioelectronics.* 128: 17–22.

138 Que, H., Yan, X., Guo, B. et al. (2019). Terminal deoxynucleotidyl transferase and rolling circle amplification induced G-triplex formation: a label-free fluorescent strategy for DNA methyltransferase activity assay. *Sensors and Actuators B: Chemical.* 291: 394–400.

139 Chen, L., Eriksson, A., Weström, S. et al. (2022). Ultra-sensitive monitoring of leukemia patients using superRCA mutation detection assays. *Nature Communications.* 13 (1): 4033.

140 Yan, H., Wen, Y., Tian, Z. et al. (2023). A one-pot isothermal Cas12-based assay for the sensitive detection of microRNAs. *Nature Biomedical Engineering.* 7 (12): 1583–1601.

141 Phillips, E.A., Silverman, A.D., Joneja, A. et al. (2023). Detection of viral RNAs at ambient temperature via reporter proteins produced through the target-splinted ligation of DNA probes. *Nature Biomedical Engineering.* 7 (12): 1571–1582.

142 Dirks, R.M. and Pierce, N.A. (2004). Triggered amplification by hybridization chain reaction. *Proceedings of the National Academy of Sciences.* 101 (43): 15275–15278.

143 Venkataraman, S., Dirks, R.M., Rothemund, P.W. et al. (2007). An autonomous polymerization motor powered by DNA hybridization. *Nature Nanotechnology.* 2 (8): 490–494.

144 Li, B., Ellington, A.D., and Chen, X. (2011). Rational, modular adaptation of enzyme-free DNA circuits to multiple detection methods. *Nucleic Acids Research.* 39 (16): e110-e.

145 Zhang, D.Y., Turberfield, A.J., Yurke, B., and Winfree, E. (2007). Engineering entropy-driven reactions and networks catalyzed by DNA. *Science.* 318 (5853): 1121–1125.

146 Xuan, F. and Hsing, I.-M. (2014). Triggering hairpin-free chain-branching growth of fluorescent DNA dendrimers for nonlinear hybridization chain reaction. *Journal of the American Chemical Society.* 136 (28): 9810–9813.

147 Bi, S., Chen, M., Jia, X. et al. (2015). Hyperbranched hybridization chain reaction for triggered signal amplification and concatenated logic circuits. *Angewandte Chemie International Edition.* 54 (28): 8144–8148.

148 Choi, H.M., Chang, J.Y., Trinh, L.A. et al. (2010). Programmable in situ amplification for multiplexed imaging of mRNA expression. *Nature Biotechnology.* 28 (11): 1208–1212.

149 Choi, H.M., Beck, V.A., and Pierce, N.A. (2014). Next-generation in situ hybridization chain reaction: higher gain, lower cost, greater durability. *ACS Nano.* 8 (5): 4284–4294.

150 Choi, H.M., Schwarzkopf, M., Fornace, M.E. et al. (2018). Third-generation in situ hybridization chain reaction: multiplexed, quantitative, sensitive, versatile, robust. *Development.* 145 (12): dev165753.

151 Yin, P., Choi, H.M., Calvert, C.R., and Pierce, N.A. (2008). Programming biomolecular self-assembly pathways. *Nature.* 451 (7176): 318–322.

152 Chen, X., Briggs, N., McLain, J.R., and Ellington, A.D. (2013). Stacking nonenzymatic circuits for high signal gain. *Proceedings of the National Academy of Sciences.* 110 (14): 5386–5391.

153 Hu, X., Fan, J., Duan, B. et al. (2018). Single-molecule catalytic hairpin assembly for rapid and direct quantification of circulating miRNA biomarkers. *Analytica Chimica Acta.* 1042: 109–115.

154 Xiong, E., Zhen, D., and Jiang, L. (2018). Homogeneous enzyme-free and entropy-driven isothermal fluorescent assay for nucleic acids based on a dual-signal output amplification strategy. *Chemical Communications.* 54 (89): 12594–12597.

155 Zhang, K., Wang, K., Zhu, X., and Xie, M. (2017). Entropy-driven reactions in living cells for assay let-7a microRNA. *Analytica Chimica Acta.* 949: 53–58.

7

CRISPR-Based Molecular Diagnostics

A Revolution in Detection and Management of Diseases

Bang-Ce Ye, Pei-Qiang Ma, and Zhen-Ping Zou

Lab of Biosystem and Microanalysis, State Key Laboratory of Bioreactor Engineering, East China University of Science and Technology, Shanghai 200237, China

CHAPTER MENU

Molecular diagnostics plays a pivotal role in the detection and management of diseases. An effective diagnostic method should possess high sensitivity, high specificity, cost-effectiveness, rapidity, portability, and user-friendliness. Conventional nucleic acid detection methodologies, predominantly reliant on polymerase chain reaction (PCR), offer remarkable sensitivity and specificity. Nonetheless, these techniques often require costly equipment, trained personnel, and considerable time and labor. These requirements may restrict their use in certain areas and real-time applications [1]. The advent of clustered regularly interspaced short palindromic repeats (CRISPR) technology has transformed the landscape of molecular diagnostics. CRISPR-associated (CRISPR/Cas) systems, recognized for their genome editing potential, have been proposed as versatile diagnostic tools. These systems align with the World Health Organization's ASSURED criteria for ideal diagnostic methods, which include the following characteristics: affordable, sensitive, specific, user-friendly, rapid/robust, equipment-free, and deliverable [2]. This alignment allows them to overcome the numerous limitations of traditional approaches. This represents a significant departure from conventional molecular detection techniques with inherent limitations.

CRISPR-based molecular diagnostics employs the distinctive capabilities of various Cas effectors. For instance, precisely targeting DNA by Cas9 and the ability of Cas13a to cleave collateral RNA have led to the development of innovative biosensing technologies. Moreover, the identification of Cas12 collateral cleavage

Diagnostic Techniques in Public Health: Analytical and Chemometric Approaches in Disease Detection and Management, First Edition. Edited by Jinghong Li and Yan Du.

of single-stranded DNA (ssDNA) has broadened the scope of detectable targets [3]. CRISPR/Cas systems can detect a wide variety of pathogens, mutations in bacteria, viruses, and genetic changes linked to cancer. They are notably effective in identifying single-nucleotide polymorphisms (SNPs), which are crucial for diagnosing genetic disorders and tailoring personalized therapies [3, 4]. By integrating CRISPR technology into diagnostic platforms, these solutions offer a cost-effective, efficient, and highly adaptable approach suitable for point-of-care testing (POCT) and ongoing disease surveillance.

With advancements in the field, CRISPR/Cas systems have become a central component of next-generation diagnostic platforms. Their capacity to provide rapid, precise, and accessible diagnostic solutions has the potential to transform healthcare, particularly in settings with limited resources and real-time disease monitoring. With ongoing innovation and refinement, CRISPR technology is poised to redefine the landscape of molecular diagnostics, advancing toward the goal of universally accessible healthcare solutions. This chapter summarizes CRISPR technology, including an examination of its mechanisms, tools, and evolution; a detailed account of the precise application of CRISPR/Cas systems for pathogen detection; and an analysis of the potential future therapeutic benefits and challenges related to its use.

7.1 Dissecting CRISPR: Its Classification, Tools, and Mechanisms

7.1.1 The Classification of CRISPR/Cas Systems

Since the discovery of CRISPR loci with short repetitive sequences in *Escherichia coli* K-12 in 1987 [5], the CRISPR/Cas system has been the subject of study for nearly 40 years. The CRISPR/Cas system is an adaptive immune defense system found in bacteria and archaea that comprises CRISPR and Cas [6–12]. The CRISPR/Cas system functions in three main phases: adaptation, expression, and interference [13–15]. In the adaptation phase, the system identifies and incorporates foreign genetic material, such as viral or plasmid DNA, into the genomes of bacteria and archaea. Subsequently, the CRISPR array was transcribed into precursor CRISPR RNA (pre-crRNA). Subsequently, this molecule is processed into mature crRNA molecules, each containing a single spacer sequence and repeat sequences, which is known as the expression phase. During the interference phase, the CRISPR/Cas system targets and cleaves foreign nucleic acids that match the sequences encoded by the crRNAs. crRNAs guide Cas effector proteins to complementary target DNA or RNA sequences, allowing for precise recognition and binding. Upon binding to a target, the Cas effector protein induces site-specific cleavage of the target nucleic acid, resulting in its degradation and neutralization. Table 7.1 presents a summary of the core features and functions of various Cas proteins. Following comprehensive identification and bioinformatic analysis, multiple types and subtypes of the CRISPR/Cas system were identified. A classification report from

Table 7.1 The core functions of CRISPR/Cas proteins.

Effector proteins	Function	References
Cas1	Cas1 possesses endonuclease activity, which is essential for cleaving both the foreign DNA and the host genome at specific sites during spacer integration. One of the key characteristics of Cas1 is its ability to form a complex with Cas2, which facilitates the integration of spacers into the CRISPR array.	[16–18]
Cas2	The primary function of Cas2 is to facilitate the integration of new spacer sequences derived from invading genetic elements into the CRISPR array, enabling the immune memory of the host organism.	[16, 19, 20]
Cas3	The primary function of Cas3 is to target and degrade foreign nucleic acids, such as viral or plasmid DNA, complementary to the spacer sequences within the CRISPR array. Cas3 achieves this through its helicase and nuclease activities, unwinding and degrading the invading DNA, ultimately leading to the inactivation of the genetic material.	[21–23]
Cas4	A component of the adaptive complex aids in the precise processing of the primary spacer and selection of protospacer adjacent motif (PAM) sequences.	[24, 25]
Cas5	Interacts with the large subunit and Cas7 subunit of the Cascade (the CRISPR-associated complex for antiviral defense) complex and binds to the 5′-handle of crRNA. In the subtype I-C system, Cas5 functions as the ribonuclease, taking over the role previously conducted by Cas6.	[26]
Cas6	A crucial role in the CRISPR/Cas immune response by recognizing and cleaving repeat sequences in precursor crRNA, generating functional crRNAs that guide interference complexes to degrade target nucleic acids.	[27–29]
Cas7	A component of the Cascade, its core function is to bind and stabilize the crRNA as part of the Cascade complex, facilitating the target recognition and interference with invading nucleic acids.	[30, 31]
Cas8	A subunit of the class 1 Cascade complex, responsible for recognizing PAM.	[30, 32]

(Continued)

Table 7.1 (Continued)

Effector proteins	Function	References
Cas9	Type II effector protein, Cas9, includes its dual ribonuclease activity, comprising both endonuclease activity for cleaving target DNA and RNA-guided DNA binding. This RNA-guided DNA binding enables Cas9 to precisely locate and bind to complementary DNA sequences specified by the single-guide RNA (sgRNA), initiating the formation of the Cas9-sgRNA complex. Cas9 proteins typically comprise multiple functional domains, including the recognition domain responsible for PAM recognition, the HNH and RuvC nuclease domains for DNA cleavage, and the RNA-binding domain for sgRNA interaction.	[33–35]
Cas10	The Cas10 protein serves as the large subunit in most type III effector complexes. Cas10 contains two catalytic domains: the histidine-aspartate (HD) domain, responsible for cleaving ssDNA, and the palm domain, which synthesizes cyclic oligoadenylates (cOA) using ATP.	[36, 37]
Cas11	The Cas11 small subunit, found in class 1 Cascade complexes, interacts with crRNA and is often fused to the large subunit Cas8 in many type I systems.	[30, 38]
Cas12	The Type V effector protein, Cas12, contains an active RuvC-like domain responsible for cleaving both strands of the target DNA. In contrast to Cas9, Cas12 produces DNA double-strand breaks with 5′ overhangs, which facilitates accurate DNA editing and modification.	[39, 40]
Cas13	Type VI effector protein, Cas13, contains two HEPN superfamily RNase domains. The primary function of Cas13 is RNA targeting and cleavage. Upon binding with a crRNA, Cas13 can identify and cut specific RNA sequences.	[41–43]

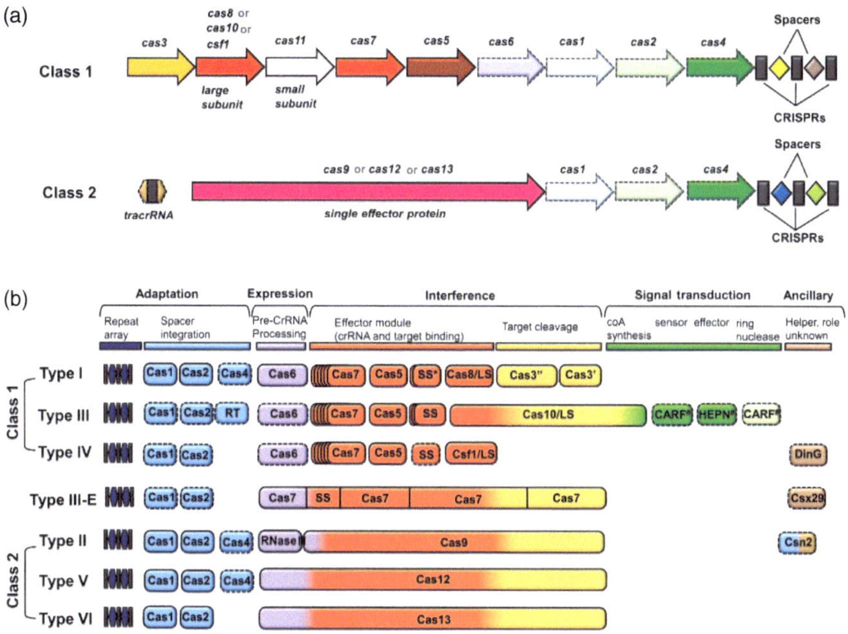

Figure 7.1 Class 1 and 2 CRISPR/Cas systems: modular organization and Cas protein nomenclature. (a) Arrangement of class 1 and 2 CRISPR/Cas loci. In class 1 systems, effector modules comprise multiple Cas proteins, forming a complex that binds to crRNA and collaborates in target binding and processing. In contrast, class 2 systems feature a single multidomain crRNA-binding protein that serves a function similar to that of the entire effector complex in class 1. Core Cas genes are labeled, whereas certain ancillary components, though common, may be absent in certain subtypes and variants, denoted by dashed outlines. (b) Functional components of CRISPR/Cas systems are depicted in the diagram, illustrating the typical genetic, structural, and functional arrangements of the six CRISPR/Cas systems. In some type I subtypes, a putative small subunit, often fused to a large subunit, is denoted by an asterisk. In addition, common ancillary components may be absent in certain subtypes and variants, as indicated by dashed outlines. Cas6 is represented by a thin solid outline in type I because of its dispensability in some systems, but not in most, and by a dashed line in type III because most of these systems appear to use the Cas6 protein provided by other CRISPR/Cas loci. *Source:* Adapted from Ref. [44].

2020 identified two classes, six types, and at least 33 subtypes of the CRISPR/Cas system (Figure 7.1) [44, 45]. Class 1 systems are typically distinguished by the presence of multi-subunit effector complexes, whereas class 2 systems comprise a single, multidomain, large nuclease, such as the well-known CRISPR/Cas9 system.

7.1.1.1 Class 1 CRISPR/Cas Systems

A defining feature of class 1 systems is the presence of multi-subunit effector complexes, which typically comprise multiple Cas proteins [46, 47]. Class 1 systems are divided into three major types (I, III, and IV) and 16 subtypes, each with a distinct effector complex and mechanism of action (Figure 7.2).

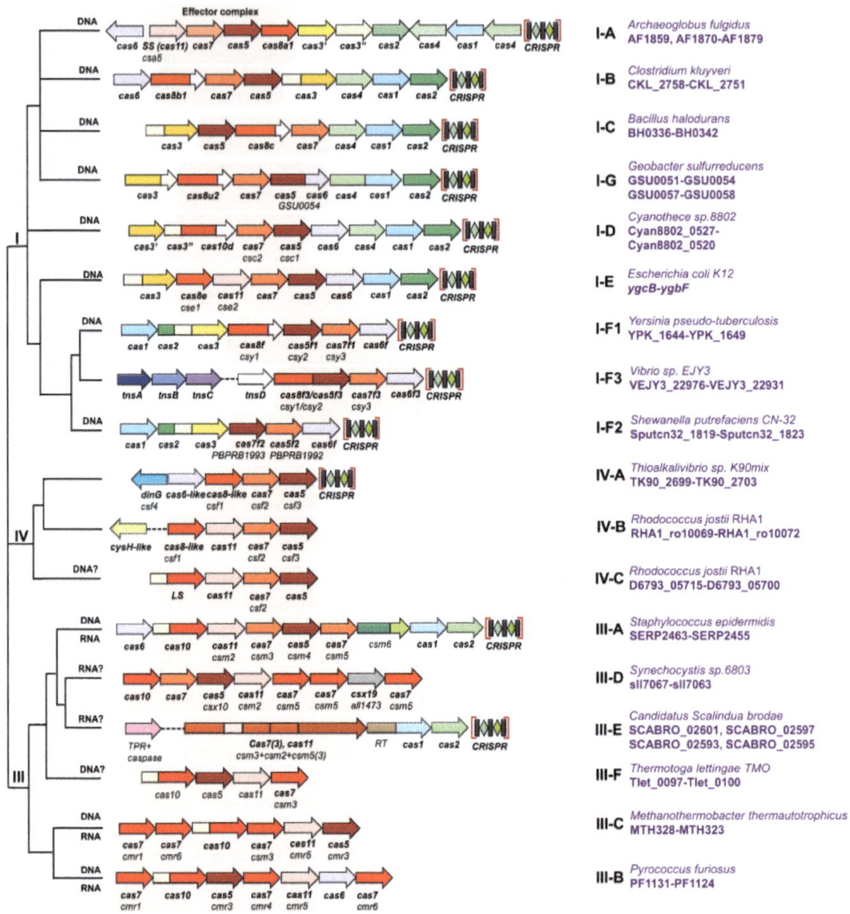

Figure 7.2 The classification of Class 1 CRISPR/Cas systems is depicted in the diagram. The dendrogram displays the likely evolutionary relationships between these types and subtypes. The right column indicates the organism and the corresponding gene range. The small subunit is encoded by various genes, such as *csm2*, *cmr5*, *cse2*, *csa5*, and others collectively designated as *cas11*. In subtypes III-A and III-B, the adaptation module genes Cas1 and Cas2 were dispensable, as indicated by the dashed lines. The cream-colored gene regions represent the histidine-aspartate (HD) nuclease domain. Genes with uncharacterized functions are depicted in gray, whereas the effector modules are shaded pink. Many loci of subtypes III-B, -C, -E, -F, IV-B, and IV-C lack CRISPR arrays and are depicted accordingly, although exceptions have been observed for each subtype of type III. CHAT denotes the protease domain of the caspase family, RT stands for reverse transcriptase, and TPR represents tetratricopeptide repeats. *Source:* Adapted from Ref. [44].

Type I CRISPR/Cas Systems Type I systems represent one of the most prevalent CRISPR/Cas systems observed in bacteria and archaea, as evidenced by studies [48, 49]. They depend on a multiprotein effector complex, Cascade [13]. The Cascade comprises various Cas proteins and crRNAs that form a large ribonucleoprotein complex. Various Cas proteins within the cascade complex perform distinct functions. Cas5 and Cas7 are involved in crRNA binding and stabilization [50], whereas

Cas8 and Cas11 facilitate DNA binding and unwinding, allowing the cascade to form stable complexes with the target DNA [51, 52]. Cas3 is the primary effector protein recruited by the Cascade complex for cleaving and degrading DNA, with dual functions of nuclease activity for cleavage and helicase activity for unwinding DNA strands [53, 54]. The integrase complex Cas1–2 is responsible for acquiring spacer sequences. The Cas5-6-7-8-11 complex searches for and binds to the target and subsequently recruits the dual-function enzyme Cas3 to cleave the target.

The type I system comprises multiple subtypes, designated I-A to I-G. The I-B system consists primarily of Cas6, Cas8, Cas7, Cas5, and Cas3. Cas8 plays a pivotal role in target DNA recognition and binding to the I-B system. In addition, Cas5 is a crucial protein within the I-B system that facilitates the binding and cleavage of target DNA [55]. The I-A system appears to be a derivative of the I-B system, where Cas3 is split into different helicase and nuclease domains, and Cas8 is divided into two parts. The I-A system regulates Cas3 nuclease activity through conformational activation rather than recruitment. Furthermore, activated Cas3 exhibits strong ssDNA trans-cleavage activity [56]. The I-C system appears to be another derivative of I-B, but with the loss of the Cas6 protein [57]. The I-D system is analogous to the I-C system, but retains the Cas6 protein. The nuclease domain responsible for cleaving target DNA is transferred from Cas3 to a subunit within the large effector complex, Cas10d [58]. The I-E system exhibits similarities to I-A, with Cas3 serving as the primary nuclease responsible for degrading the target DNA once recognized by the effector complex [59, 60]. The I-F system is distinguished by the integration of *cas2* and *cas3* genes. The I-G system is distinguished by the presence of two gene fusions, namely *cas1-cas4* and *cas5-cas6* [60, 61].

Type III CRISPR/Cas Systems Type III is the most complex subtype of the CRISPR/Cas system, relying on multiprotein effector complexes called CRISPR-associated complexes for antiviral defense (Csm in the case of the III-A system) or RNA-guided regulation (Cmr in the case of the III-B system) [62–65]. The key effector protein in the type III system is Cas10, which contains two polymerase-cyclase palm domains and is a critical protein with nucleotide cyclase and HD nuclease domains. These domains are involved in ssDNA degradation and cOA synthesis [66, 67]. Besides Cas10, the System III complex also includes other proteins that aid in RNA binding and cleavage. These include Cas5, Cas6, paralogous Cas7, and Cas11. Cas5 and Cas6 are involved in the processing and maturation of crRNAs, whereas paralogous Cas7 is involved in RNA binding and cleavage. Cas11 stabilizes the effector complex. Consequently, the type III CRISPR/Cas system can cleave the target DNA and RNA transcripts of the target gene. Nevertheless, the intricacy of System III may constrain its potential applications in biological contexts.

The type III system was initially classified into two subtypes: III-A and III-B [45]. The III-A system typically includes Cas10 (Csm1), Cas6, Csm6, Cas11 (Csm2), Cas7 (Csm3/5), and Cas5 (Csm4). The III-B system typically includes Cas10 (Cmr4), Cas6, Cas7 (Cmr1/4/6), and Cas5 (Cmr3). Compared to the III-A system, the III-B system is typically devoid of *cas1*, *cas2*, and *cas6* genes. This suggests that these genes may be shared by other CRISPR/Cas systems within the genome. Subtypes III-D and III-C are derived from III-A and III-B, respectively [62]. Subtype III-D is devoid of cas1 and cas2 and encodes a Cas10 protein devoid of the HD domain. III-C also lacks the *cas1* and *cas2* genes, but encodes a Cas10 protein with a functionally inactive cyclic domain. In addition, the type III CRISPR/Cas system encompasses two further subtypes, III-E and III-F [64].

Type IV CRISPR/Cas Systems Sequence and structural comparisons indicate that the type IV system evolved from the type III system, which contains the Cas5, Cas7, and Csf1 proteins, but not the Cas10 protein [68, 69]. Despite the absence of recognizable enzymatic domains, Csf1 has been shown to function as a large subunit of other class 1 effector complexes. The type IV system is primarily encoded on plasmids and exhibits strong plasmid-targeting bias. However, the mechanistic details of this function remain poorly understood.

The established type IV system comprises three subtypes: the three subtypes are designated IV-A, IV-B, and IV-C. Subtype IV-A and IV-B contain highly differentiated Cas7 (Csf2), Cas5 (Csf3), and Cas8-like Csf1 proteins [45, 69]. Subtype IV-A encodes a DinG-family helicase, Csf4, and a Cas6-like protein, Csf5, whereas IV-B does not. However, both IV-B and IV-C contained Cas11 protein, which was not present in IV-A. Studies have indicated that both the IV-A and IV-B systems lack recognizable nuclease domains and appear to suppress plasmid replication by blocking replication forks or transcription rather than cleaving target DNA.

7.1.1.2 Class 2 CRISPR/Cas Systems

Class 2 CRISPR/Cas systems are classified into three types—II, V, and VI (Figure 7.3) [45, 70, 71]. These systems are distinguished by the presence of a single multidomain effector protein responsible for target DNA recognition and cleavage.

Type II CRISPR/Cas systems Typically, Type II CRISPR/Cas systems comprise a CRISPR array and three to four Cas proteins, with the effector protein Cas9, a multifunctional RNA endonuclease containing RuvC and HNH nuclease domains [72]. Cas9 protein plays a pivotal role in adaptive immunity against invading genetic elements, such as bacteriophages and plasmids in bacteria. It recognizes and binds to a target DNA sequence complementary to its guide RNA (gRNA). Upon binding, the Cas9 protein generates a specific double-stranded break (DSB) at a site adjacent to the PAM near the spacer in the CRISPR array.

The identified type II CRISPR/Cas system encompasses four subtypes: II-A, II-B, II-C, and II-D subtypes have been identified [72]. The II-A system contains a functional gene, *csn2*, which encodes a protein involved in spacer integration [73]. The II-B system is distinguished by the absence of the *csn2* gene, whereas the *cas4* gene is present [74, 75]. The II-C system comprises only three functional genes: *cas1*, *cas2*, and *cas9* [76]. The II-D system contains only a small *cas9* functional gene and is unrelated to other known Cas genes [77].

Type V CRISPR/Cas Systems The effector protein of the type V CRISPR/Cas system is RNA-guided endonuclease Cas12 [78]. Similar to Cas9, Cas12 binds to a target DNA

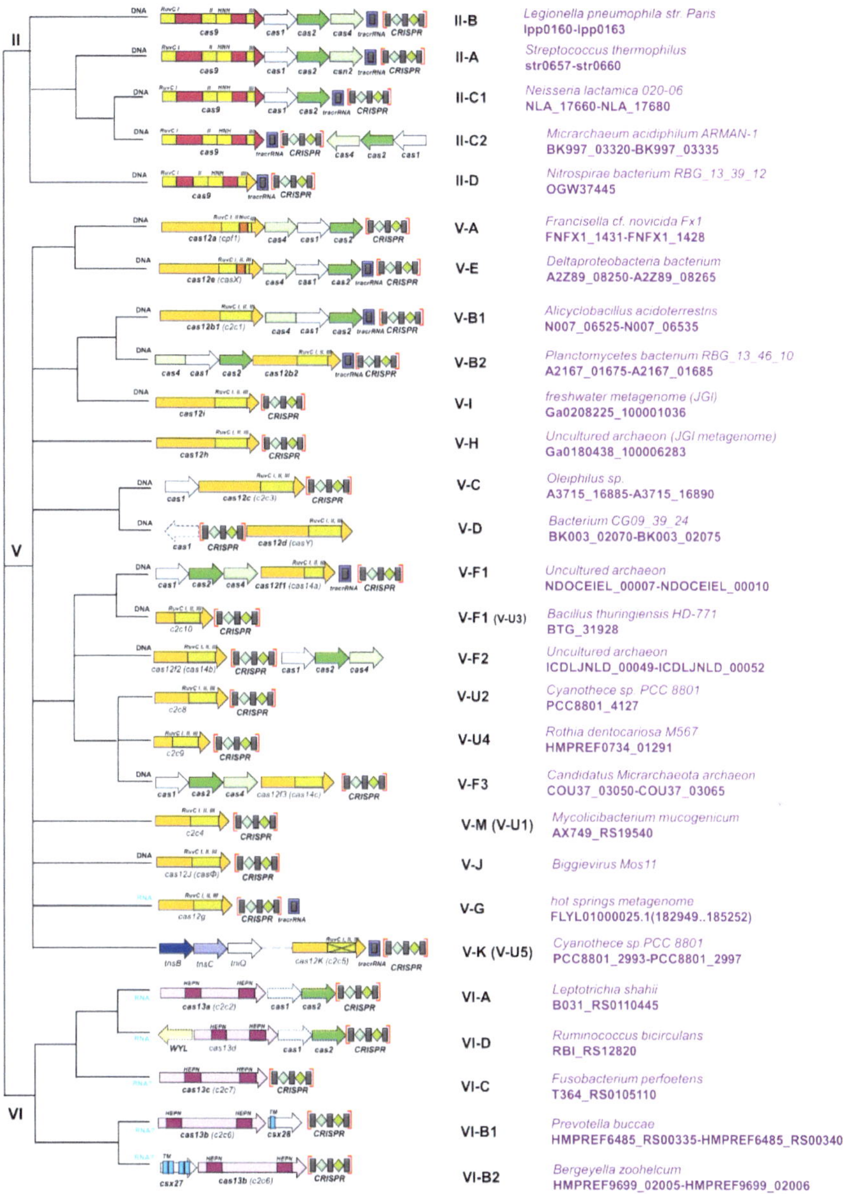

Figure 7.3 The classification of class 2 CRISPR/Cas systems is illustrated in the diagram, featuring representative loci for each subtype and distinct variants. The dendrogram on the left depicts the evolutionary relationships between these types and subtypes. The right column provides information about the organism and its corresponding gene range. The shades of gray represent the two levels of classification, depicting subtypes and variants. The adaptation module genes *cas1* and *cas2* are present only in a subset of subtype V-D, VI-A, and VI-D loci, indicated by dashed lines. Similarly, the WYL domain-encoding genes and *csx27* genes were dispensable and are displayed with dashed lines. Additional genes encoding the components of the interference module, such as tracrRNA, were also included. *Source:* Adapted from Ref. [44].

sequence complementary to its gRNA. In contrast to Cas9, Cas12 proteins, including Cas12a, possess a single RuvC-like nuclease domain, which enables them to continuously cleave both strands of the target DNA away from the PAM site [39, 79]. Notably, Cas12a exhibits intrinsic collateral cleavage activity, resulting in nonspecific degradation of nearby ssDNA molecules.

At least 12 subtypes of V-type have been identified, including V-A, V-K, and V-U (putative subtypes) [78, 80–82]. Type V effector proteins are divided into three branches. The subtypes of Cas12 include Cas12a, Cas12c, Cas12d, Cas12e, Cas12b, and those closely related to transposase B (TnpB) [83], with sizes ranging from 400 to 1500 amino acids. Initially, the V-type variants that evolved from TnpB were collectively called the V-U subtype. However, they have since been reclassified based on their functional characterization results [44], with Cas14 being renamed Cas12f as an example. The C-terminal RuvC domain of all the studied Cas12 effectors was conserved and related to the TnpB protein. This suggests that TnpB is the ancestral enzyme of different Cas12 effector proteins.

Type VI CRISPR/Cas Systems The effector protein of the VI-type CRISPR/Cas system is Cas13, which contains two Higher Eukaryotes and Prokaryotes Nucleotide-binding (HEPN) domains [84]. It is distinctive among the two classes of effectors as it exhibits RNA-guided ribonuclease activity, targeting and cleaving ssRNA instead of DNA. In addition, Cas13 exhibits distinctive trans-cleavage activity, which results in nonspecific degradation of nearby ssRNA molecules.

The identified VI-type CRISPR/Cas systems were classified into four subtypes: VI-A, -B, -C, and -D subtypes have been identified. Of these, the VI-B subtype contains two variants, VI-B1 and -B2, which have been identified [45, 85]. The effector of the VI-A system is Cas13a. In contrast to the VI-A system, the VI-D system also includes an auxiliary protein containing a tryptophan-tyrosine-leucine (ligand-binding) domain, which regulates Cas13 activity. In contrast to the VI-A system, the VI-C system lacks the *cas1* and *cas2* genes and therefore relies on other CRISPR/Cas systems to acquire spacers.

7.1.2 CRISPR/Cas Tools for Diagnostic Implications

The currently developed CRISPR/Cas-based molecular diagnostic tools primarily use simple and efficient effectors of class 2 CRISPR/Cas systems, including Cas9, Cas12, Cas13, and their variants (Figure 7.4) [86, 87]. Only a few tools employ complex effectors of Class 1 CRISPR/Cas systems, such as the Cas3 protein derived from the I-A and -E CRISPR systems, along with the associated cascade complex [56, 88, 89].

The CRISPR/Cas9 system comprises three essential components: Cas9 nuclease, crRNA, and transactivating crRNA (tracrRNA) [33, 90, 91]. These components have been engineered into sgRNAs [92, 93]. Cas9 nuclease contains a recognition domain that interacts with the gRNA and target DNA, as well as two nuclease domains, RuvC and HNH, which are responsible for inducing DSBs in the target DNA. The Cas9 protein forms a complex with sgRNA or crRNA-tracrRNA duplex, which then scans the

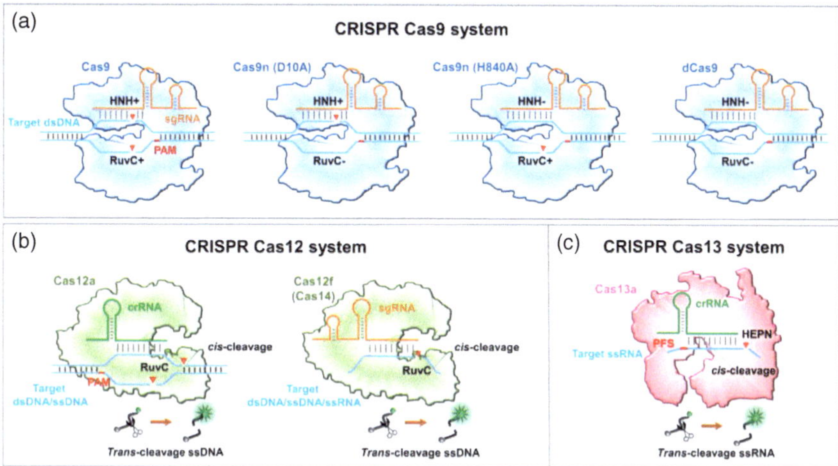

Figure 7.4 Schematic illustration of class 2 CRISPR/Cas toolbox for diagnostic implications. (a) Schematic diagram of the CRISPR/Cas9 system. (b) Schematic diagram of the CRISPR/Cas12 system. (c) Schematic diagram of the CRISPR/Cas13 system. *Source:* Adapted from Ref [86].

genomic DNA to determine the PAM sequence. Once bound to the target DNA, Cas9 nuclease induces a double-strand break of approximately three base pairs upstream of the PAM sequence (Figure 7.4a) [90].

It is also noteworthy that inactivation of either the RuvC or HNH domain results in the formation of Cas9 nickase (Cas9n), which cleaves only one strand of DNA. Specifically, inactivation of the HNH domain results in Cas9n (H840A), which cleaves the nontarget strand, whereas inactivation of the RuvC domain results in Cas9n (D10A), which cleaves the target strand (Figure 7.4a) [94]. Simultaneous inactivation of both nuclease domains results in deactivated Cas9 (dCas9), which retains its DNA-binding ability but cannot cleave target DNA (Figure 7.4b) [95, 96]. Although wild-type Cas9 exhibits robust DNA-targeting and cleavage activity, its specific PAM requirements limit its application. For example, the PAM for *Streptococcus pyogenes* Cas9 (SpCas9) is 5′-NGG-3′. Researchers have developed several engineered Cas9 variants with near-PAM-less or PAM-less characteristics to overcome the inherent limitations of wild-type Cas9. These include SpG Cas9, SpRY Cas9, SchCas9, and Nsp2Cas9 [97–100].

CRISPR/Cas12 nucleases contain a RuvC-like nuclease domain that can consecutively cleave both template and non-template strands away from the PAM site in the distal region (Figure 7.4b) [79, 101]. The enzymes exhibited efficient tracrRNA-independent ssDNA cleavage activity. Cas12a, also known as Cpf1 [39, 40, 102], requires only a single crRNA to recognize the target DNA containing PAM sites, which can be ssDNA or double-stranded DNA (dsDNA). Cas12a demonstrated highly efficient nonspecific ssDNA cleavage activity when activated, resulting in nonspecific degradation of nearby DNA molecules (Figure 7.4b). This property is important for the successful implementation of molecular diagnostics. Cas12b exhibits comparable ssDNA cleavage activity, but requires dual guidance from both crRNA and tracrRNA [103]. Cas12f, also known as Cas14 [104], is a compact Cas12 protein consisting of approximately 550 amino acids. Dual guidance from crRNA and tracrRNA is required to specifically recognize and cleave ssDNA (Figure 7.4b). This is not restricted by PAM sequences. Cas12g is a distinctive member of the Cas12 protein family that, upon target binding, can simultaneously activate ssRNase and ssDNase activities and is not constrained by PAM sequences [77].

CRISPR/Cas13 nucleases, also known as C2c2, contain two HEPN RNase domains and function as RNA-guided ribonucleases [105]. Cas13a processes pre-crRNA into mature crRNAs, which are then used to cleave the target ssRNA. Furthermore, Cas13a exhibited trans-cleavage activity, enabling it to cleave freely circulating ssRNAs in the environment (Figure 7.4c). It is crucial to highlight that the Cas13a recognition sequence is designated as the protospacer flanking site (PFS), which encompasses nucleotides A, U, and C [43, 106].

I-A CRISPR/Cas3 is one of the few type I CRISPR/Cas proteins that has been developed as a nucleic acid detection tool, as evidenced by the literature [88, 89]. In the I-C and I-F systems, Cas3 is recruited in a trans manner after the formation of an R-loop via Cascade-mediated processes. In contrast, in the I-A system, Cas3 is assembled in a cascade to form a complete effector complex that promotes R-loop formation. The nuclease activity of Cas3 is activated only by a cascade upon full R-loop formation, and it subsequently exhibits potent ssDNA trans-cleavage activity. Similarly, the I-E CRISPR/Cas3 system exhibited strong ssDNA trans-cleavage activity and has been developed for *in vitro* nucleic acid detection.

7.1.3 Detection Mechanism

Various CRISPR/Cas effector systems employ distinct mechanisms to detect target molecules. Cas9 nuclease is employed primarily for the precise detection of target DNA because of its specific recognition and cleavage abilities. The trans-cleavage abilities of Cas12 and Cas13 enzymes for ssDNA (Cas12) or ssRNA (Cas13) are advantageous, with a catalytic efficiency (K_{cat}/K_M) approaching 10^5–10^6 M^{-1} s^{-1} [107]. This high catalytic efficiency has led to the development of highly sensitive detection methods. The distinctive biological attributes and high programmability of the CRISPR/Cas system provide groundbreaking molecular diagnostic tools with exceptional sensitivity, specificity, speed, and multiplexing capability.

7.1.3.1 Nucleic Acid Detection

Nucleic Acid Detection Systems Based on CRISPR/Cas9 and its Derivatives The initial nucleic acid detection method based on the CRISPR/Cas9 system, designated nucleic acid sequence-based amplification (NASBA)-CRISPR cleavage (NASBACC), employs Cas9 protein to meticulously identify viral RNA (Figure 7.5a) [108]. The RNA sequence was amplified using NASBA with reverse primers and an attached synthetic toehold switch. When a PAM sequence is identified within the reverse-transcribed double-stranded DNA, the Cas9-sgRNA complex cleaves

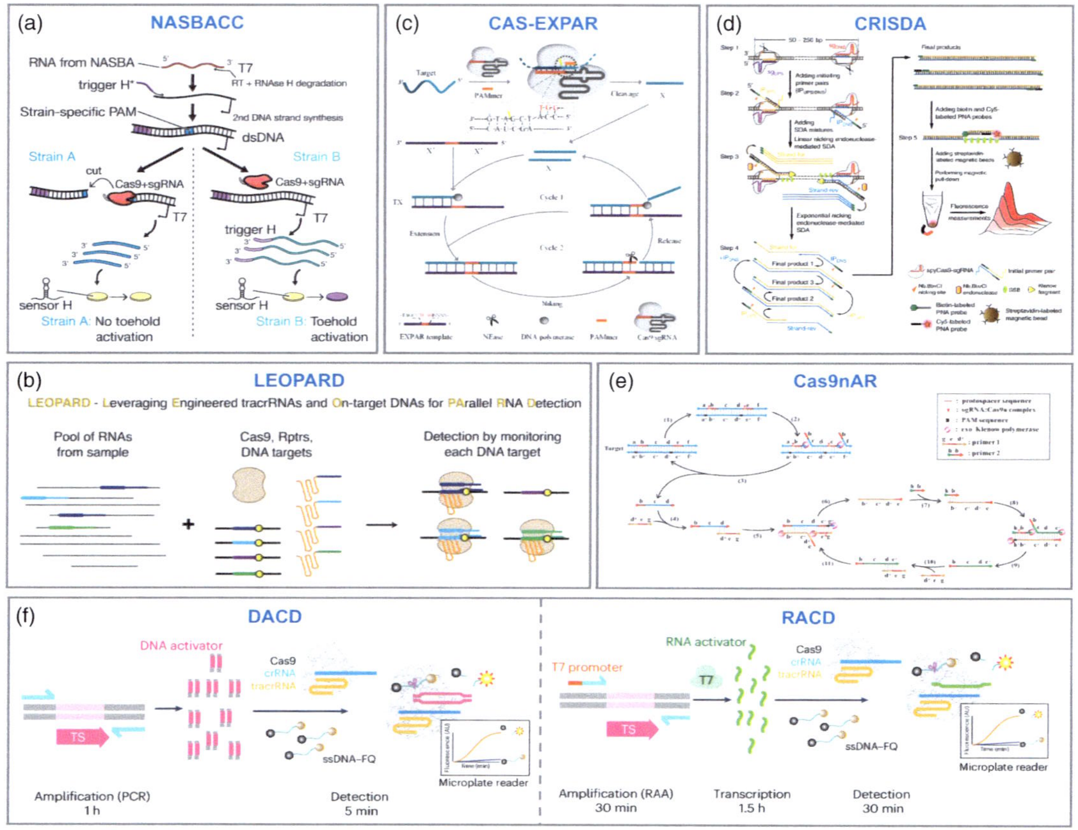

Figure 7.5 Nucleic acid detection systems based on CRISPR/Cas9 and its derivatives. (a) Schematic illustrations of NASBACC (a) (*Source:* Ref. [108] / with permission of Elsevier), LEOPARD (b) (*Source:* Ref. [109] / American Association for the Advancement of Science / CC BY 4.0), CAS-EXPAR (c) (*Source:* Ref. [110] / with permission of Elsevier), CRISDA (d) (*Source:* Ref. [111] / Springer Nature / CC BY 4.0), Cas9nAR (e) (*Source:* Ref. [112] / with permission of John Wiley & Sons), and DACD (f) (*Source:* Ref. [113] / with permission of John Wiley & Sons.)

and generates truncated RNA products devoid of a toehold switch. In the absence of a recognizable PAM sequence in reverse-transcribed dsDNA, full-length RNA products containing the toehold switch were generated, which could be activated by the toehold switch to produce a visual output signal. NASBACC can identify distinct subtypes of Zika virus (ZIKV) within three hours, with a detection limit of femtomolar (fM) levels. Jiao et al. developed an RNA detection method based on the Cas9 protein, designated LEOPARD (leveraging engineered tracrRNAs and on-target DNAs for parallel RNA detection). This method enables multiplexed detection of different RNA sequences at single-nucleotide resolution (Figure 7.5b) [109]. The system employs reprogrammed tracrRNA to convert the recognized RNA into ncrRNA, which then guides Cas9 to the target and cleaves the DNA. It is possible to simultaneously monitor multiple RNA targets in a single test when multiple DNA sequences are added to a reaction. Cas9 has also been used for DNA detection. Huang et al. developed a CRISPR/Cas9-triggered isothermal exponential amplification reaction (CAS-EXPAR) for the rapid and specific detection of target DNA [110]. The activated Cas9-sgRNA complex specifically cleaved the target DNA, which was then subjected to an EXPAR cycle amplification reaction and detected by real-time fluorescence monitoring (Figure 7.5c). The detection limit of this strategy is 0.82 attomolar (aM).

Cas9 derivatives have been successfully developed for nucleic acid detection. Cas9n is a product of deactivation of the RuvC or HNH domains of the Cas9 enzyme. Its precise cleavage enzyme activity allows its application to various isothermal amplification reactions. Zhou et al. developed a method for isothermal amplification and detection of dsDNA based on the CRISPR/Cas9-triggered nicking endonuclease (Cas9n, H840A)-mediated strand displacement amplification method (CRISDA) [111]. This method demonstrated attomolar sensitivity and single-nucleotide specificity in complex sample backgrounds (Figure 7.5d). Wang et al. developed an isothermal amplification and nucleic acid detection method, designated Cas9nAR (Cas9 nickase-mediated amplification reaction), based on another type of Cas9n (D10A). This method couples the Cas9n-sgRNA complex with the exo-Klenow polymerase to amplify the target DNA fragment (Figure 7.5e) [112]. The Cas9nAR method exhibited a zeptomolar limit of detection (LOD) within 60 minutes and single-base discrimination capability. Diagnostic technologies based on dCas9 have been developed for the detection of infectious diseases. Moon et al. reported a colorimetric viral detection method based on the dCas9 system. In this system, viral lysates and biotinylated PAMmers were added to microplate wells containing immobilized dCas9-gRNA complexes, followed by colorimetric reaction using horseradish peroxidase (HRP) and tetramethylbenzidine. This enabled naked-eye detection of SARS-CoV-2, pH1N1, and pH1N1/H275Y viruses [113]. Another design involves fusing dCas9 with split firefly luciferase, which relies on the reactivation of the luciferase enzyme owing to the proximity effect upon binding of the two fusion proteins to the target, generating a detectable biological signal [114]. Moreover, an unlabeled nucleic acid detection device, CRISPR-Chip [115], has been developed by immobilizing dCas9-gRNA complexes on a transistor, which can detect DNA targets as low as 1.7 fmol/L within 15 minutes.

Type V and VI systems exhibit trans-cleavage activity on nonspecific ssDNA or ssRNA, whereas type II systems using a sgRNA do not demonstrate this capability. However, recent studies have demonstrated that type II CRISPR/Cas9 systems directed by both crRNA and tracrRNA dual RNAs display RuvC domain-dependent trans-cleavage activities for both ssDNA and ssRNA substrates [116]. Building on these findings, Chen et al. developed novel nucleic acid detection platforms, namely DNA-activated Cas9 detection (DACD) and RNA-activated Cas9 detection (RACD). Both platforms enable molecular detection with exceptional sensitivity and the capacity to accurately differentiate individual bases. (Figure 7.5f) [116].

Nucleic Acid Detection Systems Based on CRISPR/Cas12 and CRISPR/Cas13 Although the Cas9 protein exhibits efficient cleavage activity toward the target DNA, it is often challenging to achieve secondary amplification of the signal independently, which makes it difficult to generate readable signal outputs when the target concentration is low. In contrast to Cas9, Cas12, and Cas13 exhibit distinctive collateral cleavage activities that facilitate the amplification of the output signal. In 2018, Doudna et al. reported the collateral cleavage activity of Cas12a toward ssDNA and combined this cleavage activity with recombinase polymerase amplification (RPA) to develop a method called DETECTR (DNA endonuclease-targeted CRISPR trans reporter (DETECTR)), using ssDNA labeled with a fluorophore and quencher at opposing ends as the reporting molecule (Figure 7.6a) [117]. This method exhibits attomolar sensitivity and can effectively detect and distinguish between Human Papilloma Virus (HPV) and its subtypes. Teng et al. developed a Cas12b-mediated DNA detection (CDetection) strategy based on Cas12b for the detection of HPV [119]. Doudna et al. developed Cas14-DETECTR based on miniature CRISPR/Cas14 (Cas12f) [104]. Compared to Cas12a, Cas14 is approximately half the size and can target and cleave ssDNA in the presence of tracrRNA and crRNA without the requirement of a PAM sequence. Moreover, upon target recognition, Cas14 initiates collateral cleavage of ssDNA, enabling rapid and high-fidelity SNP detection. Furthermore, Li et al. combined Cas12a with PCR to develop a rapid nucleic acid detection platform called HOLMES (a 1-h low-cost multipurpose highly efficient system) (Figure 7.6b) [101]. It is also noteworthy that HOLMES has been further developed as HOLMESv2 to address a range of application scenarios [120]. For instance, HOLMESv2 employs a single-step approach in conjunction with loop-mediated isothermal amplification (LAMP), enabling the precise differentiation of SNPs and the quantitative detection of target nucleic acids.

SHERLOCK (Specific High-Sensitivity Enzymatic Reporter UnLOCKing) represents a significant advancement in CRISPR/Cas13 detection technology (Figure 7.6c) [118, 121]. The SHERLOCK method can rapidly identify DNA and RNA with attomolar sensitivity and single-nucleotide mismatch specificity. However, when Cas13a was employed in isolation for ssRNA detection, the sensitivity was only 50 fM, which is insufficient for practical applications. Consequently, SHERLOCK employs RPA or RT-RPA for the initial round of dsDNA or ssRNA amplification, followed by T7 polymerase, to transcribe the amplified product into the ssRNA target of Cas13a and perform secondary amplification. Finally, the

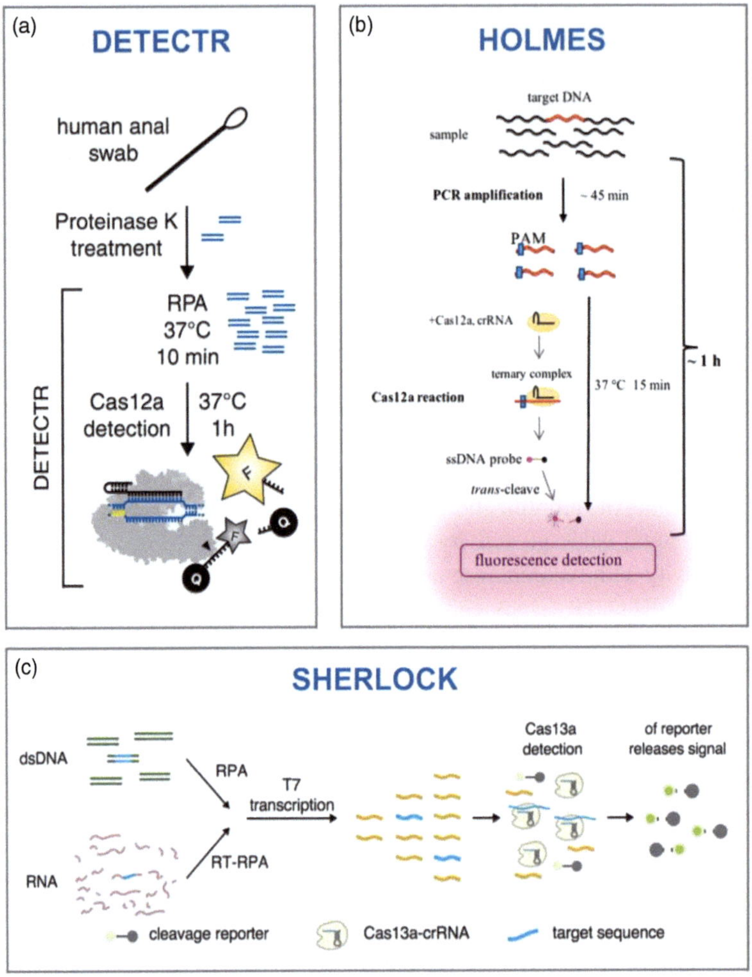

Figure 7.6 Nucleic acid detection systems based on CRISPR/Cas12 and CRISPR/Cas13. (a) Schematic illustration of DETECTR (a) (*Source:* Ref. [117] / with permission of American Association for the Advancement of Science), HOLMES (b) (*Source:* Ref. [101] / Springer Nature/CC BY 4.0), and SHERLOCK (c) (*Source:* Ref. [118] / with permission of American Association for the Advancement of Science).

collateral cleavage activity of Cas13a toward ssRNA was used to release detectable signals. Zhang et al. subsequently optimized SHERLOCK technology and developed a nucleic acid sensing platform, designated SHERLOCKv2 [122], which exhibits multiple, rapid, and portable characteristics. Compared to SHERLOCK, SHERLOCKv2 exhibits the following enhancements: optimized primer concentration to circumvent preamplification saturation, which impedes accurate quantification; utilization of the CRISPR enzyme Csm6 to amplify detection signals; incorporation of orthogonal CRISPR enzymes for multiplex detection; and achievement of lateral flow readout.

7.1.3.2 Non-nucleic Acid Targets Detection

The scope of targets for CRISPR/Cas system detection has expanded from nucleic acids to encompass other analytes, including small molecules, peptides, proteins, exosomes, and bacteria. These analytes are closely associated with human disease diagnosis and health management. Regardless of the type of target, most CRISPR/Cas-based detection systems employ precise target recognition, cleavage, and efficient collateral cleavage activity of the CRISPR/Cas effector complex toward nucleic acids. The principal distinction between the detection of diverse targets is the utilization of disparate biological transduction elements to facilitate signal conversion. Signal transduction elements and strategies include transcription factors, adaptors, riboswitches, DNAzymes, and antigen-antibody interactions [123].

Using Transcription Factors as Signal Transduction Elements Bacterial allosteric transcription factors (aTFs) typically comprise target molecule- and DNA-binding domains [124, 125]. Binding of the target molecule to an aTF often induces conformational changes in the DNA-binding domain of the aTF, weakening or enhancing its binding to DNA operators. Liang et al. developed a small-molecule detection platform, CaT-Smelor (CRISPR/Cas12a and aTF-mediated small-molecule detector) [126], based on this characteristic. This platform enables the rapid and sensitive detection of disease biomarkers and target metabolites (such as uric acid, hydroxybenzoic acid, and tetracycline) from complex clinical samples or fermentation cultures (Figure 7.7a). CaT-Smelor employs protein immobilization techniques to immobilize aTFs (CBD-aTFs) fused to cellulose-binding domains on microcrystalline cellulose. The DNA-binding domain of aTF exhibits specific binding to dsDNA containing the corresponding aTF-binding motif and PAM site. Upon binding to the target molecules, conformational changes occur in the aTF, resulting in the release of dsDNA. The released dsDNA activates the ssDNA collateral cleavage activity of Cas12a, resulting in the generation of a detectable signal. Mahas et al. developed a convenient field-deployable small-molecule detection platform by coupling aTFs, in vitro transcription systems, and CRISPR/Cas12a. Upon binding of the small-molecule ligand to an aTF, the dissociation of the aTF from the operator downstream of the T7 promoter initiates transcription. This transcript serves as a precursor for the Cas12a crRNA array, which is processed by Cas12a into mature forms [131].

Using Functional Nucleic Acids as Signal Transduction Elements Functional nucleic acids, including aptamers, deoxyribozymes, and riboswitches, are DNA or RNA sequences with specific functions [132]. Several bioreceptors and detection platforms have been developed using functional nucleic acids as signal transduction elements [132], which convert non-nucleic acid target molecules into detectable signals using the CRISPR/Cas system.

Aptamers are typically designed as short, single-stranded nucleotide sequences that bind to metal ions, small molecules, proteins, or even bacteria, with dissociation constants ranging from picomolar to nanomolar levels [133]. Samanta et al.

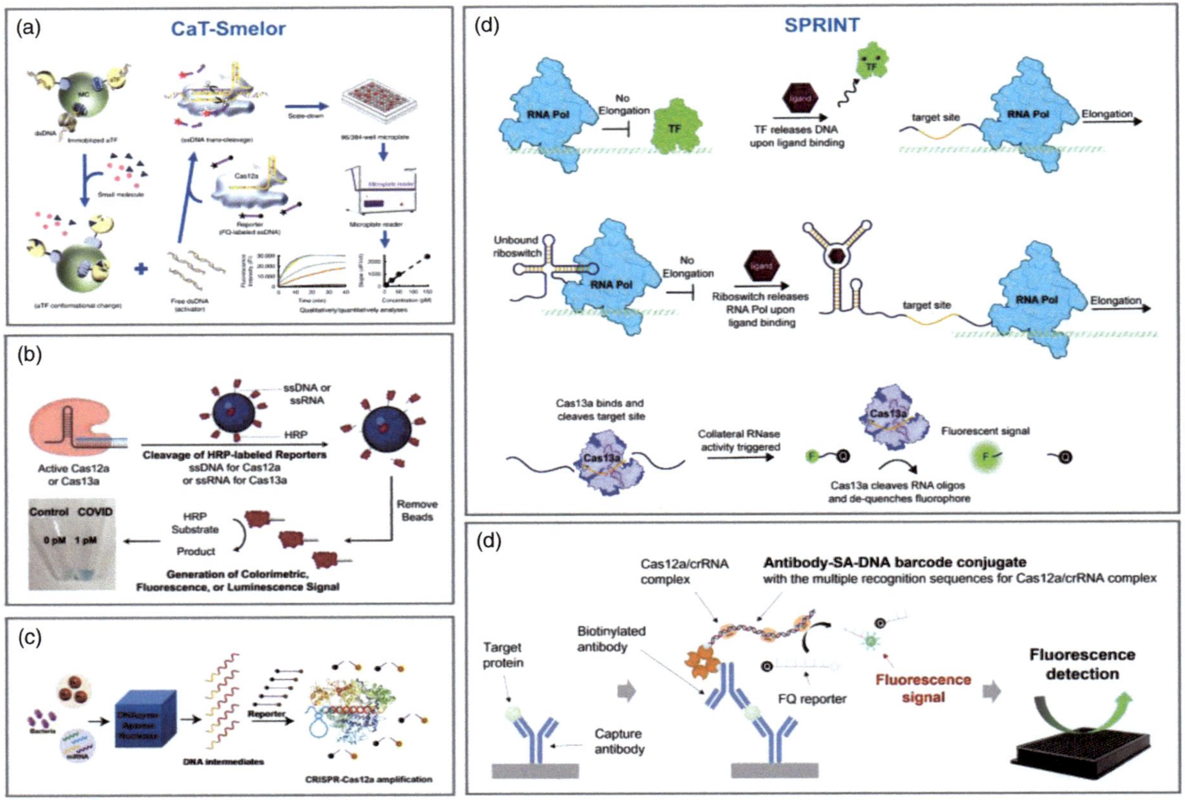

Figure 7.7 Non-nucleic acid target detection based on the CRISPR/Cas system. (a) Transcription factors are used as signal transduction elements for non-nucleic acid target detection. *Source:* Ref. [126] / Springer Nature/CC BY 4.0. (b) Aptamers are used as signal transduction elements for non-nucleic acid target detection. *Source:* Ref. [127] / with permission of American Chemical Society. (c) DNAzymes were used as signal transduction elements for non-nucleic acid target detection. *Source:* Ref. [128] / with permission of American Chemical Society. (d) Using riboswitches as signal transduction elements for non-nucleic acid target detection. *Source:* Ref. [129] / with permission of Oxford University Press. (e) Signal transduction strategies based on antigen-antibody interactions. *Source:* Ref. [130] / with permission of American Chemical Society.

developed a new CRISPR/Cas detection strategy that couples nucleic acid aptamers with HRP (Figure 7.7b) [127]. In this strategy, two ATP-binding aptamers hybridize with the ssDNA target of Cas12a to form a "locked" conformation, preventing binding to the Cas12a-gRNA complex. In the presence of ATP, it forms a complex with the aptamer, releasing the target ssDNA. The released target ssDNA then activates the Cas12a-gRNA complex to generate a detectable signal by cleaving fluorophore/quencher-tagged ssDNA reporters via its trans-cleavage activity or by further amplifying the signal through the release of HRP upon cleavage. Similar signal transduction strategies have been used for highly sensitive detection of cocaine, lipopolysaccharides, peptides, alpha-fetoprotein, thrombin, and pathogenic bacteria [123, 134].

DNAzymes are ssDNA fragments with catalytic activity that mediate various reactions, including DNA and RNA cleavage [135]. Typically, the activity of such enzymes requires the assistance of metal ions or other molecules, which provides a basis for the development of sensors for the detection of such targets [136]. The coupling of DNAzymes with CRISPR/Cas nucleases has been developed as a detection system for various metal ions such as Pb^{2+} and Na^{+}. Briefly, DNAzymes can cleave ssDNA to release DNA activators in the presence of metal ions (Figure 7.7c) [128]. DNA activators then activate the trans-cleavage activity of CRISPR/Cas12a to produce detectable signals.

CRISPR/Cas-based biosensors have made great strides in cation detection. However, the precise detection of anions, such as fluoride, remains a significant challenge [137]. Iwasaki et al. established a detection platform called SHERLOCK-based profiling of in vitro transcription (SPRINT) based on a fluoride ion-responsive riboswitch and CRISPR/Cas13a (Figure 7.7d) [129]. On this platform, the conformational change in the riboswitch upon binding fluoride leads to the release of RNA polymerase, which triggers the transcription process. The transcribed product served as an activator of the trans-cleavage activity of CRISPR/Cas13a, resulting in a detectable signal output.

Signal Transduction Strategies Based on Antigen-antibody Interactions The interaction between antigens and antibodies has been successfully coupled with the CRISPR/Cas system to detect protein biomarkers. Lee et al. developed an immuno-CRISPR assay as a novel protein detection method (Figure 7.7e) [130]. Biotinylated DNA barcodes with multiple CRISPR/Cas12a recognition sites were conjugated to Chemokine Ligand 9 (CXCL9) protein antibodies. In the presence of the target protein CXCL9, a sandwich structure with multiple Cas12a recognition sites was formed. This structure captures and activates the trans-cleavage activity of the Cas12a/crRNA complex, resulting in a detectable fluorescence signal. Compared to traditional HRP-based ELISA, this immuno-CRISPR strategy improved the detection limit for CXCL9 protein 7-fold.

7.2 Pinpoint Accuracy: Utilizing CRISPR/Cas Systems for Pathogen Detection

Recently, with the emergence and development of CRISPR/Cas toolboxes, CRISPR-based detection methods have rapidly evolved from laboratory to clinical diagnosis. Epidemiological surveillance and efficient treatment are heavily dependent on the timely and accurate diagnosis. CRISPR-based methods with high efficiency, specificity, and low cost enable accurate diagnosis for at-home, point-of-care, and other real-time tests. CRISPR-based diagnostics have primarily focused on sensing pathogens, particularly viruses, bacteria, and parasites.

7.2.1 CRISPR System for Viral Infection Diagnosis

Viruses present a formidable challenge to human health and well-being, posing a grave risk to public health. The coronavirus disease (COVID-19) pandemic has underscored the critical importance of infectious pathogenic diagnostics and prevention. Besides the primary antigen detection and SARS-CoV-2 nucleic acid detection methods, CRISPR-based diagnostic systems have been promoted for rapidly detecting COVID-19 [138].

Since the virus outbreak in early 2020, classic SHERLOCK and DETECTR methods have been reported for detecting SARS-CoV-2 [139, 140]. The SHERLOCK-based approach employs a two-stage assay consisting of an RPA-driven isothermal pre-amplification step followed by a Cas13a-mediated fluorescence reporter system [139]. This method showed 96% sensitivity with fluorescence output, 88% sensitivity with the lateral flow test, and 100% specificity with two readouts. To make it more suitable for point-of-care use, the Streamlined Highlighting of Infections to Navigate Epidemics (SHINE) was developed by combining the HUDSON method and a single-step SHERLOCK process [141], simplifying the SARS-CoV-2 viral extraction and detection (Figure 7.8a). HUDSON was improved to quickly deactivate viruses in detection samples within 10 minutes. This method achieved 90% sensitivity and 100% specificity on 50 nasopharyngeal patient samples compared with those with RT-qPCR. Another Cas13-based method eliminates the need for preamplification by the straightforward identification of SARS-CoV-2 in a swab [147]. Cas13a activation is increased by combining multiple, carefully selected crRNAs. Monitoring the fluorescence change, rather than just the fluorescence endpoint, enables a sensitivity of approximately 100 copies per microlitre in 30 minutes. Microfluidics has been used to fuse CRISPR systems for detecting 21 viruses, including SARS-CoV-2, termed microfluidic Combinatorial Arrayed Reactions for Multiplexed Evaluation of Nucleic acids (mCARMEN) [148]. The mCARMEN assay identifies six SARS-CoV-2 variants, assessed on 2,088 patient samples with virtually identical agreement to sequencing-based variant typing.

The classic Cas12a-based DETECTR assay was also improved to recognize SARS-CoV-2 by combining LAMP for the preamplification, termed SARS-CoV-2 DETECTR [140]; this assay reached a 95% positive concordance and a 100% negative concordance in comparison with PCR. (Figure 7.8b). A Cas12a-based one-pot assay

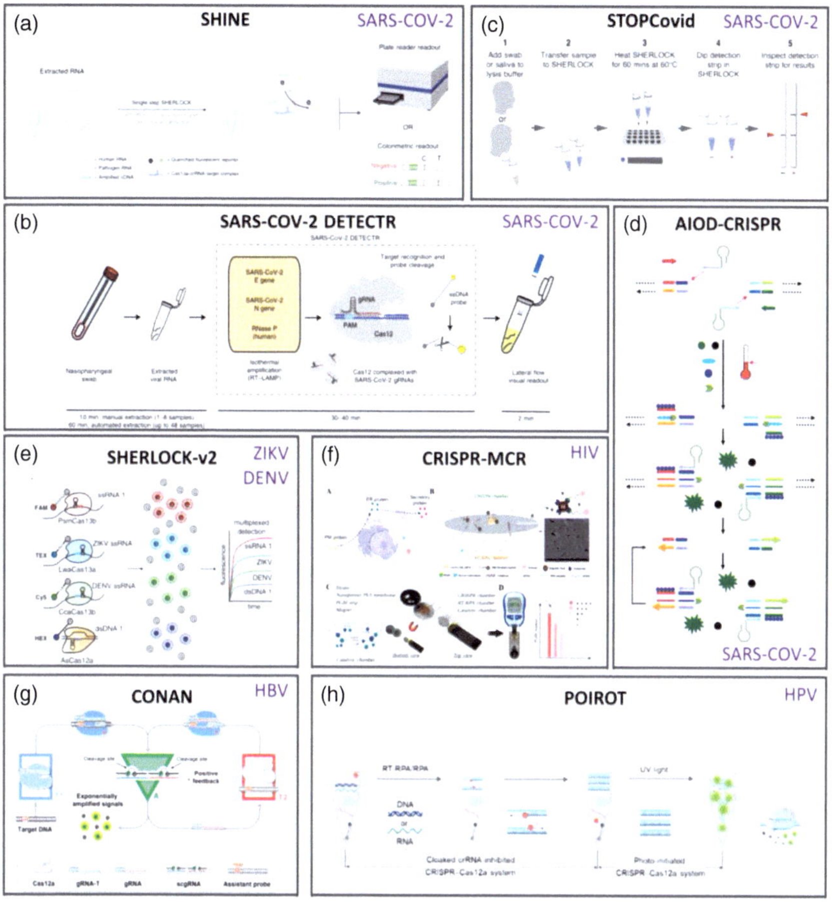

Figure 7.8 CRISPR/Cas-based system for viral infection diagnosis. (a) SHINE, combining HUDSON with a single-step SHERLOCK to detect SARS-CoV-2. *Source:* Ref. [141] / Springer Nature / CC BY 4.0. (b) DETECTR for detecting SARS-CoV-2 [140]. (c) STOPCovid for detecting SARS-CoV-2. *Source:* [142] with permission of medRxiv. (d) SARS-CoV-2 AIOD-CRISPR for detecting SARS-CoV-2. *Source:* [143] / Springer Nature / CC BY 4.0. (e) SHERLOCKv2 for detecting ZIKV and DENV. *Source:* Ref. [122] with permission of American Association for the Advancement of Science. (f) CRISPR-MCR biosensor for detecting HIV. *Source:* Ref. [144] with permission of American Chemical Society. (g) CONAN for detecting HBV. *Source:* Ref. [145] with permission of American Association for the Advancement of Science. (h) POIROT for detecting HPV. *Source:* Ref. [146] with permission of John Wily & Sons.

was reported by Zhang, referred to as SHERLOCK Testing in One Pot for Covid-19 (STOPCovid) [142]. This simplified method uses the LAMP reaction to replace the RPA for a rapid one-hour one-pot test, provides comparable sensitivity to that of RT-qPCR, and shows a 100-copy LOD per reaction in the detection of viral genomes in samples (Figure 7.8c). To simplify the method and increase the sensitivity, Zhang et al. developed STOPCovidv2 by utilizing magnetic beads to extract RNA from samples, which reduced the duration of sample extraction to 15 minutes and increased the sensitivity to 33 copies per milliliter [149]. Besides, Liu et al.

developed another Cas12a-based one-pot assay called SARS-CoV-2 all-in-one dual CRISPR–Cas12a (AIOD-CRISPR) [143], using Cas12a with two crRNAs without PAM site limitation to trigger high-efficiency dual CRISPR/Cas-based nucleic acid detection (Figure 7.8d). This method showed a low five-copies/reaction LOD for virus RNA. Many other CRISPR-based assays have been reported with high sensitivity and specificity, low cost, and a fast and simple workflow for the detection of SARS-CoV-2, such as Cas13-based ADESSO [150], DISCoVER [151], and one-pot assay including dWS-CRISPR [152] and opvCRISPR [153]. With the SARS-CoV-2-caused pandemic ending, CRISPR/Cas-based detection methods have contributed to virus detection, prevention, and control, and have provided advanced ideas and solutions for future public health emergencies.

Other than COVID-19, many viral infections threaten public health, such as the influenza virus, ZIKV, Ebola virus, African swine fever virus (ASFV), hepatitis B virus (HBV), Dengue virus (DENV), and Human Immunodeficiency Virus (HIV). In 2016, the first CRISPR-based detection method, NASBACC, was reported to detect the ZIKV in the plasma of a viremic macaque [108]. CRISPR/Cas9-mediated, PAM sequence-dependent specific cleavage combined with the NASBA-based isothermal preamplification of targets was used to identify different Zika subtypes. In 2017, the Cas13a-based SHERLOCK platform was first reported to detect DENV and ZIKV RNA, with a 2 aM sensitivity and almost 100% specificity [118]. An orthogonal approach was then used to improve SHERLOCK to SHERLOCKv2, which integrated four effectors (LwaCas13a, PsmCas13b, CcaCas13b, and AsCas12a) to generate different fluorescent reporters and achieve multiplex detection of ZIKV and DENV in a single reaction with single-molecule sensitivity (Figure 7.8e) [122].

Another RNA virus, HIV, leads to acquired immunodeficiency syndrome (AIDS), which represents one of the gravest global public health predicaments. Early diagnosis is important for the treatment of HIV patients. A nanoporous-membrane-separated cascade reaction system called CRISPR-mediated cascade reaction (CRISPR-MCR) biosensor was developed to improve the compatibility of multienzyme reactions of CRISPR/Cas-based methods (Figure 7.8f) [144]. This biosensor was integrated into a microfluidic platform and used to detect HIV with a simple glucometer, akin to at-home diabetes screening. The Cas13-based CARVER assay (Cas13-assisted restriction of viral expression and readout) has been used to detect RNA viruses in living cells [154]. The system programmed Cas13 to target and destroy almost a hundred different single-stranded RNAs with the potential to infect humans, demonstrating that the Cas13-based toolbox can be utilized for viral diagnosis and treatment.

CRISPR-based assays have also been used to detect DNA viruses, such as Herpesviridae (including cytomegalovirus [CMV] and Epstein-Barr virus [EBV] [155]) and BK polyomavirus [156]. For example, the SHERLOCK-based assay was used for the detection of CMV and BKV in clinical samples, which showed 100% sensitivity for BKV and 80% sensitivity for CMV, and 100% specificity in both cases compared to RT-qPCR assay for BKV detection in 67 HUDSON-isolated plasma and urine samples [156]. HBV constitutes one of the most formidable liver viruses, which can cause a serious liver infection known as hepatitis B. Recently, a series of CRISPR-based detection methods for HBV detection have been developed [157–162]. For instance, the CRISPR/Cas-Only Amplification Network (CONAN) method, based on the Cas12a system, was developed by engineering a Cas12a-triggered autocatalytic chemical reaction network for signal amplification [145]. Without requiring preamplification, the method achieved single-enzyme, single-step, and real-time genomic DNA detection in HBV clinical samples with high sensitivity and specificity (Figure 7.8g).

HPV16 and HPV18 are high-risk types of human papillomavirus and are major causes of cervical cancer. Thus, early diagnosis plays a vital role in timely treatment. CRISPR combined with PCR amplification methods like CRISPR-typing PCR (ctPCR) has been successfully developed for detecting HPV [163–165]. Cas9 is used to cut target sequences precisely, avoiding mutation-inducing sequences before PCR amplification, affecting amplification results, enabling the differentiation between HPV16 and HPV18 subtypes, and detecting low-frequency mutations. Liu et al. reported a high-performance multi-target detection platform (MiCaR) method, which fuses CRISPR/Cas12a and multiplex RPA on a high-performance multi-target detection platform [166]. The method showed a sensitivity of 0.26 aM and 98.1% specificity on the diagnosis of 100 potential HPV-infected samples. Recently, Yang et al. developed a universal Photo-initiated CRISPR–Cas12a system based on photo-responsive cloaked crRNA as a Robust One-pot Testing (POIROT) platform [146]. This platform integrated the trans-cleavage effect of Cas12a with RPA and showed one-copy/μL sensitivity in testing 150 clinical vaginal swab samples (Figure 7.8h).

7.2.2 CRISPR System for Bacterial Infection Diagnosis

Pathogenic bacteria, which are another major source of infection, can cause a variety of diseases and pose a formidable threat to human health. Therefore, fast and effective diagnosis of such infections is important for controlling the spread of the disease and protecting public health. CRISPR-based methods for detecting pathogenic bacteria can be broadly classified into two primary types: those that detect the genetic material (nucleic acids) of the pathogens, and those that use nucleic acid aptamers to detect the bacteria directly. These pathogenic bacteria include *Mycobacterium tuberculosis* [118, 167], *Staphylococcus aureus* [118, 122], *Listeria monocytogenes* [110], *Pseudomonas aeruginosa* [118], and *Salmonella enteric* [168].

In 2017, SHERLOCK [118] was used to distinguish bacterial strains by utilizing the V3 region of the 16S rRNA gene as the target, which included conserved flanking regions as the binding domain of RPA primers and the variable internal region as the recognizing domain (Figure 7.9a). Using five possible targeting crRNAs, SHERLOCK discriminated between genes of *P. aeruginosa* and *E. coli*. In addition, the assay also discriminated between clinical isolates of *Klebsiella pneumoniae* with two different resistance genes: Klebsiella pneumoniae carbapenemase (KPC) and New Delhi metallo-β-lactamase 1 (NDM-1). SHERLOCKv2 used

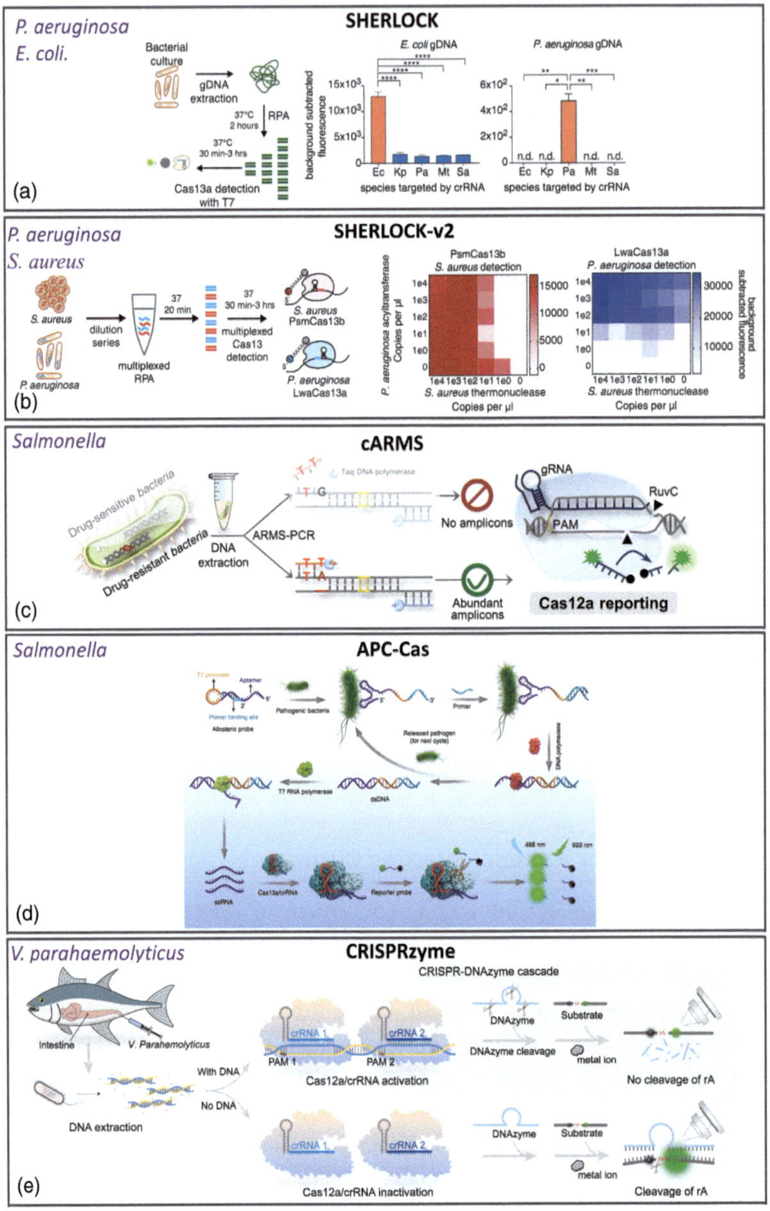

Figure 7.9 CRISPR/Cas-based system for bacterial infection diagnosis. (a) SHERLOCK for recognizing *Pseudomonas aeruginosa* and *Escherichia coli*. *Source:* Ref. [118] / with permission of American Association for the Advancement of Science. (b) SHERLOCKv2 for detecting *Pseudomonas aeruginosa* and *Staphylococcus aureus*. *Source:* Ref. [122] with permission of American Association for the Advancement of Science. (c) ARMS-qPCR-based cARMS for the detection of *Salmonella enterica*. *Source:* Ref. [169] / with permission of American Chemical Society. (d) APC-Cas for directly detecting *Salmonella enterica*. *Source:* Ref. [168] / Springer Nature/CC BY 4.0. (e) CRISPRzyme for detecting *V. parahaemolyticus* and *L. monocytogenes*. *Source:* Ref. [170] / with permission of Elsevier.

LwaCas13a and PsmCas13b to target the acyltransferase gene of *Pseudomonas aeruginosa* and the thermonuclease gene of *Staphylococcus aureus*, respectively, to achieve attomolar sensitivity in single-sample multiplexed detection (Figure 7.9b) [122].

Salmonella enterica and *Listeria monocytogenes* are foodborne pathogens that can cause severe food poisoning and illness, threatening public health. Dong et al. reported an assay called cARMS combining the amplification refractory mutation system qPCR (ARMS-qPCR) with the Cas12a signaling system to recognize the single-nucleotide-mutated *Salmonella* with antibiotic resistance [169]. The cARMS assay integrates allele-specific primed polymerization and CRISPR/Cas12 binding, endowing it with high specificity to detect single-nucleotide variations (Figure 7.9c). An electrochemical biosensor (E-CRISPR) was developed by integrating the trans-cleavage activity of Cas12a and recombinase-assisted amplification. This biosensor showed 0.68 aM LOD on genomic DNA and 26 CFU/mL sensitivity to *L. monocytogenes* in cultures. Besides nucleic acid detection, the directional detection of bacterial antigen has been developed. For example, Xing *et al.* reported a Cas13a-based allosteric probe-initiated catalysis system called APC-Cas, which combined aptamer-based allosteric probes with the Cas13a system (Figure 7.9d) [168]. The system allowed the highly selective detection of *Salmonella* with a sensitivity of one CFU in diverse samples and exhibited a performance similar to that of typical RT-qPCR.

Vibrio parahaemolyticus is a well-known pathogen responsible for causing gastroenteritis in humans, usually through the consumption of contaminated seafood, such as raw or undercooked shellfish. A preamplification-free CRISPR-DNAzyme cascade-based detection assay (CRISPRzyme assay) was developed by combining the Cas12a-based system with DNAzyme [170]. This assay showed high specificity and sensitivities of 62 CFU and 82 CFU in detecting *V. parahaemolyticus* and *L. monocytogenes*, respectively (Figure 7.9e).

7.2.3 CRISPR System for Parasites and Fungi Diagnosis

Parasites are organisms that live on or within a host and benefit at their own expense. Human parasites exist in various forms and can cause various diseases and public health issues. *Plasmodium* is a parasitic protozoan that causes malaria in humans and other organisms. CRISPR was developed to detect the *Plasmodium* species, including different *P. falciparum* strains [171] and specific distinction of *P. vivax* and *P. falciparum* [172]. In 2019, Crawford et al. reported a CRISPR/Cas9-based diagnostic assay termed Finding Low Abundance Sequences by Hybridization (FLASH) [171], coordinating with next-generation sequencing (NGS) for detecting drug-resistant *Plasmodium falciparum* and distinguishing different strains. FLASH-NGS showed an 85.6% on-target rate, far more than <0.02% on-target only with NGS (Figure 7.10a). SHERLOCK-based methods have been reported to recognize all *Plasmodium* species that cause human malaria and enable specific distinction of *P. vivax* and *P. falciparum*, the primary causes of malaria cases worldwide [172]. This assay was employed for clinical samples

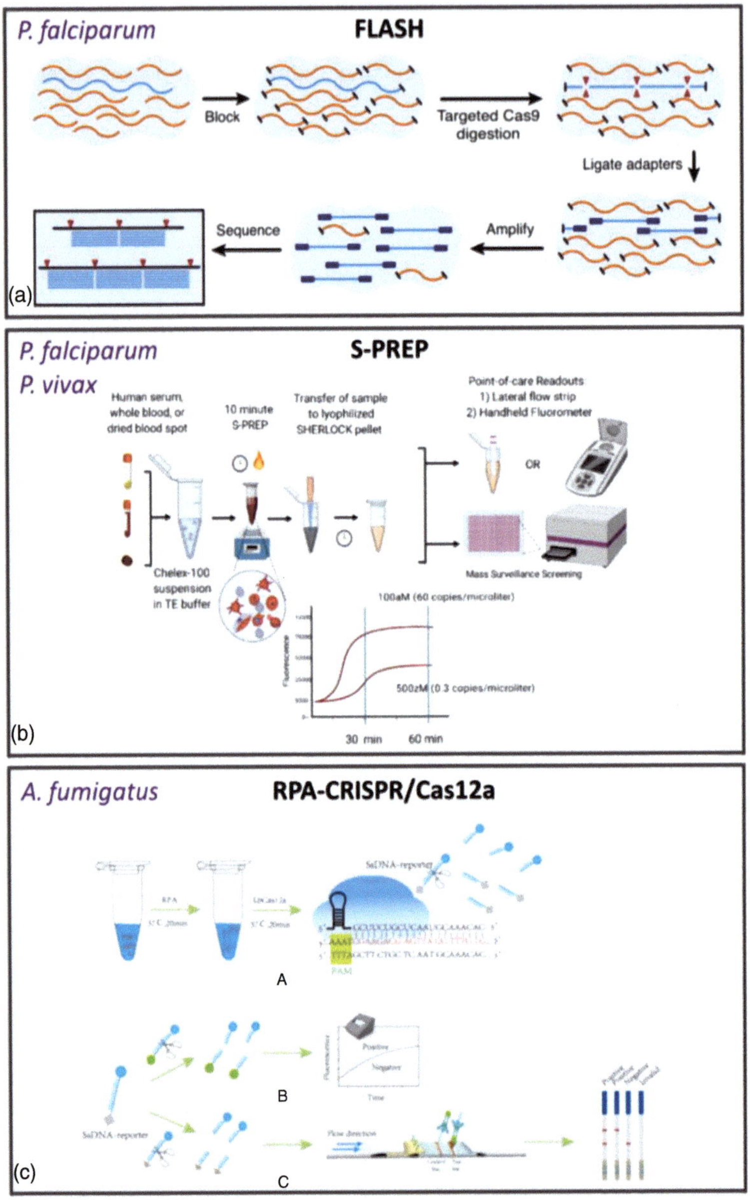

Figure 7.10 CRISPR/Cas-based system for parasites and fungi diagnosis. (a) FLASH for the detection of different *P. falciparum* strains. *Source:* Ref. [171] / with permission of Oxford University Press. (b) S-PREP for the species-specific recognition of *P. vivax* and *P. falciparum*. *Source:* Ref. [173] / National Academy of Sciences / CC BY 4.0. (c) RPA- CRISPR/Cas12a system for the detection of *Aspergillus fumigatus*. *Source:* Ref. [174] / with permission of Springer Nature.

from Uganda and Thailand and samples from an Anopheles reservoir in Thailand. It showed a 94% detection rate and 94% accuracy in identifying *falciparum*, compared to RT-qPCR. Another SHERLOCK-based method called SHERLOCK parasite rapid extraction protocol (S-PREP) uses lyophilized reagents downstream to simplify sample preparation and achieve RPA-Cas12a-based one-pot reaction [173]. This method distinguishes *P. falciparum, P. vivax, P. ovale,* and *P. malariae* in 60 minutes and shows 100% specificity and sensitivity for clinical samples infected with *P. vivax* and *P. falciparum* (Figure 7.10b).

Invasive fungal diseases (IFDs) are serious infections resulting from the invasion of tissues by fungal pathogens. These diseases are concerning in immunocompromised persons, such as those receiving chemotherapy, allograft recipients, or HIV/AIDS-infected patients. Diagnosing IFDs can be challenging because of their non-specific symptoms and the need for specialized tests to identify the pathogens accurately. A combination of clinical evaluation, imaging, and microbiological culture is the traditional clinical diagnosis method and is highly accurate, but requires a complicated and time-consuming operation. Shen et al. reported a DETECTR similar platform combining RPA with Cas12a toolbox for rapid and visual detection of *Aspergillus fumigatus*, which can cause invasive pulmonary aspergillosis in immunocompromised patients [174]. Fluorescence and a lateral flow strip were used to display the results. The method exhibited a low LOD of 100 copies/μL, 96.7% sensitivity, 100% specificity, and 98.0% consistency rate compared to fluorescence quantitative PCR (Figure 7.10c).

7.2.4 CRISPR System for Non-infectious Disease Diagnosis

Most non-infectious diseases are chronic diseases caused by environmental factors, lifestyle choices, or genetic predispositions, such as cancers, genetic disorders, diabetes, and cardiovascular diseases. CRISPR/Cas-based diagnostic assays are widely used for detecting noninfectious diseases, such as cancers [122, 175–179], genetic disorders [115, 180], and immune rejection [156].

Cancer is a collection of various complex diseases that pose a significant threat to human health and life. Precise and early diagnosis of cancer helps to take effective treatment measures before the cancer spreads or becomes more severe, increasing the cure and survival rates of the patients. Detecting cancer biomarkers in body fluids is the main non-invasive method to provide important information about cancer, increasing the chances of early detection and treatment. CRISPR-based assays have been used for sensing different kinds of cancer biomarkers, including cell-free DNA (cfDNA) [118, 122, 176, 177], microRNA (miRNA) [181–183], circulating tumor cells (CTCs) [176, 184], and exosomes [185] and proteins [186]. SHERLOCKv2 with a lateral flow test was applied to detect cfDNA in blood samples of patients with non-small cell lung cancer (NSCLC) and was used to analyze the epidermal growth factor receptor (EGFR) mutation and deletion [122], which significantly impacts treatment decisions and outcomes (Figure 7.11a). Ye et al. reported a Cas12a-based catalytic hairpin assembly (CHA) circuit for single-step microRNAs (miRNAs)

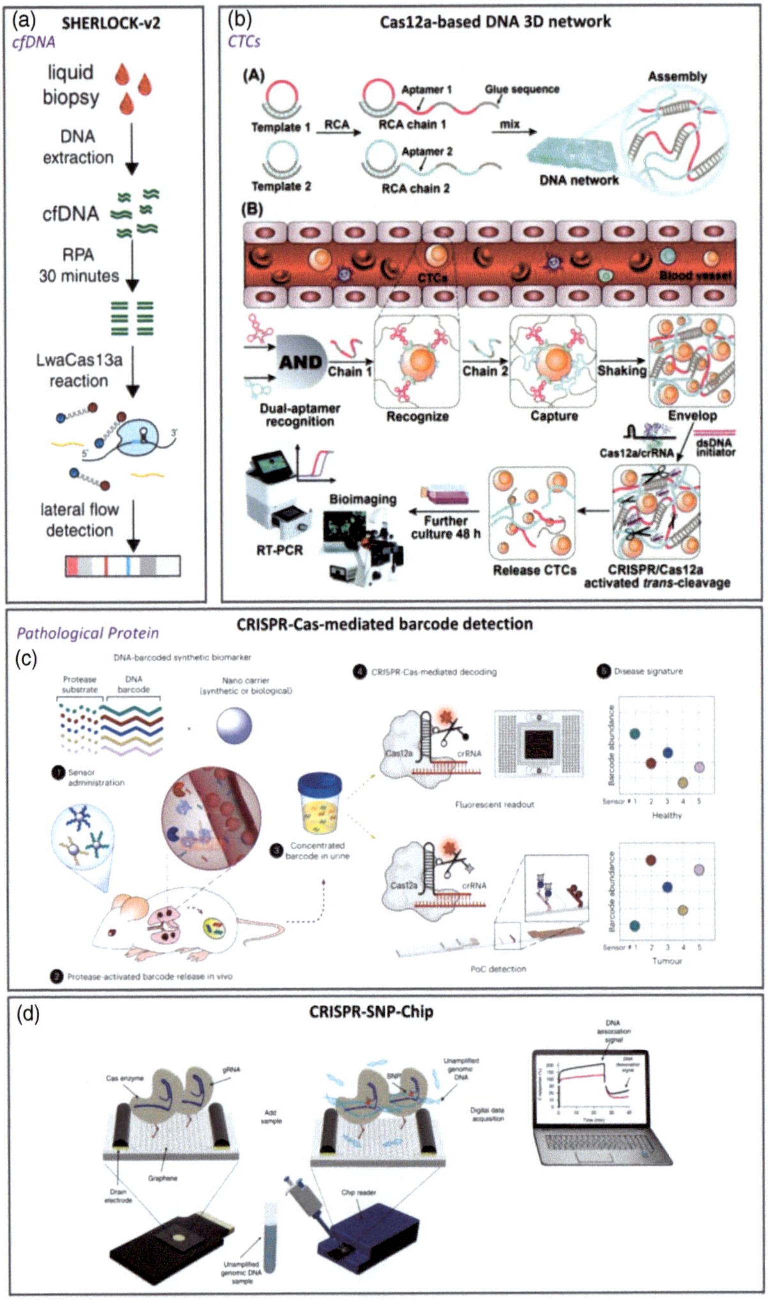

Figure 7.11 CRISPR/Cas-based system for non-infectious disease diagnosis. (a) SHERLOCKv2 for detecting cfDNA from non-small cell lung cancer (NSCLC) patients. *Source:* Ref. [122] with permission of American Association for the Advancement of Science. (b) Cas12a-responsive DNA network for detecting CTCs. *Source:* Ref. [178] / with permission of Royal Society of Chemistry. (c) CRISPR-mediated DNA barcode detection for the non-invasive detection of cancers. *Source:* Ref. [175] / Springer Nature / CC BY 4.0. (d) CRISPR-SNP-Chip for detecting SNP. *Source:* Ref. [180] / with permission of Springer Nature.

detection under isothermal conditions [181]. The approach showed a femtomolar detection limit and sensitively detected miRNA in different tumor cell lines.

Besides nucleic acid biomarkers, CTCs, exosomes, and proteins can be directly detected for cancer diagnosis. A Cas12a-responsive DNA network was developed for the specific trapping and controllable release of CTCs [178]. Double aptamers coupled with CRISPR/Cas12a enabled effective capture and encapsulation, and further controlled the release of CTCs (Figure 7.11b). Recently, a novel cancer diagnosis method was developed by using the CRISPR/Cas system to amplify urinary biomarkers [175]. This method combined chemically stable nucleic acid labels with the CRISPR/Cas enzyme and achieved disease diagnosis in urine from mouse models with different cancers by measuring tumor-related protein activity. In addition, using a microfluidic platform for CRISPR-mediated DNA barcode readout holds promise for rapidly evaluating complicated diseases and for determining treatments. (Figure 7.11c).

Genetic diseases include SNPs, deletions, polygenic, and chromosomal disorders. The diagnosis of SNPs often involves techniques such as genotyping, genome-wide association studies, and NGS, which are time-consuming and costly. The CRISPR/Cas-based SNP detection assays enable the fast, single-base specifical and cost-effective detection. For example, Aran et al. developed a CRISPR-Chip with a graphene-based field-effect transistor, which utilizes the CRISPR toolbox for digitally recognizing target sequences within genomic DNA [115]. This biosensor with a specific guide-RNA-assisted deactivated Cas9 anchored on a graphene transistor, providing a label-free, amplification-free, and rapid (within 15 minutes) detection of target DNA sequences with high sensitivity (1.7 fM), without interference from genomic contaminants. This method was used to detect mutations associated with Duchenne muscular dystrophy in clinical samples. Furthermore, Aran et al. reported another biosensor (CRISPR-SNP-Chip) that fuses Cas9 toolbox and graphene field-effect transistors for quickly and accurately detecting single-nucleotide mutations without the need for DNA amplification [180]. This technology has been demonstrated on two human disease models, sickle cell anemia (SCD) and amyotrophic lateral sclerosis (ALS), and can recognize wild and mutant genes within an hour (Figure 7.11d).

Other non-infectious diseases can also be detected using a CRISPR-based sensor platform. Stevens et al. developed the CrisprZyme assay [187] that combines CRISPR/Cas reactions with nano-enzyme-linked immunosorbent assays, enabling the quantitative detection of non-coding RNAs at room temperature without preamplification. The platform uses Cas13 enzymes to cleave reporter RNAs, modulating the binding of functionalized nano-enzymes for sensitive colorimetric readout down to the picomolar level. CrisprZyme enables the detection of disease-associated non-coding RNAs in samples from patients with acute myocardial infarction. In addition, the Cas13-based assay has been developed for ultrasensitive visualization of elevated CXCL9 mRNA levels and evaluated renal graft rejection in kidney transplant patients [156].

7.3 The CRISPR-Based Diagnosis: Opportunities and Challenges

The advent of CRISPR-based diagnostic technologies holds immense promise in tackling the pressing public health emergencies facing the world today. This revolutionary gene-editing tool has demonstrated remarkable capabilities in accurately detecting and identifying a wide range of disease-associated biomarkers, including those responsible for infectious diseases, genetic disorders, and chronic conditions. One of the key advantages of CRISPR-based diagnostics is their ability to provide rapid, non-invasive, and highly accurate testing, which can dramatically improve early disease detection and intervention. By enabling the swift identification of pathogenic agents or genetic markers, healthcare providers can quickly implement targeted treatment strategies and initiate containment measures, thereby mitigating the spread of infectious diseases and reducing the burden on overburdened health systems.

Moreover, the precision and flexibility of CRISPR/Cas technology can be leveraged to address the unique challenges posed by emerging infectious diseases and evolving pathogens. The rapid design and deployment of CRISPR-based diagnostic tests can facilitate early warning systems and enhance global disease surveillance efforts, empowering public health authorities to respond swiftly and effectively to emerging threats. Beyond infectious diseases, CRISPR-based diagnostics also hold immense potential in addressing the global burden of non-communicable diseases, such as cancer, genetic disorders, and chronic conditions. By providing personalized insights into an individual's health status and risk factors, these technologies can enable targeted prevention strategies, early intervention, and more effective disease management–ultimately realizing at-home and POCT diagnosis, improving patient outcomes, and reducing the societal and economic impact of these debilitating conditions.

However, the widespread adoption of CRISPR-based diagnostics also presents a number of formidable challenges that must be addressed. First, there is a need to discover and apply new and improved versions of CRISPR/Cas effectors that overcome the current limitations in terms of large molecular size, restricted PAM recognition sequences, and off-target effects. Emerging Cas variants like enhanced specificity Cas9 (eSpCas9) [188], high-fidelity Cas9 (Cas9-HF1) [189], and the highly specific Cas12j tools [190] provide enhanced specificity, versatility, and efficiency, which are crucial for robust CRISPR-based biosensing. Second, the development of simple, cost-effective, and convenient POCT platforms is essential for clinical and at-home diagnosis. Strategies like lyophilized CRISPR/Cas systems and integrated sample preparation and detection can enable automated or semi-automated POCT [191, 192]. Finally, the integration of CRISPR diagnostics with engineering, materials science, and artificial intelligence holds promise for advanced capabilities such as multiplexed detection and real-time monitoring of physiological conditions and disease states. Addressing these technical challenges will be key to transforming CRISPR-based diagnosis into a widely adopted diagnostic tool in both clinical and home settings.

By overcoming the challenges and embracing the opportunities presented by CRISPR-based diagnostics, the global community can unlock a future where preventable diseases are detected and managed early, health systems are more resilient and responsive, and the overall well-being of individuals and communities is significantly improved. Ultimately, the successful integration of CRISPR-based diagnostic technologies into mainstream healthcare has the potential to revolutionize the way we detect, manage, and treat a wide range of diseases, paving the way for a future where personalized, precision-driven healthcare becomes the norm and patient outcomes are significantly improved.

References

1 Kaminski, M.M., Abudayyeh, O.O., Gootenberg, J.S. et al. (2021). CRISPR-based diagnostics. *Nature Biomedical Engineering* 5: 643–656.

2 Land, K.J., Boeras, D.I., Chen, X.S. et al. (2019). REASSURED diagnostics to inform disease control strategies, strengthen health systems and improve patient outcomes. *Nature Microbiology* 4: 46–54.

3 Li, Y., Li, S., Wang, J., and Liu, G. (2019). CRISPR/Cas systems towards next-generation biosensing. *Trends in Biotechnology* 37: 730–743.

4 Li, L., Shen, G., Wu, M. et al. (2022). CRISPR-Cas-mediated diagnostics. *Trends in Biotechnology* 40: 1326–1345.

5 Ishino, Y., Shinagawa, H., Makino, K. et al. (1987). Nucleotide sequence of the iap gene, responsible for alkaline phosphatase isozyme conversion in Escherichia coli, and identification of the gene product. *Journal of Bacteriology* 169: 5429–5433.

6 Mojica, F.J., Díez-Villaseñor, C., Soria, E., and Juez, G. (2000). Biological significance of a family of regularly spaced repeats in the genomes of Archaea, bacteria and mitochondria. *Molecular Microbiology* 36: 244–246.

7 Mojica, F.J., Díez-Villaseñor, C., García-Martínez, J., and Soria, E. (2005). Intervening sequences of regularly spaced prokaryotic repeats derive from foreign genetic elements. *Journal of Molecular Evolution* 60: 174–182.

8 Mohanraju, P., Makarova, K.S., Zetsche, B. et al. (2016). Diverse evolutionary roots and mechanistic variations of the CRISPR-Cas systems. *Science* 353: aad5147.

9 Garneau, J.E., Dupuis, M., Villion, M. et al. (2010). The CRISPR/Cas bacterial immune system cleaves bacteriophage and plasmid DNA. *Nature* 468: 67–71.

10 Cong, L., Ran, F.A., Cox, D. et al. (2013). Multiplex genome engineering using CRISPR/Cas systems. *Science* 339: 819–823.

11 Hsu, P.D., Lander, E.S., and Zhang, F. (2014). Development and applications of CRISPR-Cas9 for genome engineering. *Cell* 157: 1262–1278.

12 Barrangou, R., Fremaux, C., Deveau, H. et al. (2007). CRISPR provides acquired resistance against viruses in prokaryotes. *Science* 315: 1709–1712.

13 Faure, G., Makarova, K.S., and Koonin, E.V. (2019). CRISPR-Cas: complex functional networks and multiple roles beyond adaptive immunity. *Journal of Molecular Biology* 431: 3–20.

14 Amitai, G. and Sorek, R. (2016). CRISPR-Cas adaptation: insights into the mechanism of action. *Nature Reviews. Microbiology* 14: 67–76.

15 Jackson, S.A., Mckenzie, R.E., Fagerlund, R.D. et al. (2017). CRISPR-Cas: adapting to change. *Science* 356.

16 Nuñez, J.K., Kranzusch, P.J., Noeske, J. et al. (2014). Cas1-Cas2 complex formation mediates spacer acquisition during CRISPR-Cas adaptive immunity. *Nature Structural & Molecular Biology* 21: 528–534.

17 Rollie, C., Schneider, S., Brinkmann, A.S. et al. (2015). Intrinsic sequence specificity of the Cas1 integrase directs new spacer acquisition. *eLife* 4.

18 Makarova, K.S., Wolf, Y.I., Shmakov, S.A. et al. (2020). Unprecedented diversity of unique CRISPR-Cas-related systems and Cas1 homologs in Asgard Archaea. *CRISPR Journal* 3: 156–163.

19 Samai, P., Smith, P., and Shuman, S. (2010). Structure of a CRISPR-associated protein Cas2 from Desulfovibrio vulgaris. *Acta Crystallographica. Section F, Structural Biology and Crystallization Communications* 66: 1552–1556.

20 Dixit, B., Ghosh, K.K., Fernandes, G. et al. (2016). Dual nuclease activity of a Cas2 protein in CRISPR-Cas subtype I-B of Leptospira interrogans. *FEBS Letters* 590: 1002–1016.

21 Beloglazova, N., Petit, P., Flick, R. et al. (2011). Structure and activity of the Cas3 HD nuclease MJ0384, an effector enzyme of the CRISPR interference. *The EMBO Journal* 30: 4616–4627.

22 Sinkunas, T., Gasiunas, G., Fremaux, C. et al. (2011). Cas3 is a single-stranded DNA nuclease and ATP-dependent helicase in the CRISPR/Cas immune system. *The EMBO Journal* 30: 1335–1342.

23 Huo, Y., Nam, K.H., Ding, F. et al. (2014). Structures of CRISPR Cas3 offer mechanistic insights into Cascade-activated DNA unwinding and degradation. *Nature Structural & Molecular Biology* 21: 771–777.

24 Shiimori, M., Garrett, S.C., Graveley, B.R., and Terns, M.P. (2018). Cas4 nucleases define the PAM, length, and orientation of DNA fragments integrated at CRISPR loci. *Molecular Cell* 70: 814–824.e816.

25 Lemak, S., Nocek, B., Beloglazova, N. et al. (2014). The CRISPR-associated Cas4 protein Pcal_0546 from Pyrobaculum calidifontis contains a [2Fe-2S] cluster: crystal structure and nuclease activity. *Nucleic Acids Research* 42: 11144–11155.

26 Nam, K.H., Haitjema, C., Liu, X. et al. (2012). Cas5d protein processes pre-crRNA and assembles into a cascade-like interference complex in subtype I-C/Dvulg CRISPR-Cas system. *Structure* 20: 1574–1584.

27 Carte, J., Wang, R., Li, H. et al. (2008). Cas6 is an endoribonuclease that generates guide RNAs for invader defense in prokaryotes. *Genes & Development* 22: 3489–3496.

28 Wang, R., Preamplume, G., Terns, M.P. et al. (2011). Interaction of the Cas6 riboendonuclease with CRISPR RNAs: recognition and cleavage. *Structure* 19: 257–264.

29 Niewoehner, O., Jinek, M., and Doudna, J.A. (2014). Evolution of CRISPR RNA recognition and processing by Cas6 endonucleases. *Nucleic Acids Research* 42: 1341–1353.

30 Liu, T.Y. and Doudna, J.A. (2020). Chemistry of Class 1 CRISPR-Cas effectors: binding, editing, and regulation. *The Journal of Biological Chemistry* 295: 14473–14487.

31 Mogila, I., Kazlauskiene, M., Valinskyte, S. et al. (2019). Genetic dissection of the Type III-A CRISPR-Cas system Csm complex reveals roles of individual subunits. *Cell Reports* 26: 2753–2765.e2754.

32 Brouns, S.J., Jore, M.M., Lundgren, M. et al. (2008). Small CRISPR RNAs guide antiviral defense in prokaryotes. *Science* 321: 960–964.

33 Jiang, F. and Doudna, J.A. (2017). CRISPR-Cas9 structures and mechanisms. *Annual Review of Biophysics* 46: 505–529.

34 Heler, R., Samai, P., Modell, J.W. et al. (2015). Cas9 specifies functional viral targets during CRISPR-Cas adaptation. *Nature* 519: 199–202.

35 Wei, Y., Terns, R.M., and Terns, M.P. (2015). Cas9 function and host genome sampling in Type II-A CRISPR-Cas adaptation. *Genes & Development* 29: 356–361.

36 Zhu, X. and Ye, K. (2012). Crystal structure of Cmr2 suggests a nucleotide cyclase-related enzyme in type III CRISPR-Cas systems. *FEBS Letters* 586: 939–945.

37 Samai, P., Pyenson, N., Jiang, W. et al. (2015). Co-transcriptional DNA and RNA cleavage during Type III CRISPR-Cas immunity. *Cell* 161: 1164–1174.

38 Özcan, A., Krajeski, R., Ioannidi, E. et al. (2021). Programmable RNA targeting with the single-protein CRISPR effector Cas7-11. *Nature* 597: 720–725.

39 Swarts, D.C., Van Der Oost, J., and Jinek, M. (2017). Structural basis for guide RNA processing and seed-dependent DNA targeting by CRISPR-Cas12a. *Molecular Cell* 66: 221–233.e224.

40 Fonfara, I., Richter, H., Bratovič, M. et al. (2016). The CRISPR-associated DNA-cleaving enzyme Cpf1 also processes precursor CRISPR RNA. *Nature* 532: 517–521.

41 Meeske, A.J., Nakandakari-Higa, S., and Marraffini, L.A. (2019). Cas13-induced cellular dormancy prevents the rise of CRISPR-resistant bacteriophage. *Nature* 570: 241–245.

42 Yan, W.X., Chong, S., Zhang, H. et al. (2018). Cas13d is a compact RNA-targeting type VI CRISPR effector positively modulated by a WYL-Domain-Containing Accessory Protein. *Molecular Cell* 70: 327–339.e325.

43 Abudayyeh, O.O., Gootenberg, J.S., Konermann, S. et al. (2016). C2c2 is a single-component programmable RNA-guided RNA-targeting CRISPR effector. *Science* 353: aaf5573.

44 Makarova, K.S., Wolf, Y.I., and Koonin, E.V. Section 1 chapter 2: evolutionary classification of CRISPR-Cas systems 13. In: *CRISPR, Biology and Applications*, CRISPR: Biology and Applications (ed. R. Barrangou, E.J. Sontheimer, and L.A. Marraffini). Wiley. ISBN: 978-1-683-67037-7.

45 Makarova, K.S., Wolf, Y.I., Iranzo, J. et al. (2020). Evolutionary classification of CRISPR-Cas systems: a burst of class 2 and derived variants. *Nature Reviews. Microbiology* 18: 67–83.

46 Makarova, K.S., Wolf, Y.I., Alkhnbashi, O.S. et al. (2015). An updated evolutionary classification of CRISPR-Cas systems. *Nature Reviews. Microbiology* 13: 722–736.

47 Hille, F., Richter, H., Wong, S.P. et al. (2018). The biology of CRISPR-Cas: backward and forward. *Cell* 172: 1239–1259.

48 Crawley, A.B., Henriksen, J.R., and Barrangou, R. (2018). CRISPRdisco: an automated pipeline for the discovery and analysis of CRISPR-Cas systems. *CRISPR Journal* 1: 171–181.

49 Cameron, P., Coons, M.M., Klompe, S.E. et al. (2019). Harnessing type I CRISPR-Cas systems for genome engineering in human cells. *Nature Biotechnology* 37: 1471–1477.

50 Makarova, K.S., Karamycheva, S., Shah, S.A. et al. (2019). Predicted highly derived class 1 CRISPR-Cas system in Haloarchaea containing diverged Cas5 and Cas7 homologs but no CRISPR array. *FEMS Microbiology Letters* 366.

51 Tan, R., Krueger, R.K., Gramelspacher, M.J. et al. (2022). Cas11 enables genome engineering in human cells with compact CRISPR-Cas3 systems. *Molecular Cell* 82: 852–867.e855.

52 Charpentier, E., Richter, H., Van Der Oost, J., and White, M.F. (2015). Biogenesis pathways of RNA guides in archaeal and bacterial CRISPR-Cas adaptive immunity. *FEMS Microbiology Reviews* 39: 428–441.

53 Sinkunas, T., Gasiunas, G., Waghmare, S.P. et al. (2013). in vitro reconstitution of Cascade-mediated CRISPR immunity in Streptococcus thermophilus. *The EMBO Journal* 32: 385–394.

54 Westra, E.R., Van Erp, P.B., Künne, T. et al. (2012). CRISPR immunity relies on the consecutive binding and degradation of negatively supercoiled invader DNA by Cascade and Cas3. *Molecular Cell* 46: 595–605.

55 Lu, M., Yu, C., Zhang, Y. et al. (2024). Structure and genome editing of type I-B CRISPR-Cas. *Nature Communications* 15: 4126.

56 Hu, C., Ni, D., Nam, K.H. et al. (2022). Allosteric control of type I-A CRISPR-Cas3 complexes and establishment as effective nucleic acid detection and human genome editing tools. *Molecular Cell* 82: 2754–2768.e2755.

57 O'brien, R.E., Santos, C. I, Wrapp, D. et al. (2020). Structural basis for assembly of non-canonical small subunits into type I-C cascade. *Nature Communications* 11: 5931.

58 Mcbride, T.M., Cameron, S.C., Fineran, P.C., and Fagerlund, R.D. (2023). The biology and type I/III hybrid nature of type I-D CRISPR-Cas systems. *Biochemical Journal* 480: 471–488.

59 Westra, E.R., Semenova, E., Datsenko, K.A. et al. (2013). Type I-E CRISPR-cas systems discriminate target from non-target DNA through base pairing-independent PAM recognition. *PLoS Genetics* 9: e1003742.

60 Xue, C. and Sashital, D.G. (2019). Mechanisms of Type I-E and I-F CRISPR-Cas systems in Enterobacteriaceae. *EcoSal Plus* 8.

61 Yang, L., Zhang, L., Yin, P. et al. (1931). Insights into the inhibition of type I-F CRISPR-Cas system by a multifunctional anti-CRISPR protein AcrIF24. *Nature Communications* 2022: 13.

62 Molina, R., Sofos, N., and Montoya, G. (2020). Structural basis of CRISPR-Cas Type III prokaryotic defence systems. *Current Opinion in Structural Biology* 65: 119–129.

63 Kolesnik, M.V., Fedorova, I., Karneyeva, K.A. et al. (2021). Type III CRISPR-Cas systems: deciphering the most complex prokaryotic immune system. *Biochemistry (Mosc)* 86: 1301–1314.

64 Li, X., Han, J., Yang, J., and Zhang, H. (2024). The structural biology of type III CRISPR-Cas systems. *Journal of Structural Biology* 216: 108070.

65 Stella, G. and Marraffini, L. (2024). Type III CRISPR-Cas: beyond the Cas10 effector complex. *Trends in Biochemical Sciences* 49: 28–37.

66 Niewoehner, O., Garcia-Doval, C., Rostøl, J.T. et al. (2017). Type III CRISPR-Cas systems produce cyclic oligoadenylate second messengers. *Nature* 548: 543–548.

67 Athukoralage, J.S., Rouillon, C., Graham, S. et al. (2018). Ring nucleases deactivate type III CRISPR ribonucleases by degrading cyclic oligoadenylate. *Nature* 562: 277–280.

68 Barrangou, R., Marraffini, L.A., and Sontheimer, E.J. Section 1 chapter 1: CRISPR-Cas systems: core features and common mechanisms. In: *CRISPR, Biology and Applications*, CRISPR: Biology and Applications (ed. R. Barrangou, E.J. Sontheimer, and L.A. Marraffini). Wiley. ISBN: 978-1-683-67037-7.

69 Pinilla-Redondo, R., Mayo-Muñoz, D., Russel, J. et al. (2020). Type IV CRISPR-Cas systems are highly diverse and involved in competition between plasmids. *Nucleic Acids Research* 48: 2000–2012.

70 Shmakov, S., Smargon, A., Scott, D. et al. (2017). Diversity and evolution of class 2 CRISPR-Cas systems. *Nature Reviews. Microbiology* 15: 169–182.

71 Makarova, K.S., Zhang, F., and Koonin, E.V. (2017). SnapShot: class 2 CRISPR-Cas systems. *Cell* 168: 328–328.e321.

72 Chylinski, K., Makarova, K.S., Charpentier, E., and Koonin, E.V. (2014). Classification and evolution of type II CRISPR-Cas systems. *Nucleic Acids Research* 42: 6091–6105.

73 Mosterd, C. and Moineau, S. (2020). Characterization of a Type II-A CRISPR-Cas system in Streptococcus mutans. *mSphere* 5.

74 Chylinski, K., Le Rhun, A., and Charpentier, E. (2013). The tracrRNA and Cas9 families of type II CRISPR-Cas immunity systems. *RNA Biology* 10: 726–737.

75 Bestas, B., Wimberger, S., Degtev, D. et al. (2023). A Type II-B Cas9 nuclease with minimized off-targets and reduced chromosomal translocations in vivo. *Nature Communications* 14: 5474.

76 Mir, A., Edraki, A., Lee, J., and Sontheimer, E.J. (2018). Type II-C CRISPR-Cas9 biology, mechanism, and application. *ACS Chemical Biology* 13: 357–365.

77 Altae-Tran, H., Kannan, S., Demircioglu, F.E. et al. (2021). The widespread IS200/IS605 transposon family encodes diverse programmable RNA-guided endonucleases. *Science* 374: 57–65.

78 Yan, W.X., Hunnewell, P., Alfonse, L.E. et al. (2019). Functionally diverse type V CRISPR-Cas systems. *Science* 363: 88–91.

79 Zetsche, B., Gootenberg, J.S., Abudayyeh, O.O. et al. (2015). Cpf1 is a single RNA-guided endonuclease of a class 2 CRISPR-Cas system. *Cell* 163: 759–771.

80 Koonin, E.V., Makarova, K.S., and Zhang, F. (2017). Diversity, classification and evolution of CRISPR-Cas systems. *Current Opinion in Microbiology* 37: 67–78.

81 Burstein, D., Harrington, L.B., Strutt, S.C. et al. (2017). New CRISPR-Cas systems from uncultivated microbes. *Nature* 542: 237–241.

82 Nguyen, G.T., Dhingra, Y., and Sashital, D.G. (2022). Miniature CRISPR-Cas12 endonucleases – programmed DNA targeting in a smaller package. *Current Opinion in Structural Biology* 77: 102466.

83 Karvelis, T., Druteika, G., Bigelyte, G. et al. (2021). Transposon-associated TnpB is a programmable RNA-guided DNA endonuclease. *Nature* 599: 692–696.

84 O'connell, M.R. (2019). Molecular mechanisms of RNA targeting by Cas13-containing Type VI CRISPR-Cas systems. *Journal of Molecular Biology* 431: 66–87.

85 Abudayyeh, O.O. and Gootenberg, J.S. Section 1 chapter 7: CRISPR-Cas13: biology, mechanism, and applications of RNA-guided, RNA-targeting CRISPR systems. In: *CRISPR, Biology and Applications*, CRISPR: Biology and Applications (ed. R. Barrangou, E.J. Sontheimer, and L.A. Marraffini). Wiley. ISBN: 978-1-683-67037-7.

86 Wang, T., Wang, Z., Bai, L. et al. (2023). Next-generation CRISPR-based diagnostic tools for human diseases. *TrAC Trends in Analytical Chemistry* 168: 117328.

87 Knott, G.J. and Doudna, J.A. (2018). CRISPR-Cas guides the future of genetic engineering. *Science* 361: 866–869.

88 Morisaka, H., Yoshimi, K., Okuzaki, Y. et al. (2019). CRISPR-Cas3 induces broad and unidirectional genome editing in human cells. *Nature Communications* 10: 5302.

89 Yoshimi, K., Takeshita, K., Yamayoshi, S. et al. (2022). CRISPR-Cas3-based diagnostics for SARS-CoV-2 and influenza virus. *iScience* 25: 103830.

90 Cong, L. and Zhang, F. (2015). Genome engineering using CRISPR-Cas9 system. *Methods in Molecular Biology* 1239: 197–217.

91 Doudna, J.A. and Charpentier, E. (2014). The new frontier of genome engineering with CRISPR-Cas9. *Science* 346: 1258096.

92 Jinek, M., Chylinski, K., Fonfara, I. et al. (2012). A programmable dual-RNA-guided DNA endonuclease in adaptive bacterial immunity. *Science* 337: 816–821.

93 Briner, A.E., Donohoue, P.D., Gomaa, A.A. et al. (2014). Guide RNA functional modules direct Cas9 activity and orthogonality. *Molecular Cell* 56: 333–339.

94 Ran, F.A., Hsu, P.D., Lin, C.Y. et al. (2013). Double nicking by RNA-guided CRISPR Cas9 for enhanced genome editing specificity. *Cell* 154: 1380–1389.

95 Qi, L.S., Larson, M.H., Gilbert, L.A. et al. (2013). Repurposing CRISPR as an RNA-guided platform for sequence-specific control of gene expression. *Cell* 152: 1173–1183.

96 Gilbert, L.A., Larson, M.H., Morsut, L. et al. (2013). CRISPR-mediated modular RNA-guided regulation of transcription in eukaryotes. *Cell* 154: 442–451.

97 Walton, R.T., Christie, K.A., Whittaker, M.N., and Kleinstiver, B.P. (2020). Unconstrained genome targeting with near-PAMless engineered CRISPR-Cas9 variants. *Science* 368: 290–296.

98 Christie, K.A., Guo, J.A., Silverstein, R.A. et al. (2023). Precise DNA cleavage using CRISPR-SpRYgests. *Nature Biotechnology* 41: 409–416.

99 Wang, S., Mao, H., Hou, L. et al. (2022). Compact SchCas9 recognizes the simple NNGR PAM. *Adv Sci (Weinh)* 9: e2104789.

100 Wei, J., Hou, L., Liu, J. et al. (2022). Closely related type II-C Cas9 orthologs recognize diverse PAMs. *eLife* 11.

101 Li, S.Y., Cheng, Q.X., Wang, J.M. et al. (2018). CRISPR-Cas12a-assisted nucleic acid detection. *Cell Discovery* 4: 20.

102 Yamano, T., Nishimasu, H., Zetsche, B. et al. (2016). Crystal structure of Cpf1 in complex with guide RNA and target DNA. *Cell* 165: 949–962.

103 Teng, F., Cui, T., Feng, G. et al. (2018). Repurposing CRISPR-Cas12b for mammalian genome engineering. *Cell Discovery* 4: 63.

104 Harrington, L.B., Burstein, D., Chen, J.S. et al. (2018). Programmed DNA destruction by miniature CRISPR-Cas14 enzymes. *Science* 362: 839–842.

105 East-Seletsky, A., O'connell, M.R., Knight, S.C. et al. (2016). Two distinct RNase activities of CRISPR-C2c2 enable guide-RNA processing and RNA detection. *Nature* 538: 270–273.

106 Smargon, A.A., Cox, D.B.T., Pyzocha, N.K. et al. (2017). Cas13b is a Type VI-B CRISPR-associated RNA-guided RNase differentially regulated by accessory proteins Csx27 and Csx28. *Molecular Cell* 65: 618–630.e617.

107 Huyke, D.A., Ramachandran, A., Bashkirov, V.I. et al. (2022). Enzyme kinetics and detector sensitivity determine limits of detection of amplification-free CRISPR-Cas12 and CRISPR-Cas13 diagnostics. *Analytical Chemistry* 94: 9826–9834.

108 Pardee, K., Green, A.A., Takahashi, M.K. et al. (2016). Rapid, low-cost detection of Zika virus using programmable biomolecular components. *Cell* 165: 1255–1266.

109 Jiao, C., Sharma, S., Dugar, G. et al. (2021). Noncanonical crRNAs derived from host transcripts enable multiplexable RNA detection by Cas9. *Science* 372: 941–948.

110 Huang, M., Zhou, X., Wang, H., and Xing, D. (2018). Clustered regularly interspaced short palindromic repeats/Cas9 triggered isothermal amplification for site-specific nucleic acid detection. *Analytical Chemistry* 90: 2193–2200.

111 Zhou, W., Hu, L., Ying, L. et al. (2018). A CRISPR-Cas9-triggered strand displacement amplification method for ultrasensitive DNA detection. *Nature Communications* 9: 5012.

112 Wang, T., Liu, Y., Sun, H.H. et al. (2019). An RNA-guided Cas9 Nickase-based method for universal isothermal DNA amplification. *Angewandte Chemie (International Ed. in English)* 58: 5382–5386.

113 Moon, J., Kwon, H.J., Yong, D. et al. (2020). Colorimetric detection of SARS-CoV-2 and drug-resistant pH1N1 using CRISPR/dCas9. *ACS Sensors* 5: 4017–4026.

114 Zhang, Y., Qian, L., Wei, W. et al. (2017). Paired design of dCas9 as a systematic platform for the detection of featured nucleic acid sequences in pathogenic strains. *ACS Synthetic Biology* 6: 211–216.

115 Hajian, R., Balderston, S., Tran, T. et al. (2019). Detection of unamplified target genes via CRISPR-Cas9 immobilized on a graphene field-effect transistor. *Nature Biomedical Engineering* 3: 427–437.

116 Chen, J., Chen, Y., Huang, L. et al. (2024). Trans-nuclease activity of Cas9 activated by DNA or RNA target binding. *Nature Biotechnology*.

117 Chen, J.S., Ma, E., Harrington, L.B. et al. (2018). CRISPR-Cas12a target binding unleashes indiscriminate single-stranded DNase activity. *Science* 360: 436–439.

118 Gootenberg, J.S., Abudayyeh, O.O., Lee, J.W. et al. (2017). Nucleic acid detection with CRISPR-Cas13a/C2c2. *Science* 356: 438–442.

119 Teng, F., Guo, L., Cui, T. et al. (2019). CDetection: CRISPR-Cas12b-based DNA detection with sub-attomolar sensitivity and single-base specificity. *Genome Biology* 20: 132.

120 Li, L., Li, S., Wu, N. et al. (2019). HOLMESv2: A CRISPR-Cas12b-assisted platform for nucleic acid detection and DNA methylation quantitation. *ACS Synthetic Biology* 8: 2228–2237.

121 Kellner, M.J., Koob, J.G., Gootenberg, J.S. et al. (2019). SHERLOCK: nucleic acid detection with CRISPR nucleases. *Nature Protocols* 14: 2986–3012.

122 Gootenberg, J.S., Abudayyeh, O.O., Kellner, M.J. et al. (2018). Multiplexed and portable nucleic acid detection platform with Cas13, Cas12a, and Csm6. *Science* 360: 439–444.

123 Cheng, X., Li, Y., Kou, J. et al. (2022). Novel non-nucleic acid targets detection strategies based on CRISPR/Cas toolboxes: a review. *Biosensors & Bioelectronics* 215: 114559.

124 Libis, V., Delépine, B., and Faulon, J.L. (2016). Sensing new chemicals with bacterial transcription factors. *Current Opinion in Microbiology* 33: 105–112.

125 Hicks, M., Bachmann, T.T., and Wang, B. (2020). Synthetic biology enables programmable cell-based biosensors. *ChemPhysChem* 21: 132–144.

126 Liang, M., Li, Z., Wang, W. et al. (2019). A CRISPR-Cas12a-derived biosensing platform for the highly sensitive detection of diverse small molecules. *Nature Communications* 10: 3672.

127 Samanta, D., Ebrahimi, S.B., Ramani, N., and Mirkin, C.A. (2022). Enhancing CRISPR-Cas-mediated detection of nucleic acid and non-nucleic acid targets using enzyme-labeled reporters. *Journal of the American Chemical Society* 144: 16310–16315.

128 Li, J., Yang, S., Zuo, C. et al. (2020). Applying CRISPR-Cas12a as a signal amplifier to construct biosensors for non-DNA targets in ultralow concentrations. *ACS Sensors* 5: 970–977.

129 Iwasaki, R.S. and Batey, R.T. (2020). SPRINT: a Cas13a-based platform for detection of small molecules. *Nucleic Acids Research* 48: e101.

130 Lee, I., Kwon, S.J., Sorci, M. et al. (2021). Highly sensitive immuno-CRISPR assay for CXCL9 detection. *Analytical Chemistry* 93: 16528–16534.

131 Mahas, A., Wang, Q., Marsic, T., and Mahfouz, M.M. (2022). Development of Cas12a-based cell-free small-molecule biosensors via allosteric regulation of CRISPR array expression. *Analytical Chemistry* 94: 4617–4626.

132 Liu, J., Cao, Z., and Lu, Y. (2009). Functional nucleic acid sensors. *Chemical Reviews* 109: 1948–1998.

133 Nimjee, S.M., White, R.R., Becker, R.C., and Sullenger, B.A. (2017). Aptamers as therapeutics. *Annual Review of Pharmacology and Toxicology* 57: 61–79.

134 Zhao, X., Li, S., Liu, G. et al. (2021). A versatile biosensing platform coupling CRISPR-Cas12a and aptamers for detection of diverse analytes. *Sci Bull (Beijing)* 66: 69–77.

135 Morrison, D., Rothenbroker, M., and Li, Y. (2018). DNAzymes: selected for applications. *Small Methods* 2: 1700319.

136 Khan, S., Burciu, B., Filipe, C.D.M. et al. (2021). DNAzyme-based biosensors: immobilization strategies, applications, and future prospective. *ACS Nano* 15: 13943–13969.

137 Ma, Y., Mou, Q., Yan, P. et al. (2021). A highly sensitive and selective fluoride sensor based on a riboswitch-regulated transcription coupled with CRISPR-Cas13a tandem reaction. *Chemical Science* 12: 11740–11747.

138 Soh, J.H., Balleza, E., Abdul Rahim, M.N. et al. (2022). CRISPR-based systems for sensitive and rapid on-site COVID-19 diagnostics. *Trends in Biotechnology* 40: 1346–1360.

139 Patchsung, M., Jantarug, K., Pattama, A. et al. (2020). Clinical validation of a Cas13-based assay for the detection of SARS-CoV-2 RNA. *Nature Biomedical Engineering* 4: 1140–1149.

140 Broughton, J.P., Deng, X., Yu, G. et al. (2020). CRISPR-Cas12-based detection of SARS-CoV-2. *Nature Biotechnology* 38: 870–874.

141 Arizti-Sanz, J., Freije, C.A., Stanton, A.C. et al. (2020). Streamlined inactivation, amplification, and Cas13-based detection of SARS-CoV-2. *Nature Communications* 11: 5921.

142 Joung, J., Ladha, A., Saito, M. et al. (2020). Point-of-care testing for COVID-19 using SHERLOCK diagnostics. *medRxiv*.

143 Ding, X., Yin, K., Li, Z. et al. (2020). Ultrasensitive and visual detection of SARS-CoV-2 using all-in-one dual CRISPR-Cas12a assay. *Nature Communications* 11: 4711.

144 Li, Z., Uno, N., Ding, X. et al. (2023). Bioinspired CRISPR-mediated cascade reaction biosensor for molecular detection of HIV using a Glucose Meter. *ACS Nano* 17: 3966–3975.

145 Shi, K., Xie, S., Tian, R. et al. (2021). A CRISPR-Cas autocatalysis-driven feedback amplification network for supersensitive DNA diagnostics. *Science Advances* 7.

146 Liu, P., Lin, Y., Zhuo, X. et al. (2024). Universal crRNA acylation strategy for robust photo-initiated One-Pot CRISPR-Cas12a nucleic acid diagnostics. *Angewandte Chemie (International Ed. in English)* 63: e202401486.

147 Fozouni, P., Son, S., De León, D. et al. (2021). Amplification-free detection of SARS-CoV-2 with CRISPR-Cas13a and mobile phone microscopy. *Cell* 184: 323–333.e329.

148 Welch, N.L., Zhu, M., Hua, C. et al. (2022). Multiplexed CRISPR-based microfluidic platform for clinical testing of respiratory viruses and identification of SARS-CoV-2 variants. *Nature Medicine* 28: 1083–1094.

149 Joung, J., Ladha, A., Saito, M. et al. (2020). Detection of SARS-CoV-2 with SHERLOCK One-Pot Testing. *The New England Journal of Medicine* 383: 1492–1494.

150 Casati, B., Verdi, J.P., Hempelmann, A. et al. (2022). Rapid, adaptable and sensitive Cas13-based COVID-19 diagnostics using ADESSO. *Nature Communications* 13: 3308.

151 Chandrasekaran, S.S., Agrawal, S., Fanton, A. et al. (2022). Rapid detection of SARS-CoV-2 RNA in saliva via Cas13. *Nature Biomedical Engineering* 6: 944–956.

152 Ding, X., Yin, K., Li, Z. et al. (2021). Sensitive quantitative detection of SARS-CoV-2 in clinical samples using digital warm-start CRISPR assay. *Biosensors & Bioelectronics* 184: 113218.

153 Wang, R., Qian, C., Pang, Y. et al. (2021). opvCRISPR: one-pot visual RT-LAMP-CRISPR platform for SARS-cov-2 detection. *Biosensors & Bioelectronics* 172: 112766.

154 Freije, C.A., Myhrvold, C., Boehm, C.K. et al. (2019). Programmable inhibition and detection of RNA viruses using Cas13. *Molecular Cell* 76: 826, e811–837.

155 Pickar-Oliver, A. and Gersbach, C.A. (2019). The next generation of CRISPR-Cas technologies and applications. *Nature Reviews. Molecular Cell Biology* 20: 490–507.

156 Kaminski, M.M., Alcantar, M.A., Lape, I.T. et al. (2020). A CRISPR-based assay for the detection of opportunistic infections post-transplantation and for the monitoring of transplant rejection. *Nature Biomedical Engineering* 4: 601–609.

157 Tian, Y., Fan, Z., Xu, L. et al. (2023). CRISPR/Cas13a-assisted rapid and portable HBV DNA detection for low-level viremia patients. *Emerging Microbes & Infections* 12: e2177088.

158 Zhang, X., Tian, Y., Xu, L. et al. (2022). CRISPR/Cas13-assisted hepatitis B virus covalently closed circular DNA detection. *Hepatology International* 16: 306–315.

159 Tao, Y., Yi, K., Wang, H. et al. (2022). CRISPR-Cas12a-regulated DNA adsorption and metallization on MXenes as enhanced enzyme mimics for sensitive colorimetric detection of hepatitis B virus DNA. *Journal of Colloid and Interface Science* 613: 406–414.

160 Wang, S., Li, H., Kou, Z. et al. (2021). Highly sensitive and specific detection of hepatitis B virus DNA and drug resistance mutations utilizing the PCR-based CRISPR-Cas13a system. *Clinical Microbiology and Infection* 27: 443–450.

161 Le, C., Liu, Y., Lopez-Orozco, J. et al. (2021). CRISPR technique incorporated with single-cell RNA sequencing for studying Hepatitis B infection. *Analytical Chemistry* 93: 10756–10761.

162 Chen, X., Tan, Y., Wang, S. et al. (2021). A CRISPR-Cas12b-based platform for ultrasensitive, rapid, and highly specific detection of Hepatitis B virus genotypes B and C in clinical application. *Frontiers in Bioengineering and Biotechnology* 9: 743322.

163 Wang, Q., Zhang, B., Xu, X. et al. (2018). CRISPR-typing PCR (ctPCR), a new Cas9-based DNA detection method. *Scientific Reports* 8: 14126.

164 Zhang, B., Wang, Q., Xu, X. et al. (2018). Detection of target DNA with a novel Cas9/sgRNAs-associated reverse PCR (CARP) technique. *Analytical and Bioanalytical Chemistry* 410: 2889–2900.

165 Jia, C., Huai, C., Ding, J. et al. (2018). New applications of CRISPR/Cas9 system on mutant DNA detection. *Gene* 641: 55–62.

166 Xu, Z., Chen, D., Li, T. et al. (2022). Microfluidic space coding for multiplexed nucleic acid detection via CRISPR-Cas12a and recombinase polymerase amplification. *Nature Communications* 13: 6480.

167 Ai, J.W., Zhou, X., Xu, T. et al. (2019). CRISPR-based rapid and ultra-sensitive diagnostic test for Mycobacterium tuberculosis. *Emerging Microbes & Infections* 8: 1361–1369.

168 Shen, J., Zhou, X., Shan, Y. et al. (2020). Sensitive detection of a bacterial pathogen using allosteric probe-initiated catalysis and CRISPR-Cas13a amplification reaction. *Nature Communications* 11: 267.

169 Yang, H., Yang, S., Xia, X. et al. (2022). Sensitive detection of a single-nucleotide polymorphism in foodborne pathogens using CRISPR/Cas12a-signaling ARMS-PCR. *Journal of Agricultural and Food Chemistry* 70: 8451–8457.

170 Lu, Y., Li, F., Bai, J. et al. (2022). Rapid screening of antimicrobial probiotics using CRISPR cascade. *Biosensors & Bioelectronics* 216: 114673.

171 Quan, J., Langelier, C., Kuchta, A. et al. (2019). FLASH: a next-generation CRISPR diagnostic for multiplexed detection of antimicrobial resistance sequences. *Nucleic Acids Research* 47: e83.

172 Cunningham, C.H., Hennelly, C.M., Lin, J.T. et al. (2021). A novel CRISPR-based malaria diagnostic capable of Plasmodium detection, species differentiation, and drug-resistance genotyping. *eBioMedicine* 68: 103415.

173 Lee, R.A., Puig, H., Nguyen, P.Q. et al. (2020). Ultrasensitive CRISPR-based diagnostic for field-applicable detection of Plasmodium species in symptomatic and asymptomatic malaria. *Proceedings of the National Academy of Sciences of the United States of America* 117: 25722–25731.

174 Lin, C., Zhou, J., Gao, N. et al. (2024). Establishing a pulmonary aspergillus fumigatus infection diagnostic platform based on RPA-CRISPR-Cas12a. *World Journal of Microbiology and Biotechnology* 40: 116.

175 Hao, L., Zhao, R.T., Welch, N.L. et al. (2023). CRISPR-Cas-amplified urinary biomarkers for multiplexed and portable cancer diagnostics. *Nature Nanotechnology* 18: 798–807.

176 Zhou, M., Yin, Y., Shi, Y. et al. (2022). Spherical nucleic acid reporter-based cascade CRISPR/Cas12a amplifier for stable and sensitive biosensing of circulating tumor DNA. *Chemical Communications (Cambridge)* 58: 4508–4511.

177 Yang, B., Kong, J., and Fang, X. (2022). Programmable CRISPR-Cas9 microneedle patch for long-term capture and real-time monitoring of universal cell-free DNA. *Nature Communications* 13: 3999.

178 Wang, D.X., Wang, J., Wang, Y.X. et al. (2022). A CRISPR/Cas12a-responsive dual-aptamer DNA network for specific capture and controllable release of circulating tumor cells. *Chemical Science* 13: 10395–10405.

179 Liu, Y., Chen, Y., Dang, L. et al. (2021). EasyCatch, a convenient, sensitive and specific CRISPR detection system for cancer gene mutations. *Molecular Cancer* 20: 157.

180 Balderston, S., Taulbee, J.J., Celaya, E. et al. (2021). Discrimination of single-point mutations in unamplified genomic DNA via Cas9 immobilized on a graphene field-effect transistor. *Nature Biomedical Engineering* 5: 713–725.

181 Chen, P., Wang, L., Qin, P. et al. (2022). An RNA-based catalytic hairpin assembly circuit coupled with CRISPR-Cas12a for one-step detection of microRNAs. *Biosensors & Bioelectronics* 207: 114152.

182 Zhou, T., Huang, R., Huang, M. et al. (2020). CRISPR/Cas13a powered portable electrochemiluminescence chip for ultrasensitive and specific MiRNA detection. *Advanced Scince (Weinheim)* 7: 1903661.

183 Shen, B., Li, L., Liu, C. et al. (2023). Mesoporous nanozyme-enhanced DNA tetrahedron electrochemiluminescent biosensor with three-dimensional walking nanomotor-mediated CRISPR/Cas12a for ultrasensitive detection of exosomal microRNA. *Analytical Chemistry* 95: 4486–4495.

184 Lv, Z., Wang, Q., and Yang, M. (2021). Multivalent Duplexed-Aptamer Networks regulated a CRISPR-Cas12a system for circulating tumor cell detection. *Analytical Chemistry* 93: 12921–12929.

185 Sha, L., Bo, B., Yang, F. et al. (2022). Programmable DNA-fueled electrochemical analysis of lung cancer exosomes. *Analytical Chemistry* 94: 8748–8755.

186 Ding, L., Wu, Y., Liu, L.E. et al. (2023). Universal DNAzyme walkers-triggered CRISPR-Cas12a/Cas13a bioassay for the synchronous detection of two exosomal proteins and its application in intelligent diagnosis of cancer. *Biosensors & Bioelectronics* 219: 114827.

187 Broto, M., Kaminski, M.M., Adrianus, C. et al. (2022). Nanozyme-catalysed CRISPR assay for preamplification-free detection of non-coding RNAs. *Nature Nanotechnology* 17: 1120–1126.

188 Slaymaker, I.M., Gao, L., Zetsche, B. et al. (2016). Rationally engineered Cas9 nucleases with improved specificity. *Science* 351: 84–88.

189 Kleinstiver, B.P., Pattanayak, V., Prew, M.S. et al. (2016). High-fidelity CRISPR-Cas9 nucleases with no detectable genome-wide off-target effects. *Nature* 529: 490–495.

190 Wang, Y., Qi, T., Liu, J. et al. (2023). A highly specific CRISPR-Cas12j nuclease enables allele-specific genome editing. *Science Advances* 9: eabo6405.

191 Liu, Y., Zhan, L., Qin, Z. et al. (2021). Ultrasensitive and highly specific lateral flow assays for point-of-care diagnosis. *ACS Nano* 15: 3593–3611.

192 Politza, A.J., Nouri, R., and Guan, W. (2023). Digital CRISPR systems for the next generation of nucleic acid quantification. *Trends in Analytical Chemistry* 159.

8

Precision and Speed

The Rise of High-Throughput Microfluidic Systems in Diagnostics of Infectious Diseases

Jinlu Tang[1], Yuehe Lin[2], and Zhaohui Li[1,3]

[1] *Zhengzhou University, Henan Joint International Research Laboratory of Green Construction of Functional Molecules and Their Bioanalytical Applications, Zhengzhou 450001, China*
[2] *Washington State University, School of Mechanical and Materials Engineering, Department of Chemistry, Pullman, Washington 99164, United States*
[3] *The First Affiliated Hospital of Zhengzhou University, Zhengzhou 450052, China*

CHAPTER MENU

8.1 Introduction

Nowadays, society is faced with increasingly serious public health diseases because of the rapid development of industrial science and technology, environmental pollution, drug abuse, and so on. For example, convenient transportation has accelerated the spread of epidemics, while the overuse of antibiotics has greatly accelerated the evolution of pathogens. Rapid and sensitive detection is essential to curing as well as controlling the occurrence and development of diseases. Many technologies and instruments have been invented and used for the detection and monitoring of public health diseases, such as enzyme linked immunosorbent assay (ELISA), polymerase chain reaction (PCR), in-situ hybridization (ISH), flow cytometry (FCM), chromatography, and so on [1–3]. However, although these technologies have made great achievements in the fight against human diseases, they still cannot solve the current problems caused by global public health diseases in some aspects, such as point-of-care detection, high throughput, and high specificity. In recent years, microfluidics has emerged as an integrated technology to

Diagnostic Techniques in Public Health: Analytical and Chemometric Approaches in Disease Detection and Management, First Edition. Edited by Jinghong Li and Yan Du.

solve comprehensive problems in various fields, from chemistry to physics, biology, pharmacology, and medicine [4, 5]. Microfluidic devices have exhibited many innovative features (for instance, completing separation, transport, identification, detection and reporting within a miniaturized lab chip, facilitating data collection efficiently and accurately in both high-throughput and quantitative ways) by combining different lab functions on a small scale. In this chapter, we provide a brief introduction to microfluidic systems and summarize the latest applications of microfluidic technology in the field of public infectious diseases.

8.2 The Origins and Developments of Microfluidics

Microfluidics refers to the science of manipulating fluids in systems at the micro-length scales [6]. These fluids can be gases or liquids. Systems include devices with microchannels or patterned surfaces, even specially processed paper. The span of microfluidic scales extends from 100 nm to 1 mm.

8.2.1 Inception Phase

The concept of microfluidics can be traced back to the 1950s, when scientists first achieved precise manipulation of minuscule droplets in the nanoliter and picolitre ranges. Another event that promoted its germination occurred in 1959. Richard P. Feynman delivered a visionary speech titled "There is plenty of room at the bottom" during the American Physical Society (APS) meeting at Caltech [7], which was recognized for pioneering the new era of nanotechnology. There are at least three industrial technological developments that have had profound implications for the emergence of future microfluidic technology. The first significant factor is the Silicon Valley revolution, which dates back to the invention of the first transistor by the Bell Telephone Company in 1947 [8]. The invention of the transistor promoted the exclusivity of photoetching technology, making it possible to manufacture microelements on silicon on a large scale, laying the foundation for future microelectronics manufacturing. In 1964, Jack Kilby ushered in a new era for the microelectronics manufacturing industry by discussing how many oscillating circuit components could be manufactured on silicon wafers using optical etching technology. From then on, the Silicon Valley revolution began. The second technology is the invention of inkjet printing. In 1965, R. G. Sweet designed a device that could produce submillimetric droplets by forcing ink through a nozzle with a diameter of 35 μm [9]. In this device, ink and paper were used to create high-frequency records by jetting ink droplets onto a moving chart, leaving an immediately visible and permanent record of the input signal. A crucial factor was the implementation of electrostatic control, enabling the reproducible production of large quantities of nanoliter droplets and paving the way for the development of inkjet printers. Subsequent research on this technology by Bassous in 1977 demonstrated that an array of inkjet nozzles could be fabricated on an individual silicon wafer through photolithography [10]. This not only facilitated the commercialization of inkjet printing but also demonstrated the potential of silicon as a material for the large-scale production of microfluidic devices. Another driving force is the rise of the micromanufacturing industry. In the 1970s and 1980s, with the development of micromachining processes, microelectromechanical systems (MEMS) had attained significant success, leading to the development of a wide range of commercial products to cater to the daily needs of the public [11], especially for the automotive industry. One of the most famous examples is the MEMS-based airbags that were invented in the 1980s [12]. The core technology of airbags involves building accelerometers by integrating capacitor banks onto micrometer-sized silicon wafers. There are two kinds of electrodes for the capacitor banks, one fixed and the other mobile. When there is a strong acceleration change, such as an accidental impact, the mobile electrodes produce a displacement, which in turn changes the capacitance of the capacitor, thus ejecting the air bag. This miniaturized accelerometer demonstrates extremely high sensitivity and response speed and represents the gradual maturity of lithography and related technologies in silicon-based microelectronics and MEMS [6]. This advancement directly contributes to the progress of microfluidic technology.

8.2.2 Early Research in Microfluidics

In 1975, S.C. Terry reported the first microfluidic device in his PhD thesis [13, 14], a miniaturized gas chromatograph (GC). The GC consisted of a 1.5-m-long microchannel etched onto a silicon wafer, an injector, and a thermal detector (Figure 8.1) [15]. The device was fabricated through a process involving the use of photolithography and etching techniques on a 5 cm silicon wafer, which could facilitate highly sensitive and specific analysis of the composition of the injected gas mixture. This was a milestone in the development of microfluidic technology, and the developed device was soon commercialized and widely used in fields such as cell observation, drug screening, and biomarker detection. Furthermore, it is

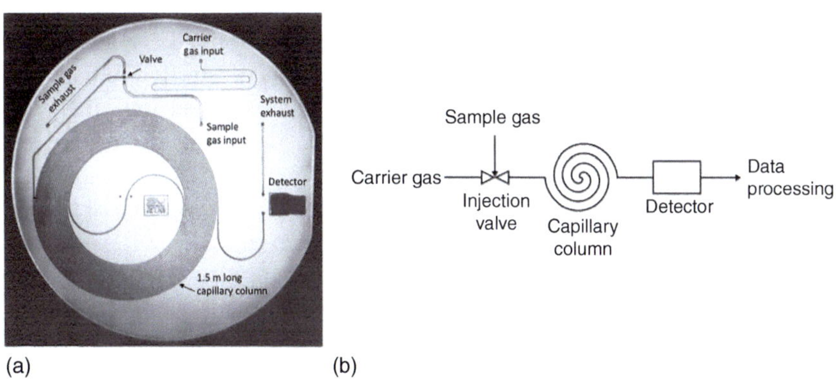

Figure 8.1 (a) Photograph of the Gas Chromatography System described by Terry et al. (b) A schematic representation illustrating all fluid handling components integrated onto a single silicon wafer. *Source:* Ref. [15] / Elsevier / CC BY 4.0.

widely acknowledged that this paper marked the true emergence of microfluidics as an independent field.

8.2.3 μTAS Concept

By 1990, Swedish scientists Manz and Widmer had proposed the rudimentary concept of microfluidic chips—Total Analysis Systems (TAS) and Micro Total Analysis Systems (μTAS) [16]. These systems encompass all necessary analytical functions such as sampling, sample transportation, and preparation involving chemical reactions and separations, along with automated detection processes. In this seminal paper, the authors outlined how the physical advantages of fluid mechanics at the microscale could result in faster and more effective analysis. Furthermore, they hypothesized that it was feasible to conduct simultaneous analysis of multiple samples on a single chip. This paper demonstrated the potential of microfluidic devices and laid the groundwork for many advances and technologies that are now commonplace in laboratories.

The introduction of the μTAS concept has sparked a surge of research activity in microfluidic technology among scientists worldwide, leading to its gradual expansion from the realm of analytical chemistry to other domains. In 1994, the inaugural μTAS conference took place in Enschede, Netherlands, attracting approximately 160 participants from diverse fields to deliberate on the burgeoning microfluidic technology. During the initial phases of microfluidic technology advancement, owing to its conceptual similarity with electronic chips in the integration of multiple analytical functions within a single device, researchers conceived the "lab on a chip" concept. This entailed fabricating various fundamental microfluidic functional units on a compact chip and integrating them to cater to diverse requirements in biology and chemistry.

8.2.4 DARPA and the Human Genome Project

Meanwhile, the establishment of two projects in the 1990s had a crucial impact on the development of microfluidic technology. One of the projects was a series of microfluidic studies for chemical and biological weapon detection initiated by the Defense Advanced Research Projects Agency (DARPA), and the other was the Human Genome Project (HGP). DARPA-related initiatives are focused on developing portable, field-deployable devices for the rapid detection of biological and chemical weapons, with the goal of identifying and preventing terrorist attacks in a timely manner, these projects have significantly expedited the transition of microfluidic technology from theoretical to practical application [6]. In the 1990s, the Human Genome Project (HGP) was collaboratively funded by the National Institutes of Health and the US Department of Energy with the objective of completing the entire human genome map within 15 years. However, due to the limitations of initial DNA sequencing technology, it was unable to meet the demands of this extensive task. As a result, researchers began to explore the potential application of microfluidic technology in DNA sequencing for realizing high-throughput, rapid, and sensitive methods for DNA sequencing. As early as 1990, Swerdlow et al. demonstrated that capillary gel electrophoresis could be used for the separation and detection of DNA sequencing samples. The separation speed of capillary gel electrophoresis is three times faster than that of traditional automated plate gel DNA sequencing instruments, along with improved resolution (2.4 times) and higher separation efficiency (5.4 times) [17]. In 1991, Manz and coworkers conducted capillary electrophoresis experiments on glass chip microchannels fabricated by microfabrication technology and performed amino acid separation and flow injection analysis, laying the foundation for the breakthrough development of microfluidic chips [18]. In 1994, Woolley and Mathies research group [19], as well as Effenhauser research group [20], respectively reported that DNA fragments could be separated at high speed by fabricating capillary electrophoresis microarrays on planar glass substrates. The establishment of capillary electrophoresis arrays effectively solved the time-consuming and labor-intensive problems faced by traditional plate gel electrophoresis-based DNA sequencing technology, taking a crucial step toward the development of the high-efficiency, parallel, and high-throughput DNA sequencing technology required for the Human Genome Project. In 1995, Woolley and Mathies achieved ultra-high-speed DNA sequencing on microfluidic chips, leading to the recognition of the commercial potential of microfluidic chips [21]. Subsequently, major corporations launched product development in this domain, significantly propelling the advancement of microfluidic technology. In 1996, PCR amplification technology (the most important technology for gene analysis, first proposed by Mullis in 1983) was successfully integrated into microfluidic chips, demonstrating the potential of micro total analysis systems in sample processing [22]. That same year, the world's first microfluidics company, Caliper, was founded in the United States. The following year, in 1996, Thomas L. Gutshall, Bill McMillan, Dr. Kurt Petersen, Dr. Greg Kovacs, Dr. Allen Northrup, and Steven Young came together in California to found Cepheid, a leading molecular diagnostics company.

8.2.5 PDMS

Despite the widespread commercialization of microfluidic technology, the realization of microfluidic devices with intricate structures and desired functionalities remains a challenge, primarily attributed to the fact that the limited materials fabrication at that time (e.g., silicon, glass, or plastic) hindered the integration of multitudes of microvalves into these devices. In 1998, a groundbreaking technology (soft lithography) was reported by Xia and Whitesides [23], allowing for the creation of microfluidic devices with lithography and soft materials. Utilizing polydimethylsiloxane (PDMS) as a typical soft material for the fabrication of microfluidic devices, as opposed to traditional materials such as silicon and glass, can significantly mitigate the complexity and duration of the fabrication process, reduce costs, as well as support an entirely clean room-independent microfabrication process once the mold is prepared [24]. Like those made from silicon, microfluidic devices constructed from PDMS can also achieve the controlled separation and detection of samples. This novel, straightforward, rapid, and efficient approach for fabricating microfluidic devices effectively decreased the costs associated with microfluidic chips and significantly reduced the barriers to entry in the field of microfluidics, enabling researchers to pursue related studies without the need for cleanrooms. This sparked a new wave of research in the microfluidics field and inspired the development and assembly of numerous innovative and resourceful microfluidic devices. It is noteworthy that this technology has also prompted a shift in business collaboration models within the microfluidic chip industry, prompting certain microfluidic chip development firms to engage in partnerships with globally recognized manufacturers of analytical instruments. For instance, the world's first commercial microfluidic analytical instrument, the Bioanalyzer 2100, was co-launched by Caliper and HP under these circumstances in 1999. Soon, in 2001, specialized journal *Lab on a Chip* in microfluidics was launched by the Royal Society of Chemistry. Another significant milestone occurred in 2002, when Quake and coworkers demonstrated that the use of PDMS deformability allowed for the integration of thousands of valves on a single device [25]. Quake et al. designed a fluidic multiplexor by mimicking electronic integrated circuits and building multi-layer PDMS structures. The operational principle of the valve leveraged the soft and elastic properties of PDMS to their fullest extent. In short, the fluidic multiplexor consisted of two layers of microchannels that were perpendicular to each other. By increasing the pressure in one type of microchannel, the channel could be deformed and bulged, thereby compressing the vertical microchannel and stopping the flow in it. Through this multiplexor system, 2,056 microvalves were integrated into a microfluidic device, which seems to have solved the valve problem mentioned earlier. This ingenious and remarkable microvalve represents an achievement beyond the capabilities of silicon or glass and has emerged as the predominant valve in the realm of microfluidics, demonstrating the transition from early single-electrophoresis separation analysis in microfluidic devices to a large-scale integrated multi-functional microfluidic platform. Concurrently, this technology directly gave rise to another major player in the microfluidics field, Fluidigm, whose worth was expected to exceed US$1 billion in the near future.

The advancement of soft lithography technology and the development of soft materials, such as PDMS, have facilitated the successful validation of a diverse array of fundamental components in microfluidic systems, including sample introduction devices, fluid manipulation and mixing techniques, detection sensors, and purification and separation apparatuses. Microfluidic devices have transitioned from the demonstration of simple theoretical concepts to practical problem solving. For example, in the previously discussed DNA sequencing process, over one million fluid operations are required. Consequently, achieving rapid and simultaneous execution of this task with minimal reagent consumption has emerged as both a challenge and an opportunity for microfluidic technology.

8.2.6 Droplet Microfluidics and Digital Microfluidics

At the turn of the century, a novel microfluidic technology emerged, distinct from prior continuous-based systems, known as droplet microfluidics [26]. Droplet microfluidics is a subfield of microfluidics that focuses on the precise generation

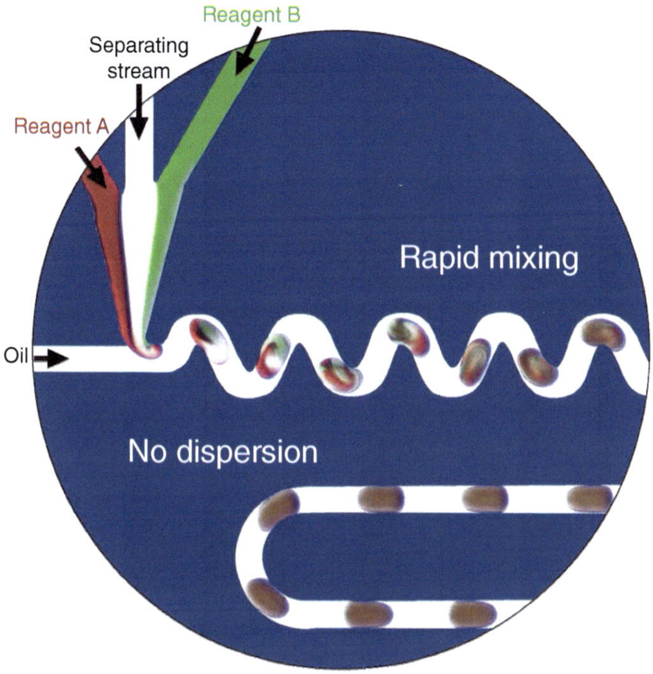

Figure 8.2 The schematic diagram of liquid droplet generation by droplet microfluidics reported by Song et al., *Source:* Ref. [27] / with permission of John Wiley & Sons.

and manipulation of discrete liquid volumes through the utilization of immiscible fluid dynamics. Normally, as shown in Figure 8.2, droplets are produced continuously and effectively by pushing and cutting two incompatible phases (like water and oil) through a narrow channel, thereby achieving control over the droplets at the nanoscale [27]. Upon initial examination, the concept of generating controlled microdroplets does not seem novel. For example, inkjet printing was already capable of carrying out this function. The uniqueness of droplet microfluidics is characterized by the precise control of droplet size through the introduction of an additional immiscible phase. The innovation lies in the ability to encapsulate reagents within sub-nanoliter microdroplets and manipulate them with precision as microreactors. Droplet microfluidics can generate millions of independent, parallel microdroplets in a limited space, which makes it possible to perform a huge number of tests with a small quantity of reagents for effectively solving the aforementioned problems. Based on these characteristics, droplet microfluidic technology has given rise to many innovative technologies and applications in the fields of biochemical reactions and analysis, such as high-throughput droplet PCR (dPCR), single-cell screening, and directed evolution of proteins. It is also widely used in the field of nanomaterial and micromaterial preparation [28].

During the identical timeframe, digital microfluidics with droplet control using electrowetting on dielectrics (EWOD) technology emerged. Unlike droplet microfluidics that manipulate droplets within closed microchannels, digital microfluidics manipulates droplets on an open array of electrodes [29]. By applying various potentials to the electrodes, it is possible to merge, mix, split, and dispense droplets individually from reservoirs. By adjusting the interfacial tension between the liquid and the electrode coated with an electrical dielectric layer, digital microfluidics could achieve precise control of the droplet volume on addressable electrodes. If the electric field in the dielectric layer only acts on one side of the droplet on the electrode array, it will cause an imbalance in the interfacial tension, which can enable the movement, division, or fusion of the droplet. Therefore, by programming the electronic control, the electrode array can be functionally segmented in different areas to perform different fluid functions, and the functions of different regions can be dynamically reassigned by changing the voltage, ultimately achieving "digital" management of the liquid. The programmable, reconfigurable, and reusable characteristics demonstrated by digital microfluidics are in perfect harmony with the idea of Labon-a-chip, creating opportunities for managing all the necessary protocols toward a specific laboratory application or a group of applications.

8.2.7 Paper Microfluidics

In 2007, a group led by G. M. Whitesides and coworkers introduced a testing platform based on patterned paper, which pioneered the development of paper microfluidics [30, 31]. The objective of paper-based microfluidics was to utilize cost-effective and readily available substrate materials as alternatives to traditional microfluidic substrates such as glass, silicon, and PDMS. The development was aimed at enhancing the accessibility and universality of corresponding biological or chemical detection methods, which are ideal for meeting point-of-care (POC) requirements and extending their applicability to resource-limited areas. Utilizing traditional lithography techniques for patterning chromatography paper, a novel concept of microfluidic paper-based analytical devices (μPADs) was proposed, featured with microscale channels and reaction areas for distinguishing it from conventional paper-based sensing devices such as pH test strips and LFA-type test strips. G. M. Whitesides and colleagues developed the 2D μPADs and used them to conduct glucose and protein analysis without reliance on external instrumentation (Figure 8.3) [30]. Subsequently, they constructed 3D μPADs by stacking multiple layers of paper, demonstrating the potential application of paper microfluidic devices in multiplexed analytical detection [32]. Since the development of μPADs, paper microfluidics, as a new branch of microfluidics, has demonstrated significant promise in the realms of medical care, biochemistry, environmental surveillance, analytical chemistry and other fields [33].

8.2.8 Open Microfluidics

An emerging field in microfluidics is referred to as open microfluidics. The term "open" indicates that the channels in the microfluidic device do not have traditional solid walls on at least one side, resulting in open boundaries for the microflows inside, such as a liquid-gas interface. The concept of "open microfluidics" was

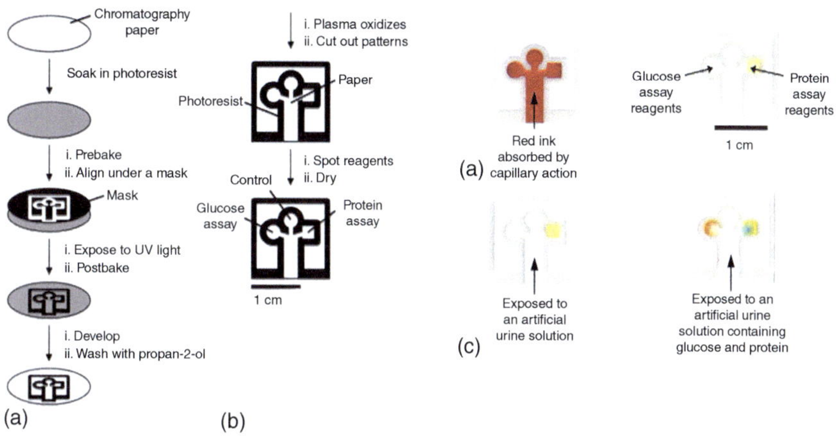

Figure 8.3 Paper microfluidics. The schematic of the fabrication process for 2D μPAD (left) and photograph of the finished product (right). *Source:* Ref. [30] / with permission of John Wiley & Sons.

initially introduced in the early 2000s to address specific inherent limitations of microfluidic technology in biomedical applications, particularly in cell culture, such as the confinement of cells within enclosed channels that hinders their interaction and subsequent manipulation [34]. In this early report, Folch and coworkers utilized a direct method, the removal of the top plate of microchannels, to construct the open microfluidic device. This method maintains the preparation approach and methodology of traditional microfluidic devices, while enabling the construction of an open microfluidic system by selectively removing specific channel walls. This approach is both straightforward and efficient, thus establishing itself as a commonly employed strategy for designing and developing open microfluidic systems. Another approach to fabricating open microfluidic systems involves the fabrication of microchannels that are then folded back on themselves, resulting in structures resembling probes. This technique was first described in 2005 by David Juncker and colleagues, who reported the development of a microfluidic probe (MFP) [35]. The MFP operates by simultaneously introducing and removing a processing liquid, generating a fluid jet between its openings that remains separate from the surrounding medium. This phenomenon is known as hydrodynamically confined flow, enabling reagents to be concentrated in a specific area of a sample without relying on enclosed channels that are prone to blockages, air bubble formation, and high hydrodynamic resistance, which can reduce device sensitivity. Utilizing these two strategies, researchers could design open microfluidic networks or open microfluidic probes, thus facilitating the precise manipulation of microfluidics and their internal substances within an open space.

8.2.9 Organ-on-a-Chip

The integration of microfabrication and tissue engineering has catalyzed the advancement of organ-on-a-chip technologies, which currently represents one of the most rapidly expanding and extensively investigated domains in microfluidics [36]. In simple terms, organ-on-a-chip refers to a novel technology that employs microfluidics and biocompatible materials to replicate the essential characteristics of in-vivo tissues or organs through the cultivation of live cells. This approach constructs a biomimetic model that more accurately mimics in-vivo pathological or physiological processes, providing a novel perspective and platform for conducting *in vitro* studies on pathological/physiological mechanisms and high-throughput drug screening. The concept of using microfluidics to simulate the physiological or pathological characteristics of human organs was first proposed by Michael Shuler and others in 2004 [37]. They successfully demonstrated the interaction between the liver and lungs on a small silicon chip by generating multicompartmental, multi-cell type bioreactor systems using a microscale cell culture analog system (μCCA), which could be considered as an upgraded version of cell culture analogue (CCA) technology developed in 1995 [38]. The term organ-on-a-chip was invented in 2010 [39]. In this landmark research, Donald Ingber's group developed a biomimetic microsystem that replicates the essential functional interface between alveoli and capillaries in human lungs by introducing cyclic biomechanical stretching forces and demonstrates a reproduction of real-life reactions [39]. More importantly, the remarkable work drew the attention of the Advancing Regulatory Science program jointly promoted by the US Food and Drug Administration (FDA) and the US National Institutes of Health (NIH), and had led to the establishment of the "Tissue Chips for Drug Screening" project funded by the FDA, NIH, and DARPA, which were two 5-year programs aimed at promoting the development of organ chips [40]. Compared to traditional two-dimensional cell culture techniques, organ-on-a-chip technology has demonstrated many advantages by integrating tissue engineering techniques with microfluidics and microelectromechanical systems, such as the ability to precisely control cells and specific tissue structures, simulation of *in vivo* chemical gradients and biomechanical actions, and the realization of vascularization or perfusion while integrating real-time tissue functional sensors for continuous monitoring [41]. With the development of advanced technologies such as induced pluripotent stem cells, gene editing, organoids, 3D printing, cell sensing technology, and microfluidics, organ-on-a-chip has seen rapid development in the past decade (Figure 8.4). One of the most significant commercial applications of organ-on-a-chip technology is the precise prediction of drug response in humans through the construction of biomimetic organs that are closest to the real situation, enabling cheap and fast drug screening [43]. In addition to providing a platform for toxicity assessment in the development of new drugs, organ-on-a-chip technology can also enhance our understanding of the physiological and pathological states of various human tissues and organs [44]. For instance, manipulating microfluidic flow rates to generate distinct oxygen gradients within a liver chip could facilitate the elucidation of hepatic metabolic zoning, thereby advancing our comprehension of its physiological functions [45]. By harnessing stem cell technology, such as iPSCs, organ-on-a-chip evolved the capacity to replicate a wide range of disease models, encompassing common diseases (e.g., cardiovascular diseases [46], urinary system diseases [47], and neural disorders [48]), tumors (e.g., glioblastoma [49],

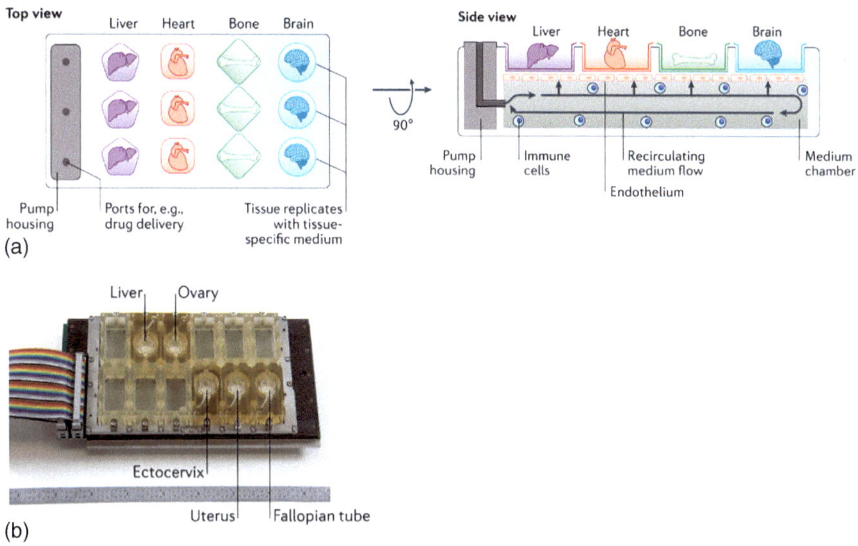

Figure 8.4 Examples of organs on a chip and related fields. (a) A simplified schematic diagram of organ coordination on multiple chips (illustrating a multiorgan chip system for studying the organ toxicity of doxorubicin, coordinating the liver, heart, bones, and brain). *Source:* Ref. [40] / with permission of Springer Nature. (b) A photo of a female reproductive system chip, including five organ chips of the ovary, cervix, uterus, fallopian tube, and liver, which can be used to simulate the hormonal conditions of a woman's menstrual cycle and pregnancy. *Source:* Ref. [42] / Springer Nature / CC BY 4.0.

lung cancer [50], and ovarian cancer [51]), and even rare disorders (such as Barth syndrome [52]). In these contexts, organ chips have developed a variety of branches with different intentions and functions, such as "patient-on-a-chip" [53] and "body on a chip" [54]. These platforms hold promise for elucidating the specific pathogenic mechanisms underlying the aberrant conditions and facilitating the development of tailored therapeutic strategies for individual patients, thereby advancing the field of personalized medicine [55].

8.2.10 3D-Printed Microfluidics

As mentioned above, significant advancements in microfluidics have been closely related to the emergence of innovative technologies. For instance, advancements in soft lithography technology had resulted in the emergence of diverse new microfluidic devices, further promoting the widespread application of microfluidics in fields such as analytical science, drug screening, and separation of biologically active substances. The emergence of 3D printing and iterative advancements in the technology have significantly broadened the repertoire of fabrication techniques for microfluidic devices, ushering in a pivotal new frontier in the realm of microfluidic technology development—3D-printed microfluidics. 3D printing technology, also referred to as additive manufacturing, was initially researched and published by Hideo Kodama of the Nagoya Municipal Industrial Research Institute in 1981 [56].

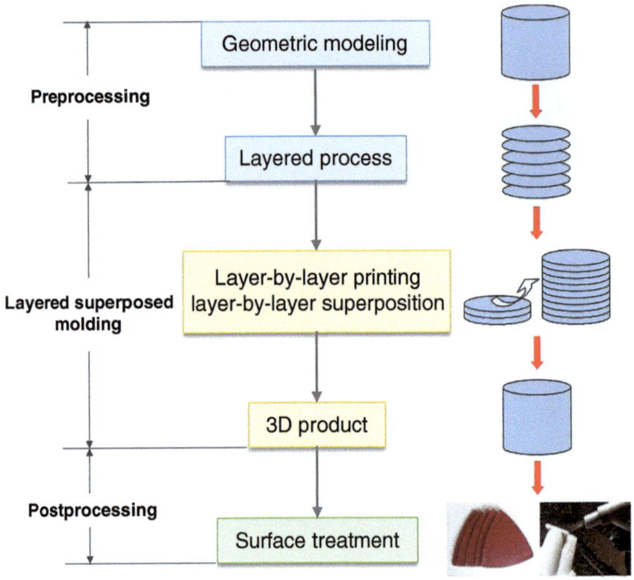

Figure 8.5 3D printing process. *Source:* Ref. [57]/with permission of John Wiley & Sons.

In contrast to conventional manufacturing methods, the fundamental principle of 3D printing technology lies in its "layer-by-layer" approach to production, which entails the sequential deposition of material layers to fabricate intricate components within a single manufacturing cycle (Figure 8.5) [58]. Then, in the late 1990s, 3D printing technology for microfluidic devices began to emerge [59]. In 2002, McDonald et al. reported pioneering work, in which the group utilized a solid-object printer to directly fabricate a masker for PDMS microfluidic devices without the need for any other masks [60]. Even though the resolution (which was limited to 250 μm at the lowest) of this technology was lower than that of photolithography, it provided a brand-new perspective for developing microfluidic chips with complex structures and multilevel features. Meanwhile, microfluidic devices directly fabricated using 3D printing technology were not reported until 2011. Moore et al. utilized fused deposition modeling, a type of 3D printing technology, to fabricate disc-based microfluidics (centrifugal microfluidics) and demonstrated that the centrifugal microfluidics produced using 3D printing technology exhibited comparable operability to those fabricated using other microfabrication technologies [61]. Compared to the current popular PDMS-molded microfluidic systems, 3D-printed microfluidics circumvents several challenges encountered by the former (such as the high probability of leakage at connection points when multiple components are assembled, the requirement for highly skilled professional operators for PDMS automation processes, and the slow or expensive PDMS molding process) [62] and exhibits many additional advantages: (i) the remarkable capacity to produce shapes and structures for microfluidic devices in three dimensions using a method that was not achievable before. Therefore, it is feasible to manufacture robust and standardized connection ports to effectively address the issue of leakage in integrated devices [63]. For example, commercial-grade threaded adapters can be fabricated using 3D printing technology for microfluidic devices, allowing for easy and reliable connection to external tubing [64]. It is also suggested that the process of commercialization is relatively straightforward. (ii) The material used is closer to commercial thermoplastic than PDMS, enabling the 3D printing preparation of simple microfluidic devices with faster speed (10–15 minutes) and lower cost. Therefore, it also supports agile iterative design and rapid prototyping, enabling researchers to embrace a strategy of "rapid and frequent experimentation" for further accelerating the innovation and development of 3D-printed microfluidics. (iii) As a 3D digital manufacturing process, it enables rapid modification of device characteristics with each print through the adjustment of the CAD software design. Scientists can effortlessly reproduce devices by distributing the design files in various research facilities [57]. (iv) 3D printing has the potential to overcome topological limitations and revolutionize the manufacturing of various types of valves or micro-pumps in microfluidic devices, thereby enabling fully automated fluid handling [65]. Currently, there are several methods of 3D printing used for the fabrication of microfluidic devices, such as fused deposition modeling (FDM), selective laser sintering (SLS), stereolithography (SLA), PolyJet printing, and direct laser writing (DLW). Each 3D printing technology possesses distinct characteristics, which are complemented by traditional soft lithography techniques to shape the thriving microfluidics market of today, catering to the demand from multifunctional, high-throughput microfluidic applications in biological and chemical fields (see Table 8.1).

8.2.11 CRISPR-based Microfluidics

Over the past decade, a series of groundbreaking technologies have emerged, significantly influencing scientific research and propelling advancements in microfluidic technology. Each year, thousands of studies related to microfluidics are published [84]. While an exhaustive discussion of these technologies exceeds the scope of this chapter, we will succinctly introduce several promising emerging innovations. Notably, we'd like to highlight CRISPR technology [85]. As discussed in the previous chapter, CRISPR has established itself as a pivotal tool in biological analysis, molecular biology, and clinical diagnostics, substantially enhancing the repertoire for detecting disease-related biomarkers such as nucleic acids, proteins, and even certain small molecules [86]. Furthermore, it has catalyzed new avenues in microfluidics, which can be referred to as CRISPR-based microfluidics [87]. By integrating CRISPR technology with microfluidics platforms, CRISPR-based microfluidics is expected to effectively address the defects of high cost and potential cross-contamination in CRISPR detection methods, achieving high-throughput, parallel, and quantitative detection of analytes [88]. For example, Welch et al. developed a virus and variant detection platform named microfluidic Combinatorial Arrayed Reactions for Multiplexed Evaluation of Nucleic acids (mCARMEN) by combining CRISPR diagnostic technology with the commercialized Fluidigm microfluidic technology platform [89]. This innovative platform

Table 8.1 Comparison of the soft lithography, fused deposition modeling (FDM), selective laser sintering (SLS), stereolithography (SLA), PolyJet printing, and direct laser writing (DLW) methods in terms of microfluidics fabrication [59, 62, 66–79].

General Characteristics	Soft Lithography	FDM	SLS	SLA	PolyJet	DLW
Principle	Non-photolithographic strategy, through an elastomeric stamp with patterned relief structures to deposit ink on a substrate	Extrusion-based 3D printing technique, using a nozzle gun to construct objects by stacking softened or melted wires, and subsequently solidifying them through the process of cooling	Powder bed fusion-like technology, a laser beam is employed to create layered structures through the process of melting and bonding powdered materials	Vat photopolymerization process, using an ultraviolet (UV) laser beam to produce objects layer by layer	Inkjet-based 3D printing technique, jetting numerous photopolymer droplets onto an integrated base and curing them using a UV light emitter	Laser-based 3D printing technique, utilizing focused laser light to trigger spatially targeted two-photon (or multi-photon) polymerization of photosensitive materials at designated positions in the 3D space through a point-by-point and/or layer-by-layer process
Resolution	5 μm or less	~100 μm, special case: 40 μm [80]	~50 μm [81]	~300 μm, special case: 10 μm [73]	~250 μm, 54 μm [82]	~1 μm, special case: 0.1 μm [83]

Material	Photoresists, unsensitized polymers (PDMS, PMMA, PE, PVC, etc.), precursor polymers, colloidal materials, sol-gel materials, biological macromolecules	PLA, ABS, polypropylene, and other thermoplastics	Powdered plastic, metal, ceramic, PC, acrylic styrene, PVC, ABS wax, etc.	Liquid photosensitive resin, acrylate resin, polymer resins, modified resin	Photopolymer resins, acrylate-based polymer	photosensitive material, liquid-phase poly (ethylene glycol) diacrylate (PEGDA), propylene glycol methyl ether acetate (PGMEA)
Speed	Very low	Moderate	Moderate to low	Moderate to low	Fast	Fast
Cost-Effectiveness	Affordable	Affordable	High initial costs	High initial costs	Very high initial investment	High initial costs
Biocompatibility	Biocompatible and bio-functional	Depends on the material utilized, many thermoplastics are biocompatible	Depends on the material utilized	Depends on the material utilized, many resins are cytotoxic and not biocompatible, PEG-DA is biocompatible	Depends on the material utilized	Depends on the material utilized
Multi-material fabrication	Possible	Easily	Easily	Easily	Easily	Easily

enabled high-throughput surveillance of 21 viruses (including SARS-CoV-2, other coronaviruses, and influenza strains) and 6 SARS-CoV-2 variants, while also being able to quantitatively measure the number of viral genomic copies. By integrating various types of microfluidic devices, such as paper-based microfluidics [90], centrifugal microfluidics [91], and digital microfluidics [92], CRIPSR-based microfluidics have successfully been used for high-specificity, high-sensitivity detection of disease-related markers such as viruses, bacteria, and miRNAs. Although these two technologies are yet to be seamlessly integrated, the future prospects of CRIPSR-based microfluidic technology are promising.

8.2.12 AI-Assisted Microfluidics

Another technology that has had a significant impact on modern society and science is artificial intelligence (AI). AI is a developing technology aimed at enhancing the theories, methodologies, techniques, and practical applications related to human intelligence. Its purpose is to allow computer systems to carry out tasks that typically necessitate human cognitive abilities, including learning, reasoning, perception, and language comprehension [93]. The field of AI has made swift advancements, resulting in notable developments in areas such as biomedicine, agriculture, material synthesis, and environmental monitoring [94–99]. New AI methodologies are transforming data processing and analysis, particularly serving as effective tools for detecting, quantifying, screening, or even forecasting large data sets derived from microfluidics [100]. A typical application is the unparalleled efficiency and reliability of AI in single-cell profiling, where an AI-assisted detection system can swiftly and accurately analyze tens of thousands of single-cell data sets obtained from high-throughput microfluidic platforms, thereby providing researchers with invaluable insights. Relying on this unique characteristic, Yellen et al. employed a trapezoidal microfluidic network to capture images of 100,000 cell clone characteristics in a single operation and utilized machine learning and computer vision techniques to assess the phenotypic parameters within these images, quantifying the responses of acute myeloid leukemia clones derived from single cells to targeted treatments while uncovering infrequent resistance and morphological characteristics at frequencies as low as 0.05% [101]. Meanwhile, Soteriou et al. had applied the technology in solid tissue biopsy [102]. In this study, researchers employed microfluidic-based real-time deformability cytometry to capture morphological images of individual free-floating cells isolated from tissues in a high-throughput way. Subsequently, they applied unsupervised dimensionality reduction and logistic regression analysis to interpret the data, ultimately achieving differentiation among cell subpopulations in various tissues by extracting physiological phenotypic parameters from images of single cells within 30 minutes. Current research indicates that AI could significantly improve microfluidic systems in terms of biomedical detection and drug screening, enhancing accuracy, efficiency, and workflow optimization [100, 103–105]. Furthermore, we believe that in the near future, AI and microfluidic synergistic systems will be applied in more fields of research.

8.3 Applications of Microfluidic Technology in the Detection of Public Infectious Diseases

8.3.1 SARS-CoV-2 Detection

The emergence of COVID-19 has consistently presented significant risks to both public health and the global economy. According to the World Health Organization (WHO), by September 2024, the pandemic had resulted in over 776 million cases and caused 7.05 million deaths [106]. The pathogen responsible for COVID-19 is the severe acute respiratory syndrome coronavirus 2 (SARS-CoV-2). SARS-CoV-2 is one of the seven known coronaviruses capable of infecting humans, alongside HCoV-229E, HCoV-OC43, HCoV-NL63, HCoV-HKU1, SARS-CoV, and MERS-CoV [107]. It belongs to the *Sarbecovirus* subfamily within the beta genus of coronaviruses and exhibits a pronounced capacity for human-to-human transmission via respiratory droplets and contact [108]. Currently, the diagnostic techniques for SARS-CoV-2 could be broadly divided into three categories: nucleic acid detection techniques targeting the viral RNA genome, immunoassay techniques focused on viral antigens or host-generated antibodies, and imaging techniques for detecting physiological alterations in the patient's lungs [109–111]. Particularly, nucleic acid detection techniques and immunoassay techniques demonstrate a remarkable ability to accurately identify COVID-19 in the early stages at the molecular level, which holds substantial implications for mitigating disease transmission and preventing large-scale outbreaks of infectious diseases [112]. Both these technologies are grounded in the molecular biological characteristics of SARS-CoV-2, which include a single-stranded RNA genome approximately 30 kb in length and four structural proteins: nucleocapsid (N) protein, spike (S) protein, membrane (M) protein, and envelope (E) protein, as shown in Figure 8.6. In detail, nucleic acid detection technology primarily identified gene targets indicative of viral presence, including the N, S, and E genes, as well as open reading frame 1ab *(ORF 1ab)*.

Nucleic acid detection technology is commonly referred to as nucleic acid amplification test (NAAT), as it frequently employs various amplification strategies,

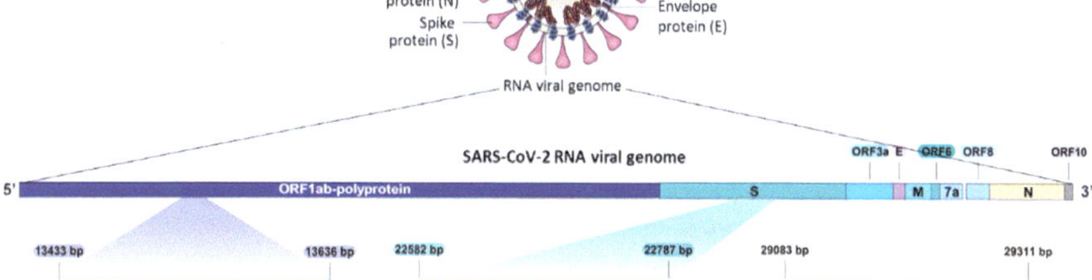

Figure 8.6 Schematic representation of the genomes, genes, and proteins of SARS-CoV-2 virus. *Source:* Ref. [113] / with permission of American Chemical Society.

including reverse transcription-polymerase chain reaction (RT-PCR), hybridization chain reaction (HCR), loop-mediated isothermal amplification (LAMP), and the CRISPR/Cas system [114–116]. NAAT exhibits inherent high sensitivity and specificity, facilitating the detection of low-copy numbers of viral RNA even at minimal viral loads prior to the onset of COVID-19 symptoms (Figure 8.7), thereby often serving as the gold standard for diagnosing viral infections [114]. However, NAATs have demonstrated indications of being overwhelmed in their real-world application for mitigating the spread of COVID-19 during global outbreaks, with several current limitations becoming increasingly evident. Firstly, the effectiveness of NAAT is limited by the sample pre-processing process [117], and secondly, NAAT requires trained and skilled personnel, expensive instruments, and time-consuming and complex operational steps. These factors significantly constrain the capacity of resource-limited regions to implement rapid and large-scale screening of the virus, thereby hindering the effective management of widespread COVID-19 transmission [118]. Meanwhile, immunoassay technology, which detects the N protein or S protein as viral antigens, is an important auxiliary technique for the rapid diagnosis of SARS-CoV-2. In comparison to nucleic acid amplification tests (NAAT), immunoassay tests exhibit a lower complexity in sample preparation, offer greater operational convenience, and provide rapid test results. These attributes render them particularly suitable for diverse applications, including the fast screening of asymptomatic individuals and urgent responses during large-scale disease outbreaks [119]. Consequently, most of the commercially available COVID-19 *in vitro* diagnostic devices are indeed based on immunoassays [120]. However, as virus-related antigens do not form until two to three weeks after infection (as shown in Figure 8.7), immunoassays frequently exhibit low sensitivity and a high false-negative rate during the early stages of infection [121]. Moreover, some conventional immunoassays still depend on expensive instruments like microplate readers and require specialized complex procedures, which hinders rapid disease

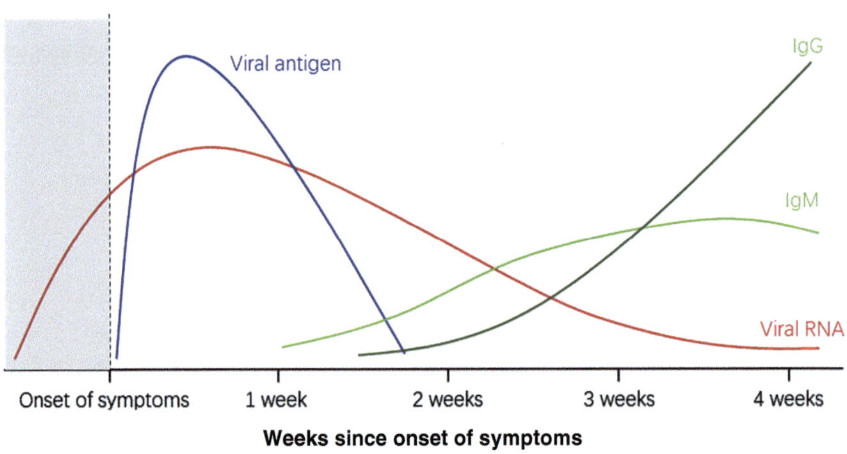

Figure 8.7 The relative expression levels and timeline of virus RNA, viral antigens, and immune antibodies during SARS-CoV-2 infection.

prevention and control in resource-limited areas. Therefore, developing new technologies characterized by high throughput, portability, low cost, and even the capability for point-of-care testing (POCT) is of great importance for rapid COVID-19 screening and timely control of the spread of the epidemic.

Microfluidic technology holds significant promise for high-throughput, rapid analysis, and detection, particularly in the domain of point-of-care testing (POCT) for disease diagnosis, food safety assessment, and environmental monitoring [122–126]. The characteristics of microfluidic technology—namely miniaturization, seamless integration, cost-effectiveness, and high throughput—are particularly well-aligned with the immediate detection and sample-to-answer demands of contemporary COVID-19 testing and control strategies. Microfluidic technology has made significant contributions to COVID-19 screening work in the past [127]. To date, a variety of innovative microfluidic technologies have been developed to support efforts in combating the pandemic and enhancing global public health. These innovative microfluidic technologies seamlessly integrate established analytical principles, including nucleic acid amplification techniques and immunoassay methods, with advanced functionalities inherent to microfluidic devices, such as one-step sample processing and automated analysis. This integration significantly enhances diagnostic testing by promoting high sensitivity, specificity, and throughput [128]. For instance, Ackerman et al. developed a scalable multiplexed pathogen detection platform known as Combinatorial Arrayed Reactions for Multiplexed Evaluation of Nucleic acids (CARMEN) by integrating CRISPR-based nucleic acid amplification technology (NAAT) with droplet microfluidics [92]. This innovative platform leveraged the high efficiency and rapid operation of droplet microfluidics to generate tens of thousands of nanoliters of liquid droplets while harnessing the superior nucleic acid amplification potential of the CRISPR/Cas13 system to simultaneously detect 169 human-associated viruses, including a rapid assessment for COVID-19. The miniaturization afforded by droplet microfluidics enabled a reduction in reagent costs per test by over 300-fold. Furthermore, its intrinsic multiplexing and high-throughput functionalities were anticipated to transform diagnostic and monitoring efforts during global pandemics from the targeted detection of priority samples to comprehensive screening across large sample volumes, thereby significantly enhancing global public health outcomes. In this section, we conduct a concise review of the most recent advancements in microfluidic nucleic acid tests and microfluidic immunoassays for COVID-19 diagnostics. We will further analyze the distinctive characteristics of these technologies through the lens of various types of microfluidic devices, including paper-based microfluidics, centrifugal microfluidics, droplet microfluidics, and integrated microfluidics. Finally, we will address the challenges encountered in this domain and explore prospects for future development.

8.3.1.1 Paper Microfluidic Devices for the Detection of SARS-CoV-2

Paper was first invented in the Han Dynasty of China around CE 105 [129]. Since then, it has been extensively utilized in various scientific fields, including chemistry and medicine. Applications range from litmus paper for pH testing to paper chromatography for material separation, as well as the development of blood glucose test strips, pregnancy test strips, and even paper-based devices for diagnosing pathogens and infectious diseases [130]. Currently, paper-based microfluidic devices are extensively employed to detect SARS-CoV-2 for facilitating COVID-19 screening. These devices leverage passive fluid transport through capillary action or suction mechanisms, eliminating the need for external power sources. Their affordability, portability, mass-producibility, and straightforward waste disposal render them particularly suitable for efficient, rapid, and on-site testing in resource-limited areas [131].

In the field of COVID-19, the most widely used paper-based microfluidic devices are the lateral flow test (LFT) kits. In 2022, the WHO strongly supported the use of antigen LFTs for the self-testing of COVID-19. Currently, hundreds of commercialized products based on LFTs are being used for the detection of COVID-19 [132]. Most of the LFTs are designed for the detection of antigens or antibodies and can provide qualitative visual testing results within 5–30 minutes. In addition to using 1D LFTs for SARS-CoV-2 detection, researchers have also explored the application of other types of paper-based microfluidic devices for COVID-19 detection. For example, Yakoh et al. reported the development of an electrochemical sensor based on a 2D microfluidic paper-based analytical device (2D μPAD) for COVID-19 diagnosis [133]. As shown in Figure 8.8a, the 2D μPAD consisted of a foldable paper substrate featuring three functional components (a working ePAD, a counter ePAD, and a closing ePAD), in which a spike protein receptor-binding domain (SP RBD) was selectively and chemically fixed with oxidized graphene (GO)-EDC/NHS on the hydrophilic region of the working ePAD. The detection of SARS-CoV-2 could be accomplished within 30 minutes, with a detection limit of 1 ng/mL. Clinical results from blood samples indicated that the device displayed acceptable sensitivity and specificity rates of 100% and 90%, respectively. Ehzari et al. reported the development of a 2D μPAD that employed cost-effective plant-derived antibodies in conjunction with an electrochemical platform for the label-free detection of SARS-CoV-2 [139]. The device utilized the electrocatalytic activity of nickel-iron-layered double hydroxide, and demonstrated a detection limit of 0.23 fg/mL for SARS-CoV-2.

However, the high cost of antibodies, coupled with the stringent conditions required for their storage and transportation, brings significant challenges for paper-based immunoassays. Consequently, researchers began to investigate the development of antibody-free paper-based detection methods for the diagnosis of COVID-19. Bordbar et al. designed a colorimetric sensor array on 2D μPAD to diagnose COVID-19 by tracking metabolites in urine samples [134]. The paper-based microfluidic pattern consisted of 16 small circles surrounding a central large circle (Figure 8.8b). The 16 detection areas were populated with colorimetric compounds, such as gold and silver nanoparticles, porphyrin dyes, metal ion complexes, and organic dyes that were modified with reagents sensitive to carbonyl species. By introducing urine samples into the central circular area, the sensing molecules located in the 16 surrounding detection zones could identify the corresponding metabolites and generate visual signals within 10 minutes. For COVID-19 patients, the detection sensitivity was about 73%.

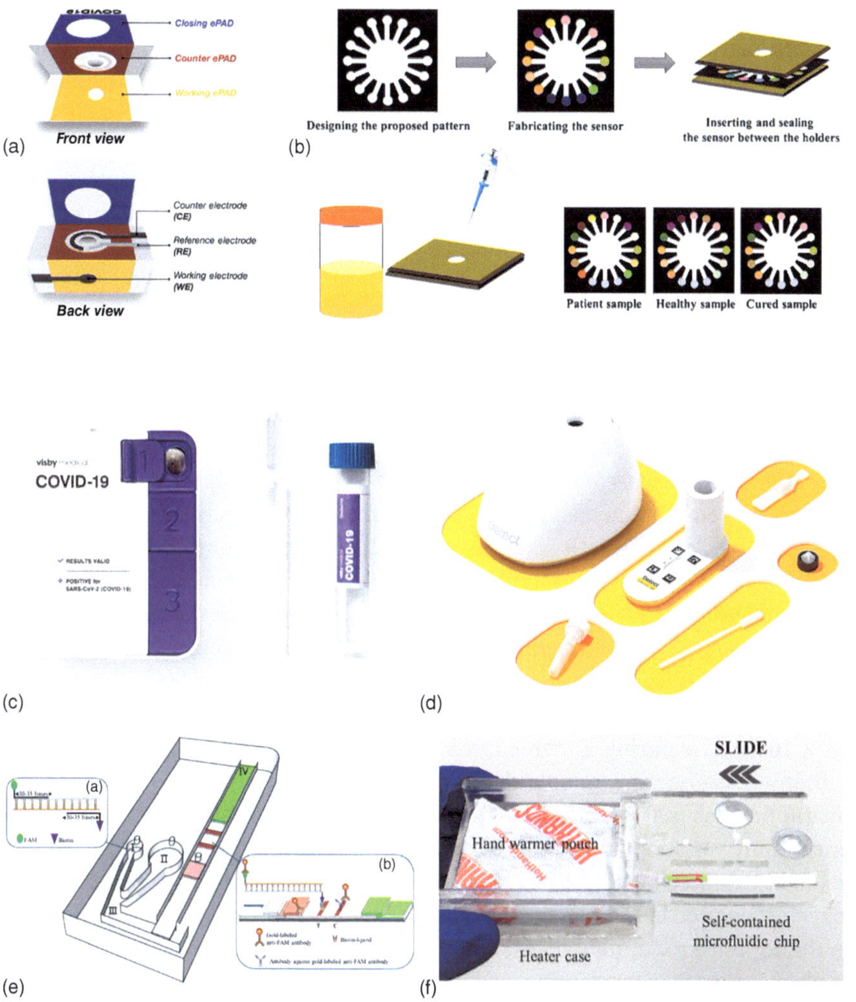

Figure 8.8 Strategies of paper-based microfluidic devices for the detection of SARS-CoV-2. (a) An electrochemical sensor based on 2D μPAD. *Source:* Ref. [133] / with permission of Elsevier. (b) A colorimetric sensor based on 2D μPAD. *Source:* Ref. [134]/with permission of Springer Nature. (c) The Flu-COVID product developed by Visby Medical, integrating PCR reaction. *Source:* Ref. [135] / United States Patent Office. (d) The Detect product developed by DETECT, INC, integrating LAMP reaction. *Source:* Ref. [136]/United States Patent Office. (e) Schematic illustration of the MI-IF-RPA system for COVID-19 detection by integrating LFAs and RPA. *Source:* Ref. [137] / with permission of Royal Society of Chemistry. (f) self-contained paper-based microfluidic system for COVID-19 detection by integrating LFAs, RPA and CRISPR amplification. *Source:* Ref. [138] / with permission of Elsevier.

However, conventional paper-based microfluidic devices frequently encounter challenges related to inadequate sensitivity in the detection of antigens or metabolite-associated molecules, primarily due to their inability to facilitate signal amplification. As previously noted, nucleic acid tests, especially NAATs, can provide direct measurement of viral genomes and are regarded as the gold standard in virological diagnostics. However, traditional NAATs exhibit several inherent limitations, including complex operational procedures, time-consuming and labor-intensive workflows, and the need for costly instruments and equipment. These factors constrained their applicability in large-scale sample screening and hindered their rapid and comprehensive deployment in epidemic prevention. Microfluidic devices, which integrate a variety of functional components such as meticulously designed microchannels, microreactors, micropumps, and intelligent microheaters, can significantly minimize manual intervention, decrease detection time and costs, and enhance amplification efficiency. In recent years, numerous microfluidic NAATs utilizing various types of microfluidic devices have been developed for the detection of SARS-CoV-2. These innovations demonstrate enhanced sensitivity, high throughput, reduced detection time and low cost, significantly augmenting the practicality of NAATs. Aided by nucleic acid detection technology, paper-based microfluidic strategies are particularly advantageous for the early screening and diagnosis of COVID-19.

To date, several commercial paper-based microfluidic products have been successfully developed and received emergency use authorization from the FDA for global bedside testing or home self-testing of COVID-19. For instance, the Flu-COVID product developed by Visby Medical, a leading company in the field of *in vitro* diagnostics, combined the gold standard PCR technology with microfluidic technology for achieving precise diagnosis of COVID-19 in just 30 minutes [135]. This product was the world's first commercialized handheld PCR device, with a size of about 7.6×10.2 cm, featuring a "three-in-one" function of sample processing, nucleic acid amplification, and sample detection (Figure 8.8c). The device incorporated microfluidic technology, employing elongated channels to facilitate the necessary temperature variations across distinct thermal zones during PCR. The PCR product would be captured by nucleic acid hybridization on a lateral flow strip, followed by enzyme-catalyzed colorimetric detection. The user performed a wash-off operation on the collected nasal/pharyngeal swab and placed it in the sample hole, followed by plugging in the power supply and pressing the 1, 2, and 3 buttons in sequence. Within 30 minutes, the result could be observed in the indicator window. However, this disposable three-in-one PCR testing device required additional accessories for integrating the auxiliary temperature cycling process, leading to its complex design and operation, as well as higher costs. Consequently, numerous companies are engaged in the development of products utilizing isothermal amplification technology as an alternative to the gold standard PCR for viral detection. DETECT, INC. developed a paper-based microfluidic device named Detect, utilizing LAMP technology for at-home COVID-19 testing [136]. The device featured a streamlined design, comprising a reaction tube for LAMP and a test card for LFA (Figure 8.8d). Guided by the mobile application, users could perform nasal swab sampling and sample insertion, obtaining results within one hour. Experimental data indicated that the detection limit of the Detect product could achieve 313 copies/swab, demonstrating an overall consistency of 97.3% when compared to high-sensitivity PCR detection, with a sensitivity of 90.9% and specificity of 100%. Furthermore, its price was under $50.

Similarly, professor Yang's team developed the microfluidic-integrated lateral flow recombinase polymerase amplification (MI-IFRPA) method for the rapid and sensitive detection of SARS-CoV-2 by integrating reverse transcription recombinase polymerase amplification (RPA) with a universal lateral flow assay strip detection system on a microfluidic chip [137]. As shown in Figure 8.8e, the chip was small in size, simple in design, easy to process, and mass-producible. It was made up of three layers of PMMA (designed using AutoCAD software and processed by laser), with a total size of 70 mm × 35 mm × 5 mm. It is worth noting that the entire testing process was very simple and fast, requiring only nucleic acid extraction and loading, followed by incubation to obtain the results. The total duration of this process was approximately 30 minutes, significantly shorter than the time (around two hours) required for conventional PCR technology. The performance of the chip was assessed with clinically confirmed COVID-19 cases, demonstrating a sensitivity of 97% and a specificity of 100%, which were closely aligned with those obtained from current reverse transcription-polymerase chain reaction (RT-PCR) diagnostic tests. Li et al. integrated isothermal amplification, CRISPR cleavage, and lateral flow detection into a microfluidic platform, and utilized a hand warmer as the heat source to develop a simple, cost-effective, and self-contained paper-based microfluidic system for non-invasive visual detection of SARS-CoV-2 (Figure 8.8f) [138]. By integrating RPA and CRISPR amplification, the device could detect as low as 100 copies of SARS-CoV-2 RNA within 60 minutes, demonstrating a sensitivity of 94.1% for clinical nasopharyngeal swab samples, alongside a specificity of 100% and an accuracy of 95.8%.

Although NAAT significantly enhances analytical sensitivity, nucleic acid amplification methods are hindered by their time-consuming nature and the necessity for additional heating devices. These factors contribute to increased design complexity and cost in practical applications, ultimately diminishing their universality. The development of amplification-free nucleic acid analysis technologies can effectively address these challenges. To achieve this objective, Zhang et al. developed a colorimetric detection method for SARS-CoV-2 by employing foldable paper strips that utilize nucleic acid strand displacement reactions in conjunction with Ag(I) ions to modulate urease activity [140]. As shown in Figure 8.9, this method involved the meticulous design of a double-stranded DNA containing cytosine-cytosine mispairing for the loading of Ag(I) ions. In the presence of target viral RNA, the double-stranded DNA structure was disrupted by a competitive chain displacement reaction, leading to the release of Ag(I) ions. These released Ag(I) ions served as urease inhibitors to attenuate the amplification effect associated with urea hydrolysis, which affected the production of $NH4^+$ and thus the pH value of the solution. Ultimately, SARS-CoV-2 could be detected visually using a pH indicator. This method effectively utilized the enzymatic amplification properties of urease, thereby simplifying the complex procedures and eliminating the need for heating equipment in nucleic acid amplification. This approach significantly reduced detection costs (approximately $0.3 per test) while delivering results within 30 minutes. Notably, this method was capable of distinguishing SARS-CoV-2 variants, demonstrating 100% concordance with real-time quantitative PCR.

In addition to amplifying the detection signal by using proteases, the sensitivity could also be increased by increasing the number of signal molecules. Similarly, Wang et al. developed an amplification-free nucleic acid immunoassay leveraging the unique ability of the S9.6 monoclonal antibody to specifically recognize DNA-RNA hybrid chains. The method employed fluorescent nanoparticles to modify both the DNA probe and the S9.6 antibody, utilizing fluorescence from a test strip as the signal output. It enabled accurate detection of SARS-CoV-2 RNA within one hour. In clinical evaluations involving pharyngeal swabs and sputum samples, the assay demonstrated 100% sensitivity and 99% specificity when compared to real-time quantitative RT-PCR.

In conclusion, paper-based microfluidic devices for nucleic acid detection offered significant advantages such as portability, rapidity, and cost-effectiveness, demonstrating considerable potential for the rapid diagnosis of COVID-19. However, they also brought several challenges. First, their sensitivity was constrained when analyzing complex biological samples; thus, improvements in reliability and accuracy were essential. False-positive results may inadvertently expose healthy individuals to infected areas, leading to serious medical incidents. Second, progress in the development of these devices for multiplex detection was sluggish; multichannel and multilabel technologies remained underdeveloped. Additionally, cross-contamination among multiple targets poses a critical issue that should be addressed while navigating the numerous mutations of the COVID-19 virus. Finally, the level of intelligent integration remained relatively rudimentary. Achieving real-time monitoring capabilities and enabling data transmission to cloud platforms would significantly enhance epidemic analysis and governmental response strategies, thereby facilitating more effective prevention and treatment measures against COVID-19.

8.3.1.2 Droplet/Digital Microfluidic Devices for the Detection of SARS-CoV-2

Given the highly transmissible nature of SARS-CoV-2, large-scale population screening is crucial for the effective management of the COVID-19 pandemic. Consequently, there is an urgent need to develop high-throughput, rapid, and cost-effective diagnostic methods to safeguard global public health. Droplet microfluidic technology and digital microfluidic technology enable the rapid generation and manipulation of tens of thousands of sub-nanoliter droplets, demonstrating significant utility in high-throughput experiments such as drug screening and gene sequencing, while markedly reducing both the time and cost associated with data collection. In these contexts, numerous researchers are committed to employing these technologies to facilitate the rapid and high-throughput detection of SARS-CoV-2.

For example, Park et al. developed a droplet microfluidic system based on surface-enhanced Raman scattering (SERS) for the rapid and sensitive detection of SARS-CoV-2 [142]. In the PDMA microfluidic device, SARS-CoV-2 lysate was mixed with SERS nanotags composed of gold nanoparticles conjugated monoclonal antibodies and antibody-conjugated magnetic beads, and then flowed through a T-junction channel to generate droplets in an oil phase (Figure 8.10a). As these droplets continuously passed through the laser confocal detection area, the

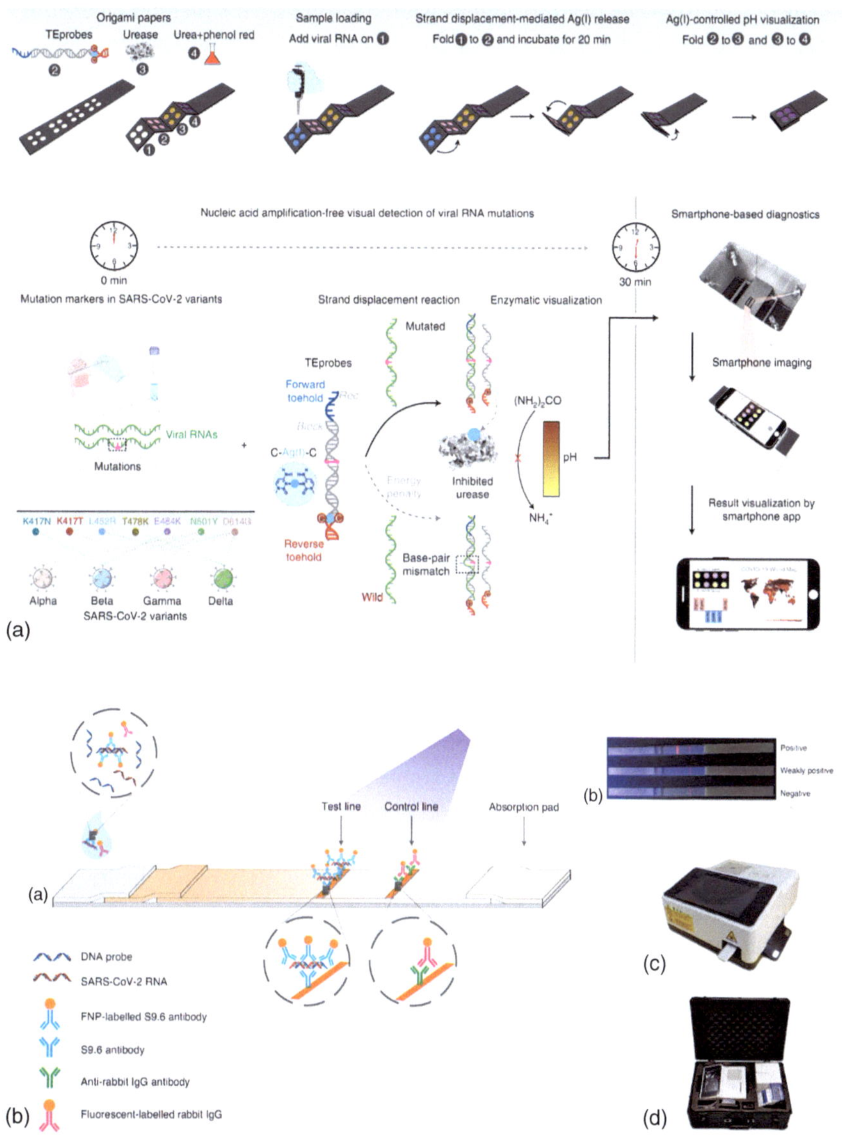

Figure 8.9 Amplification-free nucleic acid analysis strategies of paper-based microfluidic devices for the detection of SARS-CoV-2. (a) A colorimetric strategy employing foldable paper strips that utilize nucleic acid strand displacement reactions and Ag(I) ions-modulated urease activity. *Source:* Ref. [140] / with permission of Springer Nature. (b) A nanoparticle-enhanced fluorescence strategy integrating amplification-free nucleic acid immunoassay and LFAs. *Source:* Ref. [141] / with permission of Springer Nature.

Raman signals from the SERS nanotags were measured. This method enabled the completion of the entire assay within 10 minutes, achieving a detection limit of 0.22 PFU/mL and demonstrating excellent reproducibility with a coefficient of variation of 1.79%. However, generating uniform and appropriately sized droplets for laser confocal counting requires complex and precise flow rate regulation along with repeated testing, thereby increasing operational complexity. Digital microfluidic technology facilitates the automatic generation and manipulation of droplets through electrowetting, offering enhanced operational convenience. As shown in Figure 8.10b, Zeng et al. reported an Aggregation-Induced Emission Fluorescence (AIEgens)-enhanced immunofluorescence analysis platform based on digital microfluidics for rapid and sensitive detection of SARS-CoV-2 nucleocapsid (N) protein [143]. The platform featured a straightforward operational procedure and enabled the high-sensitivity quantification of SARS-CoV-2 nucleocapsid (N) protein in small-volume samples (less than 1 μL) within 25 minutes, as well as achieving a detection limit of 8.91 pg/mL.

Droplets not only facilitate a revolutionary enhancement in experimental throughput—achieving an order of magnitude increase compared to conventional microplate methods—but also function as independent microreactors, allowing for precise and rapid control over reaction temperature, duration, and other parameters. These characteristics align exceptionally well with the requirements of nucleic acid amplification tests (NAATs), heralding unprecedented advancements in the field of nucleic acid detection [28]. For example, droplet digital PCR (ddPCR), utilizing droplet-based microfluidics, had demonstrated significant efficacy in the domains of rare mutation detection, large-scale single-cell sequencing, and other nucleic acid-related assays [144]. Moreover, ddPCR had received regulatory approval from the US Food and Drug Administration for its application in monitoring the treatment of chronic myeloid leukemia [28]. During the COVID-19 pandemic, ddPCR facilitated the rapid detection of SARS-CoV-2 in environmental samples such as air and wastewater [145, 146], making substantial contributions to epidemiological research and the analysis of viral transmission pathways. Numerous studies have revealed that ddPCR possesses significant potential for application in the diagnosis and prognostic monitoring of COVID-19 patients, with sensitivity and specificity comparable to the gold standard RT-qPCR, and it might even outperform RT-qPCR in quantifying SARS-CoV-2 RNA level [147–151]. ddPCR effectively addresses the inherent limitations of conventional qPCR in quantification, which depends on standard curves, and theoretically facilitates absolute quantification of target analytes. However, the turnaround time associated with ddPCR is comparatively longer, which somewhat constrains its practical applications. To address this issue, Yin et al. developed a rapid ddPCR technology for the highly sensitive and precise detection of SARS-CoV-2 RNA [152]. As shown in Figure 8.11a, the rapid ddPCR system presented here could accomplish 40 cycles of thermal amplification of RNA within five minutes and simultaneously detect the ORF1ab gene, N gene, and RNase P gene in just 15 minutes. Clinical samples from both positive and control cases demonstrated that the testing results were consistent with qPCR, but the false-positive rate for low-concentration samples was lower than that observed with qPCR.

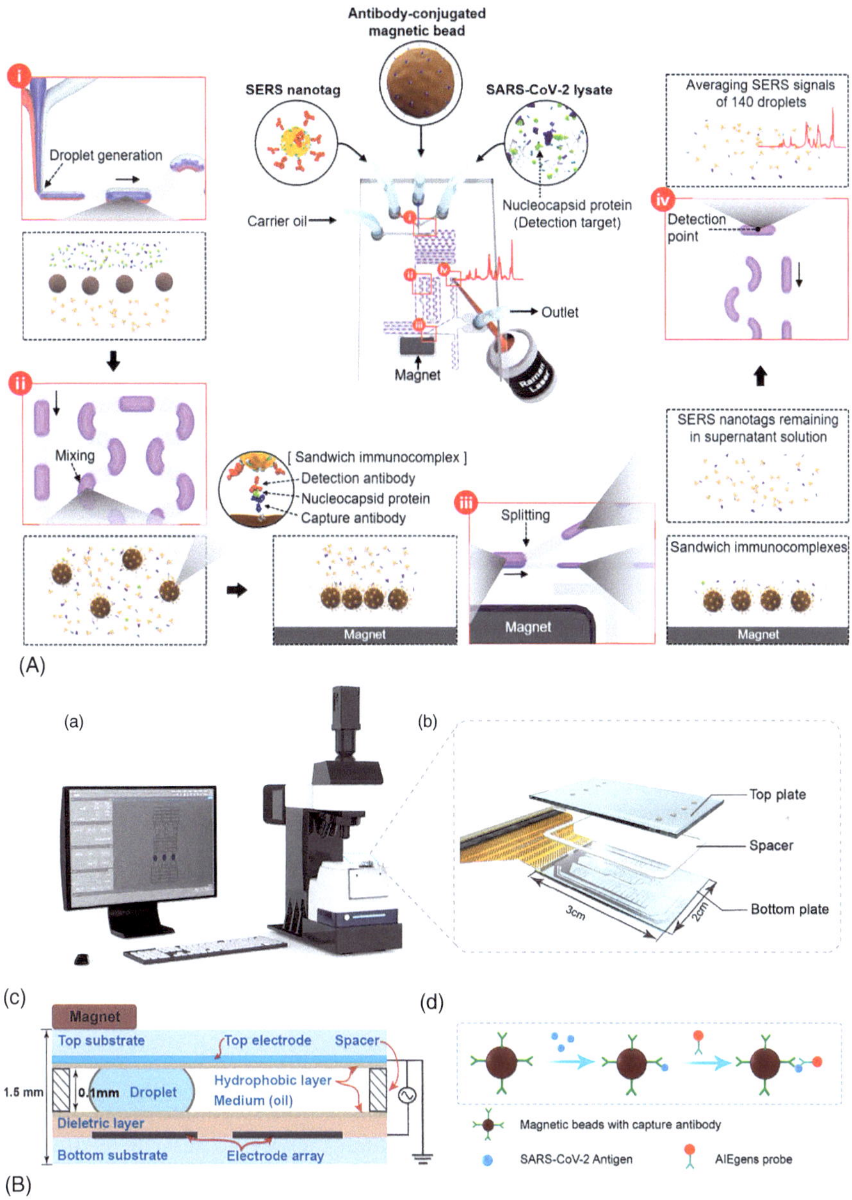

Figure 8.10 Strategies of droplet/digital microfluidic devices for immunoassay detection of SARS-CoV-2. (a) SERS-based sensor using droplet microfluidic. *Source:* Ref. [142] / with permission of Elsevier. (b) Digital microfluidics platform based on AIEgens-enhanced immunofluorescent assay. *Source:* Ref. [143] / with permission of Elsevier.

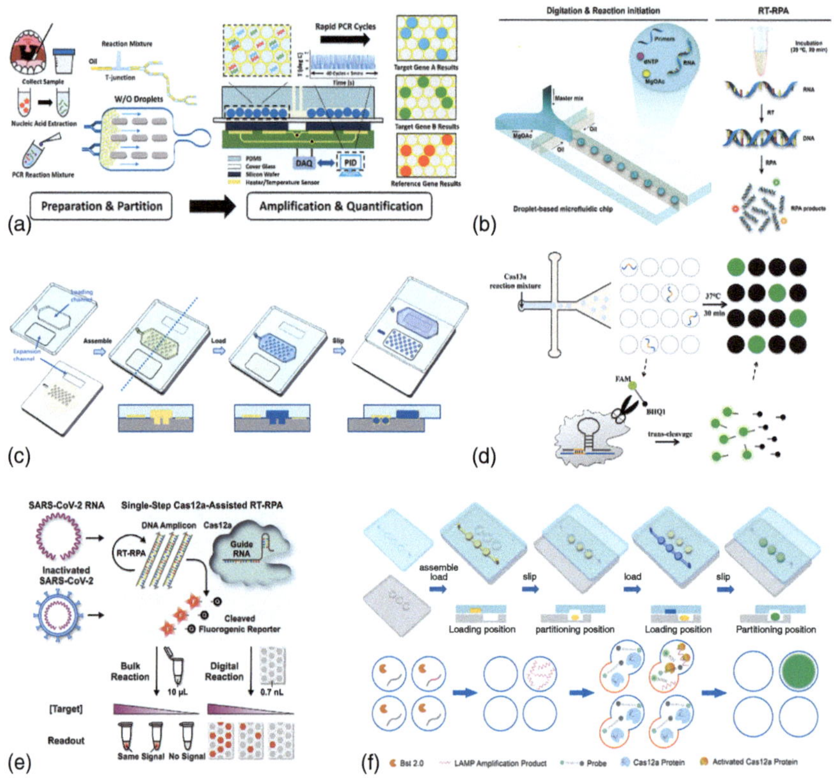

Figure 8.11 Strategies of droplet/digital microfluidic devices via nucleic acid amplification technology for the detection of SARS-CoV-2. (a) Rapid ddPCR technology. *Source:* Ref. [152]/with permission of Elsevier. (b) Digital RT-RPA technology. *Source:* Ref. [153] / with permission of Royal Society of Chemistry. (c) Digital LAMP technology. *Source:* Ref. [154] / with permission of Royal Society of Chemistry. (d) Digital CRISPR-Cas system technology. *Source:* Ref. [155] / with permission of American Chemical Society. (e) Digital CRISPR-Cas system combined with RT-PCR technology. *Source:* Ref. [156] / with permission of John Wiley & Sons. (f) Digital LAMP-CRISPR technology. *Source:* Ref. [157] / Royal Society of Chemistry / CC BY 4.0.

Concurrently, researchers are investigating the incorporation of alternative thermocycling technologies into droplet microfluidic devices to mitigate reliance on intricate temperature control systems in PCR technology, thereby facilitating more efficient and rapid COVID-19 testing. Choi et al. developed an innovative digital droplet repeatable polymerase amplification (ddRPA) technology for the detection of COVID-19 by integrating recombinase polymerase amplification (RPA) into a droplet digital microfluidics system (Figure 8.11b) [153]. This method enabled the quantitative detection of SARS-CoV-2 RNA within 30 minutes, with a minimum detection limit of 2.23 copies/μL. The results from clinical sample analyses indicated that ddRPA was in complete agreement with RT-qPCR assay outcomes. Shen's group developed a digital loop-mediated isothermal amplification (dLAMP) technology for the quantitative detection of SARS-CoV-2 RNA [154]. Utilizing SilpChip technology, the researchers generated a droplet array, thereby streamlining the operational process (Figure 8.11c). The dLAMP results demonstrated a high consistency with the commercial digital PCR outcomes from Stilla. The combination of the CRISPR-Cas system and droplet microfluidics has also been explored for the detection of COVID-19. Shan et al. integrated droplet microfluidics with CRISPR-Cas13a to establish a digital detection method, achieving a detection of SARS-CoV-2 within 30 minutes, with a detection limit of 470 aM (Figure 8.11d) [155]. Various nucleic acid multiple amplification strategies have also been explored for the detection of COVID-19. For example, Park et al. established the first digital CRISPR-Cas system combined with RT-PCR for the detection of SARS-CoV-2 [156]. This system integrated commercial digital microfluidic chips, RT-PCR, and CRISPR-Cas12a, facilitating qualitative and quantitative detection of SARS-CoV-2 RNA within 15 and 30 minutes, with a minimum detection limit of 1 copy/mL (Figure 8.11e). Yu et al. further integrated LAMP and CRISPR technologies on a sliding droplet microfluidic platform to develop dLAMP-CRISPR technology for the ultra-sensitive detection of SARS-CoV-2, achieving a theoretical detection limit of 400 copies/mL [157].

There is no doubt that droplet microfluidic technology had exhibited exceptional analytical performance in the realm of COVID-19 detection, such as similar clinical efficacy and even lower false-negative rate when compared to RT-PCR. However, most droplet microfluidic technologies remained at the laboratory stage, making their application for the on-site diagnosis of SARS-CoV-2 challenging. This limitation primarily emerged from their reliance on complex instrumentation, the intricacies involved in droplet generation, and the high overall costs. Some problems were being actively researched and addressed. For example, Chen et al. successfully prepared thousands of droplets in just five minutes using an elliptical pipette and combined it with a portable thermal cycler, a portable light source, and a smartphone to achieve highly sensitive detection of SARS-CoV-2 RNA with a detection limit of 3.8 copies/20 μL [158]. We anticipate that, with sustained research efforts, droplet and digital microfluidic technologies will excel in the realm of disease detection. In particular, ddPCR technology is poised to supplant qPCR for nucleic acid-related assays in the future, thereby assuming an increasingly vital role in life analysis and related fields.

8.3.1.3 Other Integrated Microfluidic Platforms for the Detection of SARS-CoV-2

While numerous microfluidic-based devices have demonstrated efficacy in COVID-19 detection, particularly lateral flow test strips, the majority of currently developed microfluidic technologies do not fully align with the WHO's guidelines for diagnostic tools aimed at infectious diseases in resource-limited settings, referred to as "REASSURED." The criteria include real-time connectivity, ease of specimen collection, affordable, sensitive, specific, user-friendly, rapid and robust, equipment-free, and delivery to those who need it [159]. For example, microfluidic devices can integrate fluid, electronic, mechanical, and chemical components into a single device, but the signal capture is often dependent on large laboratory instruments such as fluorescence spectrophotometers or fluorescence microscopes.

On the other extreme, the output may only depend on subjective visual identification. In response to these issues, researchers have attempted to integrate more novel technologies or tools to introduce more compatible microfluidic platforms for the diagnosis of infectious diseases that meet the "REASSURED" standards. For example, the application of 3D printing technology facilitated the rapid and cost-effective fabrication of microfluidic chips [76]. Nguyen et al. employed light-curing 3D printing technology to fabricate an integrated 3D-printed dark box, which was coupled with a portable detection device for enabling the detection of SARS-CoV-2 RNA within 70 minutes [160]. As illustrated in Figure 8.12a, this 3D-printed apparatus comprised a separation chamber for RNA extraction and a microfluidic chip designed for RNA LAMP amplification and detection, offering advantages such as straightforward fabrication and cost-effectiveness. It was suitable for deployment in remote and resource-limited settings, thereby contributing to the optimization of the medical supply chain.

In addition to fulfilling essential evaluation criteria, including sensitivity, specificity, ease of use, and cost-effectiveness, the development of low-instrumentation, non-instrumented, or even platform-free microfluidic systems that require no additional equipment holds considerable practical significance. The development of centrifugal microfluidic discs can minimize reliance on external equipment [165]. For instance, Li et al. reported an innovative centrifugal microfluidic system capable of automating sample loading, nucleic acid extraction, purification, and collection within 30 minutes, offering advantages such as rapidity, independence from specialized personnel, and user-friendliness (Figure 8.12b) [161]. However, this system is yet to integrate nucleic acid amplification units and signal detection equipment; thus, it continues to rely on real-time quantitative PCR technology for the detection of SARS-CoV-2 RNA. Teixeira et al. introduced an enhanced approach by integrating a commercial DVD drive with a centrifugal microfluidic disk to develop an all-in-one detection device for multiplex immunochemical assays targeting respiratory infectious diseases [166]. This device was capable of performing semi-automated testing on six samples within 30 minutes, demonstrating commendable diagnostic sensitivity (98–100%) and specificity (84–98%).

An alternative promising strategy involved utilizing smartphones to substitute for any supplementary equipment that may be required for on-site COVID-19 testing [167]. Smartphones are characterized by their light weight and portable design, as well as communication and data processing capabilities, enabling them to replace traditional instruments and integrate seamlessly with microfluidic devices for a variety of detection functions [168]. Smartphones can collect and analyze field data through local communication networks, ultimately providing medical personnel with access, offering advantages of cost-effectiveness and time-efficiency. As a typical example, Bayin et al. developed a lateral flow analysis platform utilizing superparamagnetic nanoparticles and a giant magnetoresistance sensing system, enabling the quantitative detection of IgM and IgG through immunoassay methods [169]. This platform could transmit detection results to smartphones via Bluetooth, facilitating remote analysis of COVID-19 patients' status by medical personnel. Samper et al. integrated the communication capabilities of mobile phones with

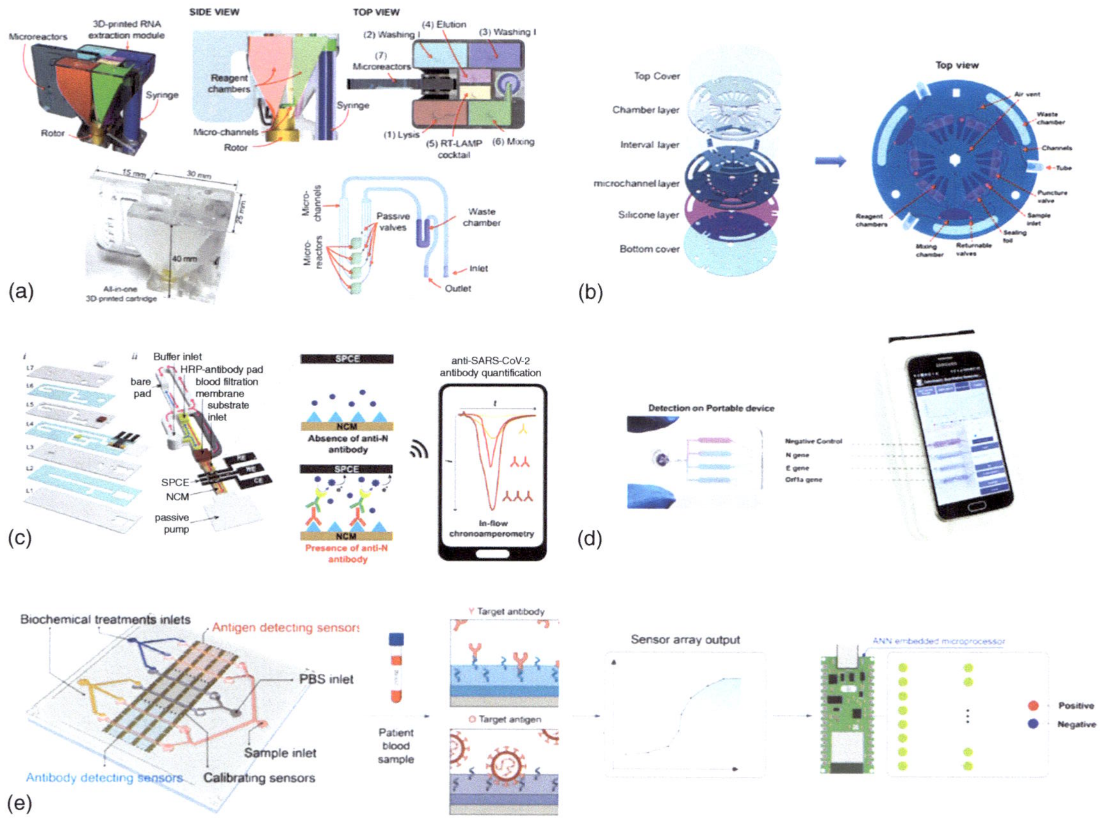

Figure 8.12 Strategies of integrated microfluidic devices for the detection of SARS-CoV-2. (a) An integrated 3D-printed microfluidic device. *Source:* Ref. [160] / with permission of Elsevier. (b) Centrifugal microfluidic device. *Source:* Ref. [161] / with permission of Royal Society of Chemistry. (c) Smartphone-assisted capillary-flow microfluidic device. *Source:* Ref. [162] / with permission of American Chemical Society. (d) Smartphone-assisted 3D-printed integrated microfluidic chip. *Source:* Ref. [163] / with permission of Elsevier. (e) Artificial intelligence-assisted on-chip device. *Source:* Ref. [164] / John Wiley & Sons / CC BY 4.0.

capillary microfluidic devices to establish a platform based on electrochemical immunoassay for the on-site detection of anti-SARS-CoV-2 nucleocapsid protein antibodies [162]. As shown in Figure 8.12c, this platform eliminated the need for sample pretreatment and enabled direct IgG detection in serum within 20 minutes, achieving a low detection limit of 5 ng/mL. Furthermore, the device could be utilized alongside a near-field communication potentiometer controlled by a smartphone, providing immediate quantitative electrochemical readings without requiring professional interpretation of results. The system was user-friendly and cost-effective, demonstrating significant potential for point-of-care diagnostics.

Smartphones possess direct imaging capabilities that can substitute for large imaging equipment by utilizing integrated cameras or in conjunction with specific external microscopic devices. This functionality enables the direct visualization of COVID-19 test results and the conversion of image signals into electronic data, thereby superseding traditional visual inspection methods while demonstrating enhanced sensitivity and precision. Kim et al. presented a handheld portable device comprising a smartphone and an integrated microscope [170]. This device, in conjunction with an antibody-conjugated sub-micron particle-modified paper-based microfluidic chip, was capable of directly capturing liquid droplets and aerosols containing SARS-CoV-2 from the air for detection purposes. The system employed the fluorescence microscope on the smartphone to quantify particles on the chip, facilitating SARS-CoV-2 detection within 30 minutes. An improved strategy for SARS-CoV-2 detection without additional microscopes was reported by Sun et al. [171]. The system employed LAMP to amplify specific nucleic acid sequences on a silicon-based microfluidic chip and utilized a smartphone camera to directly image the resultant fluorescence at the conclusion of the reaction, enabling results to be obtained within 30 minutes, with a detection limit comparable to that of PCR. By digitally analyzing the colorimetric results obtained through smartphones, it was possible to effectively avoid subjective differences in color perception during naked eye readings, thereby significantly enhancing the recognizability and detection sensitivity of the results. Yin et al. reported a strategy that directly employed smartphones to analyze colorimetric signals on 3D-printed microfluidic chips, facilitating high-sensitivity detection of SARS-CoV-2 in wastewater within one hour, with a detection limit of 100 copies/mL (Figure 8.12d) [163]. This platform was both portable and cost-effective, requiring no complex equipment while enabling intelligent, connected, and on-site detection.

Smartphones can effectively manage the detection process through the implantation of meticulously designed programs and can be combined with integrated microfluidic devices to facilitate the automation of sample preparation, detection, data collection and analysis, and result reporting and storage. This system can reduce reliance on specialized personnel and eliminates the need for complex instruments, thus significantly reducing detection costs and improving detection efficiency, which is particularly beneficial for comprehensive disease control in resource-limited settings. Yedire et al. had reported a fully automated microfluidic platform controlled by a smartphone [172]. The platform consisted of a 3D-printed microfluidic chip for LAMP reactions, a portable imaging system, and an intelligent module for control and data analysis. Using a smartphone, users could remotely control sample collection, lysis, heating, and amplification, and achieve imaging and digital analysis of the final colorimetric result. The system could detect SARS-CoV-2 RNA in 13 minutes, demonstrating certain clinical feasibility.

The intelligent analysis facilitated by artificial intelligence (AI) algorithms enables the rapid processing of substantial data sets, significantly enhancing both the speed and accuracy of medical diagnostics. Strategies combining multiple integrated microfluidic technologies with artificial intelligence have been reported for COVID-19 testing. For example, Bae et al. developed a biosensor utilizing indium gallium zinc oxide in conjunction with microfluidic chips and artificial neural networks (Figure 8.12e), enabling the detection of the SARS-CoV-2 spike protein in under 20 minutes, with a detection limit of 1 pg/mL [164]. Sun et al. developed an artificial intelligence-assisted chip technology by integrating μPAD with deep learning, aimed at analyzing RNA associated with the SARS-CoV-2 ORF1ab gene in qPCR results [173]. This method enhanced the efficiency of traditional PCR testing by approximately 67% and possessed the capability to predict the final output of qPCR.

8.3.2 Microfluidic Systems for Zika Virus and Dengue Virus Detection

Zika virus (ZIKV) and dengue virus (DENV) are RNA viruses classified within the Flaviviridae family, capable of infecting humans through hematophagous arthropods such as mosquitoes or ticks, leading to severe diseases. ZIKV was first isolated in Uganda in 1947, subsequently identified in Asia in 1966, and began to spread throughout the Americas in 2015. Between 2014 and 2016, it triggered outbreaks in French Polynesia, a significant epidemic in Brazil, and an outbreak in Singapore. In February 2016, the WHO declared ZIKV as a Public Health Emergency of International Concern [174]. ZIKV is primarily transmitted to humans through the bite of an infected mosquito belonging to the *Aedes* genus [175]. To date, ZIKV infections have been reported in 89 countries and regions [176]. ZIKV infections can lead to serious complications, such as microcephaly and Guililan-Barré syndrome, posing a huge burden on public health. ZIKV infection can be confirmed through laboratory assays, which include nucleic acid testing and serological analysis of blood, urine, and other bodily fluids. Dengue is a viral infection transmitted to humans by mosquitoes, predominantly occurring in tropical and subtropical regions. While the majority of infections are either asymptomatic or exhibit mild symptoms, a subset of patients may progress to severe manifestations, which can be fatal in critical cases. In recent years, the global incidence of dengue fever has escalated dramatically, surpassing 6.5 million reported cases and more than 7,300 dengue-related deaths as of 2023. The development of methods for the rapid and specific detection of these pathogens, particularly point-of-care diagnostic strategies, holds significant scientific implications. Currently, a variety of commercial diagnostic technologies and platforms are available for the detection of ZIKV, DENV, and other related vector-borne diseases [177]. Some of these technologies utilize microfluidic systems. For example, the ViroTrack™ system developed by Blusense Diagnostics in Denmark can be used for the rapid detection of biomarkers related to dengue and Zika fevers, delivering results within a mere 15 minutes. Microfluidic chips are integral to this system and play a pivotal role in its operation by facilitating the separation of plasma from whole blood through centrifugation. This process ensures that the analysis concentrates solely on the relevant components necessary for accurate biomarker detection. An optical magnetic readout method, which exploits light-scattering phenomena induced by rotating nanoparticle chains—commonly referred to as chip-based immunomagnetic assays—is employed for signal detection. Furthermore, the ViroTrack system possesses multiplexing capabilities, enabling simultaneous detection of multiple targets within a single sample [178]. Clinical studies have demonstrated that the system has robust bedside diagnostic accuracy, achieving a positive predictive value of 97.5%. This capability could streamline dengue fever screening and offer valuable insights for enhancing disease monitoring and outbreak response related to mosquito-borne illnesses, including DENV and ZIKV infection [179]. An additional example of a commercialized product was the VereFever™ developed by Veredus Laboratories in Singapore using the VerePLEX™ Biosystem platform. It exemplified a multiplex test that integrates PCR and Lab-On-Chip (LOC) technologies into a cohesive testing workflow. This system was intended for the qualitative detection, differentiation, and identification of multiple vector-borne pathogens (including DENV and ZIKV) within a single assay [180].

8.3.2.1 Paper Microfluidic Devices for the Detection of DENV or ZIKV

Lateral flow-based paper microfluidic devices have garnered significant attention in the realm of rapid diagnostics, with several commercialized lateral flow immunoassay (LFA) kits currently available for the diagnosis of DENV [181]. These commercial LFAs primarily employ immunoassay techniques to detect DENV nonstructural protein 1 (NS1) or IgG/IgM antibodies, thereby facilitating the rapid identification of DENV [182, 183]. In recent years, numerous novel paper-based microfluidic devices have been reported to improve the detection capabilities for DENV and ZIKV. For example, Seok et al. integrated microfluidic technology with LAMP techniques to develop a paper-based laboratory platform for the molecular diagnosis of ZIKV, DENV, and chikungunya virus [184]. The entire diagnostic workflow—including sample preparation, nucleic acid extraction, amplification, and fluorescence detection—was executed on this lateral flow-type paper chip (Figure 8.13a). This system enabled simultaneous detection of three distinct viruses within 60 minutes and could quantitative assay ZIKV RNA in the range of 5–5,000 copies. However, the use of fluorescence signals necessitates specific light sources and detectors, which may increase the complexity and cost of the device. A color change-based paper sensor enabling the visual detection of ZIKV effectively addresses the demand for low-cost diagnostic methods [188]. Based on this concept, Kaarj et al. developed a paper-based microfluidic chip for the detection of ZIKV RNA, utilizing a colorimetric approach [185]. As shown in Figure 8.13b, the chip was fabricated using wax jet printing technology and effectively filtered biological samples, including tap water, urine, and plasma. Following RT-LAMP amplification, the presence of ZIKV RNA was indicated by changes in acidity with a pH indicator solution, allowing for subsequent quantitative analysis via smartphone imaging. This platform enabled the detection of ZIKV RNA within 15 minutes, with a limit of detection as low as 1 copy/mL. An innovative paper-based microfluidic device for the detection of DENV RNA was implemented on a folded paper strip [186]. The platform integrated RT-LAMP technology with colorimetry, and interfaced with a simple, portable integrated system capable of simultaneously conducting nucleic acid separation, isothermal amplification, chromatographic analysis, and result interpretation (Figure 8.13c). Within 30 minutes, this platform could efficiently detect all four serotypes of DENV from approximately 50 μL of human serum, demonstrating exceptional specificity and sensitivity. From another perspective, the direct detection of complete virus particles can reduce the dependence on antibody cross-reactivity in immunoassays and provide a more intuitive reflection of the infection status, which may become a superior alternative. Draz et al. integrated paper-based microfluidic technology with platinum nanoprobes to develop a nanoparticle-enhanced impedance biosensor for the detection of ZIKV lysates [187]. The utilization of metal nanoprobes for viral capture significantly enhanced detection sensitivity, achieving a 10-fold increase compared to the direct detection of virus lysates (Figure 8.13d). The paper chip could be fabricated within 40 minutes using a layer-by-layer assembly approach. This impedance-based biosensing method leveraging paper-based microfluidics enabled the detection of ZIKV in approximately 75 minutes, with a limit of detection of 10 copies/μL.

8.3.2.2 Other Integrated Microfluidic Platforms for the Detection of DENV or ZIKV

In addition to cost-effective paper-based microfluidic chips, several innovative microfluidic devices have been developed for the detection of DENV and ZIKV. For example, Hin et al. developed a platform called FeverDisk based on centrifugal microfluidic technology for detecting multiple potential viruses and bacteria that cause unexplained fevers [189]. The FeverDisk pre-stored all reagents required for real-time detection, including magnetic particles for the in-situ extraction of pathogens and enzymes required for LAMP reactions. By simply adding the sample to the disk, this fully automated integrated system could deliver detection results within two hours. Furthermore, the FeverDisk was capable of distinguishing four distinct subtypes of DENV and could simultaneously detect up to 12 different infectious fever pathogens. Another fully automatic detection system based on centrifugal microfluidic technology was reported by Zhang et al., in which recombinase-aided amplification (RAA) and CRISPR-Cas13a were integrated for detecting various viruses, such as DENV and ZIKV [190]. As shown in Figure 8.14A, the system was designed with an automatically rotating chip and a thermally triggered wax valve, and the required reagents were stored within the chip. It could simultaneously detect 10 virus species within 45 minutes through multiple channels, showing a detection limit of 5 copies per reaction for simulated plasma samples. The results of detecting clinical DENV specimens confirmed that the system's sensitivity and specificity were comparable to those of RT-PCR. Sharma et al. developed a fully automated Lab-on-a-Chip detection system for the rapid identification of ZIKV

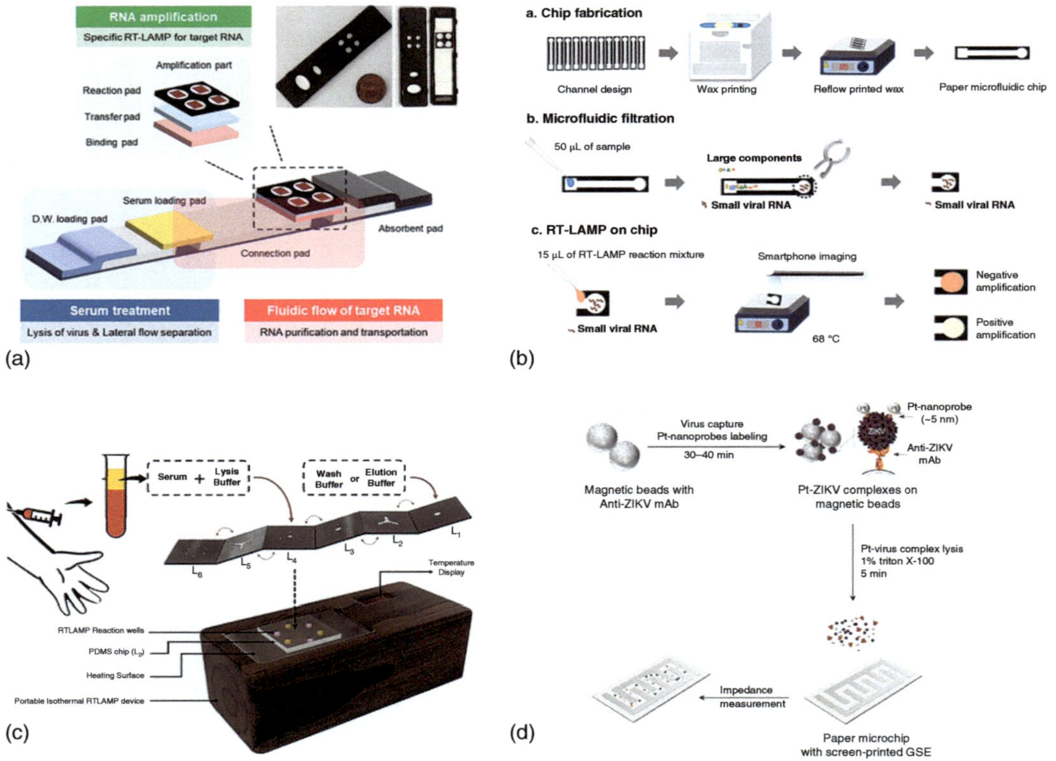

Figure 8.13 Strategies of paper microfluidic devices for the detection of DENV or ZIKV. (a) A fluorescence-based paper microfluidic device based on LAMP technology for the detection of DENV and ZIKV. *Source:* Ref. [184] / with permission of Elsevier. (b) A paper microfluidic device based on LAMP technology for the colorimetric assays of ZIKV. *Source:* Ref. [185] / Springer Nature / CC BY 4.0. (c) An origami-based microfluidic device using LAMP technology for the colorimetric assays of DENV. *Source:* Ref. [186] / with permission of American Chemical Society. (d) Nanoparticle-enhanced electrical sensing of ZIKV on a paper microchip. *Source:* Ref. [187] / with permission of Royal Society of Chemistry.

utilizing a RT-LAMP technique based on colorimetric assays [194]. This system employed magnetic drive technology to precisely control the movement of magnetic beads within the microfluidic chip, thereby facilitating automation and enabling the visual detection of ZIKV within 40 minutes. Powell et al.'s study showed that building winding microchannels with silicon-based chips and combining them with qPCR technology could achieve rapid and highly sensitive detection of ZIKV in just 25 minutes [195]. ddPCR was also explored for the detection of ZIKV [196]. The study by Hu et al. demonstrated that ddPCR could detect ZIKV at a concentration of 1 copy/μL, exhibiting exceptional accuracy and sensitivity in samples with low ZIKV concentrations, thereby rendering it particularly suitable for the early diagnosis of ZIKV. Kasputis et al. employed dLAMP technology to facilitate the rapid detection of ZIKV [191]. As illustrated in Figure 8.14b, they implemented a vacuum lung design strategy to construct a microfluidic chip and further integrated digital technology with RT-LAMP, enabling rapid and reliable detection of ZIKV within just one hour, with a detection limit as low as 100 copies/μL.

Mobile phone-based strategies for detecting DENV or ZIKV are also receiving widespread attention from researchers. For example, Ganguli et al. developed a hands-free smartphone-based microfluidic diagnostics platform for the real-time detection of ZIKV, DENV, and other viruses in whole blood [197]. This microfluidic platform employed RT-LAMP as the nucleic acid amplification strategy, utilizing a smartphone to capture real-time images of the amplification reaction and provide digital visual readouts. The platform demonstrated high specificity and clinically relevant sensitivity, capable of detecting ZIKV at a concentration of 1.56×10^5 PFU/mL in whole blood. Zhu et al. developed a chitosan-modified capillary microfluidic chip, integrating it with in-situ PCR amplification for the detection of ZIKV [192]. As shown in Figure 8.14c, the authors employed a smartphone as the signal acquisition terminal, successfully detecting ZIKV in saliva within 30 minutes. Recently, Jankelow et al. reported a smartphone-based microfluidic integrated system for the rapid detection of ZIKV in whole blood [193]. This system comprised a microfluidic chip featuring six parallel reaction chambers and a clip-on detection device that amplified ZIKV RNA via RT-LAMP while employing the smartphone to monitor the dynamic fluorescence of LAMP reports within the reaction chambers (Figure 8.14d). The system could provide detection results in approximately 30 minutes and detect ZIKV RNA at a concentration as low as 2.70×10^2 copies/μL.

8.3.3 Microfluidic Systems for Ebola Virus Detection

The Ebola virus (EBOV) is a single-stranded negative-sense RNA virus, which belongs to the filovirus family and possesses the feature of high infectivity and lethality [198]. It was first identified in Sudan and Zaire (now the Democratic Republic of the Congo) in 1976 [199]. Ebola virus disease, formerly referred to as Ebola hemorrhagic fever, can lead to exudation at the site of venipuncture, melena, rectal bleeding, and hematemesis [200]. Due to its exceptionally high infectivity, this virus has the potential to cause large-scale outbreaks, epidemics, and pandemics. Intermittent outbreaks of Ebola have been reported across Africa; notable instances

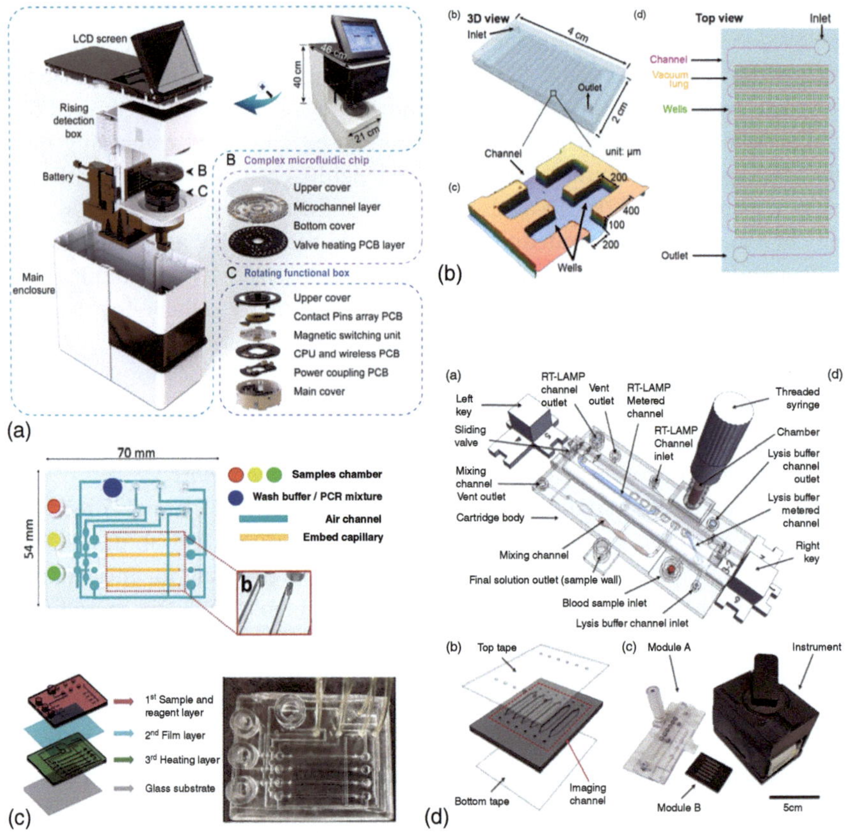

Figure 8.14 Strategies of various microfluidic devices for the detection of DENV or ZIKV. (a) A portable all-in-one centrifugal microfluidic device for the detection of DENV and ZIKV. *Source:* Ref. [190] / with permission of Royal Society of Chemistry. (b) A self-powered digital LAMP microfluidic chip for the detection of ZIKV. *Source:* Ref. [191] / with permission of Royal Society of Chemistry. (c) An on-chip device integrated with chitosan-modified capillaries for the detection of ZIKV. *Source:* Ref. [192] / MDPI / CC BY 4.0. (d) A hands-free smartphone-based microfluidic diagnostics platform for the detection of ZIKV, DENV, and other viruses in whole blood. *Source:* Ref. [193] / with permission of Royal Society of Chemistry.

include the outbreak in Kikwit, Democratic Republic of the Congo in 1995, outbreaks in Uganda from 2000 to 2001, and the West African epidemic from 2014 to 2016 [201]. More recently, an outbreak occurred in the Democratic Republic of the Congo between 2018 and 2020. According to data from the WHO, approximately 3470 confirmed and suspected cases were reported during this period, resulting in around 2,280 fatalities [202]. Such infectious diseases exert devastating impacts on global health systems as well as economic and social structures [203]. Therefore, early diagnosis coupled with clinical suspicion is essential for reducing mortality rates and preventing further outbreaks due to the high infectivity and lethality associated with the EBOV [204].

Currently, several techniques have been established for the laboratory diagnosis of EBOV, including viral genome detection, viral antigen detection, and assessment of host immune responses [200]. However, these methods are often time-consuming and necessitate controlled laboratory environments, typically relying on trained personnel and specialized equipment. For instance, the implementation of RT-PCR—the gold standard for EBOV detection—requires a dedicated clean room and the effective inactivation of EBOV prior to testing, which poses significant challenges for its widespread adoption in the resource-limited regions of Africa [205]. Therefore, the development of novel multi-functional rapid detection tools for on-site and real-time identification of EBOV is of great significance [206]. Microfluidic chip technology can utilize physical and chemical properties at the micro and nanoscale to build POC sensors, possessing significant potential for early and rapid diagnosis of infectious viruses [207]. For example, BioFire's microfluidic chip-based product FilmArray® BioThreat-E was granted emergency authorization for Ebola virus detection in West Africa in 2014. The chip used nested multiplex PCR analysis technology, which could detect up to 24 pathogens in the same blood sample in a single test. Additionally, the detection process was significantly faster than traditional PCR or RT-PCR methods, taking only about one hour, making it well-suited for early, rapid screening of infectious diseases [204]. A clinical evaluation report indicated that FilmArray® BioThreat-E demonstrated exceptional performance in whole blood EBOV detection, achieving sensitivity and specificity rates of 95.7% and 100%, respectively [208]. Furthermore, this product could detect EBOV in saliva and urine samples.

8.3.3.1 Paper Microfluidic Devices for the Detection of EBOV

Paper-based sensors are increasingly recognized for their notable cost-effectiveness, particularly in the rapid detection of infectious diseases [209]. However, paper-based sensors are often limited by low sensitivity and prolonged detection times. In contrast, paper microfluidic technology, which integrates sample separation, enrichment, and signal amplification within a single microreactor framework, holds significant promise for addressing these limitations and demonstrates considerable application potential [210]. In the field of EBOV detection, numerous lateral flow strips have been developed and applied in commercial products, including SD Q Line Ebola Zaire Ag test [211], OraQuick Ebola Rapid Antigen Test [212], QuickNavi-Ebola [213], DEDIATEST Ebola [211], and EBOLA Ag K-SeT [214]. Meanwhile, many laboratory methods have also been reported. For example, Magro et al. employed RT-RPA technology for the detection of EBOV RNA on μPADs [215]. This system enabled the detection of EBOV within 20 minutes with a portable device, which also achieved a sensitivity of 90% while significantly reducing costs and streamlining operational procedures. Another approach employed 3D μPADs in conjunction with immunoassay techniques for the colorimetric detection of anti-EBOV IgG [216]. Couturier et al. developed an immunoassay method that integrated immunochromatography with silver amplification technology for the detection of EBOV [217]. This method could identify trace levels of EBOV glycoprotein in the plasma from monkey cohorts, with a detection limit of 2.2×10^4 genome copies/mL. Clinical performance evaluation indicated that this method exhibited 100% specificity and 85.7% sensitivity. Nanomaterials can effectively enhance the detection performance of paper-based microfluidic devices, including LFTs and μPADs [218]. Meanwhile, the strategy of using nanoparticles to enhance the signal for lateral flow detection of EBOV has also been explored [219]. Hu and his colleagues employed a multifunctional nanoball incorporating quantum dots and gold nanoparticles for the detection of EBOV glycoprotein on a lateral flow assay (Figure 8.15a) [220]. Utilizing lateral flow immunoassay, this platform could achieve high-sensitivity quantitative detection of EBOV glycoprotein within 20 minutes, with a detection limit of 0.18 ng/mL. Moreover, the platform was applicable for detecting EBOV in urine, plasma, and tap water. Notably, this platform featured dual signal reading modes that facilitated not only quantitative detection via fluorescence but also qualitative visual assessment through colorimetry. In contrast to the direct application of nanoparticles as signaling entities, Yan's research group integrated the enzyme-mimicking catalytic properties of Fe_3O_4 magnetic nanoparticles with lateral flow assay technology to develop a nanozyme-based immunoassay strip for the rapid detection of EBOV glycoprotein [221]. As shown in Figure 8.15b, Fe_3O_4 magnetic nanoparticles exhibited pseudo-peroxidase activity, enabling them to catalyze the corresponding substrate (e.g., 3,3′-diaminobenzidine) in the presence of hydrogen peroxide, thereby facilitating colorimetric detection. This approach capacitated high-sensitivity visual detection of EBOV glycoprotein with a detection limit of 1 ng/mL. Furthermore, this method allowed for the detection of EBOV in clinical samples within 30 minutes and demonstrated sensitivity and accuracy comparable to that of ELISA.

8.3.3.2 Other Microfluidic Devices for the Detection of EBOV

Digital microfluidic technology has significant advantages such as fast response, parallel detection, high integration, automation, and miniaturization. These features enable the precise execution of multiple detection steps within microliter-scale samples while demonstrating enhanced sensitivity for the efficient identification of early infections or low-concentration pathogens. Consequently, this technology is particularly well-suited for the development of real-time diagnostic devices targeting infectious diseases [226]. For EBOV detection, Piraino et al. developed a multiplexed microfluidic platform capable of detecting multiple anti-EBOV IgG antibodies in human serum [227]. This platform integrated digital and analog detection technologies, specifically, molecular interaction-induced mechanical capture (MITOMI), to create a portable, integrated hybrid microfluidic device with high sensitivity and a wide dynamic range for biomarker detection. The system could identify anti-EBOV IgG at concentrations as low as 1 pM in serum. Furthermore, the team had progressively combined this microfluidic device with cost-effective, portable control systems and a fluorescence USB microscope to develop an innovative portable microfluidic device that could detect 100 pM of anti-EBOV IgG in ultra-low-volume whole blood samples. Recently, digital RCA technology was specifically developed by Ciftci et al. for the detection of EBOV [222]. In this study, the authors developed a range of hook-and-loop RCA systems for the specific detection of various EBOV genes and

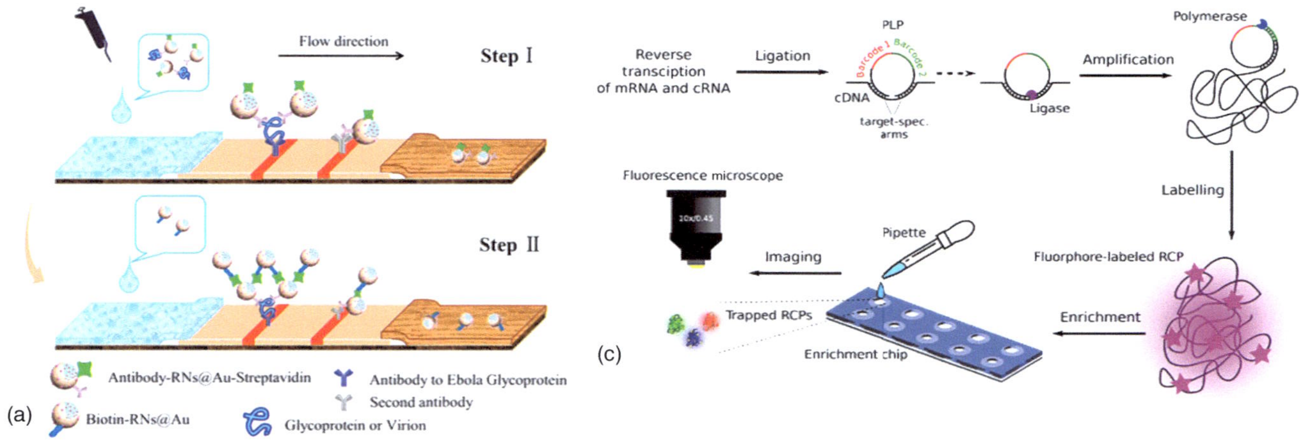

Figure 8.15 Strategies of various microfluidic devices for the detection of EBOV. (a) Nanomaterial-assisted lateral flow assay for the detection of EBOV. *Source:* Ref. [220] / with permission of American Chemical Society. (b) Nanozyme-based immunochromatographic strip for the detection of EBOV. *Source:* Ref. [221]/with permission of Elsevier. (c) Digital RCA-based platform for the detection of EBOV. *Source:* Ref. [222] / with permission of Elsevier. (d) qRT-PCR based on-chip device for the detection of EBOV. Reprinted from reference [223], Copyright 2017 by Springer Nature. (e) RT-LAMP based disc microfluidic chip for the detection of EBOV. *Source:* Ref. [224] / MDPI / CC BY 4.0. (f) CRISPR-Cas13a based automated microfluidic chip for the detection of EBOV. *Source:* Ref. [225] / with permission of American Chemical Society.

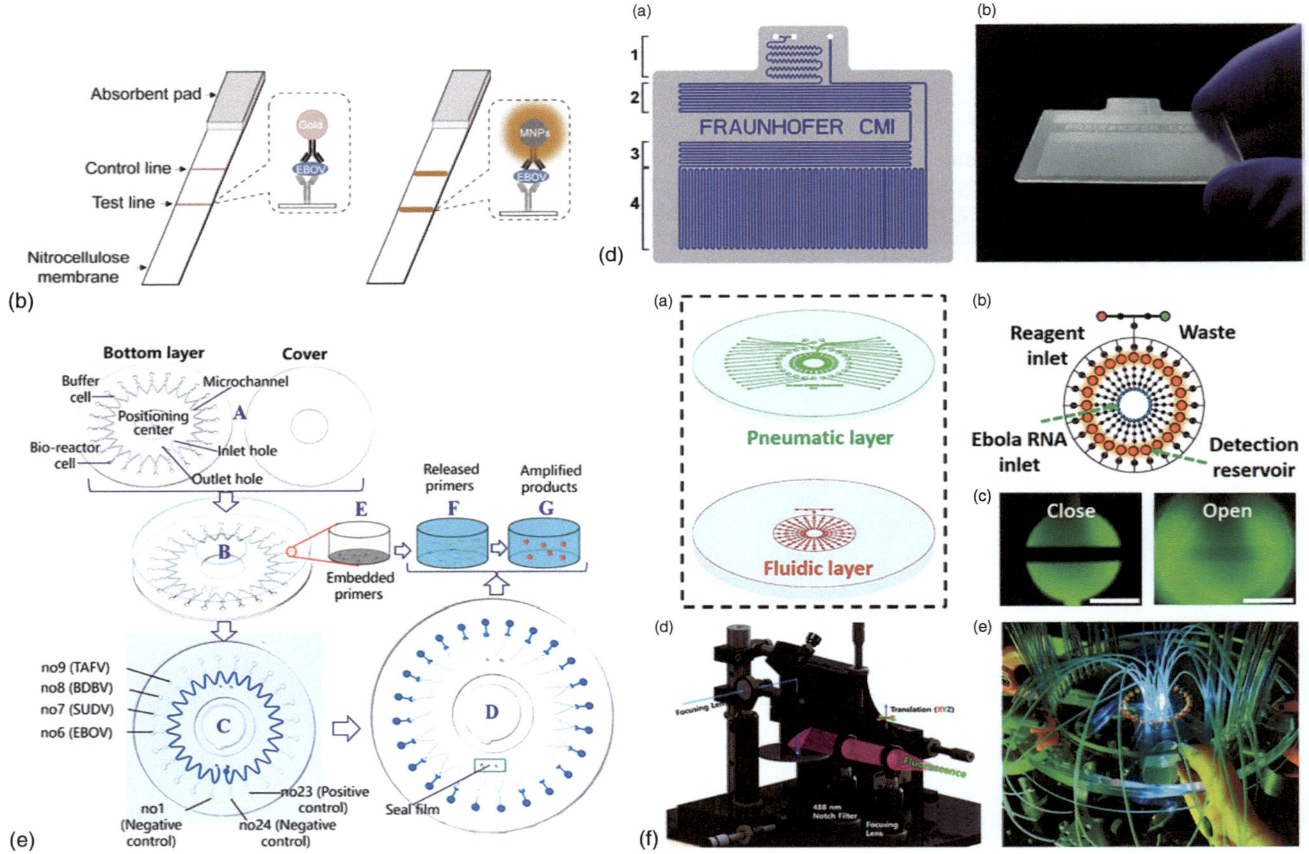

Figure 8.15 (*Continued*)

performed enrichment and readout of the RCA products on a pump-free digital microfluidic chip (Figure 8.15c). This strategy enabled the simultaneous detection of EBOV, ZIKV, and DENV, demonstrating high specificity, sensitivity, and multiplexing capability.

Various nucleic acid amplification techniques have been developed for the detection of EBOV. For instance, Na et al. developed a microbead RCA strategy for the rapid detection of EBOV [228]. Unlike digital microfluidics, this strategy utilized the DNA hydrogels formed by RCA to create spatial steric hindrance between thousands of microbeads, blocking fluid pathways between microbeads. By injecting staining fluid, the visual detection of the target pathogen could be achieved based on the length of fluid flow. This strategy was simple and fast and could complete the detection of EBOV in 15 minutes, while integrating multi-channel design to achieve simultaneous detection of multiple pathogens. Fernandez-Carballo et al. realized on-chip detection of EBOV RNA using qRT-PCR technology [223]. The authors employed a roll-to-roll embossing technique to fabricate a disposable microfluidic chip with an extended microfluidic channel, which incorporated distinct temperature zones for the qRT-PCR analysis of EBOV RNA (Figure 8.15d). This approach enabled the high-sensitivity detection of EBOV RNA within 30–50 minutes, with a detection limit of 10 RNA copies/μL. Lin et al. reported an RT-LAMP-based method using centrifugal microfluidics for detecting EBOV [224]. As shown in Figure 8.15e, this system could perform four parallel RT-LAMP reactions for EBOV simultaneously on a disc-shaped microfluidic chip, with results obtained within 50 minutes and a detection limit as low as 10 copies per reaction. A study utilizing a CRISPR-Cas13a-based disc microfluidic device demonstrated that the platform could implement the specific detection of EBOV RNA within five minutes, with a detection limit of 5.45×10^4 copies/μL (Figure 8.15f) [225].

8.4 Limitations, Current Trends, and Future Prospects

Microfluidic systems possess the capability to rapidly and efficiently acquire and process data from low-volume samples, while simultaneously integrating multiple functionalities into a single device [229]. They enable result output without necessitating specialized personnel or large-scale instruments. These distinctive capabilities align seamlessly with the development of real-time diagnostic systems for medical applications [230]. As previously stated, microfluidic devices are employed extensively in the four predominant central laboratory techniques—blood chemistry analysis, immunoassays, nucleic acid amplification tests, and flow cytometry [229]. Furthermore, they have significantly contributed to the detection of infectious viruses, bacteria, parasites, and other pathogens by integrating both traditional and innovative molecular detection methodologies [231]. In addition to applications in diagnostics, microfluidics have garnered significant attention across various fields, including drug screening, biological development, and aerospace, with a diverse array of innovative microfluidic devices and systems continually emerging [4, 232, 233]. However, microfluidic devices which actually apply to practical problems have not been as widely used as expected, and only a small fraction of microfluidic devices have been commercialized in contrast to the extensive theoretical demonstrations present in literature [234]. This indicates that, from a technical perspective, microfluidic technology is progressively gaining strength and maturity; however, several critical challenges remain to be addressed.

The primary challenge lies in effectively integrating a comprehensive array of experimental processes—including sample preparation, separation, extraction, target identification, signal generation, output, and result analysis—into a single chip to realize a true "lab-on-a-chip." To accomplish this objective, contemporary microfluidic systems frequently depend on substantial external auxiliary apparatus. For example, the delivery of fluids in microfluidic devices typically requires additional fluid pumps. While passive transport by capillary action can be used to replace fluid pumps in some microfluidic systems like paper-based microfluidic systems [31], it can't meet the needs of other types of microfluidic systems. In the realm of sample detection, particularly concerning microfluidic devices that utilize nucleic acids, there is a predominant reliance on standardized PCR technology (such as GeneXpert) [235]. While a limited number of microfluidic devices employ isothermal amplification techniques such as LAMP, these systems still necessitate additional temperature control components. Consequently, it is imperative to further assess the efficacy and practicality of amplification technologies like CRISPR-based cutting systems that are compatible with room temperature or exhibit broad temperature tolerance within microfluidic frameworks [88]. Looking ahead, there exists an urgent demand for the development of innovative high-sensitivity detection methodologies that obviate the need for amplification. In terms of result readout, it is recommended to minimize the use of large instruments such as fluorescence microscopes. The adoption of compact, portable handheld devices can significantly streamline the overall design and facilitate equipment integration. Depending on the detection strategy employed, various portable devices may be selected, for instance, those utilizing colorimetric or fluorescence methods can be photographed and digitally processed using a smartphone. For electrochemical detection, these devices can be integrated with commercially available glucose meters and digital multimeters. Furthermore, the integration of microfluidic platforms encompasses various dimensions, including standardization within the broader field, advancements in industrial technologies, and the evolving needs of customers.

An additional aspect that merits investigation in the advancement of microfluidic technology is its cost-effectiveness. As previously noted, microfluidic technology aligns closely with the demand for rapid detection equipment in public health within resource-limited settings. Consequently, to facilitate the widespread adoption of microfluidic devices, it is essential that their costs are lower than those of traditional laboratory methods and that their accessibility is significantly enhanced. However, the cost-effectiveness of microfluidic devices is still significantly lower than that of traditional laboratory systems at present [236]. One of the primary factors is the diversity of materials and manufacturing processes, which resulted in academic research environments utilizing materials that are typically suited for small-scale production and face challenges in transitioning seamlessly to large-scale commercial manufacturing. For example, the academic community extensively employs PDMS as a material to fabricate microfluidic devices, for developing various innovative applications, due to its superior optical transmittance and ease of processing. However, PDMS exhibits permeability defects concerning small molecules, which leads to the evaporation, permeation, and absorption of liquid samples [237]. Moreover, PDMS encounters significant challenges in mass production since devices made from PDMS cannot be fabricated using high-throughput techniques such as injection molding, rolling, or embossing [230]. Additionally, the cost of PDMS is relatively higher compared to many other polymers, thereby imposing considerable limitations on its commercial applications. This situation is likely to extend the translational timeline between academic research and commercial products while further escalating development costs. Based on the current advancements in technology, paper-based microfluidics demonstrate significant potential. However, in order to meet the growing social needs, there is an urgent need to develop more novel materials. Another concern is that the chips utilized for transporting samples and reagents in microfluidic systems are frequently challenging to clean and are predominantly designed as disposable products, which significantly elevates costs compared to the permanent equipment employed in traditional laboratories. One viable approach to address this challenge is the development of modular, reconfigurable microfluidic devices [238] that employ a detection mode akin to portable glucose meters, allowing for the replacement of only those modules that are contaminated by samples or reagents, thereby reducing the overall costs associated with microfluidic systems.

The standardization of microfluidics represents a critical yet underexplored area of research. The establishment of globally recognized standards has the potential to significantly enhance the comprehensive advancement of microfluidic technologies [239]. Although several standards have been proposed, including guidelines and recommendations for microfluidic interconnections, their practical efficacy remains limited [234]. Professionals across various sectors within the microfluidics domain should collaborate to convene workshops, aiming to establish consensus standards that enhance the reliability and practicality of microfluidic devices. These unified standards must encompass fundamental specifications and validation criteria pertaining to materials and equipment manufacturing, fluid manipulation and transport, detection and analytical methodologies, as well as biological sample handling and processing. The ultimate goal is to improve the reliability of microfluidic systems, minimize maintenance costs, promote compatibility and interoperability among diverse devices, facilitate the large-scale production of microfluidic components, and propel the industrialization of this technology [6, 234, 240, 241].

Microfluidic technology holds significant potential for application in public health, particularly in the realm of rapid diagnostics [242]. Future advancements in innovative detection methodologies [88, 241, 243], novel functional materials [241, 244], machine learning [104], wearable technologies [126], and artificial intelligence [99, 100] are poised to further enhance the integration of microfluidics within public health initiatives. Portable handheld devices are expected to see increased utilization in the design of cost-effective microfluidic systems, which will facilitate large-scale, rapid, and on-site pathogen detection and regional epidemic screening. The convergence of Internet of Things (IoT) technologies with smartphones will enable real-time monitoring and seamless transmission of diagnostic results. AI-driven algorithms will refine the analysis and decision-making processes associated with extensive data sets, while wearable microfluidic devices can continuously monitor multiple biomarkers non-invasively and provide real-time feedback. By integrating big data analytics with AI, comprehensive personal health records can be established to support effective health management strategies. In summary, microfluidic technology is set to become an indispensable component driving advancements in global public health and heralding a new era in disease detection and monitoring [245].

References

1 Shakeri Hossein Abad, Z., Kline, A., Sultana, M. et al. (2021). Digital public Health surveillance: A systematic scoping review. *npj Digit. Med.* 4 (1): 1–13. https://doi.org/10.1038/s41746-021-00407-6.

2 Cassedy, A., Parle-McDermott, A., and O'Kennedy, R. (2021). Virus detection: A review of the current and emerging molecular and immunological methods. *Front. Mol. Biosci.* 8. https://doi.org/10.3389/fmolb.2021.637559.

3 Mao, K., Zhang, H., Pan, Y., and Yang, Z. (2021). Biosensors for wastewater-based epidemiology for monitoring public Health. *Water Research* 191: 116787. https://doi.org/10.1016/j.watres.2020.116787.

4 Tabeling, P. (2023). *Introduction to Microfluidics*, 2nde. Oxford University PressOxford. https://doi.org/10.1093/oso/9780192845306.001.0001.

5 Lu, D., Shao, G., Du, D. et al. (2011). Enzyme entrapped Nanoporous scaffolds formed through flow-induced gelation in a microfluidic filter device for sensitive biosensing of Organophosphorus compounds. *Lab Chip* 11 (3): 381–384. https://doi.org/10.1039/C0LC00337A.

6 Whitesides, G.M. (2006). The origins and the future of Microfluidics. *Nature* 442 (7101): 368–373. https://doi.org/10.1038/nature05058.

7 Toumey, C. (2009). Plenty of room. *Plenty of History. Nature Nanotech* 4 (12): 783–784. https://doi.org/10.1038/nnano.2009.357.

8 Warner, R.M. (2001). Microelectronics: its unusual origin and personality. *IEEE Transactions on Electron Devices* 48 (11): 2457–2467. https://doi.org/10.1109/16.960368.

9 Sweet, R.G. (1965). High frequency recording with electrostatically deflected ink jets. *Review of Scientific Instruments* 36 (2): 131–136. https://doi.org/10.1063/1.1719502.

10 Bassous, E., Taub, H.H., and Kuhn, L. (1977). Ink jet printing nozzle arrays etched in silicon. *Applied Physics Letters* 31 (2): 135–137. https://doi.org/10.1063/1.89587.

11 Judy, J.W. (2001). Microelectromechanical systems (MEMS): fabrication, design and applications. *Smart Mater. Struct.* 10 (6): 1115. https://doi.org/10.1088/0964-1726/10/6/301.

12 Ramakrishnan, J., Gaurav, P.T.R., Chandar, N.S., and Sudharsan, N.M. (2021). Structural design, analysis and DOE of MEMS-based capacitive accelerometer for automotive airbag application. *Microsyst Technol* 27 (3): 763–777. https://doi.org/10.1007/s00542-020-04979-3.

13 Terry, S.C. (1975). *A Gas Chromatography System Fabricated on A Silicon Wafer Using Integrated. CIRCUIT TECHNOLOGY.*; Stanford University.

14 Terry, S.C., Jerman, J.H., and Angell, J.B. (1979). A gas chromatographic air Analyzer fabricated on a silicon wafer. *IEEE Transactions on Electron Devices* 26 (12): 1880–1886. https://doi.org/10.1109/T-ED.1979.19791.

15 Convery, N. and Gadegaard, N. (2019). 30 years of Microfluidics. *Micro and Nano Engineering* 2: 76–91. https://doi.org/10.1016/j.mne.2019.01.003.

16 Manz, A., Graber, N., and Widmer, H.M. (1990). Miniaturized Total chemical analysis systems: A novel concept for chemical sensing. *Sensors and Actuators B: Chemical* 1 (1): 244–248. https://doi.org/10.1016/0925-4005(90)80209-I.

17 Swerdlow, H. and Gesteland, R. (1990). Capillary gel electrophoresis for rapid, high resolution DNA sequencing. *Nucleic Acids Research* 18 (6): 1415–1419. https://doi.org/10.1093/nar/18.6.1415.

18 Harrison, D.J., Fluri, K., Seiler, K. et al. (1993). Micromachining a miniaturized capillary electrophoresis-based chemical analysis system on a Chip. *Science* 261 (5123): 895–897. https://doi.org/10.1126/science.261.5123.895.

19 Woolley, A.T. and Mathies, R.A. (1994). Ultra-high-speed DNA fragment separations using microfabricated capillary Array electrophoresis chips. *Proceedings of the National Academy of Sciences* 91 (24): 11348–11352. https://doi.org/10.1073/pnas.91.24.11348.

20 Effenhauser, C.S., Paulus, A., Manz, A., and Widmer, H.M. (1994). High-speed separation of antisense oligonucleotides on a micromachined capillary electrophoresis device. *Anal. Chem.* 66 (18): 2949–2953. https://doi.org/10.1021/ac00090a024.

21 Woolley, A.T. and Mathies, R.A. (1995). Ultra-high-speed DNA sequencing using capillary electrophoresis chips. *Anal. Chem.* 67 (20): 3676–3680. https://doi.org/10.1021/ac00116a010.

22 Woolley, A.T., Hadley, D., Landre, P. et al. (1996). Functional integration of PCR amplification and capillary electrophoresis in a microfabricated DNA analysis device. *Anal. Chem.* 68 (23): 4081–4086. https://doi.org/10.1021/ac960718q.

23 Xia, Y. and Whitesides, G.M. (1998). Soft lithography. *Angewandte Chemie International Edition* 37 (5): 550–575. https://doi.org/10.1002/(SICI)1521-3773(19980316)37:5<550::AID-ANIE550>3.0.CO;2-G.

24 Duffy, D.C., McDonald, J.C., Schueller, O.J.A., and Whitesides, G.M. (1998). Rapid prototyping of microfluidic Systems in Poly(Dimethylsiloxane). *Anal. Chem.* 70 (23): 4974–4984. https://doi.org/10.1021/ac980656z.

25 Thorsen, T., Maerkl, S.J., and Quake, S.R. (2002). Microfluidic large-scale integration. *Science* 298 (5593): 580–584. https://doi.org/10.1126/science.1076996.

26 Nan, L., Zhang, H., Weitz, D.A., and Shum, H.C. (2024). Development and future of droplet Microfluidics. *Lab Chip* 24 (5): 1135–1153. https://doi.org/10.1039/D3LC00729D.

27 Song, H., Tice, J.D., and Ismagilov, R.F. (2003). A microfluidic system for controlling reaction networks in time. *Angewandte Chemie International Edition* 42 (7): 768–772. https://doi.org/10.1002/anie.200390203.

28 Ding, Y., Howes, P.D., and deMello, A.J. (2020). Recent advances in droplet Microfluidics. *Anal. Chem.* 92 (1): 132–149. https://doi.org/10.1021/acs.analchem.9b05047.

29 Choi, K., Ng, A.H.C., Fobel, R., and Wheeler, A.R. (2012). Digital Microfluidics. *Annual Rev. Anal. Chem.* 5 (1): 413–440. https://doi.org/10.1146/annurev-anchem-062011-143028.

30 Martinez, A.W., Phillips, S.T., Butte, M.J., and Whitesides, G.M. (2007). Patterned paper as a platform for inexpensive, Low-volume, portable bioassays. *Angewandte Chemie International Edition* 46 (8): 1318–1320. https://doi.org/10.1002/anie.200603817.

31 Microfluidics, P. (2019). *Theory and Applications; Bhattacharya, S., Kumar, S., Agarwal, A. K., Eds.; Advanced Functional Materials and Sensors.* Singapore: Springer Singapore. https://doi.org/10.1007/978-981-15-0489-1.

32 Martinez, A.W., Phillips, S.T., and Whitesides, G.M. (2008). Three-dimensional microfluidic Devices fabricated in layered paper and tape. *Proceedings of the National Academy of Sciences* 105 (50): 19606–19611. https://doi.org/10.1073/pnas.0810903105.

33 Liu, B., Du, D., Hua, X. et al. (2014). Paper-based electrochemical biosensors: from test strips to paper-based Microfluidics. *Electroanalysis* 26 (6): 1214–1223. https://doi.org/10.1002/elan.201400036.

34 Hsu, C.-H., Chen, C., and Folch, A. (2004). "Microcanals" for micropipette access to single cells in microfluidic environments. *Lab Chip* 4 (5): 420–424. https://doi.org/10.1039/B404956J.

35 Juncker, D., Schmid, H., and Delamarche, E. (2005). Multipurpose Microfluidic Probe. *Nature Mater* 4 (8): 622–628. https://doi.org/10.1038/nmat1435.

36 Leung, C.M., de Haan, P., Ronaldson-Bouchard, K. et al. (2022). A guide to the organ-on-a-Chip. *Nat Rev Methods Primers* 2 (1): 1–29. https://doi.org/10.1038/s43586-022-00118-6.

37 Sin, A., Chin, K.C., Jamil, M.F. et al. (2004). The design and fabrication of three-chamber microscale cell culture Analog Devices with integrated dissolved oxygen sensors. *Biotechnology Progress* 20 (1): 338–345. https://doi.org/10.1021/bp034077d.

38 Sweeney, L.M., Shuler, M.L., Babish, J.G., and Ghanem, A. (1995). A cell culture analogue of rodent physiology: application to naphthalene toxicology. *Toxicology in vitro* 9 (3): 307–316. https://doi.org/10.1016/0887-2333(95)00007-U.

39 Huh, D., Matthews, B.D., Mammoto, A. et al. (2010). Reconstituting organ-level lung functions on a Chip. *Science* 328 (5986): 1662–1668. https://doi.org/10.1126/science.1188302.

40 Low, L.A., Mummery, C., Berridge, B.R. et al. (2021). Organs-on-chips: into the next decade. *Nat Rev Drug Discov* 20 (5): 345–361. https://doi.org/10.1038/s41573-020-0079-3.

41 Zhao, Y., Landau, S., Okhovatian, S. et al. (2024). Integrating Organoids and organ-on-a-Chip Devices. *Nat Rev Bioeng* 2 (7): 588–608. https://doi.org/10.1038/s44222-024-00207-z.

42 Xiao, S., Coppeta, J.R., Rogers, H.B. et al. (2017). A microfluidic culture model of the human reproductive tract and 28-day menstrual cycle. *Nat Commun* 8 (1): 14584. https://doi.org/10.1038/ncomms14584.

43 Zhang, B. and Radisic, M. (2017). Organ-on-a-Chip Devices advance to market. *Lab Chip* 17 (14): 2395–2420. https://doi.org/10.1039/C6LC01554A.

44 Wang, H., Ning, X., Zhao, F. et al. (2024). Human Organoids-on-chips for biomedical research and applications. *Theranostics* 14 (2): 788–818. https://doi.org/10.7150/thno.90492.

45 Lee-Montiel, F.T., George, S.M., Gough, A.H. et al. (2017). Control of oxygen tension recapitulates zone-specific functions in human liver microphysiology systems. *Exp Biol Med (Maywood)* 242 (16): 1617–1632. https://doi.org/10.1177/1535370217703978.

46 Hinson, J.T., Chopra, A., Nafissi, N. et al. (2015). Titin mutations in iPS cells define sarcomere insufficiency as a cause of dilated cardiomyopathy. *Science.* https://doi.org/10.1126/science.aaa5458.

47 Cruz, N.M., Song, X., Czerniecki, S.M. et al. (2017). Organoid Cystogenesis reveals a critical role of microenvironment in human polycystic kidney disease. *Nature Mater* 16 (11): 1112–1119. https://doi.org/10.1038/nmat4994.

48 Amartumur, S., Nguyen, H., Huynh, T. et al. (2024). *Neuropathogenesis-on-Chips for Neurodegenerative Diseases. Nat Commun* 15 (1): 2219. https://doi.org/10.1038/s41467-024-46554-8.

49 Yi, H.-G., Jeong, Y.H., Kim, Y. et al. (2019). A bioprinted human-Glioblastoma-on-a-Chip for the identification of patient-specific responses to Chemoradiotherapy. *Nat Biomed Eng* 3 (7): 509–519. https://doi.org/10.1038/s41551-019-0363-x.

50 Sontheimer-Phelps, A., Hassell, B.A., and Ingber, D.E. (2019). Modelling cancer in microfluidic human organs-on-chips. *Nat Rev Cancer* 19 (2): 65–81. https://doi.org/10.1038/s41568-018-0104-6.

51 Dabbagh Moghaddam, F., Dadgar, D., Esmaeili, Y. et al. (2023). Microfluidic platforms in diagnostic of ovarian cancer. *Environmental Research* 237: 117084. https://doi.org/10.1016/j.envres.2023.117084.

52 Wang, G., McCain, M.L., Yang, L. et al. (2014). Modeling the mitochondrial cardiomyopathy of Barth syndrome with induced pluripotent stem cell and heart-on-Chip Technologies. *Nat Med* 20 (6): 616–623. https://doi.org/10.1038/nm.3545.

53 Saiding, Q., Ma, J., Ke, C., and Cui, W. (2022). From "Organs on a Chip" to "patient on a Chip". *Innovation* 3 (5): https://doi.org/10.1016/j.xinn.2022.100282.

54 Li, Z.A. and Tuan, R.S. (2022). Towards establishing human body-on-a-Chip Systems. *Stem Cell Research & Therapy* 13 (1): 431. https://doi.org/10.1186/s13287-022-03130-5.

55 Ingber, D.E. (2022). Human organs-on-chips for disease modelling, drug development and personalized medicine. *Nat Rev Genet* 23 (8): 467–491. https://doi.org/10.1038/s41576-022-00466-9.

56 Hartings, M.R. and Ahmed, Z. (2019). Chemistry from 3D printed objects. *Nat Rev Chem* 3 (5): 305–314. https://doi.org/10.1038/s41570-019-0097-z.

57 He, Y., Wu, Y., Fu, J. et al. (2016). Developments of 3D printing Microfluidics and applications in chemistry and biology: A review. *Electroanalysis* 28 (8): 1658–1678. https://doi.org/10.1002/elan.201600043.

58 MacDonald, E. and Wicker, R. (2016). Multiprocess 3D printing for increasing component functionality. *Science* 353 (6307): aaf2093. https://doi.org/10.1126/science.aaf2093.

59 Rupal, B.S., Garcia, E.A., Ayranci, C., and Qureshi, A.J. (2019). 3D printed 3D-Microfluidics: recent developments and design challenges. *JID* 22 (1): 5–20. https://doi.org/10.3233/jid-2018-0001.

60 McDonald, J.C., Chabinyc, M.L., Metallo, S.J. et al. (2002). Prototyping of microfluidic Devices in poly(Dimethylsiloxane) using solid-object printing. *Anal. Chem.* 74 (7): 1537–1545. https://doi.org/10.1021/ac010938q.

61 Moore, J.L., McCuiston, A., Mittendorf, I. et al. (2011). Behavior of capillary valves in centrifugal microfluidic Devices prepared by three-dimensional printing. *Microfluid Nanofluid* 10 (4): 877–888. https://doi.org/10.1007/s10404-010-0721-1.

62 Au, A.K., Huynh, W., Horowitz, L.F., and Folch, A. (2016). 3D-printed Microfluidics. *Angew Chem Int Ed* 55 (12): 3862–3881. https://doi.org/10.1002/anie.201504382.

63 Chen, C., Fu, Y., Sparks, S.S. et al. (2024). 3D-printed flexible microfluidic Health monitor for in situ sweat analysis and biomarker detection. *ACS Sens.* 9 (6): 3212–3223. https://doi.org/10.1021/acssensors.4c00528.

64 Lee, K.G., Park, K.J., Seok, S. et al. (2014). 3D printed modules for integrated microfluidic Devices. *RSC Adv.* 4 (62): 32876–32880. https://doi.org/10.1039/C4RA05072J.

65 Sochol, R.D., Sweet, E., Glick, C.C. et al. (2016). 3D printed microfluidic circuitry via multijet-based additive manufacturing. *Lab Chip* 16 (4): 668–678. https://doi.org/10.1039/C5LC01389E.

66 Young, O.M., Xu, X., Sarker, S., and Sochol, R.D. (2024). Direct laser writing-enabled 3D printing strategies for microfluidic applications. *Lab Chip* 24 (9): 2371–2396. https://doi.org/10.1039/D3LC00743J.

67 Song, Y., Ghafari, Y., Asefnejad, A., and Toghraie, D. (2024). An overview of selective laser sintering 3D printing Technology for Biomedical and Sports Device Applications: processes, materials, and applications. *Optics & Laser Technology* 171: 110459. https://doi.org/10.1016/j.optlastec.2023.110459.

68 Lin, W.-Z.S., Bostic, W.K.V., and Malmstadt, N. (2024). 3D-printed microfluidic device for high-throughput production of lipid nanoparticles incorporating

SARS-CoV-2 spike protein mRNA. *Lab Chip* 24 (2): 162–170. https://doi.org/10.1039/D3LC00520H.

69 Duarte, L.C., Figueredo, F., Chagas, C.L.S. et al. (2024). A review of the recent achievements and future trends on 3D printed microfluidic Devices for bioanalytical applications. *Analytica Chimica Acta* 1299: 342429. https://doi.org/10.1016/j.aca.2024.342429.

70 Deka, M., Sinha, N., Das, R. et al. (2024). A review on the surface modification of materials for 3D-printed diagnostic Devices. *Anal. Methods* 16 (4): 485–495. https://doi.org/10.1039/D3AY01742G.

71 Su, R., Wang, F., and McAlpine, M.C. (2023). 3D printed Microfluidics: advances in strategies, integration, and applications. *Lab Chip* 23 (5): 1279–1299. https://doi.org/10.1039/D2LC01177H.

72 Griffin, K. and Pappas, D. (2023). 3D printed Microfluidics for bioanalysis: A review of recent advancements and applications. *TrAC Trends in Analytical Chemistry* 158: 116892. https://doi.org/10.1016/j.trac.2022.116892.

73 Xu, Y., Qi, F., Mao, H. et al. (2022). In-situ transfer vat Photopolymerization for transparent microfluidic device fabrication. *Nat Commun* 13 (1): 918. https://doi.org/10.1038/s41467-022-28579-z.

74 Gonzalez, G., Roppolo, I., Pirri, C.F., and Chiappone, A. (2022). Current and emerging trends in polymeric 3D printed microfluidic Devices. *Additive Manufacturing* 55: 102867. https://doi.org/10.1016/j.addma.2022.102867.

75 Prabhakar, P., Sen, R.K., Dwivedi, N. et al. (2021). 3D-printed Microfluidics and potential biomedical applications. *Front. Nanotechnol.* 3: 609355. https://doi.org/10.3389/fnano.2021.609355.

76 Nielsen, A.V., Beauchamp, M.J., Nordin, G.P., and Woolley, A.T. (2020). 3D printed microfluidics. *Annual Review of Analytical Chemistry* 13: 45–65. https://doi.org/10.1146/annurev-anchem-091619-102649.

77 Waheed, S., Cabot, J.M., Macdonald, N.P. et al. (2016). 3D printed microfluidic Devices: enablers and barriers. *Lab Chip* 16 (11): 1993–2013. https://doi.org/10.1039/C6LC00284F.

78 Bhattacharjee, N., Urrios, A., Kang, S., and Folch, A. (2016). The upcoming 3D-printing revolution in Microfluidics. *Lab Chip* 16 (10): 1720–1742. https://doi.org/10.1039/C6LC00163G.

79 Ho, C.M.B., Ng, S.H., Li, K.H.H., and Yoon, Y.-J. (2015). 3D printed Microfluidics for biological applications. *Lab Chip* 15 (18): 3627–3637. https://doi.org/10.1039/C5LC00685F.

80 Nelson, M.D., Ramkumar, N., and Gale, B.K. (2019). Flexible, transparent, Sub-100 μm microfluidic channels with fused deposition Modeling 3D-printed thermoplastic polyurethane. *J. Micromech. Microeng.* 29 (9): 095010. https://doi.org/10.1088/1361-6439/ab2f26.

81 Roy, N.K., Behera, D., Dibua, O.G. et al. (2019). A novel microscale selective laser sintering (μ-SLS) process for the fabrication of microelectronic parts. *Microsyst Nanoeng* 5 (1): 1–14. https://doi.org/10.1038/s41378-019-0116-8.

82 Castiaux, A.D., Pinger, C.W., Hayter, E.A. et al. (2019). PolyJet 3D-printed enclosed microfluidic channels without Photocurable supports. *Anal. Chem.* 91 (10): 6910–6917. https://doi.org/10.1021/acs.analchem.9b01302.

83 Goodling, A.E., Nagelberg, S., Kaehr, B. et al. (2019). Colouration by Total internal reflection and interference at microscale concave interfaces. *Nature* 566 (7745): 523–527. https://doi.org/10.1038/s41586-019-0946-4.

84 Kumar, A.S., Venkatesalu, S., Dilliyappan, S. et al. (2024). Microfluidics as diagnostic tools. *Clinica Chimica Acta* 556: 117841. https://doi.org/10.1016/j.cca.2024.117841.

85 Pacesa, M., Pelea, O., and Jinek, M. (2024). Past, present, and future of CRISPR genome editing technologies. *Cell* 187 (5): 1076–1100. https://doi.org/10.1016/j.cell.2024.01.042.

86 Wang, J.Y. and Doudna, J.A. (2023). CRISPR technology: A decade of genome editing is only the beginning. *Science* 379 (6629): eadd8643. https://doi.org/10.1126/science.add8643.

87 Li, P., Xiong, H., Yang, B. et al. (2022). Recent Progress in CRISPR-based microfluidic assays and applications. *TrAC Trends in Analytical Chemistry* 157: 116812. https://doi.org/10.1016/j.trac.2022.116812.

88 Avaro, A.S. and Santiago, J.G. (2023). A critical review of microfluidic systems for CRISPR assays. *Lab on a Chip* 23 (5): 938–963. https://doi.org/10.1039/D2LC00852A.

89 Welch, N.L., Zhu, M., Hua, C. et al. (2022). Multiplexed CRISPR-based microfluidic platform for clinical testing of respiratory viruses and identification of SARS-CoV-2 variants. *Nat Med* 28 (5): 1083–1094. https://doi.org/10.1038/s41591-022-01734-1.

90 Tang, Y., Qi, L., Liu, Y. et al. (2022). CLIPON: A CRISPR-enabled strategy that turns commercial pregnancy test strips into general point-of-need test Devices. *Angewandte Chemie International Edition* 61 (12): e202115907. https://doi.org/10.1002/anie.202115907.

91 Chen, Y., Zong, N., Ye, F. et al. (2022). Dual-CRISPR/Cas12a-assisted RT-RAA for ultrasensitive SARS-CoV-2 detection on automated centrifugal Microfluidics. *Anal. Chem.* 94 (27): 9603–9609. https://doi.org/10.1021/acs.analchem.2c00638.

92 Ackerman, C.M., Myhrvold, C., Thakku, S.G. et al. (2020). Massively multiplexed nucleic acid detection with Cas13. *Nature* 582 (7811): 277–282. https://doi.org/10.1038/s41586-020-2279-8.

93 Messeri, L. and Crockett, M.J. (2024). Artificial intelligence and illusions of understanding in scientific research. *Nature* 627 (8002): 49–58. https://doi.org/10.1038/s41586-024-07146-0.

94 Varghese, C., Harrison, E.M., O'Grady, G., and Topol, E.J. (2024). Artificial intelligence in surgery. *Nat Med* 30 (5): 1257–1268. https://doi.org/10.1038/s41591-024-02970-3.

95 El Jarroudi, M., Kouadio, L., Delfosse, P. et al. (2024). Leveraging edge artificial intelligence for sustainable agriculture. *Nat Sustain* 7 (7): 846–854. https://doi.org/10.1038/s41893-024-01352-4.

96 Bahadir, C.D., Omar, M., Rosenthal, J. et al. (2024). Artificial intelligence applications in histopathology. *Nat Rev Electr Eng* 1 (2): 93–108. https://doi.org/10.1038/s44287-023-00012-7.

97 Song, A.H., Jaume, G., Williamson, D.F.K. et al. (2023). Artificial intelligence for digital and computational pathology. *Nat Rev Bioeng* 1 (12): 930–949. https://doi.org/10.1038/s44222-023-00096-8.

98 Gregoire, J.M., Zhou, L., and Haber, J.A. (2023). Combinatorial synthesis for AI-driven materials discovery. *Nat. Synth* 2 (6): 493–504. https://doi.org/10.1038/s44160-023-00251-4.

99 Brownstein, J.S., Rader, B., Astley, C.M., and Tian, H. (2023). Advances in artificial intelligence for infectious-disease surveillance. *N Engl J Med* 388 (17): 1597–1607. https://doi.org/10.1056/NEJMra2119215.

100 Zhou, J., Dong, J., Hou, H. et al. (2024). High-throughput microfluidic systems accelerated by artificial intelligence for biomedical applications. *Lab Chip* 24 (5): 1307–1326. https://doi.org/10.1039/D3LC01012K.

101 Yellen, B.B., Zawistowski, J.S., Czech, E.A. et al. (2021). Massively parallel quantification of phenotypic heterogeneity in single-cell drug responses. *Science. Advances* 7 (38): eabf9840. https://doi.org/10.1126/sciadv.abf9840.

102 Soteriou, D., Kubánková, M., Schweitzer, C. et al. (2023). Rapid single-cell physical Phenotyping of mechanically dissociated tissue biopsies. *Nat. Biomed. Eng* 7 (11): 1392–1403. https://doi.org/10.1038/s41551-023-01015-3.

103 Su, Z., He, J., Zhou, P. et al. (2020). A high-throughput system combining microfluidic hydrogel droplets with deep learning for screening the Antisolvent-crystallization conditions of active pharmaceutical ingredients. *Lab Chip* 20 (11): 1907–1916. https://doi.org/10.1039/D0LC00153H.

104 Lashkaripour, A., Rodriguez, C., Mehdipour, N. et al. (2021). Machine learning enables design automation of microfluidic flow-focusing droplet generation. *Nat Commun* 12 (1): 1–14. https://doi.org/10.1038/s41467-020-20284-z.

105 Ren, Z., Zhang, Z., Wei, J. et al. (2022). Wavelength-multiplexed hook Nanoantennas for machine learning enabled mid-infrared spectroscopy. *Nat Commun* 13 (1): 3859. https://doi.org/10.1038/s41467-022-31520-z.

106 *COVID-19 cases | WHO COVID-19 dashboard*. datadot. https://data.who.int/dashboards/covid19/cases (accessed 10 September 2024).

107 Cui, J., Li, F., and Shi, Z.-L. (2019). Origin and evolution of pathogenic coronaviruses. *Nat Rev Microbiol* 17 (3): 181–192. https://doi.org/10.1038/s41579-018-0118-9.

108 Chan, J.F.-W., Yuan, S., Kok, K.-H. et al. (2020). A familial cluster of pneumonia associated with the 2019 novel coronavirus indicating person-to-person transmission: A study of a family cluster. *The Lancet* 395 (10223): 514–523. https://doi.org/10.1016/S0140-6736(20)30154-9.

109 Tegally, H., Wilkinson, E., Giovanetti, M. et al. (2021). Detection of a SARS-CoV-2 variant of concern in South Africa. *Nature* 592 (7854): 438–443. https://doi.org/10.1038/s41586-021-03402-9.

110 Mina, M.J. and Andersen, K.G. (2021). COVID-19 Testing: One Size Does Not Fit All. *Science* 371 (6525): 126–127. https://doi.org/10.1126/science.abe9187.

111 García-Fiñana, M. and Buchan, I.E. (2021). Rapid Antigen Testing in COVID-19 Responses. *Science* 372 (6542): 571–572. https://doi.org/10.1126/science.abi6680.

112 Tromberg, B.J., Schwetz, T.A., Pérez-Stable, E.J. et al. (2020). Rapid scaling up of Covid-19 diagnostic testing in the United States — the NIH RADx initiative. *N Engl J Med* 383 (11): 1071–1077. https://doi.org/10.1056/NEJMsr2022263.

113 Zhang, L., Wang, H., Yang, S. et al. (2024). High-throughput and integrated CRISPR/Cas12a-based molecular diagnosis using a deep learning enabled microfluidic system. *ACS Nano*. https://doi.org/10.1021/acsnano.4c05734.

114 Drain, P.K. (2022). Rapid diagnostic testing for SARS-CoV-2. *N Engl J Med* 386 (3): 264–272. https://doi.org/10.1056/NEJMcp2117115.

115 Greenleaf, M.N., Damhorst, G.L., Ku, D.N. et al. (2022). Designing for simplicity: lessons from Mesa biotech for microfluidic entrepreneurs and early-stage companies. *Lab Chip* 22 (8): 1469–1473. https://doi.org/10.1039/D2LC00081D.

116 Li, M., Ge, H., Sun, Z. et al. (2022). A loop-mediated isothermal amplification-enabled analytical assay for the detection of SARS-CoV-2: A review. *Front. Cell. Infect. Microbiol.* 12. https://doi.org/10.3389/fcimb.2022.1068015.

117 Zhang, H., Cao, L., Brodsky, J. et al. (2024). Quantitative or digital PCR? A comparative analysis for choosing the optimal one for biosensing applications. *TrAC Trends in Analytical Chemistry* 174: 117676. https://doi.org/10.1016/j.trac.2024.117676.

118 Smyrlaki, I., Ekman, M., Lentini, A. et al. (2020). Massive and rapid COVID-19 testing is feasible by extraction-free SARS-CoV-2 RT-PCR. *Nat Commun* 11 (1): 4812. https://doi.org/10.1038/s41467-020-18611-5.

119 Grossberg, A.N., Koza, L.A., Ledreux, A. et al. (2021). A multiplex Chemiluminescent immunoassay for serological profiling of COVID-19-positive symptomatic and asymptomatic patients. *Nat Commun* 12 (1): 740. https://doi.org/10.1038/s41467-021-21040-7.

120 *COVID-19 in vitro Diagnostic Medical Devices | COVID-19 in vitro Diagnostic Devices and Test Methods Database*. https://covid-19-diagnostics.jrc.ec.europa.eu/devices?loggedIn=0&device_id=&manufacturer=&text_name=&marking=Yes&method=&rapid_diag=1&target_type=&search_method=AND (accessed 15 September 2024).

121 Cai, Q., Mu, J., Lei, Y. et al. (2021). Simultaneous detection of the spike and Nucleocapsid proteins from SARS-CoV-2 based on ultrasensitive single molecule assays. *Anal Bioanal Chem* 413 (18): 4645–4654. https://doi.org/10.1007/s00216-021-03435-z.

122 Zhang, X., Zhu, D., Yang, X. et al. (2024). Nanozyme-enabled microfluidic biosensors: A promising tool for on-site food safety analysis. *Trends in Food Science & Technology* 148: 104486. https://doi.org/10.1016/j.tifs.2024.104486.

123 Wang, Y., Chen, J., Yang, Z. et al. (2024). Advances in nucleic acid assays for infectious disease: the role of microfluidic technology. *Molecules* 29 (11): 2417. https://doi.org/10.3390/molecules29112417.

124 Wang, N., Zhang, J., Xiao, B., and Chen, A. (2024). Microfluidic-assisted integrated nucleic acid test strips for POCT. *Talanta* 267: 125150. https://doi.org/10.1016/j.talanta.2023.125150.

125 Shanehband, N. and Naghib, S.M. (2024). Recent advances in Nano/Microfluidics-based cell isolation techniques for cancer diagnosis and treatments. *Biochimie* 220: 122–143. https://doi.org/10.1016/j.biochi.2024.01.001.

126 Chen, S., Qiao, Z., Niu, Y. et al. (2023). Wearable flexible microfluidic sensing technologies. *Nat Rev Bioeng* 1 (12): 950–971. https://doi.org/10.1038/s44222-023-00094-w.

127 Cavalera, S., Colitti, B., Rosati, S. et al. (2021). A multi-target lateral flow immunoassay enabling the specific and sensitive detection of Total antibodies to SARS COV-2. *Talanta* 223: 121737. https://doi.org/10.1016/j.talanta.2020.121737.

128 Zarean Mousaabadi, K., Talebi Vandishi, Z., Kermani, M. et al. (2023). Recent developments toward microfluidic point-of-care diagnostic sensors for viral infections. *TrAC Trends in Analytical Chemistry* 169: 117361. https://doi.org/10.1016/j.trac.2023.117361.

129 Gao, Z.Y. (2011). A research on early spread of traditional hand-made paper in East Asia countries. *Advanced Materials Research* 174: 537–540. https://doi.org/10.4028/www.scientific.net/AMR.174.537.

130 Lee, J.H. (ed.) (2021). Paper-based medical diagnostic devices: as a part of bioanalysis-advanced materials, methods, and devices. In: *Bioanalysis*, vol. 10. Singapore: Springer Singapore 10.1007/978-981-15-8723-8.

131 Budd, J., Miller, B.S., Weckman, N.E. et al. (2023). Lateral flow test engineering and lessons learned from COVID-19. *Nat Rev Bioeng* 1 (1): 13–31. https://doi.org/10.1038/s44222-022-00007-3.

132 US Food and Drug Administration. in vitro Diagnostics EUAs - antigen diagnostic tests for SARS-CoV-2. https://www.fda.gov/medical-devices/covid-19-emergency-use-authorizations-medical-devices/in-vitro-diagnostics-euas-antigen-diagnostic-tests-sars-cov-2 (accessed 26 September 2024).

133 Yakoh, A., Pimpitak, U., Rengpipat, S. et al. (2021). Paper-based electrochemical biosensor for diagnosing COVID-19: detection of SARS-CoV-2 antibodies and antigen. *Biosensors and Bioelectronics* 176: 112912. https://doi.org/10.1016/j.bios.2020.112912.

134 Bordbar, M.M., Samadinia, H., Sheini, A. et al. (2022). Non-invasive detection of COVID-19 using a microfluidic-based colorimetric sensor Array sensitive to urinary metabolites. *Microchim Acta* 189 (9): 316. https://doi.org/10.1007/s00604-022-05423-1.

135 Andreyev, B.; Moravick, K. E.; Ciopyk, B.; Briones, V.; Loney, G.; Zerda, A. D. L.; Ching, J.; Kelly, C.; Chu, S.; Swenson, D.; Huang, H. Devices and methods for molecular diagnostic testing. US9623415B2, 18 April 2017. https://patents.google.com/patent/US9623415B2en (accessed 24 September 2024).

136 Roswech, T.; Rothberg, J. M.; Bean, I.; Caravias, M. F.; Dyer, M. Rapid diagnostic test. US20220146508A1, 12 May, 2022. https://patents.google.com/patent/US20220146508A1en (accessed 23 September 2024).

137 Liu, D., Shen, H., Zhang, Y. et al. (2021). A microfluidic-integrated lateral flow Recombinase polymerase amplification (MI-IF-RPA) assay for rapid COVID-19 detection. *Lab Chip* 21 (10): 2019–2026. https://doi.org/10.1039/D0LC01222J.

138 Li, Z., Ding, X., Yin, K. et al. (2022). Instrument-free, CRISPR-based Diagnostics of SARS-CoV-2 using self-contained microfluidic system. *Biosensors and Bioelectronics* 199: 113865. https://doi.org/10.1016/j.bios.2021.113865.

139 Ehzari, H., Safari, M., Hallaj, R., and Amiri, M. (2024). Electrochemical microfluidic paper-based analytical device for label-free detection of SARS-CoV-2 antigen by LDH redox probe. *Microchemical Journal* 197: 109779. https://doi.org/10.1016/j.microc.2023.109779.

140 Zhang, T., Deng, R., Wang, Y. et al. (2022). A paper-based assay for the colorimetric detection of SARS-CoV-2 variants at single-nucleotide resolution. *Nat. Biomed. Eng* 6 (8): 957–967. https://doi.org/10.1038/s41551-022-00907-0.

141 Wang, D., He, S., Wang, X. et al. (2020). Rapid lateral flow immunoassay for the fluorescence detection of SARS-CoV-2 RNA. *Nat Biomed Eng* 4 (12): 1150–1158. https://doi.org/10.1038/s41551-020-00655-z.

142 Park, S., Su Jeon, C., Choi, N. et al. (2022). Sensitive and reproducible detection of SARS-CoV-2 using SERS-based microdroplet sensor. *Chemical Engineering Journal* 446: 137085. https://doi.org/10.1016/j.cej.2022.137085.

143 Zeng, Y., Gan, X., Xu, Z. et al. (2024). AIEgens-enhanced rapid sensitive Immunofluorescent assay for SARS-CoV-2 with digital Microfluidics. *Analytica Chimica Acta* 1298: 342398. https://doi.org/10.1016/j.aca.2024.342398.

144 Hou, Y., Chen, S., Zheng, Y. et al. (2023). Droplet-based digital PCR (ddPCR) and its applications. *TrAC Trends in Analytical Chemistry* 158: 116897. https://doi.org/10.1016/j.trac.2022.116897.

145 Chirizzi, D., Conte, M., Feltracco, M. et al. (2021). SARS-CoV-2 concentrations and virus-laden aerosol size distributions in outdoor air in north and south of Italy. *Environment International* 146: 106255. https://doi.org/10.1016/j.envint.2020.106255.

146 Graham, K.E., Loeb, S.K., Wolfe, M.K. et al. (2021). SARS-CoV-2 RNA in wastewater settled solids is associated with COVID-19 cases in a large urban Sewershed. *Environ. Sci. Technol.* 55 (1): 488–498. https://doi.org/10.1021/acs.est.0c06191.

147 Huang, H., Huang, K., Sun, Y. et al. (2022). A digital microfluidic RT-qPCR platform for multiple detections of respiratory pathogens. *Micromachines* 13 (10): 1650. https://doi.org/10.3390/mi13101650.

148 Oranger, A., Manzari, C., Chiara, M. et al. (2021). Accurate detection and quantification of SARS-CoV-2 genomic and subgenomic mRNAs by ddPCR and meta-Transcriptomics analysis. *Commun Biol* 4 (1): 1–10. https://doi.org/10.1038/s42003-021-02748-0.

149 Deiana, M., Mori, A., Piubelli, C. et al. (2020). Assessment of the direct quantitation of SARS-CoV-2 by droplet digital PCR. *Sci Rep* 10 (1): 18764. https://doi.org/10.1038/s41598-020-75958-x.

150 Ishak, A., AlRawashdeh, M.M., Esagian, S.M., and Nikas, I.P. (2021). Diagnostic, prognostic, and therapeutic value of droplet digital PCR (ddPCR) in

COVID-19 patients: A systematic review. *Journal of Clinical Medicine* 10 (23): 5712. https://doi.org/10.3390/jcm10235712.

151 Ho, K.-L., Liao, H.-Y., Liu, H.M. et al. (2022). Digital microfluidic qPCR cartridge for SARS-CoV-2 detection. *Micromachines* 13 (2): 196. https://doi.org/10.3390/mi13020196.

152 Yin, H., Wu, Z., Shi, N. et al. (2021). Ultrafast multiplexed detection of SARS-CoV-2 RNA using a rapid droplet digital PCR system. *Biosensors and Bioelectronics* 188: 113282. https://doi.org/10.1016/j.bios.2021.113282.

153 Choi, J.W., Seo, W.H., Kang, T. et al. (2023). Droplet digital Recombinase polymerase amplification for multiplexed detection of human coronavirus. *Lab Chip* 23 (10): 2389–2398. https://doi.org/10.1039/D3LC00025G.

154 Lyu, W., Zhang, J., Yu, Y. et al. (2021). Slip formation of a high-density droplet Array for nucleic acid quantification by digital LAMP with a random-access system. *Lab Chip* 21 (16): 3086–3093. https://doi.org/10.1039/D1LC00361E.

155 Shan, X., Gong, F., Yang, Y. et al. (2023). Nucleic acid amplification-free digital detection method for SARS-CoV-2 RNA based on droplet Microfluidics and CRISPR–Cas13a. *Anal. Chem.* 95 (45): 16489–16495. https://doi.org/10.1021/acs.analchem.3c02007.

156 Park, J.S., Hsieh, K., Chen, L. et al. (2021). Digital CRISPR/Cas-assisted assay for rapid and sensitive detection of SARS-CoV-2. *Advanced Science* 8 (5): 2003564. https://doi.org/10.1002/advs.202003564.

157 Yu, Z., Xu, L., Lyu, W., and Shen, F. (2022). Parallel multistep digital analysis SlipChip demonstrated with the quantification of nucleic acid by digital LAMP-CRISPR. *Lab Chip* 22 (16): 2954–2961. https://doi.org/10.1039/D2LC00284A.

158 Chen, L., Yadav, V., Zhang, C. et al. (2021). Elliptical pipette generated large microdroplets for POC visual ddPCR quantification of Low viral load. *Anal. Chem.* 93 (16): 6456–6462. https://doi.org/10.1021/acs.analchem.1c00192.

159 Land, K.J., Boeras, D.I., Chen, X.-S. et al. (2019). REASSURED Diagnostics to inform disease control strategies, strengthen Health systems and improve patient outcomes. *Nat Microbiol* 4 (1): 46–54. https://doi.org/10.1038/s41564-018-0295-3.

160 Nguyen, H.Q., Nguyen, V.D., Phan, V.M., and Seo, T.S. (2024). A novel point-of-care platform for rapid SARS-CoV-2 detection utilizing an all-in-one 3D-printed microfluidic cartridge and IoT technology. *Sensors and Actuators B: Chemical* 410: 135632. https://doi.org/10.1016/j.snb.2024.135632.

161 Li, S., Wan, C., Xiao, Y. et al. (2024). Multiple on-line active valves based centrifugal Microfluidics for dynamic solid-phase enrichment and purification of viral nucleic acid. *Lab Chip* 24 (12): 3158–3168. https://doi.org/10.1039/D4LC00074A.

162 Samper, I.C., Sánchez-Cano, A., Khamcharoen, W. et al. (2021). Electrochemical capillary-flow immunoassay for detecting anti-SARS-CoV-2 Nucleocapsid protein antibodies at the point of care. *ACS Sens.* 6 (11): 4067–4075. https://doi.org/10.1021/acssensors.1c01527.

163 Yin, K., Ding, X., Xu, Z. et al. (2021). Multiplexed colorimetric detection of SARS-CoV-2 and other pathogens in wastewater on a 3D printed integrated microfluidic Chip. *Sensors and Actuators B: Chemical* 344: 130242. https://doi.org/10.1016/j.snb.2021.130242.

164 Bae, B., Baek, Y., Yang, J. et al. (2023). Near-sensor computing-assisted simultaneous viral antigen and antibody detection via integrated label-free biosensors with Microfluidics. *InfoMat* 5 (10): e12471. https://doi.org/10.1002/inf2.12471.

165 Malic, L., Brassard, D., Fonte, D.D. et al. (2022). Automated sample-to-answer centrifugal microfluidic system for rapid molecular Diagnostics of SARS-CoV-2. *Lab Chip* 22 (17): 3157–3171. https://doi.org/10.1039/D2LC00242F.

166 Teixeira, W., Pallás-Tamarit, Y., Juste-Dolz, A. et al. (2022). An all-in-one point-of-care testing device for multiplexed detection of respiratory infections. *Biosensors and Bioelectronics* 213: 114454. https://doi.org/10.1016/j.bios.2022.114454.

167 Li, Y., Zhang, J., Chen, J. et al. (2024). Detection of SARS-CoV-2 based on artificial intelligence-assisted smartphone: A review. *Chinese Chemical Letters* 35 (7): 109220. https://doi.org/10.1016/j.cclet.2023.109220.

168 Kanchi, S., Sabela, M.I., Mdluli, P.S., and Inamuddin; Bisetty, K. (2018). Smartphone based bioanalytical and diagnosis applications: A review. *Biosensors and Bioelectronics* 102: 136–149. https://doi.org/10.1016/j.bios.2017.11.021.

169 Bayin, Q., Huang, L., Ren, C. et al. (2021). Anti-SARS-CoV-2 IgG and IgM detection with a GMR based LFIA system. *Talanta* 227: 122207. https://doi.org/10.1016/j.talanta.2021.122207.

170 Kim, S., Akarapipad, P., Nguyen, B.T. et al. (2022). Direct capture and smartphone quantification of airborne SARS-CoV-2 on a paper microfluidic Chip. *Biosensors and Bioelectronics* 200: 113912. https://doi.org/10.1016/j.bios.2021.113912.

171 Sun, F., Ganguli, A., Nguyen, J. et al. (2020). Smartphone-based multiplex 30-minute nucleic acid test of live virus from nasal swab extract. *Lab Chip* 20 (9): 1621–1627. https://doi.org/10.1039/D0LC00304B.

172 Yedire, S.G., Hosseini, I.I., Shieh, H. et al. (2023). Additive manufacturing leveraged microfluidic setup for sample to answer colorimetric detection of pathogens. *Lab Chip* 23 (18): 4134–4145. https://doi.org/10.1039/D3LC00429E.

173 Sun, H., Xiong, L., Huang, Y. et al. (2022). AI-aided on-Chip nucleic acid assay for smart diagnosis of infectious disease. *Fundamental Research* 2 (3): 476–486. https://doi.org/10.1016/j.fmre.2021.12.005.

174 Musso, D., Ko, A.I., and Baud, D. (2019). Zika virus infection — after the pandemic. *N Engl J Med* 381 (15): 1444–1457. https://doi.org/10.1056/NEJMra1808246.

175 Maslow, J.N. and Roberts, C.C. (2020). Zika virus: a brief history and review of its pathogenesis rediscovered, *Zika Virus: Methods and Protocols*. In: (ed. G. Kobinger and T. Racine), 1–8. New York, NY: Springer US. https://doi.org/10.1007/978-1-0716-0581-3_1.

176 Zika virus. https://www.who.int/news-room/fact-sheets/detail/zika-virus (accessed 5 October 2024).

177 Mitsakakis, K., D'Acremont, V., Hin, S. et al. (2018). Diagnostic tools for tackling febrile illness and enhancing patient management. *Microelectron Eng* 201: 26–59. https://doi.org/10.1016/j.mee.2018.10.001.

178 Liao, T., Wang, X., Donolato, M. et al. (2020). Evaluation of ViroTrack Sero Zika IgG/IgM, a new rapid and quantitative Zika serological diagnostic assay. *Diagnostics (Basel)* 10 (6): 372. https://doi.org/10.3390/diagnostics10060372.

179 Chong, Z.L., Soe, H.J., Ismail, A.A. et al. (2021). Evaluation of the diagnostic accuracy of a new biosensors-based rapid diagnostic test for the point-of-care diagnosis of previous and recent dengue infections in Malaysia. *Biosensors (Basel)* 11 (5): 129. https://doi.org/10.3390/bios11050129.

180 Varghese, J., De Silva, I., and Millar, D.S. (2023). *Latest Advances in Arbovirus Diagnostics. Microorganisms* 11 (5): 1159. https://doi.org/10.3390/microorganisms11051159.

181 Raafat, N., Blacksell, S.D., and Maude, R.J. (2019). A review of dengue Diagnostics and implications for surveillance and control. *Trans R Soc Trop Med Hyg* 113 (11): 653–660. https://doi.org/10.1093/trstmh/trz068.

182 Pal, S., Dauner, A.L., Mitra, I. et al. (2014). Evaluation of dengue NS1 antigen rapid tests and ELISA kits using clinical samples. *PLOS ONE* 9 (11): e113411. https://doi.org/10.1371/journal.pone.0113411.

183 Luo, R., Fongwen, N., Kelly-Cirino, C. et al. (2019). Rapid diagnostic tests for determining dengue Serostatus: A systematic review and key informant interviews. *Clin Microbiol Infect* 25 (6): 659–666. https://doi.org/10.1016/j.cmi.2019.01.002.

184 Seok, Y., Batule, B.S., and Kim, M.-G. (2020). Lab-on-paper for all-in-one molecular Diagnostics (LAMDA) of Zika, dengue, and Chikungunya virus from human serum. *Biosensors and Bioelectronics* 165: 112400. https://doi.org/10.1016/j.bios.2020.112400.

185 Kaarj, K., Akarapipad, P., and Yoon, J.-Y. (2018). Simpler, faster, and sensitive Zika virus assay using smartphone detection of loop-mediated isothermal amplification on paper microfluidic chips. *Sci Rep* 8 (1): 12438. https://doi.org/10.1038/s41598-018-30797-9.

186 Biswas, P., Mukunthan Sulochana, G.N., Banuprasad, T.N. et al. (2022). All-serotype dengue virus detection through multilayered origami-based paper/polymer Microfluidics. *ACS Sens.* 7 (12): 3720–3729. https://doi.org/10.1021/acssensors.2c01525.

187 Draz, M.S., Venkataramani, M., Lakshminarayanan, H. et al. (2018). Nanoparticle-enhanced electrical detection of Zika virus on paper microchips. *Nanoscale* 10 (25): 11841–11849. https://doi.org/10.1039/C8NR01646A.

188 Pardee, K., Green, A.A., Takahashi, M.K. et al. (2016). Rapid, Low-cost detection of Zika virus using programmable biomolecular components. *Cell* 165 (5): 1255–1266. https://doi.org/10.1016/j.cell.2016.04.059.

189 Hin, S., Lopez-Jimena, B., Bakheit, M. et al. (2021). Fully automated point-of-care differential diagnosis of acute febrile illness. *PLOS Neglected Tropical Diseases* 15 (2): e0009177. https://doi.org/10.1371/journal.pntd.0009177.

190 Zhang, Y., Guo, Y., Liu, G. et al. (2024). Portable all-in-one microfluidic system for CRISPR–Cas13a-based fully integrated multiplexed nucleic acid detection. *Lab Chip* 24 (14): 3367–3376. https://doi.org/10.1039/D4LC00326H.

191 Kasputis, T., Yeh, P.-C., Liu, L. et al. (2024). Development of a self-powered digital LAMP microfluidic Chip (SP-dChip) for the detection of emerging viruses. *Lab Chip* 24 (14): 3490–3497. https://doi.org/10.1039/D4LC00265B.

192 Zhu, X., Zhao, J., Hu, A. et al. (2020). A novel microfluidic device integrated with chitosan-modified capillaries for rapid ZIKV detection. *Micromachines* 11 (2): 186. https://doi.org/10.3390/mi11020186.

193 Jankelow, A.M., Lee, H., Wang, W. et al. (2022). Smartphone clip-on instrument and microfluidic processor for rapid sample-to-answer detection of Zika virus in whole blood using spatial RT-LAMP. *Analyst* 147 (17): 3838–3853. https://doi.org/10.1039/D2AN00438K.

194 Sharma, S., Kabir, M.A., and Asghar, W. (2020). Lab-on-a-chip Zika detection with reverse transcription loop-mediated isothermal amplification–based assay for point-of-care settings. *Archives of Pathology & Laboratory Medicine* 144 (11): 1335–1343. https://doi.org/10.5858/arpa.2019-0667-OA.

195 Powell, L., Wiederkehr, R.S., Damascus, P. et al. (2018). Rapid and sensitive detection of viral nucleic acids using silicon microchips. *Analyst* 143 (11): 2596–2603. https://doi.org/10.1039/C8AN00552D.

196 Hui, Y., Wu, Z., Qin, Z. et al. (2018). Micro-droplet digital polymerase chain reaction and real-time quantitative polymerase chain reaction technologies provide highly sensitive and accurate detection of Zika virus. *Virol. Sin.* 33 (3): 270–277. https://doi.org/10.1007/s12250-018-0037-y.

197 Ganguli, A., Ornob, A., Yu, H. et al. (2017). Hands-free smartphone-based Diagnostics for simultaneous detection of Zika, Chikungunya, and dengue at point-of-care. *Biomed Microdevices* 19 (4): 73. https://doi.org/10.1007/s10544-017-0209-9.

198 Jacob, S.T., Crozier, I., Fischer, W.A. et al. (2020). Ebola virus disease. *Nat Rev Dis Primers* 6 (1): 13. https://doi.org/10.1038/s41572-020-0147-3.

199 Khan, S., Muhammad, null, Rauf, A. et al. (2018). Comprehensive review on Ebola (EBOV) virus: future prospects. *Infect Disord Drug Targets* 18 (2): 96–104. https://doi.org/10.2174/1871526517666170817100828.

200 Feldmann, H., Sprecher, A., and Geisbert, T.W. (2020). Ebola. *New England Journal of Medicine* 382 (19): 1832–1842. https://doi.org/10.1056/NEJMra1901594.

201 Malvy, D., McElroy, A.K., de Clerck, H. et al. (2019). Ebola Virus Disease. *The Lancet* 393 (10174): 936–948. https://doi.org/10.1016/S0140-6736(18)33132-5.

202 Keita, A.K., Koundouno, F.R., Faye, M. et al. (2021). Resurgence of Ebola virus in 2021 in Guinea suggests a new paradigm for outbreaks. *Nature* 597 (7877): 539–543. https://doi.org/10.1038/s41586-021-03901-9.

203 null, null. (2022). Randomized trial of vaccines for Zaire Ebola virus disease. *New England Journal of Medicine* 387 (26): 2411–2424. https://doi.org/10.1056/NEJMoa2200072.

204 Bettini, A., Lapa, D., and Garbuglia, A.R. (2023). Diagnostics of Ebola virus. *Front. Public Health* 11. https://doi.org/10.3389/fpubh.2023.1123024.

205 Coarsey, C.T., Esiobu, N., Narayanan, R. et al. (2017). Strategies in Ebola virus disease (EVD) Diagnostics at the point of care. *Crit Rev Microbiol* 43 (6): 779–798. https://doi.org/10.1080/1040841X.2017.1313814.

206 Quick, J., Loman, N.J., Duraffour, S. et al. (2016). Real-time, portable genome sequencing for Ebola surveillance. *Nature* 530 (7589): 228–232. https://doi.org/10.1038/nature16996.

207 Nasrollahi, F., Haghniaz, R., Hosseini, V. et al. (2021). Micro and Nanoscale Technologies for Diagnosis of viral infections. *Small* 17 (45): 2100692. https://doi.org/10.1002/smll.202100692.

208 Gay-Andrieu, F., Magassouba, N., Picot, V. et al. (2017). Clinical evaluation of the BioFire FilmArray® BioThreat-E test for the diagnosis of Ebola virus disease in Guinea. *Journal of Clinical Virology* 92: 20–24. https://doi.org/10.1016/j.jcv.2017.04.015.

209 Pardee, K., Green, A.A., Ferrante, T. et al. (2014). Paper-based synthetic gene networks. *Cell* 159 (4): 940–954. https://doi.org/10.1016/j.cell.2014.10.004.

210 Nishat, S., Jafry, A.T., Martinez, A.W., and Awan, F.R. (2021). Paper-based Microfluidics: simplified fabrication and assay methods. *Sensors and Actuators B: Chemical* 336: 129681. https://doi.org/10.1016/j.snb.2021.129681.

211 Wonderly, B., Jones, S., Gatton, M.L. et al. (2019). Comparative performance of four rapid Ebola antigen-detection lateral flow immunoassays during the 2014-2016 Ebola epidemic in West Africa. *PLOS ONE* 14 (3): e0212113. https://doi.org/10.1371/journal.pone.0212113.

212 Mukadi-Bamuleka, D., Bulabula-Penge, J., De Weggheleire, A. et al. (2022). Field performance of three Ebola rapid diagnostic tests used during the 2018–20 outbreak in the eastern Democratic Republic of the Congo: A retrospective, multicentre observational study. *The Lancet Infectious Diseases* 22 (6): 891–900. https://doi.org/10.1016/S1473-3099(21)00675-7.

213 Makiala, S., Mukadi, D., De Weggheleire, A. et al. (2019). Clinical evaluation of QuickNaviTM-Ebola in the 2018 outbreak of Ebola virus disease in the Democratic Republic of the Congo. *Viruses* 11 (7): 589. https://doi.org/10.3390/v11070589.

214 Colavita, F., Biava, M., Mertens, P. et al. (2018). EBOLA Ag *K*-SeT rapid test: field evaluation in Sierra Leone. *Clinical Microbiology and Infection* 24 (6): 653–657. https://doi.org/10.1016/j.cmi.2017.10.019.

215 Magro, L., Jacquelin, B., Escadafal, C. et al. (2017). Paper-based RNA detection and multiplexed analysis for Ebola virus Diagnostics. *Sci Rep* 7 (1): 1347. https://doi.org/10.1038/s41598-017-00758-9.

216 Murray, L.P., Govindan, R., Mora, A.C. et al. (2021). Antibody affinity as a driver of signal generation in a paper-based immunoassay for Ebola virus surveillance. *Anal Bioanal Chem* 413 (14): 3695–3706. https://doi.org/10.1007/s00216-021-03317-4.

217 Couturier, C., Wada, A., Louis, K. et al. (2020). Characterization and analytical validation of a new antigenic rapid diagnostic test for Ebola virus disease detection. *PLOS Neglected Tropical Diseases* 14 (1): e0007965. https://doi.org/10.1371/journal.pntd.0007965.

218 Ge, X., Asiri, A.M., Du, D. et al. (2014). Nanomaterial-enhanced paper-based biosensors. *TrAC Trends in Analytical Chemistry* 58: 31–39. https://doi.org/10.1016/j.trac.2014.03.008.

219 Zhao, F., Wu, W., Zhao, M. et al. (2023). Enzyme-like Nanomaterials-integrated microfluidic Technology for Bioanalysis. *TrAC Trends in Analytical Chemistry* 158: 116833. https://doi.org/10.1016/j.trac.2022.116833.

220 Hu, J., Jiang, Y.-Z., Wu, L.-L. et al. (2017). Dual-signal readout Nanospheres for rapid point-of-care detection of Ebola virus glycoprotein. *Anal. Chem.* 89 (24): 13105–13111. https://doi.org/10.1021/acs.analchem.7b02222.

221 Duan, D., Fan, K., Zhang, D. et al. (2015). Nanozyme-strip for rapid local diagnosis of Ebola. *Biosensors and Bioelectronics* 74: 134–141. https://doi.org/10.1016/j.bios.2015.05.025.

222 Ciftci, S., Neumann, F., Abdurahman, S. et al. (2020). Digital rolling circle amplification–based detection of ebola and other tropical viruses. *The Journal of Molecular Diagnostics* 22 (2): 272–283. https://doi.org/10.1016/j.jmoldx.2019.10.014.

223 Fernández-Carballo, B.L., McBeth, C., McGuiness, I. et al. (2018). Continuous-flow, microfluidic, qRT-PCR system for RNA virus detection. *Anal Bioanal Chem* 410 (1): 33–43. https://doi.org/10.1007/s00216-017-0689-8.

224 Lin, X., Jin, X., Xu, B. et al. (2019). Fast and parallel detection of four Ebola virus species on a microfluidic-Chip-based portable reverse transcription loop-mediated isothermal amplification system. *Micromachines* 10 (11): 777. https://doi.org/10.3390/mi10110777.

225 Qin, P., Park, M., Alfson, K.J. et al. (2019). Rapid and fully microfluidic Ebola virus detection with CRISPR-Cas13a. *ACS Sens.* 4 (4): 1048–1054. https://doi.org/10.1021/acssensors.9b00239.

226 Xu, X., Cai, L., Liang, S. et al. (2023). Digital Microfluidics for biological analysis and applications. *Lab Chip* 23 (5): 1169–1191. https://doi.org/10.1039/D2LC00756H.

227 Piraino, F., Volpetti, F., Watson, C., and Maerkl, S.J. (2016). A digital–analog microfluidic platform for patient-centric multiplexed biomarker diagnostics of ultralow volume samples. *ACS Nano* 10 (1): 1699–1710. https://doi.org/10.1021/acsnano.5b07939.

228 Na, W., Nam, D., Lee, H., and Shin, S. (2018). Rapid molecular diagnosis of infectious viruses in Microfluidics using DNA hydrogel formation. *Biosensors and Bioelectronics* 108: 9–13. https://doi.org/10.1016/j.bios.2018.02.040.

229 Yager, P., Edwards, T., Fu, E. et al. (2006). Microfluidic diagnostic Technologies for Global Public Health. *Nature* 442 (7101): 412–418. https://doi.org/10.1038/nature05064.

230 Sackmann, E.K., Fulton, A.L., and Beebe, D.J. (2014). The present and future role of Microfluidics in biomedical research. *Nature* 507 (7491): 181–189. https://doi.org/10.1038/nature13118.

231 Lehnert, T., Gijs, M., and M. A. (2024). Microfluidic Systems for Infectious Disease Diagnostics. *Lab on a Chip* 24 (5): 1441–1493. https://doi.org/10.1039/D4LC00117F.

232 Ayuso, J.M., Virumbrales-Muñoz, M., Lang, J.M., and Beebe, D.J. (2022). A role for microfluidic systems in precision medicine. *Nat Commun* 13 (1): 3086. https://doi.org/10.1038/s41467-022-30384-7.

233 Chauhan, R., Minocha, N., Coliaie, P. et al. (2023). Emerging microfluidic platforms for crystallization process development. *Chemical Engineering Research and Design* 197: 908–930. https://doi.org/10.1016/j.cherd.2023.08.021.

234 Reyes, D.R., Heeren, H.v., Guha, S. et al. (2021). Accelerating innovation and commercialization through standardization of microfluidic-based medical Devices. *Lab Chip* 21 (1): 9–21. https://doi.org/10.1039/D0LC00963F.

235 Taly, V., Viovy, J.-L., and Descroix, S. (ed.) (2017). Microchip diagnostics: methods and protocols. In: *Methods in Molecular Biology*, vol. 1547. New York, NY: Springer New York. https://doi.org/10.1007/978-1-4939-6734-6.

236 Verma, N. and Pandya, A. (2022). Challenges and opportunities in micro/Nanofluidic and lab-on-a-Chip. In: *Progress in Molecular Biology and Translational Science*, vol. 186, 289–302. Elsevier. https://doi.org/10.1016/bs.pmbts.2021.07.016.

237 Regehr, K.J., Domenech, M., Koepsel, J.T. et al. (2009). Biological implications of Polydimethylsiloxane-based microfluidic cell culture. *Lab Chip* 9 (15): 2132–2139. https://doi.org/10.1039/B903043C.

238 Paratore, F., Bacheva, V., Bercovici, M., and Kaigala, G.V. (2021). Reconfigurable Microfluidics. *Nat Rev Chem* 6 (1): 70–80. https://doi.org/10.1038/s41570-021-00343-9.

239 Chin, C.D., Linder, V., and Sia, S.K. (2012). Commercialization of microfluidic point-of-care diagnostic Devices. *Lab Chip* 12 (12): 2118–2134. https://doi.org/10.1039/C2LC21204H.

240 Hajam, M.I. and Khan, M.M. (2024). Microfluidics: A concise review of the history, principles, design, applications, and future outlook. *Biomater. Sci.* 12 (2): 218–251. https://doi.org/10.1039/D3BM01463K.

241 Nielsen, J.B., Hanson, R.L., Almughamsi, H.M. et al. (2020). Microfluidics: innovations in materials and their fabrication and functionalization. *Anal Chem* 92 (1): 150–168. https://doi.org/10.1021/acs.analchem.9b04986.

242 Rasheed, S., Kanwal, T., Ahmad, N. et al. (2024). Advances and challenges in portable optical biosensors for onsite detection and point-of-care diagnostics. *TrAC Trends in Analytical Chemistry* 173: 117640. https://doi.org/10.1016/j.trac.2024.117640.

243 Lu, Y., Zhang, J., Lu, X., and Liu, Q. (2024). Isothermal nucleic acid amplification based microfluidic "lab-on-a-Chip" for the detection of pathogenic bacteria and viruses in Agri-foods. *Trends in Food Science & Technology* 148: 104482. https://doi.org/10.1016/j.tifs.2024.104482.

244 Xue, X., Kim, Y.S., Ponce-Arias, A.-I. et al. (2024). A patterned human neural tube model using microfluidic gradients. *Nature* 628 (8007): 391–399. https://doi.org/10.1038/s41586-024-07204-7.

245 Gurkan, U.A., Wood, D.K., Carranza, D. et al. (2024). Next generation microfluidics: fulfilling the promise of lab-on-a-Chip Technologies. *Lab on a Chip* 24 (7): 1867–1874. https://doi.org/10.1039/D3LC00796K.

9

The Future of Detection

Instrumentation and Techniques on the Horizon

Tingting Zheng, Limin Zhang, and Yang Tian

East China Normal University, School of Chemistry and Molecular Engineering, Department of Chemistry, Shanghai Key Laboratory of Green Chemistry and Chemical Processes, Dongchuan Road 500, 200241 Shanghai, China

CHAPTER MENU

9.1 The Electrifying World of Detection: New-Age Electrochemical Methods

9.1.1 Highly Selective Electrochemical Strategies Using Double-Recognition Molecular Probes

Electrochemical techniques are a classic method for analyzing chemical signals, and they have been applied across various fields to monitor the concentrations of chemical substances, thanks to their advantages of low cost, real-time monitoring, and high spatiotemporal resolution. This method is especially suitable for applications requiring precise and timely data, providing reliable quantitative analysis support. These advantages make electrochemical techniques a crucial tool in areas such as environmental monitoring, biomedical research, and industrial process control [1]. The primary challenge for electrochemical sensors is enhancing their selectivity to meet the demanding requirements of measurements in complex environments. This involves improving the sensor's ability to distinguish the target analyte from coexisting substances, ensuring accurate and reliable detection in complex systems. Such enhancement is crucial for optimizing sensor performance in real-world applications. Benefiting from the natural specific recognition, enzyme-based sensors have been one of the most popular approaches to enhance the selectivity of electrochemical sensors. However, the restricted number of available enzymes greatly limits the potential and performance of these electrochemical sensors.

Diagnostic Techniques in Public Health: Analytical and Chemometric Approaches in Disease Detection and Management, First Edition. Edited by Jinghong Li and Yan Du.

To tackle this challenge, a dual-recognition strategy has been proposed to realize the selective detection through designing and synthesizing the molecular probe with the specific ligands for target analytes, and the response group with electrochemical activity. Utilizing this dual-recognition strategy, several electrochemical probes have been synthesized to specifically detect a range of targets including metal ions, amino acids, reactive oxygen species (ROS) (Figure 9.1a), and sulfides [2–7]. For instance, Chai et al. synthesized a copper-free derivative of bovine erythrocyte copper-zinc superoxide dismutase (E_2Zn_2SOD) capable of the selective recognition of Cu^{2+} ions. This approach leverages the specific interaction between E_2Zn_2SOD and Cu^{2+}, facilitating precise chemical detection [6]. Faradic current generated by Cu^{2+} captured by E_2Zn_2SOD located at a distinguished potential, which established the basis of electrochemical analysis for Cu^{2+}. Given the importance of the

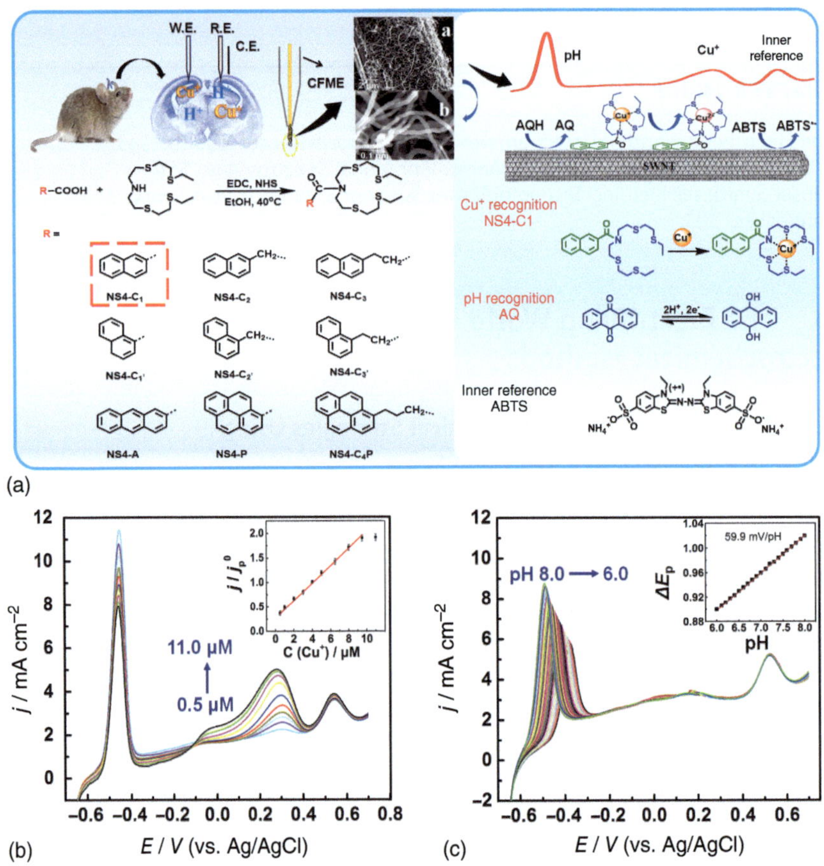

Figure 9.1 (a) Schematic diagram of an electrochemical sensor for simultaneous detection of Cu^+ and pH. (b) Electrochemical response obtained at CFME/SWCNT/AQ + NS4-C1 + ABTS for different concentrations of Cu^+, and the corresponding linear plot between Cu^+ concentration and current density. (c) Electrochemical responses at CFME/SWCNT/AQ + NS4-C1 + ABTS for different pH and the corresponding plot between ΔE_p and pH. *Source:* [2] / with permission of John Wiley & Sons.

redox couple Cu^{+}/Cu^{2+} in biological systems, Liu et al. synthesized nine *N,N*-bis (2-[2-(ethylthio)ethyl])-based derivatives as a Cu^{+} recognition probe. Their aim was to explore the roles of Cu^{+} in brain functions (Figure 9.1a) [2]. In a comprehensive optimization for the electrochemical detection of Cu^{+}, the parameters of the alkyl chain length, polyaromatic structure, and substitution group site of NS4 were systematically evaluated. Among the tested derivatives, NS4-C1 emerged as the optimal candidate for Cu^{+} detection due to its most negative potential and the highest current density, making it superior for this application. Meanwhile, a selective pH recognition of 9, 10-anthraquinone was co-assembled at the electrode surface. This single biosensor, featuring dual-signal outputs, can simultaneously determine Cu^{+} amounts from current signals and pH values from half-wave potentials (Figure 9.1b–c).

Furthermore, an alternative double recognition-sensing strategy was constructed to obtain the variation of polysulfide (H_2S_n) without electrochemical activity through rationally transferring specific chemical interaction into a selective faradic current signal [8]. Dong et al. first designed and synthesized a specific recognition molecule for H_2S_n, 3,4-bis([2-fluoro-5-nitrobenzoyl]oxy)-benzoic acid (M_{PS-1}) (Figure 9.2) [9, 10]. To differentiate H_2S_n from H_2S and to elucidate their functional differences in physiological processes, they additionally developed a recognition probe for H_2S. Upon interaction with their targets, M_{PS-1} and M_{HS-1} generate a quinone derivative via a specific chemical reaction, producing two distinct anodic

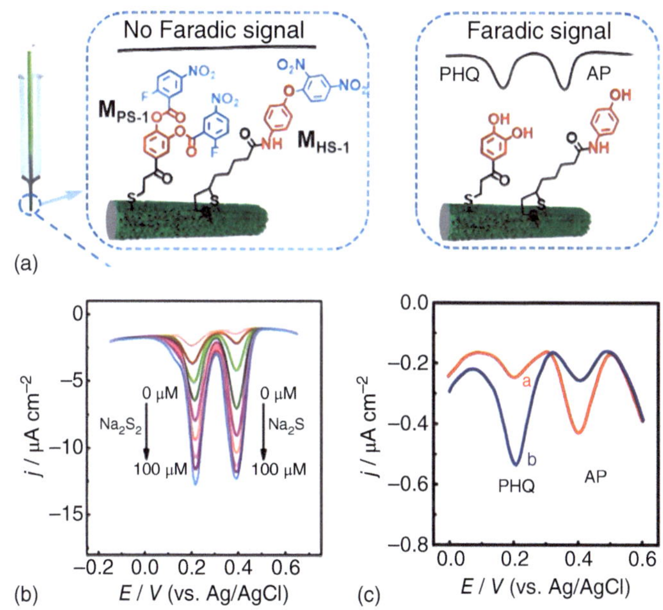

Figure 9.2 (a) Schematic diagram for the double-recognition strategy for selective detection of H_2S and H_2S_n. (b) Electrochemical responses obtained at CFME/mAu/M_{PS-1} + M_{HS-1} for simultaneous determination of Na_2S_2 and Na_2S. (c) Electrochemical responses at CFME/mAu/M_{PS-1} + M_{HS-1} in the mice brains before (a) and after (b) MCAO 2.0h. *Source:* Ref. [9] / with permission of John Wiley & Sons.

peaks. When these probes are assembled on a mesoporous gold membrane-coated carbon fiber microelectrode (CFME), the resulting microsensor exhibits high selectivity for H_2S and H_2S_n. With this advanced electrochemical sensor, researchers can accurately monitor changes in H_2Sn and H_2S levels within mouse brains, providing valuable insights into their physiological roles.

Probe reversibility is crucial and necessary for recording continuous signal fluctuation. To further develop an electrochemical method which can monitor the reversible variation of Ca^{2+} concentrations with high selectivity, Liu et al. confined three Ca^{2+} ligands with different affinities for specific recognition at the electrode surface (Figure 9.3a) [11]. Through experiments assessing the responses of three types of molecules to Ca^{2+} in artificial cerebrospinal fluid (aCSF), it was observed that the electrode modified by MBAPTA (Figure 9.3b) not only possesses the highest affinity for Ca^{2+}, but also has the highest sensitivity. However, this electrode is also demonstrated to be the poorest in reversibility. On the other hand, the electrode modified with METH showed superior reversibility, making it more suitable for repeated measurements (Figure 9.3b). These observations demonstrated that the relatively weak affinity of METH was beneficial for reversible Ca^{2+} identification and provided high selectivity for Ca^{2+} quantification. To further investigate these dynamics, they constructed an array of seven microelectrodes integrated into microfiber arrays, enabling continuous tracking and quantification of extracellular Ca^{2+} changes during cerebral ischemia–reperfusion in seven brain regions (Figure 9.3c). The results showed a progressive decrease in extracellular Ca^{2+} levels from the superficial to deeper brain areas during ischemia (Figure 9.3d).

9.1.2 More Accurate Electrochemical Sensors with Self-Calibration Ability

Due to the differences in environments between in vitro conditions and complex samples, another challenge in developing practical electrochemical devices for analysis is the quantitative determination of target molecules in real-world settings. Electrochemical sensors frequently experience baseline drift when used in complex biological samples. In order to avoid this problem, Li et al. reported a dual-reporter approach to realize the baseline drift correction during the practical measurement using the aptamer-based electrochemical sensors [12]. The sensor incorporates two redox reporters on the aptamer: one serves as the target-responsive sensor, while the other, which reports at a distinct, nonoverlapping redox potential, acts as a drift-correcting reference. Taking the difference in their relative signals largely eliminates the drift observed in these sensors when used with flowing, undiluted whole blood. This approach reduces drift from up to 50% to less than 2% over several hours via continuous operation under these complex conditions.

This self-calibration strategy is demonstrated to still be efficient for pH determination in a complex brain environment. The predominant method for pH determination is dependent on the Nernst equation, which defines a linear relationship between the measured potential and pH. In contrast to the low solution resistance in 0.1 M dilute electrolyte solutions (approximately 100 Ω), the resistance

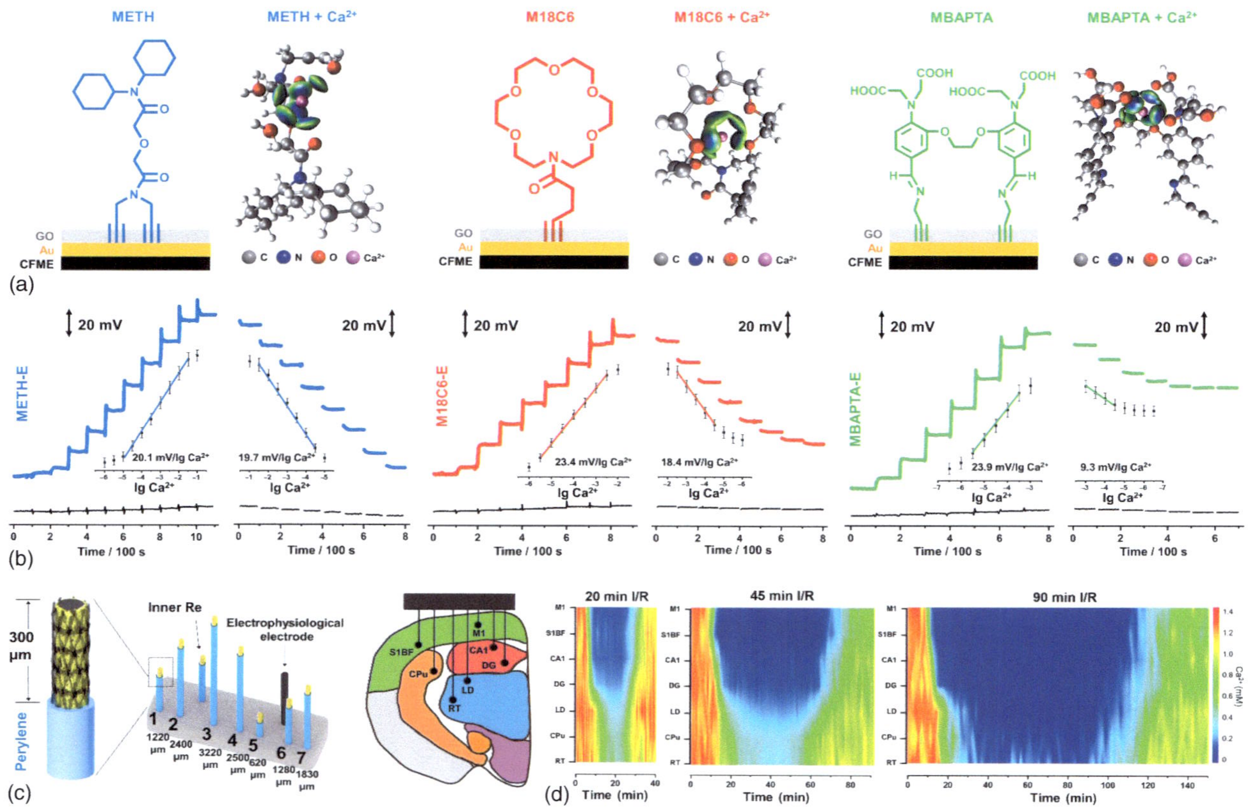

Figure 9.3 (a) Molecular structures of Ca^{2+} probes and DFT simulations for their interaction with Ca^{2+}. (b) Open-circuit potentiometric curves obtained from METH-E, M18C6-E, and MBAPTA-E with repeatable testing of Ca^{2+}. (c) Schematic diagram for the METH-E microarray. (d) Heat maps tested at METH-E microarray in recording the extracellular Ca^{2+} concentration changes in seven brain regions upon ischemia and reperfusion for different times. *Source:* Ref. [11] / with permission of John Wiley & Sons.

in cerebrospinal fluid can be substantially higher, reaching up to 3000 Ω [13]. This increased resistance leads to a significant IR drop for microelectrodes. For example, a microelectrode generating a current signal of ~1 μA will experience an IR drop of approximately 3–5 mV in the brain environment. This effect must be considered to ensure accurate pH measurements under these challenging conditions. Even a minor pH variation of 0.2–0.5 pH units in living organisms can signify various severe cellular issues and diseases. Consequently, pH measurements relying solely on potential variations without built-in corrections can cause errors of ~20–50% in brain tissue. To overcome this difficulty, a two-channel electrochemical pH sensor was developed through decorating 1,2-naphthoquinone (1,2-NQ) as the specific recognition for pH, and the FcHT (CFME/Au/1,2-NQ + FcHT) as an inner reference at one electrode (Figure 9.2b) [14, 15]. The pH calibration curve data were derived from the difference in half-wave potential ($E_{1/2}$) between the response and reference channels ($\Delta E_{1/2}$). Since the $\Delta E_{1/2}$ value is almost not influenced by the potential of the outer reference electrode and potential shifts induced by IR drop, this method obviously improves the accuracy of in vivo pH monitoring. Using this accurate sensing model, the pH variations were successfully detected in the rat brain, followed by ischemia at different times.

9.1.3 Highly Stable Assembled Surfaces with Anti-Biofouling for Long-Term Measurement

Stable molecular assembly on the electrode surface is critical for ensuring reliable and long-term performance of electrochemical sensors in practical systems. Thiol-functionalized gold interfaces represent one of the most prevalent models for achieving such stable assemblies and have been extensively studied in electrochemical sensor applications. Despite their popularity, a significant drawback of thiol ligands is their tendency to be displaced by abundant biological thiols, such as glutathione. This displacement can result in detection distortions and an unstable assembly of molecular probes on the gold surface, thereby compromising the accuracy and reliability of the sensor.

To enhance the stability of molecular assemblies on electrode surfaces, bidentate thiols were designed as surface linker groups, leading to the development of a novel, stable electrochemical sensor for H_2S_n (Figure 9.4A–B) [16]. The use of molecules with bidentate thiols significantly improved the resistance to interference from glutathione (GSH). Even after extended periods of measurement in a 5 mM GSH solution, these molecules maintained their stability and functionality, demonstrating their robustness against biological thiols. Zhang et al. made a systematic study of electrochemical sensors assembled using Au—S, Au—Se, and Au—C≡C bonds (Figure 9.4C) [17], focusing on key parameters such as the formal potential ($E^{0\prime}$), peak current density (jp), and electron transfer rate constant (ks). Their results indicated that the Au—C≡C-based surface offered superior performance, with the highest bond energy, the most negative Gibbs free energy, and the greatest surface assembly efficiency (Figure 9.4D–E). These findings highlight the advantages of using Au—C≡C bonds for creating robust and efficient biosensing platforms.

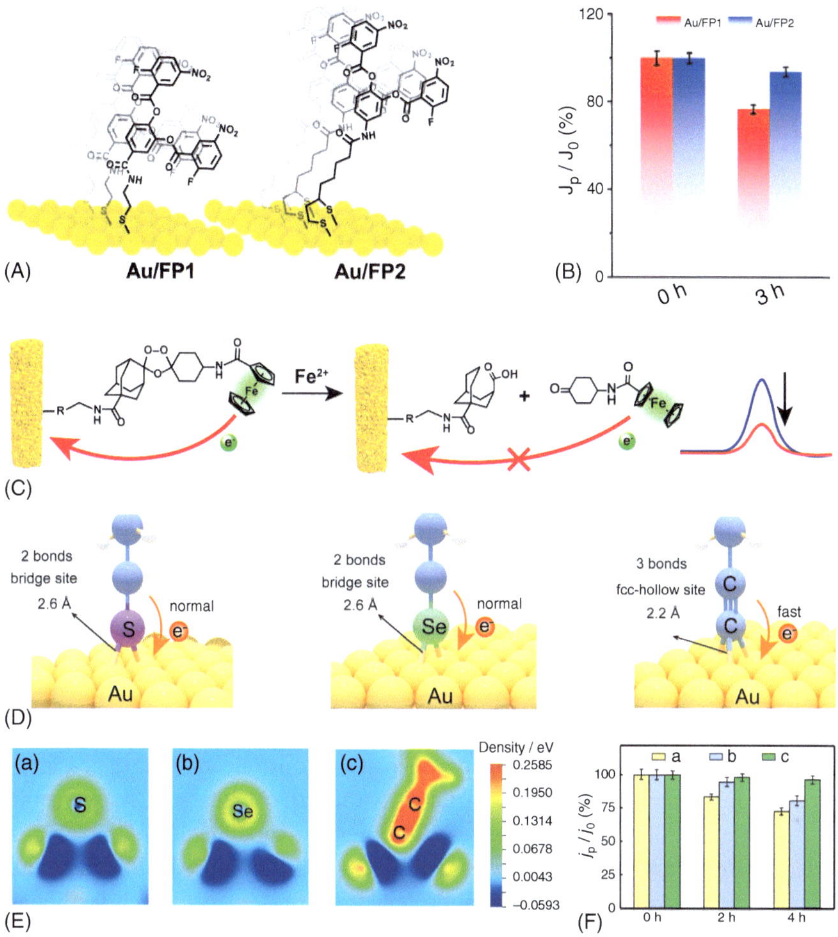

Figure 9.4 (A) Scheme of an electrochemical sensor for H_2S_n assembled by Au—S bond (Au/FP1) and bidentate thiols (Au/FP2) on Au surfaces. (B) Interferences from the different electrodes in 5 mM GSH solution. *Source:* Ref. [16] / with permission of American Chemical Society. (C) Schematic diagram for the working principle of Fe^{2+} electrochemical sensor. (D) Schematic diagram for the optimized gold surfaces assembled via Au—S, Au—Se, and Au—C≡C bonds. (E) Stability of the ligands assembled via Au—S, Au—Se, and Au—C≡C bonds at the electrode surface by immersing in 5 mM GSH solution for different times. (F) DFT calculation of differential charge density map obtained from Au—S (a), Au—Se (b), and Au—C≡C (c) bonds modified at the electrode surface. *Source:* Ref. [17] / with permission of John Wiley & Sons.

Following a four-hour immersion in a 5 mM GSH solution, the developed biosensors assembled using three different bonds exhibited varying degrees of signal reduction. Specifically, the biosensor utilizing Au—C≡C bonding showed only *a* ~5.3% reduction in signal, while those using Au—S and Au—Se bonds experienced reductions of ~38.3% and ~19.8%, respectively (Figure 9.4F). Subsequently, six-channel microelectrode arrays were fabricated to monitor Fe^{2+} changes with long-term stability in mouse models of Alzheimer's disease. It was discovered that the main pathways of Fe^{2+} entry into cells differed significantly between the hippocampal region and the cortex and striatum.

Recently, the self-assembly interface of *N*-heterocyclic carbene (NHC) on gold has attracted more and more attention and has become a potential self-assembly mode. Unlike typical carbenes with high reactivity but limited stability, NHCs generally have two heteroatoms adjacent to the carbene carbon to afford the electron-rich π-system(N-C=C-N) and a σ-electronegativity influence, which enhances the strength of the NHC-gold bond. Considering that the strength of Au—NHC bond (160 kJ/mol) has been measured to be stronger than the Au—S bond (125 kJ/mol), NHC is expected to be an extremely valuable ligand for protecting and functionalizing gold and other metal surfaces. Crudden's group reported a straightforward deposition method to construct self-assembled NHC monolayers on gold, and deposited NHC on a clean gold SPR chip, which provides more stable protein sensing results versus thiol analogs under extremes of pH (5–10) and high thermal stability, surviving heating at 65 °C in air for 24 hours. Other self-assembly interfaces of NHC on gold are also emerging to meet the needs of more fields. However, the application scope of NHC-based interfaces in fields other than materials science and homogeneous catalysis remains limited and urgently needs to be broadened. The application of NHC on various kinds of harsh and extreme conditions represents a unique synthesis challenge that, if overcome, could significantly expand the usefulness of NHC-based interfaces.

For long-term measurements with high stability in vivo, another key challenge is that the proteins coexisting in vivo are prone to adsorb onto electrode surfaces, leading to a great loss of electric signals, particularly the current response. To overcome this challenge, one of the main strategies is the chemical modification of the sensing interface using hydrophilic coatings. A notable approach proposed by Zhang et al. employs mussel adhesive protein-inspired polydopamine (PDA), which facilitates the interfacial attachment of zwitterionic sulfobetaine methacrylate molecules through nucleophilic addition. PDA, acting as a strong adhesive primer, can be securely anchored to various substrates, creating a robust and antifouling electrode surface [18].

Recently, Liu et al. developed a new microelectrode array with good biofouling resistance [11]. To achieve this, they electrodeposited graphene oxide microstrips onto the surface of a gold particle-decorated CFME (EDACE) to prevent the adsorption of protein (Figure 9.5a–b). The graphene oxide surface with various oxygen-containing functional groups exhibited excellent hydrophilicity and carried negative charges (Figure 9.5c–d), which are critical for its anti-biofouling properties (Figure 9.5e). Further modification of the electrode surface with a Ca^{2+} ligand via an Au—C≡C bond resulted in a biosensor that retained approximately 92% sensitivity over 60 consecutive days of continuous in vivo measurements (Figure 9.5f). This highlights the long-term stability of EDACE for real-time in vivo Ca^{2+} monitoring. Using this advanced tool, the researchers were the first to observe slow, progressive increases in extracellular Ca^{2+} concentrations in the days following cerebral hemorrhage.

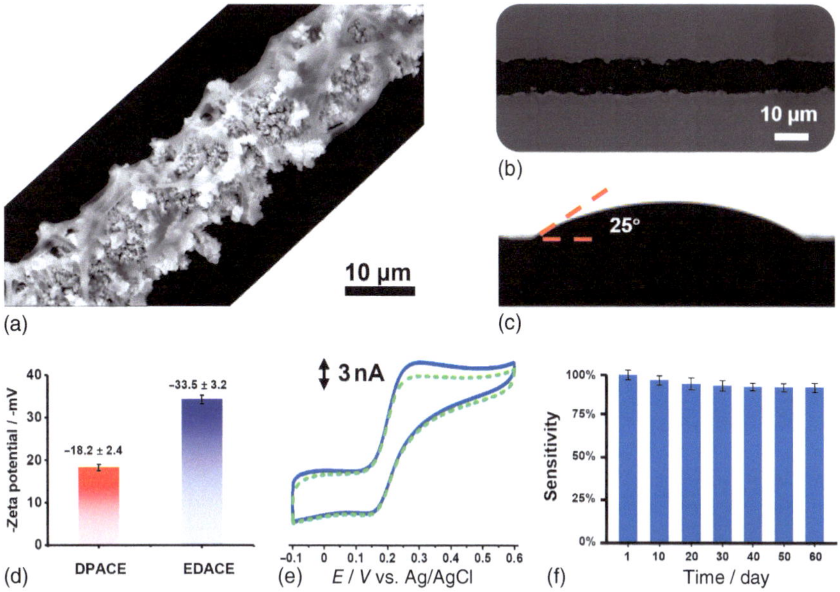

Figure 9.5 (a) SEM images of EDACE. (b) Fluorescence images of EDACE after being immersed in FITC-BSA solution for 2 h. (c) Contact angles of electrodeposition of graphene oxide. (d) Zeta potentials tested at DPACE and EDACE surfaces. (e) Voltammetric curves recorded at EDACE in 10 μM DA before (solid curve) and after (dashed curve) the electrodes were put into BSA solution for 2 h. (f) Post-calibration sensitivities of EDACE functionalized by Ca^{2+} ligand after being implanted in the brain for different days. *Source:* Ref. [11] / with permission of John Wiley & Sons.

9.1.4 Iontronics Based on Electrochemistry at Liquid/Liquid Interfacial

Although the electrochemical methods based on faradic processes have been investigated in broad fields, this type of sensing strategy is still challenging to directly determine electrochemically inactive species based on faradic signal because of the redox reactions occurring beyond the potential window. Different from that of a traditional solid/electrolyte interface, the current signal can be acquired when an ion transfer reaction occurs at a liquid/liquid (L/L) interface, referring to interfaces composed of two immiscible electrolyte solutions. As a consequence, electrochemical platforms based on the L/L interface have emerged as a powerful analytical strategy for the determination of ions. However, ion transfer at the L/L interface is generally thermodynamically unfavorable and lacks selectivity, leading to weak current response and potential interference. To solve these problems, ionophores or specific ligands are added in the aqueous phase or the organic phase to facilitate the ion transfer through two phases via a specific chemical interaction between the ion and ligand. Based on this strategy, some liquid interfacial sensors were proposed to realize the determination of ions relevant to neuroscience. For instance, Zhang et al. employed for the first time a 1,2-DCE/water interfacial sensor to detect choline in the brain, demonstrating

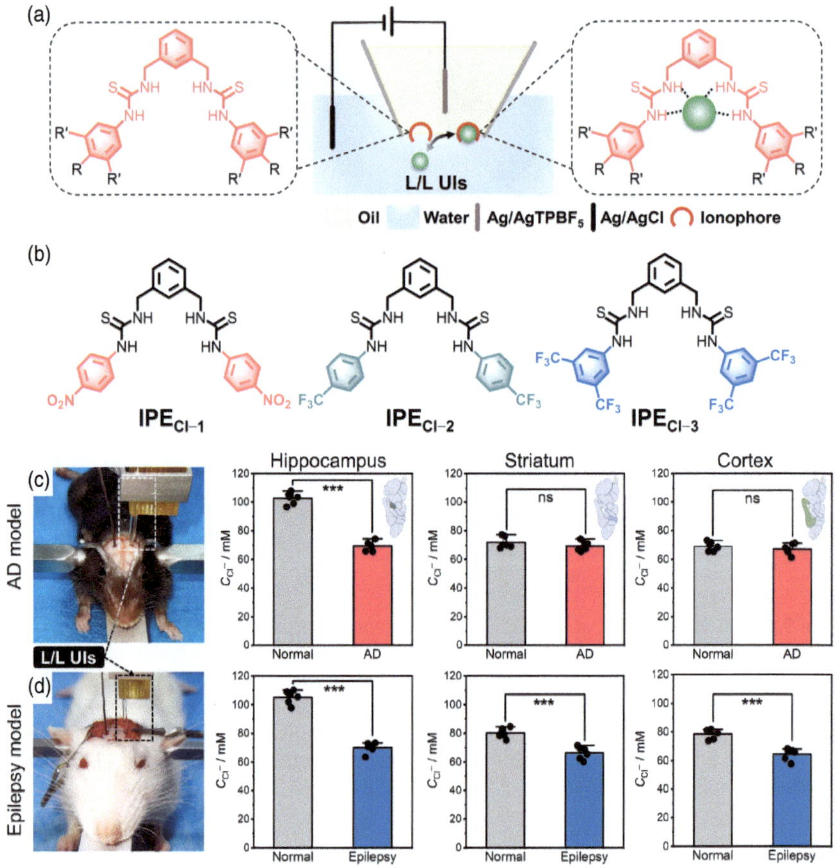

Figure 9.6 (a) Working principle of molecularly tailored L/L interface supported at micro-pipettes for facilitating Cl^- transfer process. (b) Molecular structure of designed bis-thioureas ionophores. (c–d) In vivo determination of Cl^- in different brain regions of with (c) AD model, and (d) the epilepsy model. *Source:* Reproduced with permission from Ref. [20] / American Association for the Advancement of Science - AAAS.

its potential feasibility for in vivo analysis [19]. However, the weak mechanical stability of the L/L interface generally poses a risk of organic solvent leakage, which can lead to biological toxicity affecting brain functions. Recently, Gu et al. reported an efficient L/L interfacial iontronics by using polyvinylchloride (PVC) to enhance the interfacial stability of the organic phase (Figure 9.6a–b) [20]. Furthermore, lipophilic bis-thioureas ionophores were incorporated into the organic gel phase to enhance the Cl^- transfer reactions at an L/L ultramicrointerface. Taking advantage of this molecularly tailored L/L interfacial ultramicro iontronics, they dynamically track the cerebral chloride regulation in vivo. These studies clearly show that the L/L interface serves as an alternative and effective electrochemical platform to the liquid/solid interface, allowing for the analysis of more charged ions, especially non-electroactive ions.

9.1.5 New In Vivo Electrochemical Measurement without External Power Support

For in vivo neurochemical sensing, electrochemical sensors have emerged as a crucial tool due to their high spatiotemporal resolution. This feature is vital for obtaining a real-time chemical imaging of the living brain, allowing researchers to monitor dynamic changes in neurochemical concentrations with precision and accuracy [21]. To date, in vivo electrochemical monitoring of brain chemistry has primarily utilized voltammetry and amperometry. Although these techniques differ in performance, they share a common requirement for external electric fields to drive electrode reactions. This dependence on externally powered sensing can introduce concerns regarding potential electrical disturbances caused by applied electric fields and leaky currents from potentiostats or batteries, which may impact neuronal activity. Galvanic redox potentiometry (GRP) is a novel sensing technique based on the principle of spontaneous bipolarization [22]. GRP sensors operate without the need for external power sources and instead measure the open-circuit voltage at the equilibrium state of a galvanic redox cell implemented on a single bipolar carbon fiber. The near-zero circuit current ensures minimal potential drops within brain tissue and significantly reduces the generation of electric fields. This design effectively avoids any electrical interference with adjacent neurons, making GRP sensors highly advantageous for in vivo neurochemical monitoring. More importantly, it indicates a minimal extent of electrode reaction, effectively reducing the accumulation of reaction products on the electrode surface and preventing the passivation of reactive sites. These characteristics are important for maintaining high sensitivity and long-term stability, ensuring reliable performance over extended periods. Using this sensing strategy, a new concept of enzymatic GRP has been proposed [23]. This developed approach displays rapid time response with high selectivity characteristic of enzyme electrodes, while also benefiting from the excellent biocompatibility and reliability originated from GRP sensors. In this case, glucose dehydrogenase was selected as the recognition element of glucose, and carbon black was taken as the efficient relay station to improve the response time of the chemical reaction. The GRP biosensor was demonstrated to possess high selectivity toward O_2, and other neurochemicals coexisting in the brain environment. Moreover, the good reversibility and high sensitivity endow the GRP biosensor with the ability to monitor the dynamic changes of glucose in the rat brain. In recent years, Lu et al. constructed a fouling-free 5-HT GRP sensor [24]. The GRP system utilizes bipolar electrochemistry, containing an anodic indicating electrode that is sensitive to 5-HT and a cathodic reference electrode immersed in K_2IrCl_6 solution. As 5-HT is added to the solution, the bipolar device forms a circuit driven by a negative Gibbs free energy change, resulting in the spontaneous generation of a concentration-dependent open-circuit potential response. Compared with the conventional amperometric 5-HT sensors with a sensitivity loss of more than 90% during 30 minutes, this GRP 5-HT sensor exhibits excellent anti-biofouling performance, which shows a less than 3% change of signal in 2 hours of testing. Very recently, Mao coworkers reported a H_2S sensor by combining a single-atom catalyst and GRP to solve this issue (Figure 9.7a) [25]. In this study, they apply atomically dispersed NiN_4 active sites on the sensing interface to facilitate electrochemical oxidation of H_2S at low potentials, enabling the spontaneous bipolarization of a single carbon fiber. Under open-circuit conditions, bias-free potentiometric sensing remarkably reduces sulfur accumulation on the electrode surface, thereby significantly enhancing the stability and sensitivity of electrochemical devices. The NiN_4-SACs-based GRP sensor provides sensitive detection of H_2S concentration variations in a broad range over three orders of

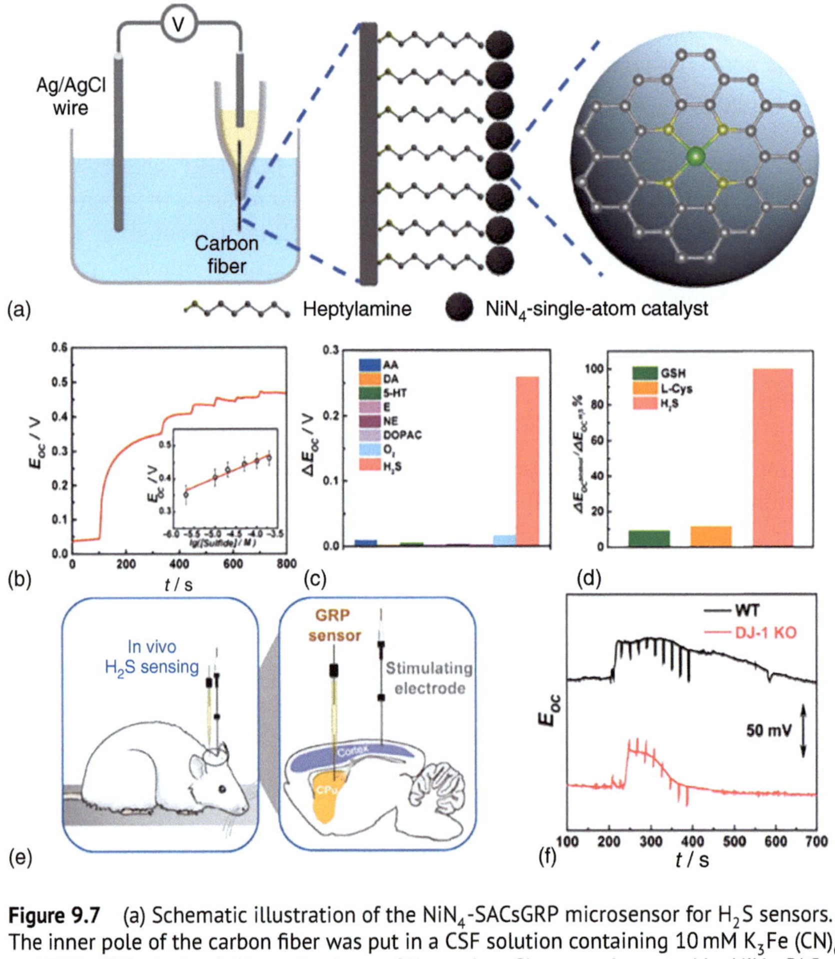

Figure 9.7 (a) Schematic illustration of the NiN_4-SACsGRP microsensor for H_2S sensors. The inner pole of the carbon fiber was put in a CSF solution containing 10 mM $K_3Fe\ (CN)_6$ and $K_4Fe\ (CN)_6$ (yellow). The outer layer of the carbon fiber was decorated by NiN_4-SACs to catalyze the electrochemical oxidation of H_2S in bulk media (blue). (b) Typical responses of NiN_4-SACs-GRP toward NaHS injections into bulk media. Inset, linear plot of E_{OC} outputs versus sulfide concentration logarithm. (c) E_{OC} changes in response to active substances commonly existing in the brain. (d) Percentage representation of biothiols caused E_{OC} disturbances to H_2S signals. (e) Schematic illustration of the cortex-striatum electrical stimulation model and GRP sensor implantation. (f) Real-time monitoring of striatum H_2S upon continuous electrical stimulation in the cortex of WT (black curve) and DJ-1 KO (red curve) mice.

magnitude, exceptional stability, and high selectivity (Figure 9.7b–d). Additionally, they demonstrated the successful quantification of H_2S release behaviors in the brains of mice under both normal and Parkinson's disease conditions employing the GRP sensor (Figure 9.7e–f). This developed H_2S sensor can efficiently avoid the potential risk of the conventional H_2S sensors, the low stability and sensitivity loss resulting from electrode passivation caused by sulfur poisoning.

Besides GRP, a more facile sensing model without external power support is constructing the electrochemical sensor on the basis of open-circuit potentiometry. In order to establish the relationship between electrical signal and chemical reaction, the open-circuit potentiometry as a signal output to replace the faradic current signal can enable more reliable in vivo measurement. Zhao et al. synthesized and optimized specific recognition ionophores for the detection of K^+, Ca^{2+}, Na^+, and pH. Meanwhile, a reference electrode was adopted to avoid interferences from the brain environment [26]. This method, without applied electric fields, allows simultaneous analysis of chemical and electrical signals in the living brain without crosstalk (Figure 9.8c–d), and can realize the quantification of the above four kinds of ions (Figure 9.8a–9.7b). Using this microarray, the extracellular K^+ with a low amount is found to have an important influence on epileptic seizures occurring in freely moving rats (Figure 9.8e).

The photoelectrochemical (PEC) technique is an innovative and rapidly developing analytical tool that leverages incident light as the excitation source and electrical current as the output signal [27–29]. One of its key advantages is the formation of an internal electric field at the interface between the photoelectrode and the electrolyte. This internal electric field effectively negates the need for external electrical bias, simplifying the device design and enhancing its operational efficiency [29]. Thereby, PEC biosensors represent a promising alternative to traditional electrical stimulation methods for in vivo detection. By using near-infrared (NIR) light or micro-optical fibers, these sensors have been proposed and initially tested for in vivo applications. This approach not only mitigates the need for external electrical bias but also enhances the versatility and applicability of biosensors in complex biological environments. Xu et al. developed an innovative reverse design strategy to tailor photoelectrodes based on the unique molecular structures of target analytes, focusing particularly on antibiotics (Figure 9.9a) [30]. Taking tetracycline (TET) as a model analyte, they designed and synthesized a Cu_1-TiO_2-x-based composite photoelectrode. In this design, the dicarbonyl group and adjacent dihydroxyl group of TET form stable coordination bonds with Cu and interact with Ti sites in TiO_2-x, respectively (Figure 9.9b–c). The atomic dispersion of Cu atoms on the TiO_2-x substrate (Cu_1-TiO_2-x) creates synergistic anchoring sites that enhance the efficiency and selectivity of the TET molecule capture. These molecular interactions enabled the development of implantable PEC sensors featuring a Cu_1-TiO_2-x photoelectrode encapsulated within an antifouling layer. These sensors allow for the selective capture and identification of TET molecules and facilitate long-term monitoring of TET metabolism in real biological environments (Figure 9.9d–e). This approach not only enhances the sensitivity and specificity of the biosensors but also ensures their robust performance over extended periods.

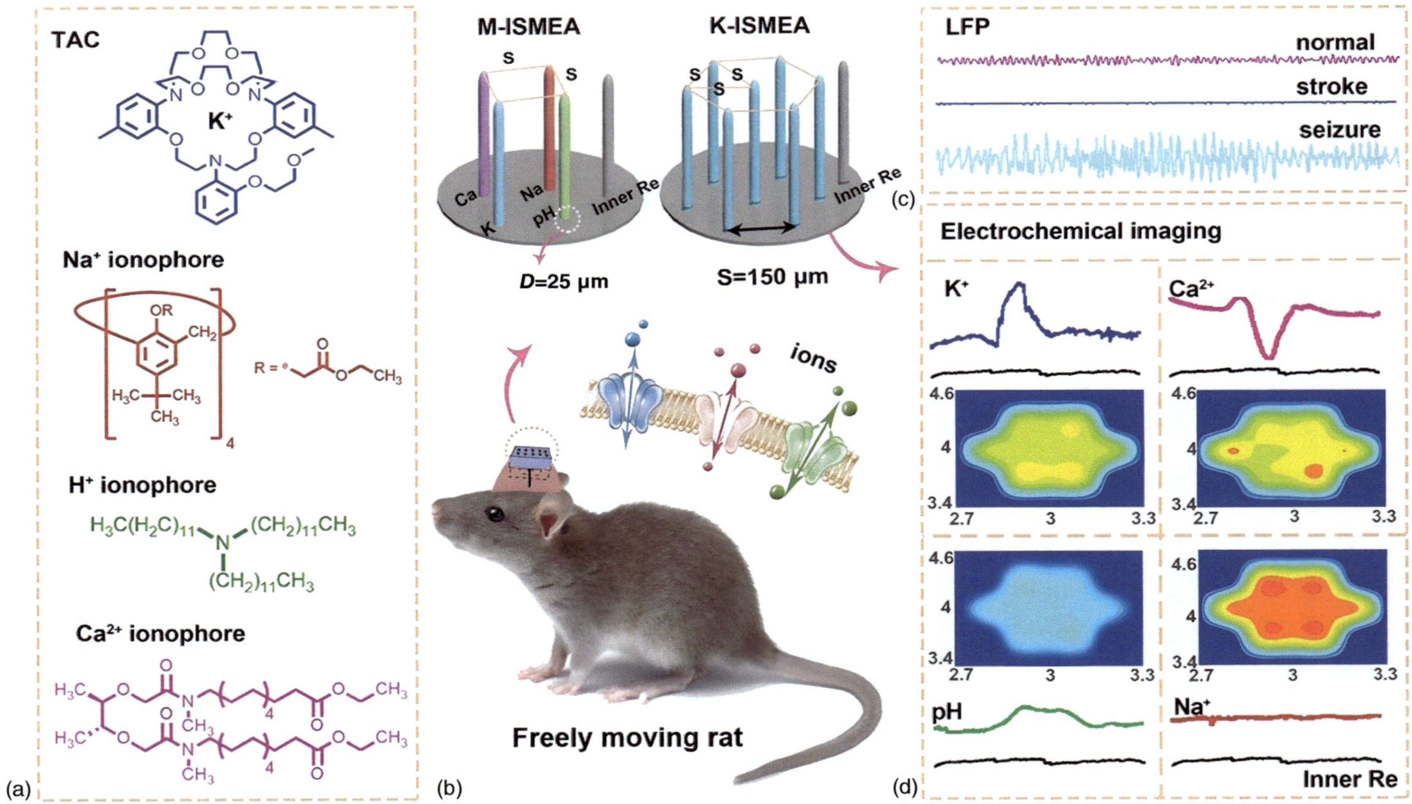

Figure 9.8 (a) Molecular structures of the specific ligands for K^+, Na^+, H^+, and Ca^{2+}. (b) Schematic illustration for the developed microarrays, (c) LFP signals recorded in the rat brain during seizure. (d) Real-time imaging and determination of K^+, Na^+, Ca^{2+}, and pH levels in freely moving rat brains. *Source:* Reproduced with permission from Ref. [26] / John Wiley & Sons.

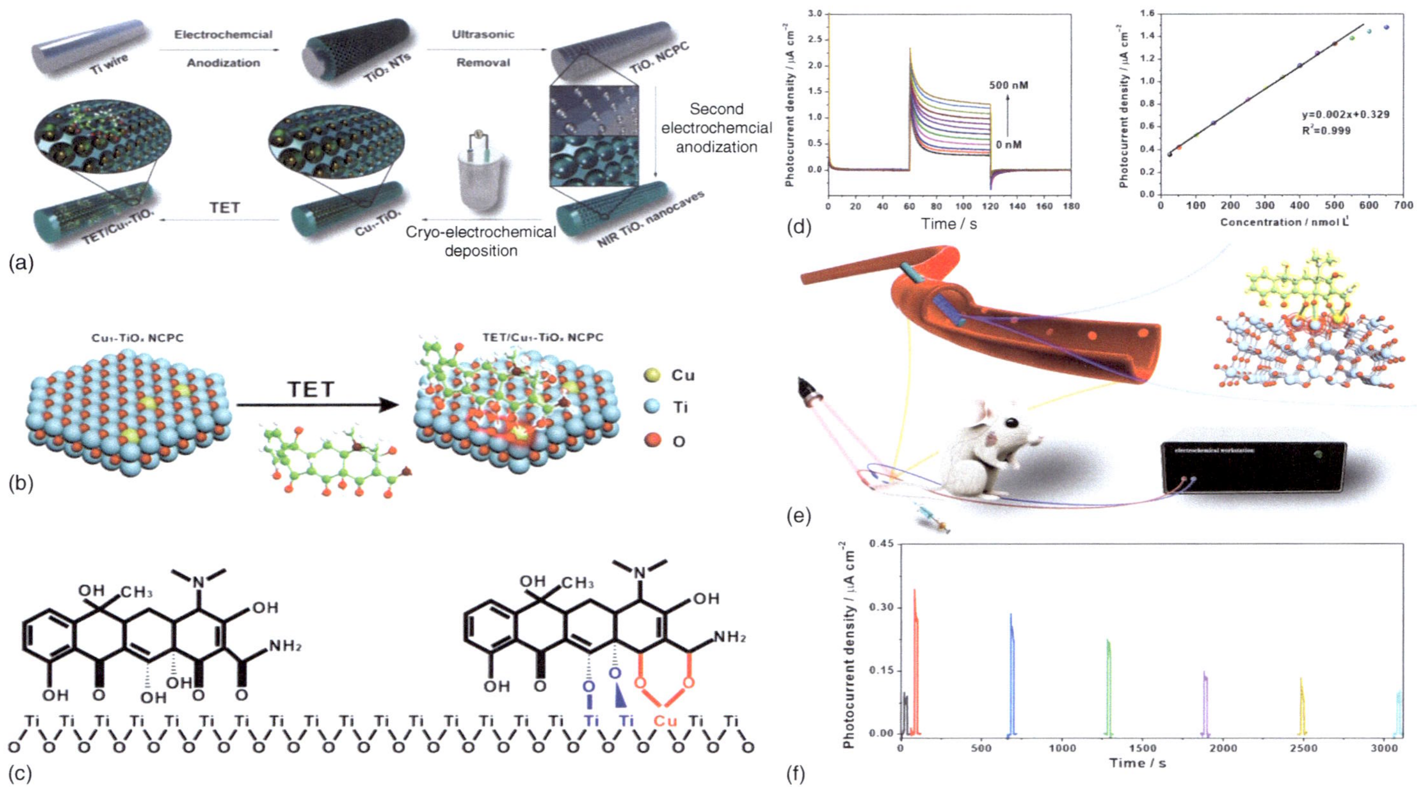

Figure 9.9 (a) Fabrication processes of Cu_1-TiO_2-x. (b) Schematic illustration of Cu_1-TiO_2-x for TET recognition with triple sites. (c) bonding a dicarbonyl group with Cu_1 and a neighboring dual-hydroxyl group with Ti sites. (d) PEC detection of TET with different concentrations, and a linear calibration curve for PEC TET detection. (e) Schematic PEC detection of TET on Cu_1-TiO_2-x/Z-MOF in a mouse. (f) PEC in vivo tracking TET metabolism in the living system. *Source:* Ref. [30] / Springer Nature.

9.1.6 Summary and Outlook

The development of electrochemical methods with efficient performance for selective and accurate acquisition and quantification of chemical information is important in an intricate determination environment. A double-recognition strategy was proposed to achieve accurate measurements of chemical species by designing specific molecular probes capable of chemical recognition of target molecules, generating distinguished electric signals. The affinity and recognition mechanisms of specific recognition ligands were carefully modulated to facilitate long-term tracking and dynamic quantification. This optimization ensures reliable and continuous monitoring over extended periods, with precise detection and quantification of target molecules. Moreover, to meet the requirements for real-time recognition in a long-term way, a new reversible recognition ligand based on moderate binding free energy is needed for real-time monitoring of Ca^{2+}. In addition, the improved stability and resistance to contamination of the electrodes were also efficiently established. In order to reduce the risk of external powered or triggered sensing ways to bring up the potential electrical disturbance to neurons, the electrochemical microarrays based on self-powered electrochemical systems based on open-circuit potentiometry were established for real-time monitoring of chemical signals in living systems.

Unfortunately, there were still lots of challenges for creating next-generation in terms of high specificity recognition, ultra-high response speed, and multi-channeling in vivo sensors. Emerging technologies such as molecular simulation design aided by AI techniques and thermodynamic calculations are poised to provide critical guidance for the continued development of electrochemical sensors, which enable carrying out complex molecular recognition with rapid response. Advances in microfabrication and lithography techniques promise to enhance the channel density of electrochemical arrays. These improvements will be advantageous for the recording of intricate chemical signal networks within biological systems. Consequently, we believe that the design strategies and methods outlined in this review can be extended to encompass a broader spectrum of molecules and more specialized contexts, enhancing our ability to study and understand complex processes.

9.2 Probing the Minuscule: SERS In Enhanced Vibrational Spectroscopy

9.2.1 Raman Instruments

9.2.1.1 The Basic Principles of Raman Scattering

The Raman scattering phenomenon was discovered by Indian physicist Raman in 1928. It provides a spectroscopic technique for studying the vibration and rotation of molecules in matter. Due to their features such as rapid detection, simplicity of operation, high sensitivity, and portability, Raman spectrometers play a crucial role in sudden public health emergencies [31–33].

When light shines on a material, most photons only change direction without altering their frequency. This type of scattering is called Rayleigh scattering. However, a small portion of photons not only change direction but also change frequency. This frequency-altered scattering is known as Raman scattering. Raman scattering can be divided into two types: Stokes Raman scattering and anti-Stokes Raman scattering. Stokes scattering refers to scattered light having a lower frequency than the incident light, while anti-Stokes scattering refers to scattered light having a higher frequency. Stokes scattering is the most common type of Raman scattering, occurring when photons interact with the vibrational modes within molecules, resulting in energy absorption and scattering at a slightly lower frequency and energy. Anti-Stokes scattering occurs when molecules are initially in an excited vibrational state and transfer part of their vibrational energy to the incident photon during scattering, resulting in the emission of a higher-energy photon.

The frequency difference $\Delta\nu$ between the incident light and the scattered light is referred to as the Raman shift. Generally, the intensity of the Stokes scattered light is significantly greater than that of the anti-Stokes scattered light, which is the reason of that Stokes' scattered light is typically measured in Raman spectroscopy analysis. For the same type of molecule, as the frequency of the incident light changes, the frequency of the Raman lines also changes; however, the Raman shift $\Delta\nu$ remains constant, independent of the frequency of the incident light. It is related to the vibrational and rotational energy levels of the molecule (Figure 9.10).

9.2.1.2 Laser Raman Spectrometer

The experimental setup for Raman spectroscopy typically consists of a laser light source, a sample cell, a monochromator, and a detection system (Figure 9.11).

Laser Light Source The laser light source for commercial instruments is typically provided by continuous gas lasers such as helium–neon lasers, argon ion lasers, and krypton ion lasers.

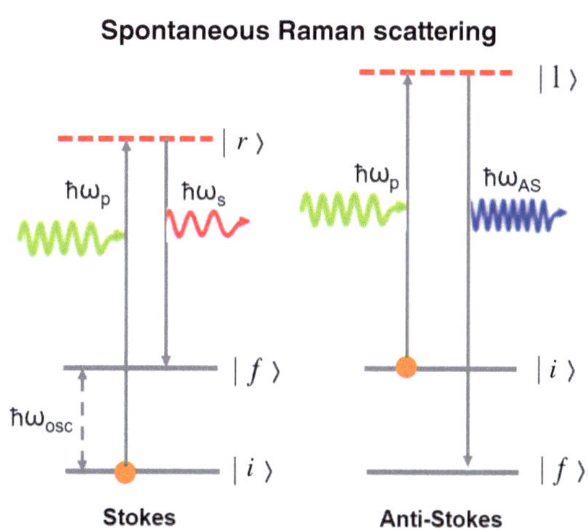

Figure 9.10 Spontaneous Raman scattering is based on inelastic scattering of light leading to a frequency redshift (Stokes) or blueshift (anti-Stokes). *Source:* Ref. [34] / Royal Society of Chemistry.

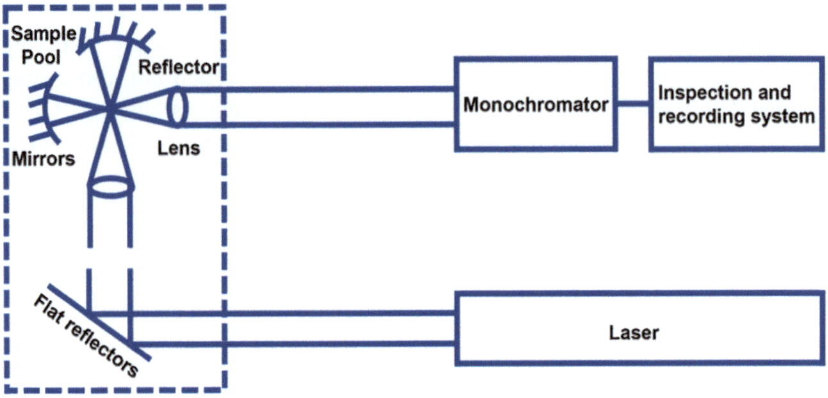

Figure 9.11 Experimental setup for laser Raman spectroscopy.

As a light source for laser Raman spectroscopy, the following three requirements must be met: First, the single-line output power typically ranges from several milliwatts to tens of milliwatts, depending on the sample and optical configuration. Second, the power stability should be of high quality, with variations not exceeding 1%. Third, the device should possess a long operational lifespan, enduring for a minimum of 1,000 hours.

Sample Device The most important aspect of the sample apparatus is to consider how to effectively irradiate the sample and focus scattered radiation. The most typical device can be seen in the upper left part of the diagram. The lens below focuses the laser beam on the sample; the lens on the right optimally focuses the scattered light at the appropriate solid angle, matching it with the solid angle of the dispersion system. The mirror above allows the laser beam to pass through the focal point at the center of the sample multiple times; the mirror on the left doubles the solid angle entering the dispersion system. This setup enhances the effective intensity of both incident and scattered light. The type of sample chamber depends on the quantity and nature of the sample. For trace samples, whether liquid, solution, or solid powder, fine crystals can be placed in capillaries of different diameters. For larger samples, they can be placed in conventional containers such as reagent bottles or vials. In order to prevent the decomposition of the sample, the most effective method is to employ a rotation technique. The focus of the laser beam moves relative to the surface of the sample to avoid local overheating.

Monochromator Because of the weak Raman scattering, it is necessary to minimize stray light interference in order to improve the signal-to-noise ratio and sensitivity. Therefore, in Raman spectroscopy analysis, not only is it necessary to effectively suppress Rayleigh scattering and stray light while maintaining sufficient excitation intensity, but also a monochromator in front of the detection system is essential. Typically, a double-grating monochromator is used. Monochromators should have good imaging quality, high resolution, and minimal stray light.

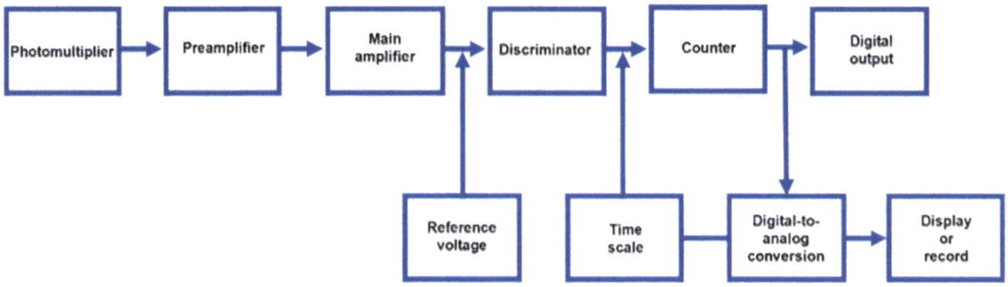

Figure 9.12 Photon counter schematic diagram.

Detection and Documentation Systems For Raman scattered light falling within the visible light range, a photomultiplier tube can be used as a detector. Because Raman light is weak, it is necessary for the photomultiplier tube to have high quantum efficiency and low dark current. To reduce dark current, cooling measures are often employed.

The weaker Raman signals can be detected using a photon counter. The schematic diagram of the photon counter is shown in the Figure 9.12.

The output voltage or current signal of a photon counter is proportional to the photon count rate (photons/second) received by the photomultiplier tube, independent of the energy of the photon. The photon counting range is usually $10 \sim 10^6$ photons/second, so for strong Raman scattered light, it is not suitable to use a photon counter. When a photon is detected by a photomultiplier tube, it is output as a photoelectron pulse. After the pulse signal is amplified by the preamplifier and the main amplifier, it is compared with a reference voltage signal by a discriminator, and the signal greater than the reference voltage is selected and sent to the counter. The counter outputs a digital signal at specified intervals (time scales) and can also be displayed or recorded by a digital-to-analog converter to obtain an analog signal.

9.2.1.3 Portable Raman Spectrometer

The portable Raman spectrometer is small in size, lightweight, and easy to carry, making it suitable for rapid analysis in the field [35–37].

The characteristics of Raman light are that it diverges in all directions in space, and the designed collection angle is small. This requires higher excitation power density, improved collection efficiency, or longer integration times to generate sufficient Raman signals. The reduced Raman scattering efficiency necessitates longer integration time or stronger laser power, which imposes higher stability and repeatability requirements on the spectrometer's detection performance.

9.2.1.4 Stimulated Raman Scattering Microscopy

Similar to the photon emission process, Raman scattering can also be divided into spontaneous Raman scattering and stimulated Raman scattering (SRS) [38]. Stimulated Raman scattering (SRS) microscopy offers chemical specificity without exogenous labels, noninvasiveness, high spatial resolution, and real-time monitoring capability [39].

In spontaneous Raman scattering, a laser beam characterized by a frequency ωp irradiates the sample. Due to inelastic scattering, the signal is generated at the Stokes and anti-Stokes frequencies, which are respectively denoted as ω_S and ω_{AS}. In contrast, for stimulated Raman scattering (SRS), two laser beams, namely one at frequency ωp and the other at frequency ωS, converge on the sample (as depicted in Figure 9.13). When the difference frequency, $\Delta\omega = \omega_p - \omega_S$, commonly known as the Raman shift, aligns with a particular molecular vibrational frequency Ω, the amplification of the Raman signal is accomplished by means of stimulated excitation.

In SRS microscopy, the two laser beams are overlapped both spatially and temporally. Subsequently, they are focused collinearly into a diffraction-limited point within the sample. At the microscope's tight focus, the stimulation factor can attain a maximum of 10^8, which endows SRS bio-imaging with high sensitivity and speed.

The detection approach of SRS exhibits uniqueness and distinguishes itself from most other optical microscopy techniques. Based on the principle of energy conservation, every vibrational excitation results in the loss of one photon in the pump beam (known as stimulated Raman loss, SRL) and the gain of one photon in the Stokes beam (termed stimulated Raman gain, SRG). Consequently, detection of the SRS signal relies on measuring the change in the relative intensity of the incident laser. ($\Delta I/I \sim 10^{-3} \sim 10^{-7}$, Figure 9.14). In order to achieve sensitive detection that surpasses the low-frequency laser background noise, a modulation at a high frequency (megahertz range) is imposed on one of the laser beams, with the subsequent transfer of this modulation to the other beam being quantified. Subsequently, a radio-frequency (RF) lock-in amplifier is utilized to extract the SRS signal [40].

By leveraging the combined advantages of chemical specificity and detection sensitivity, the single-molecule Raman sensitivity in the full far field has been achieved for the first time. Moreover, the multicolor surface-enhanced Raman forward scattering (SREF) imaging of isotopologues has been demonstrated, breaking the color barrier of fluorescence microscopy. The idler beam from a commercial optical parametric oscillator (OPO) laser source is doubled and then recombined with the OPO signal beam to form a new pair of pump and Stokes beams (as depicted in Figure 9.14), thus opening up a new SREF excitation range of around 600 nm. Benefiting from the large number of commercial fluorescent dyes within this visible range, the versatility of SREF can be fully tested [41].

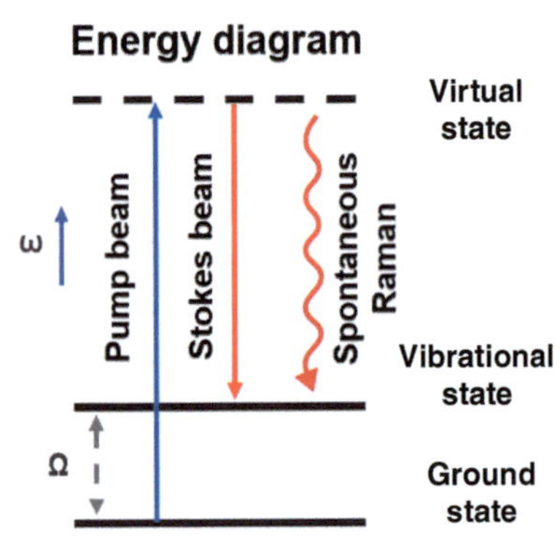

Figure 9.13 Energy diagram for SRS.

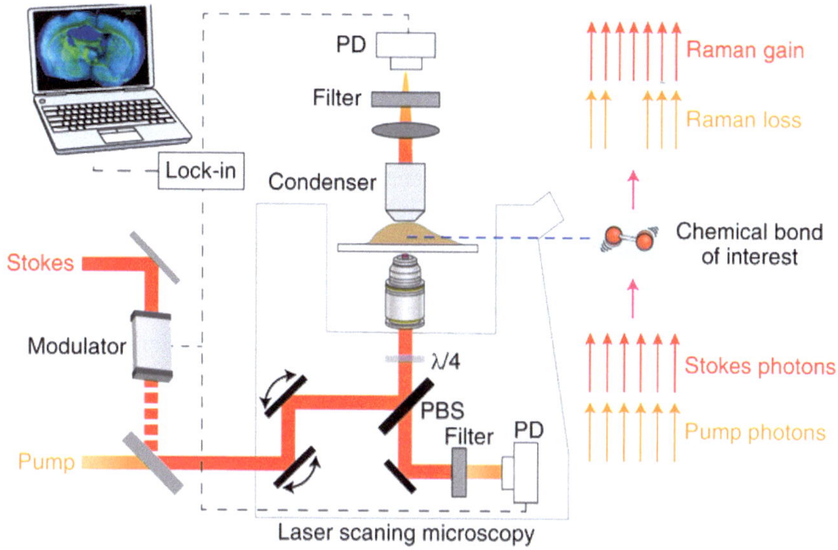

Figure 9.14 New stimulated Raman excited fluorescence (SREF) laser microscopy system opening the excitation band near 600 nm. (a) Energy diagram for the SREF process. (b) Wavelength of SHG of OPO idler (red) and corresponding resonance Raman shift (blue) as a function of OPO signal wavelength when OPO signal is set as the Stokes beam and the SHG is set as the pump beam. c System setup for SREF excitation band around 600 nm. PPLN, periodically poled lithium niobate; QWP, quarter-wave plate; DM, dichroic mirror; PBS, polarization beam splitter; BS, beam splitter; SPCM, single-photon counting module; DAC, digital-to-analog converter; ADC, analog-to-digital converter; EOM, electro-optical modulator; APD, avalanched photodiode; PD, photodiode. *Source:* Ref. [41] / with permission of American Chemical Society.

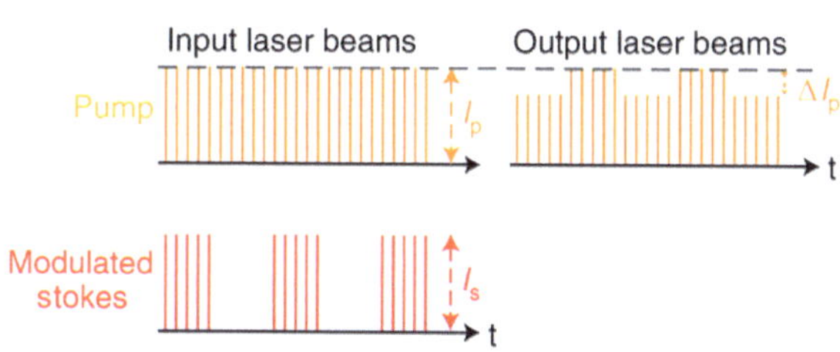

Figure 9.15 SRS detection scheme using high-frequency modulation/demodulation. Temporal profiles of the input and output pump and Stokes pulse trains. *Source:* Reproduced from Ref. [40] with permission.

9.2.2 The Principles of SERS-Based Detection

When a beam of light strikes the surface of a material, the interaction between the light and the molecules of the material essentially involves collisions, which can be either elastic or inelastic. During this process, if a photon excites a molecule and the molecule subsequently returns to its ground state without a net change in energy, it is considered an elastic collision. During this process, the photon's kinetic energy may vary, but its total energy remains unchanged, a phenomenon referred to as

Rayleigh scattering. Conversely, in inelastic collisions, the molecule transitions to an excited state, resulting in changes to both the photon's kinetic energy and total energy. Inelastic light scattering, which includes Stokes scattering ($\omega_s = \omega_0 - \omega_r$) and anti-Stokes scattering ($\omega_s = \omega_0 + \omega_r$), also known as Raman scattering, occurs for only one photon out of every 10^6–10^9 photons [42].

Various physical and chemical enhancement mechanisms have been proposed to understand the SERS mechanisms. Electromagnetic enhancement (EM) and chemical enhancement (CE) are the two most widely accepted mechanisms [43]. EM arises from the amplification of the local electromagnetic field on the rough surface of a metal-based SERS-active substrate, leading to an increased SERS signal. The CE mechanism, on the other hand, involves electron transitions within the adsorbed molecules. These transitions occur as electrons move from the Fermi level of the metal to new excited states or from these excited states back to the metal nanoparticles, resulting in charge transfer. Both mechanisms act simultaneously during Raman scattering, generating a strong SERS effect on the active substrate and metal surface. This synergistic interaction greatly enhances the sensitivity of SERS detection. In this section, the basic concepts and principles of EM and CE will be presented, which are crucial for conducting sensitive and reliable analytical SERS.

9.2.2.1 Electromagnetic Enhancement

Electromagnetic enhancement, also known as the surface plasmon resonance (SPR) model [44], describes a physical optical phenomenon involving the interaction between free electron clouds at the metal surface and within the semiconductor. When incident light resonates with the free electron clouds on the surface of metal nanoparticles, it triggers the surface plasmon resonance effect. This resonance generates significant energy in the oscillating molecules. The surface plasmon resonance model is illustrated in Figure 9.16. The fundamental conditions for SPR to occur are as follows: (i) the presence of surface plasmons, (ii) an appropriate excitation source, and (iii) specific excitation conditions that must be satisfied between the excitation source and the plasmons. The resonance effect becomes localized to the nanoscale, known as localized surface plasmon resonance (LSPR), when noble metals exhibit nanometer-scale roughness. Consequently, this LSPR effect results in a significant amplification of the Raman scattering signal because of the enhancement of the local electric field [45]. The following expression represents the relationship:

$$I_{\mathrm{SERS}} = |E(\omega_{\mathrm{inc}})|^4 \tag{9.1}$$

where I_{SERS} represents the enhancement in the intensity of SERS, and $E(\omega_{\mathrm{inc}})$ denotes the enhancement of the electric field at the Stokes shifted frequency ω_s.

9.2.2.2 Chemical Enhancement

The charge transfer model in CE has attracted considerable attention from researchers. This model involves the transfer of charge between the metal substrate and the adsorbed substance. When a molecule is adsorbed onto the metal surface, the resulting charge transfer excites electrons from around the Fermi level to higher

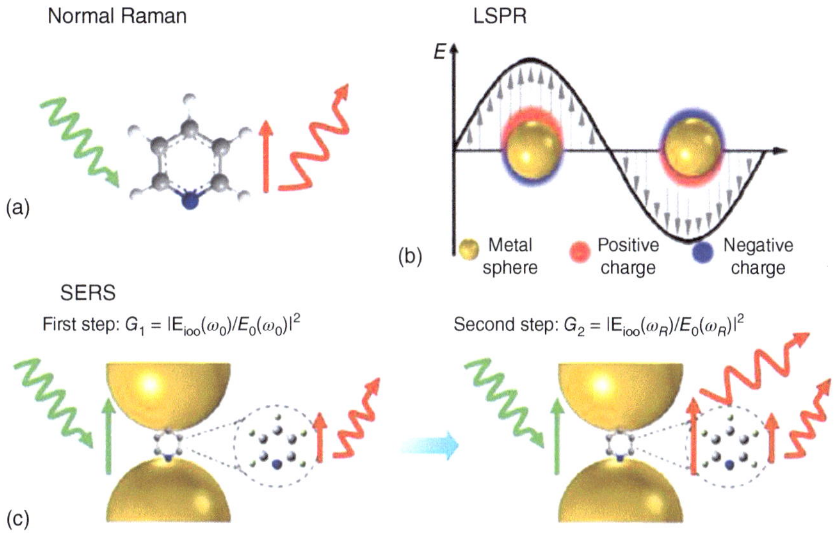

Figure 9.16 Schematics of (a) normal Raman, (b) localized surface plasmon resonance (LSPR), and (c) electromagnetic enhancement mechanism in SERS. *Source:* Adapted from Ref. [46].

Figure 9.17 Illustration of four interrelated processes for resonance-like Raman scattering or chemical enhancement. *Source:* Ref. [48] / with permission of American Chemical Society.

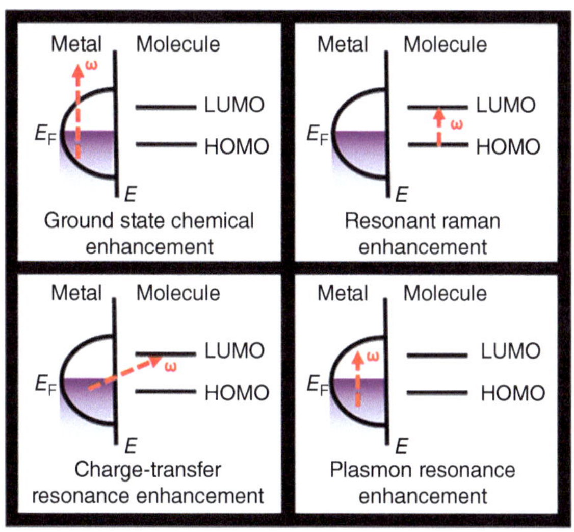

energy states. After a brief relaxation period, the electrons return to the ground state, which corresponds to the metal surface. During this process, the electrons on the metal surface and the adsorbate undergo a resonant transition, thereby generating the SERS effect. As illustrated in Figure 9.17, there are four interconnected processes that contribute to the CE: (i) an augmentation in molecular polarizability upon adsorption onto a metal surface, (ii) resonance in charge transfer within the molecule-metal complex, (iii) intensification resulting from resonance scattering, and (iv) amplification due to plasmon resonance [47].

9.2.2.3 SERS Substrate

As a surface-enhanced spectroscopic technique, the advancement of SERS is inseparable from the application and development of substrates. Since the discovery of the surface-enhanced Raman spectroscopy phenomenon, research on SERS-active substrates has been a key focus and hotspot in the field of SERS studies, serving as a prerequisite for achieving high SERS enhancement signals. An ideal SERS substrate should possess the following advantages: (i) high SERS enhancement intensity for excellent detection sensitivity, (ii) high stability to ensure consistent enhancement effects across the substrate, (iii) strong anti-interference capability, (iv) ease of preparation and long-term storage, and (v) compatibility with other techniques like tip-enhanced Raman spectroscopy (TERS) and in-situ electrochemical SERS [49]. These characteristics have driven the continuous discovery of SERS-active substrates, broadening the applications of SERS technology across various fields.

Plasmonic metals with nanostructured morphologies, including Au, Ag, and Cu, exhibiting robust and adjustable LSPR absorption within the visible spectrum, have been fabricated for utilization as substrates in SERS. Dong et al. enhanced the SERS activity through various surface electrochemical roughening methods, achieving a signal enhancement of 1 to 3 orders of magnitude on electrodes made of metals such as Pt, Pd, Rh, Ru, Fe, Co, and Ni. They also invented the shell-isolated nanoparticle-enhanced Raman spectroscopy (SHINERS), which overcame the issue of substrate material universality [50]. In addition to plasmonic metals, various other materials have been explored for the development of SERS substrates, encompassing semiconductors, metal–organic frameworks (MOFs), graphene, as well as their composites with plasmonic metals. Das et al. introduced a strategy for fabricating nitrogen-doped graphene quantum dots (N-GQDs) using a modified Hummers' method to enhance their SERS performance [51]. To enhance their properties further, specific functional groups were introduced onto the N-GQDs via vacuum annealing. When excited by a 488 nm laser, SERS performance can achieve a detection limit of 1.0×10^{-10} mol/L for rhodamine B (RhB). The primary reason for this improvement was due to two mechanisms: facilitated charge transfer (CT) and the Förster resonance energy transfer (FRET) effect, both of which were facilitated by robust π-π stacking interactions between the N-GQDs and RhB.

9.2.3 Advancements in SERS Technology for Public Health Emergency Response

SERS technology, a highly sensitive spectral detection technique capable of single-molecule level analysis, leverages the attributes of spontaneous Raman spectroscopy for molecular fingerprinting [52]. It is extensively applied in medical diagnosis, food safety, biological analysis, and other fields due to its high sensitivity, multiplexing capabilities, and quantitative detection [53]. Particularly, SERS technology excels in detecting pathogens, which include disease-causing microorganisms such as bacteria, viruses, and parasites [54]. Viruses and bacteria pose the greatest risk to human health, as they can rapidly multiply within the human body, endangering lives [55]. The ongoing refinement of SERS technology is crucial for the swift identification of early bacterial and viral infections, enabling prompt medical responses to significant public health crises [56].

9.2.3.1 SERS Technology for Detection of RNA Viruses

Virus families are classified based on whether their genetic material is DNA or RNA. RNA viruses, also referred to as ribonucleoviruses, contain RNA as their genetic material. They are more prone to variability compared to DNA viruses, which contributes to their potential to cause a broad spectrum of severely infectious diseases that impact global health [57]. In light of this, the focus of this section is on the application of SERS technology for detecting these critical pathogens.

In recent years, the extensive and swift spread of the novel coronavirus SARS-CoV-2 has led to a significant number of severe pneumonia instances. Meanwhile, due to its immune escape ability and high variability, respiratory infectious diseases like the novel coronavirus and influenza have become key and difficult issues in the field of virus detection, posing challenges to traditional detection approaches [58]. The SERS technology, with its advantages such as rapid response, high sensitivity, and simple operation, can be applied for the rapid and precise identification of SARS-CoV-2 at different periods, including the early control, mid-term diagnosis, and late defense of large-scale infections. A gold nanostructure, functionalized with human angiotensin-converting enzyme 2 (ACE2), serves as a "virus trap" effectively selecting, capturing, and rapidly detecting the coronavirus S protein [59]. This advanced SERS biosensor exhibits remarkable sensitivity, achieving a 10^6-fold enrichment of viruses due to the high affinity between ACE2 and the S protein, combined with the structure's oblique gold nanoneedles that act as traps. Furthermore, it boasts a 10^9-fold enhancement of Raman signals, stemming from the multicomponent SERS effects (Figure 9.18a). A self-assembled novel Cu_2O nanoarray, boasting a free carrier density of 1.78×10^{21} cm^{-3}, which rivaled that of noble metals, had set a new record for enhancement factors among semiconductor substrates, achieving an impressive EF of 3.19×10^{10} [33]. Subsequently, this Cu_2O nanoarray was engineered into a highly sensitive and selective SERS chip, capable of quantifying SARS-CoV-2 RNA without the need for enzymes or amplification, with a remarkable detection limit of 60 copies/mL in just five minutes (Figure 9.18b). A point-of-care SERS detection platform had been introduced, specifically designed to target the SARS-CoV-2 spike protein, which enabled a streamlined one-step process for identifying the SARS-CoV-2 antigen by utilizing aptamer-specific recognition to capture substrates and detection probes [60]. Within five minutes, a handheld Raman spectrometer was able to detect SARS-CoV-2 virus and its variants without requiring any pretreatment. Furthermore, this test has the capability to uniquely identify the SARS-CoV-2 antigen, ensuring no cross-reactivity with antigens from other coronaviruses or influenza A (Figure 9.18c). Therefore, the platform holds significant promise for utilization in swift point-of-care diagnostic tests for detecting SARS-CoV-2.

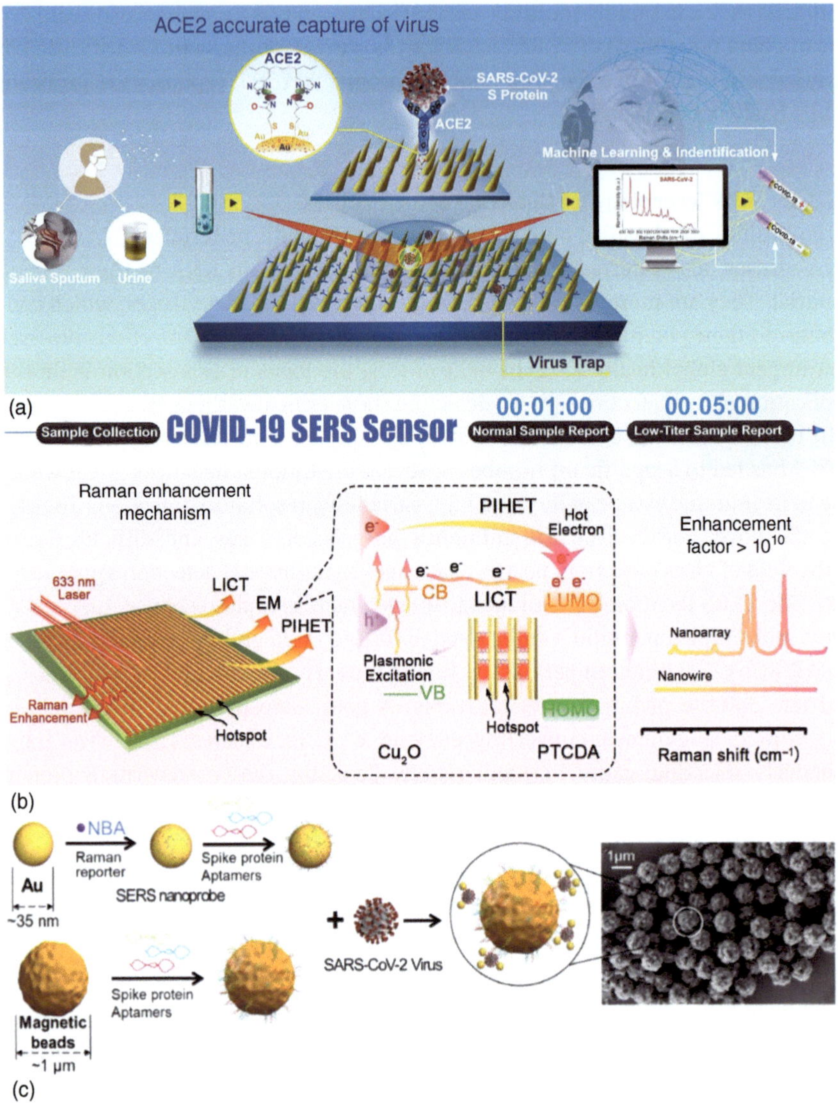

Figure 9.18 SERS technology is used for the detection of COVID-19. (a) Schematic diagram of the design and operation process of the COVID-19 SERS sensor. *Source:* Ref. [59] / with permission of Springer Nature. (b) Raman enhancement mechanism of Cu_2O nanoarray with a significant enhancement factor. *Source:* Ref. [33] / with permission of John Wiley & Sons. (c) Raman enhancement mechanism of Cu_2O nanoarray with a significant enhancement factor. *Source:* Reproduced with permission from Ref. [60] / American Chemical Society.

9.2.3.2 SERS Technology for Detection of DNA Viruses

DNA viruses constitute a significant class of viral pathogens that utilize DNA as their genetic material. These viruses replicate through the action of DNA-dependent DNA polymerase, enabling them to concurrently target human cells and disrupt metabolic processes during infection, thereby complicating the disease dynamics [61]. In response to the challenges posed by the infection and transmission of DNA viruses, the employment of SERS technology has been instrumental in enhancing the speed and precision of detection methodologies. This technological advancement is anticipated to furnish robust support for the timely diagnosis and intervention of viral infections [62].

Monkeypox (MPX), previously known as monkeypox, is a zoonotic disease caused by the monkeypox virus (MPXV), a member of the *Orthopoxvirus* genus, which is part of the Poxviridae family. This virus is characterized by its ability to induce a disease with clinical manifestations similar to those of smallpox. In 2022, the World Health Organization (WHO) declared it a Public Health Emergency of International Concern (PHEIC), highlighting the need for a coordinated international response to contain the spread of the disease [63]. A research endeavor to capture the distinctive genetic fingerprints of the MPXV genome and various antigenic proteins, without the necessity of designing tailored probes [64]. This method boasted a minimum detection threshold of 100 copies/mL, accompanied by impressive reproducibility and a favorable signal-to-noise ratio. Consequently, the correlation between the intensity of characteristic peaks and the concentration of proteins and nucleic acids can facilitate the establishment of a concentration-dependent spectral line, exhibiting a strong linear relationship. Furthermore, principal component analysis (PCA) enabled the differentiation of SERS spectra of four distinct MPXV proteins within serum samples. (Figure 9.19a). Another study reported a dual-signal, colorimetric and SERS immunochromatographic assay (ICA) that offered flexible, highly sensitive, and precise MPXV testing in diverse sophisticated samples [65]. This method involved coating a 2D MoS_2 nanosheet with a thin 1 nm layer of polyethyleneimine, enabling two compact layers of 30 nm gold nanoparticles (AuNPs) through electrostatic forces. This approach improved the colorimetric

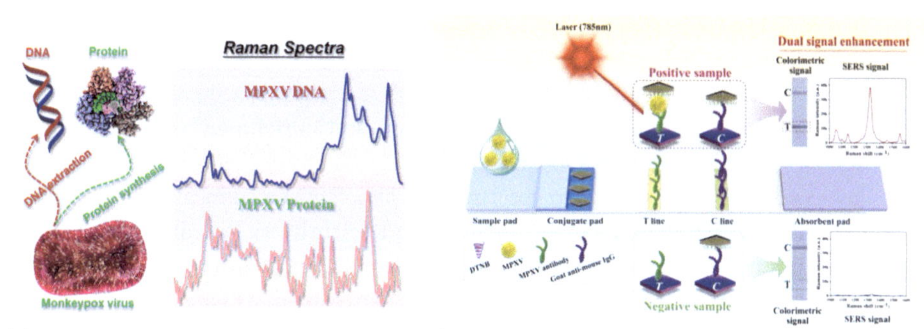

(a) (b)

Figure 9.19 SERS technique is used for the detection of monkeypox virus. (a) Rapid detection of the monkeypox virus genome and antigen proteins based on surface-enhanced Raman spectroscopy. *Source:* Ref. [64] / with permission of American Chemical Society. (b) The principle of MPXV detection by colorimetric-SERS dual-mode ICA. *Source:* Ref. [65] / with permission of Elsevier.

performance while also creating numerous effective SERS hotspots. (Figure 9.19b) Additionally, the film-like dual-signal tag (MoS_2@Au–Au) experiences a substantial enhancement in its SERS activity due to the synergistic effect of chemical amplification from MoS_2 sheets and electromagnetic enhancement from Au–Au hotspots. This assay demonstrated rapid detection capabilities, completing analysis in under 20 minutes, and exhibited high precision in real samples with an RSD of less than 9.89%. Its sensitivity surpassed traditional AuNP-based colorimetric ICA by over 500 times, making it a highly promising candidate for field applications.

9.2.3.3 SERS Technology for the Detection of Bacteria

Globally, bacterial infections are among the foremost causes of mortality. Traditional diagnostic methodologies for bacterial infections necessitate sample cultivation to ascertain the bacterial species and their susceptibility to antimicrobial agents, a process that is notably time-consuming, often extending over several days. SERS technology has garnered significant interest because of its distinctive advantages, including label-free detection, the absence of complex pretreatment requirements, noninvasiveness, and expedited detection times.

Tian et al. reported on the fabrication of a SERS substrate based on BNC, which was produced through a gravity-assisted filtration technique. The considerably lower surface roughness of BNC compared to conventional filter paper resulted in excellent uniformity of SERS activity across the entire substrate. The smooth surface of the BNC-based SERS substrate renders it an optimal platform for gathering, detecting, and identifying bacteria. The three-dimensional plasmonic composites featuring BNC, as presented here, exhibit significant potential for diverse applications spanning sensing, catalysis, and energy harvesting [66]. Kim et al. introduce a novel fiber-optic SERS probe (SERS-on-a-tip) fabricated via a simple two-step protocol using readily available components and materials, exhibiting high controllability and repeatability [67]. This is the first reported use of a fiber-optic SERS probe for detecting live, unlabeled bacteria (Figure 9.20). Lin et al. presented a highly sensitive SERS filter-like substrate, comprising AuNPs embedded within mesoporous silica (AuNPs @ MS), that was synthesized through a straightforward one-step process. This substrate was designed for the filtration and enrichment of analytes from aqueous samples (Figure 9.20) [68]. Recently, this group demonstrated a simple, low-cost, and rapid optical fingerprinting technology for clinical microorganism detection using AgNPs-decorated filter-like SERS substrates combined with pattern recognition techniques, achieving 66%–100% prediction accuracy for three bacteria species through LOOCV, offering a basic and cost-effective tool for rapid bacterial detection, quantification, and classification at species levels [69]. Wang et al. reported a plasmonic filter fabricated from silver-coated self-assembled photonic crystal for discriminating nosiheptide product sources by filtrating, capturing, and identifying streptomycete spores via SERS, outperforming traditional methods and showing potential for food safety, environmental, and clinical applications (Figure 9.20c) [70].

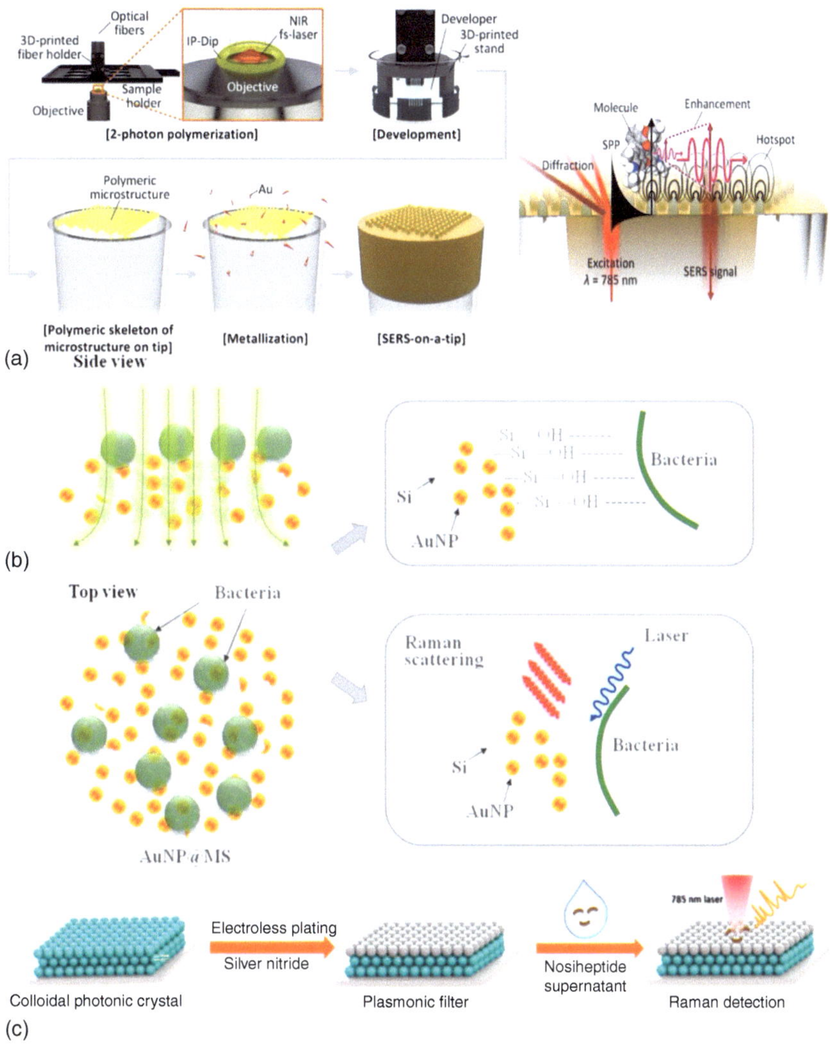

Figure 9.20 SERS technique is used for the detection of bacteria. (a) Schematic diagram, preparation process, and working principle of the SERS-on-a-tip probe. *Source:* Ref. [67] / John Wiley & Sons / CC BY 4.0. (b) SERS enhancement produced by Au NPs is used for bacterial detection. *Source:* Ref. [68] / with permission of Elsevier. (c) Schematic of plasmonic filter design and Streptomyces spores discrimination. *Source:* Ref. [70] / with permission of American Chemical Society.

9.2.4 Summary and Outlook

The rampant spread of pathogenic epidemics has profoundly affected human life, necessitating an urgent technological solution for the efficient identification and detection of pathogens. SERS spectral analysis technology presents significant potential in this regard, leveraging its molecular "fingerprint" information, absence of standardization requirements, nondestructive nature, and in situ detection capabilities. The practical utility of SERS in biosensing is commendable and is progressing toward practical application, enhanced by the convergence of microfluidic chips, CRISPR technology, and the burgeoning fields of machine learning and artificial intelligence. The advancement in energy technology further bolsters the applicability of SERS sensors in field detection. It is anticipated that an increasing array of SERS products will emerge, genuinely applicable to biochemical and molecular sensing and disease diagnostics.

9.3 Sensor Synthesis: Merging Technologies for Superior Diagnostic Proficiency

SERS has emerged as a groundbreaking analytical technique in recent years, revolutionizing the field of molecular detection and analysis. By combining the molecular fingerprinting capability of Raman spectroscopy with the signal enhancement provided by plasmonic nanostructures, SERS has opened new possibilities in sensing and diagnostic applications. The integration of SERS with other analytical technologies has further expanded its capabilities, creating powerful hybrid platforms that offer unprecedented sensitivity, selectivity, and reliability.

The fundamental principles of SERS technology rely on two primary enhancement mechanisms. The electromagnetic enhancement occurs through LSPR, while chemical enhancement takes place through charge transfer between analytes and substrate. Advancements in nanofabrication methodologies have facilitated the production of increasingly intricate SERS substrates, yielding enhancement factors surpassing the threshold of 10^{10}. This remarkable enhancement has pushed detection limits to single-molecule levels, making SERS an invaluable tool for trace analysis.

Substrate innovation has played a crucial role in advancing SERS technology. Modern fabrication techniques have enabled the development of hierarchical nanostructures that combine multiple plasmonic materials, maximizing enhancement effects while maintaining stability. Self-assembled monolayers with precise molecular spacing have improved reproducibility and quantification capabilities. Three-dimensional architectures have been designed to maximize hotspot density, significantly increasing detection sensitivity. Additionally, the emergence of stimuli-responsive substrates has enabled dynamic signal modulation, adding a new dimension to SERS analysis.

9.3.1 Electrochemical-SERS Hybrid Systems

The integration of SERS with electrochemical techniques has revolutionized analytical capabilities by creating powerful hybrid systems capable of simultaneous spectroscopic and electrochemical measurements. These EC-SERS platforms provide both structural information about molecular species and insights into electron transfer processes, enabling comprehensive characterization of complex biological systems. Modern EC-SERS instruments feature sophisticated in situ cells with precise potential control, enabling real-time monitoring of electrochemical reactions. The development of flow-through systems has further enhanced the capability for dynamic measurements, while miniaturization efforts have led to portable platforms suitable for field applications. In the diagnostic realm, EC-SERS has demonstrated remarkable potential, particularly in detecting disease biomarkers with unprecedented sensitivity and specificity. The synergistic combination leverages SERS's molecular fingerprinting capabilities with electrochemical techniques' rapid and quantitative nature, allowing for multimodal analysis of biological samples. Recent developments have focused on improving stability and reproducibility through optimized nanostructure design and surface chemistry, while new biocompatible materials have expanded the range of applicable biological samples. The technology has shown promise, particularly in cancer diagnostics, enabling early detection through ultrasensitive measurement of tumor markers and circulating tumor cells. Similarly, applications in pathogen detection have demonstrated rapid identification of bacteria and viruses, with potential for antimicrobial susceptibility testing. Jiang et al. designed an electrochemically regenerated SERS sensor, intended as a versatile tool for detecting metabolites associated with hydrogen peroxide (H_2O_2). Consequently, this SERS sensor emerges as a novel platform for the detection of H_2O_2-related metabolites in biological fluids, potentially facilitating early diagnosis of metabolic disorders [71] (Figure 9.21).

An innovative SERS sensor, developed by Wang et al., has been introduced for the quantitative assessment of miRNA-21-5p and miRNA-let-7a levels in serum and cerebrospinal fluid (CSF). This approach enables in vivo quantitative detection and sustained monitoring of these miRNAs. It holds promise as a potent tool for biomolecule analysis and may offer insights into the diagnosis and treatment of numerous diseases [72] (Figure 9.22).

Real-time monitoring capabilities have proven valuable for tracking therapeutic responses and disease progression. Despite these advances, challenges remain in

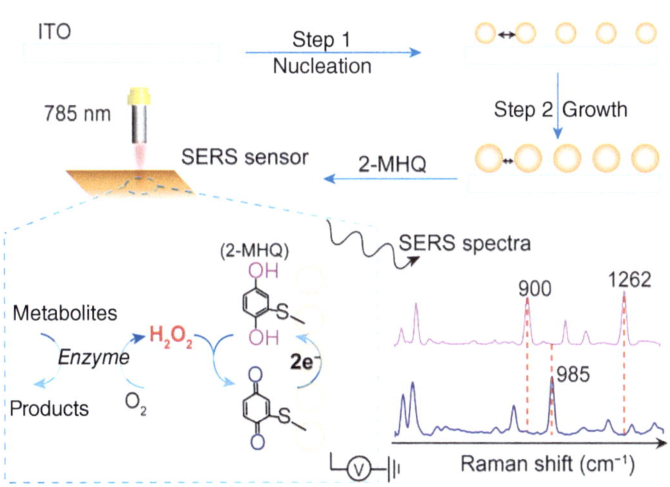

Figure 9.21 Illustration of the electrochemically renewable SERS sensor for H_2O_2-involved metabolites detection. *Source:* Ref. [71] / with permission of Elsevier.

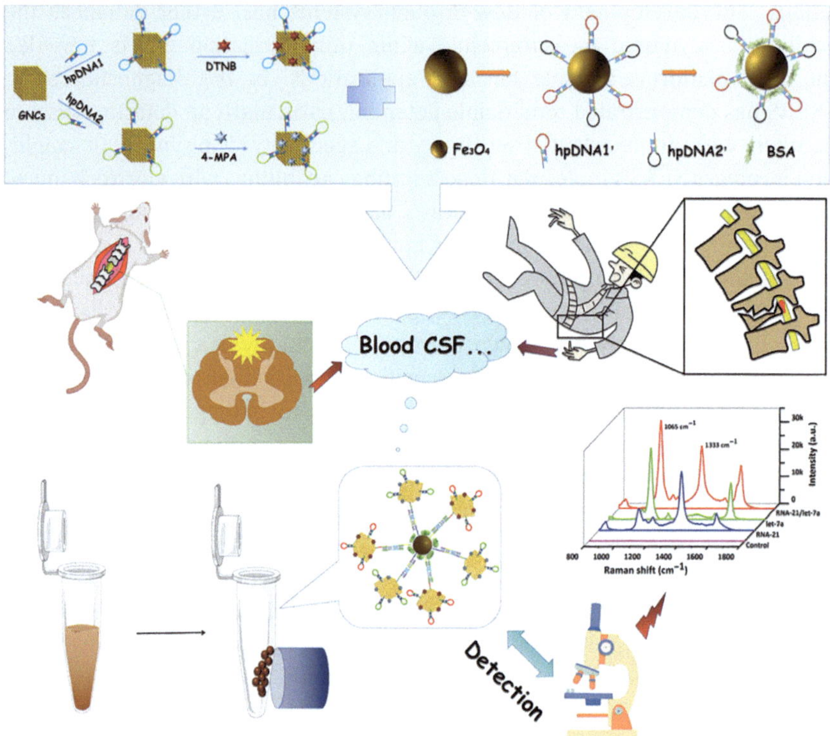

Figure 9.22 Flow diagram of preparing SERS signal probe by coupling hpDNA nanoparticles and quantitatively detecting target miRNA in rat models and clinical samples. *Source:* Ref. [72] / with permission of American Chemical Society.

standardization, cost reduction, and clinical validation. Ongoing research aims to address these limitations through improved fabrication techniques, novel nanomaterial designs, and extensive clinical studies. The continued evolution of EC-SERS technology, coupled with advances in data analytics and device miniaturization, positions it as a powerful tool for next-generation diagnostic platforms, particularly in personalized medicine and telemedicine applications. Future developments may focus on expanding multiplexing capabilities, enhancing sensitivity through advanced plasmonic materials, and developing more robust calibration methods for quantitative analysis. The integration of artificial intelligence and cloud-based platforms could further enhance diagnostic capabilities, potentially enabling real-time disease monitoring and predictive diagnostics.

9.3.2 Microfluidic Integration in SERS

The integration of microfluidics represents another significant advancement in SERS technology. By combining SERS with microfluidic platforms, researchers have addressed several limitations of traditional SERS analysis. These integrated systems provide improved sample handling, precise volume control, and enhanced mixing capabilities. The development of paper-based microfluidic SERS devices has further simplified sample processing while significantly reducing costs. These innovations have made SERS more suitable for point-of-care diagnostics and field-testing applications. Microfluidic SERS platforms offer several key advantages, including automated sample preparation, reduced reagent consumption, and controlled analyte delivery to hotspots. Advanced microfluidic designs incorporate multiple functional elements such as mixing chambers, separation channels, and concentration zones, enabling complex sample processing workflows. The integration of droplet-based microfluidics has enabled single-cell analysis and high-throughput screening capabilities. Novel fabrication techniques, including 3D printing and soft lithography, have facilitated the development of sophisticated microfluidic architectures with optimized SERS detection zones. The incorporation of gradient generators allows for automated concentration-dependent measurements, while integrated valves and pumps enable precise flow control. Recent developments in optofluidic systems have enhanced light–matter interactions within microchannels, leading to improved SERS enhancement factors. Paper-based platforms have emerged as cost-effective alternatives, offering advantages such as capillary-driven flow, natural filtering capabilities, and compatibility with colorimetric readouts. Zhou et al. clinically validated an sEV Glycan PHenotype (EV-GLYPH) test, leveraging a microfluidic system coupled with SERS for comprehensive profiling of small extracellular vesicle (sEV) glycans in non-small cell lung cancer (NSCLC). Their findings underscore the potential of sEV glycan profiling for noninvasive diagnostic and prognostic applications, paving the way for promising clinical utilities and enhancing our understanding of sEV glycosylation in lung cancer [73] (Figure 9.23).

Ho et al. introduced a novel droplet microfluidic platform integrated with SERS-based aptasensor for the rapid and sensitive detection of HER2-positive exosomes from breast cancer cells [74] (Figure 9.24).

These advances in microfluidic SERS technology have significantly expanded its applications in biomedical diagnostics, environmental monitoring, and food safety testing. Ongoing research focuses on improving device reproducibility, developing standardized fabrication protocols, and enhancing detection sensitivity through optimized microfluidic designs. The continued evolution of integrated microfluidic-SERS platforms promises to further advance the field of rapid, sensitive, and user-friendly analytical devices.

9.3.3 SERS-Fluorescence Dual-Mode Detection

The combination of SERS with fluorescence spectroscopy has created powerful dual-mode detection platforms. This integration leverages the complementary strengths of both techniques: while SERS provides detailed structural information at the molecular level, fluorescence offers excellent sensitivity and dynamic range. The dual-mode approach enhances reliability through orthogonal validation of results. Recent innovations include switchable platforms that can alternate between SERS and fluorescence modes based on specific molecular triggers, offering

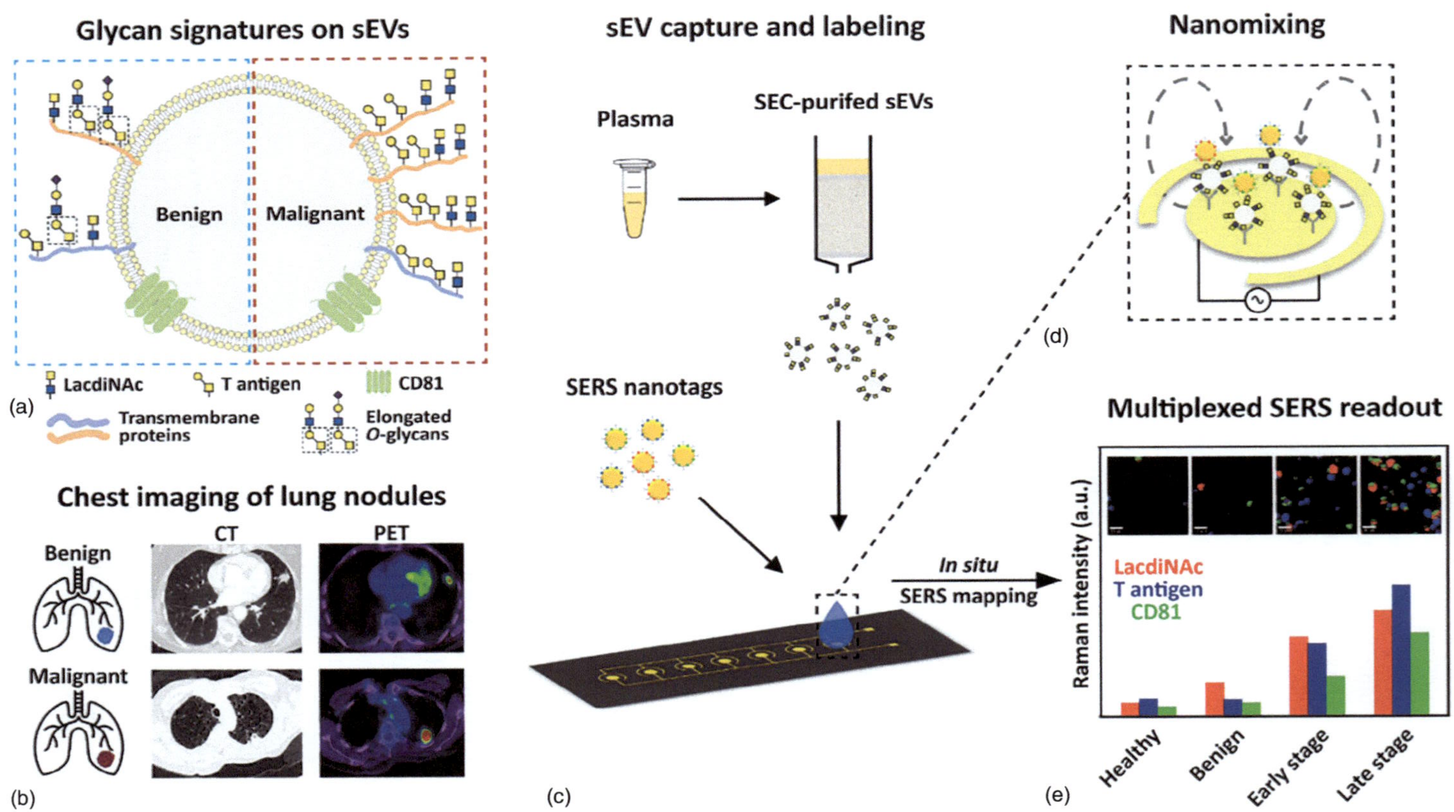

Figure 9.23 Principle of EV-GLYPH assay for early-stage NSCLC identification. (a) Glycan signatures on sEVs derived from patients with benign and malignant lung diseases. (b) CT and PET images of benign and malignant lung nodules. (c) Working scheme of EV-GLYPH assay. SEC-purified sEVs from plasma are captured by anti-MUC1 antibody immobilized on an electrode and subsequently labeled with SERS nanotags against LacdiNAc, T antigen, and CD81. (d) A nanomixing fluid flow is generated on the electrode surface by the applied ac-EHD field. (e) With in situ SERS mapping, the average Raman intensities and false-color SERS spectral images (insets) are established based on the characteristic Raman signals of three SERS nanotags (WFA-MBA, 1,075 cm^{-1}, red; PNA-TFMBA, 1,375 cm^{-1}, blue; anti-CD81-DTNB, 1,335 cm^{-1}, green), representing the expression of LacdiNAc, T antigen, and CD81, respectively. *Source:* Reproduced with permission from Ref. [73] / John Wiley & Sons / CC BY 4.0.

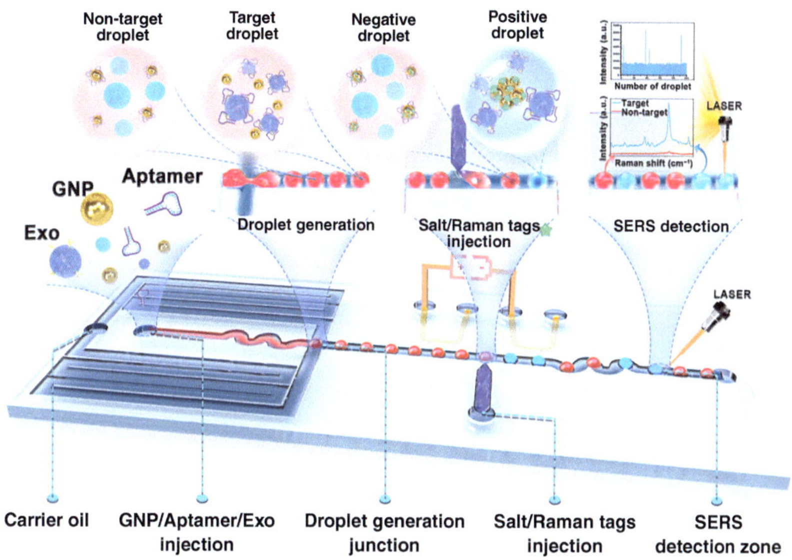

Figure 9.24 Illustration of the process of SERS-based droplet microfluidic platform for detecting HER2-positive exosomes. *Source:* Ref. [74] / with permission of American Chemical Society.

unprecedented flexibility in analytical applications. The synergistic integration of these techniques has enabled multiplexed detection strategies where SERS and fluorescence signals can be simultaneously monitored, providing comprehensive molecular fingerprinting. Advanced plasmonic nanostructures have been developed to enhance both SERS and fluorescence signals through optimized near-field effects. These engineered materials often incorporate specific surface chemistries to maintain optimal distances between fluorophores and metal surfaces, preventing quenching while maximizing enhancement. The development of rationally designed molecular probes that can generate both SERS and fluorescence signals has further expanded the capabilities of these dual-mode systems. Huang et al. introduced a dual-mode biosensor combining SERS and fluorescence (FL) for the detection of hepatocellular carcinoma (HCC)-associated microRNA (miR-224), utilizing a well-organized gold nanoarray (Au NA) substrate integrated with a target-catalyzed hairpin assembly (CHA) approach. This innovative biosensor offers novel pathways for early prediction, alerting, monitoring, and staging of HCC [75] (Figure 9.25).

Li et al. reported the construction of highly stable SERS/fluorescent dual-mode peptide probes, facilitating precise collagen fingerprint profiling for hepatic fibrosis staging. These dual-mode probes, SF-I, SF-IV, and SF-D, comprise silver nanoparticles (Ag NPs), Raman reporters, and FAM-labeled collagen targeting peptides (F-CTP), stabilized by electrostatic repulsion between Ag NPs and F-CTP. SF-I, SF-IV, and SF-D demonstrate strong binding affinity to type I, type IV, and denatured

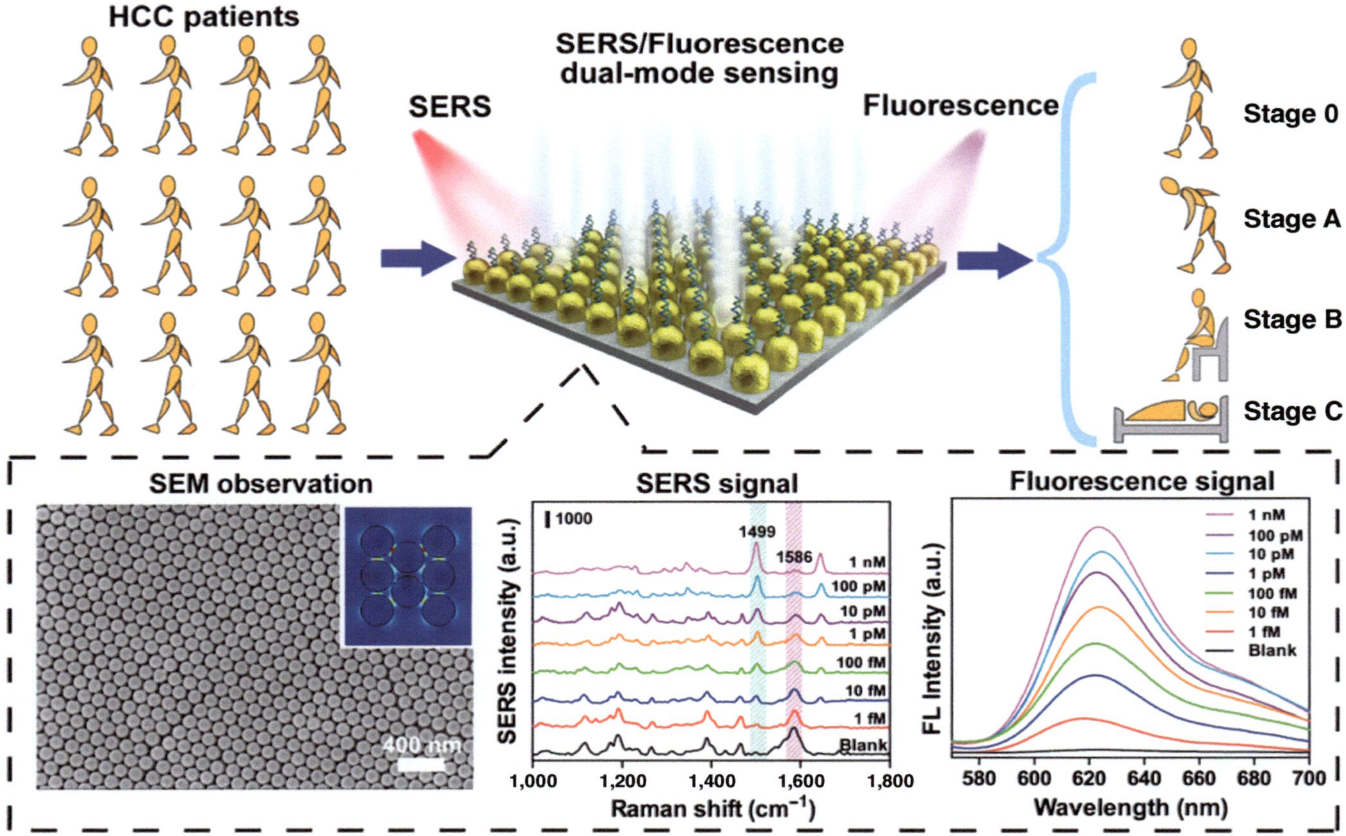

Figure 9.25 Schematic Illustration of procedures of the Au NA substrate-based SERS/FL dual-mode sensing and application for clinical staging. *Source:* Reproduced with permission from Ref. [75] / American Chemical Society.

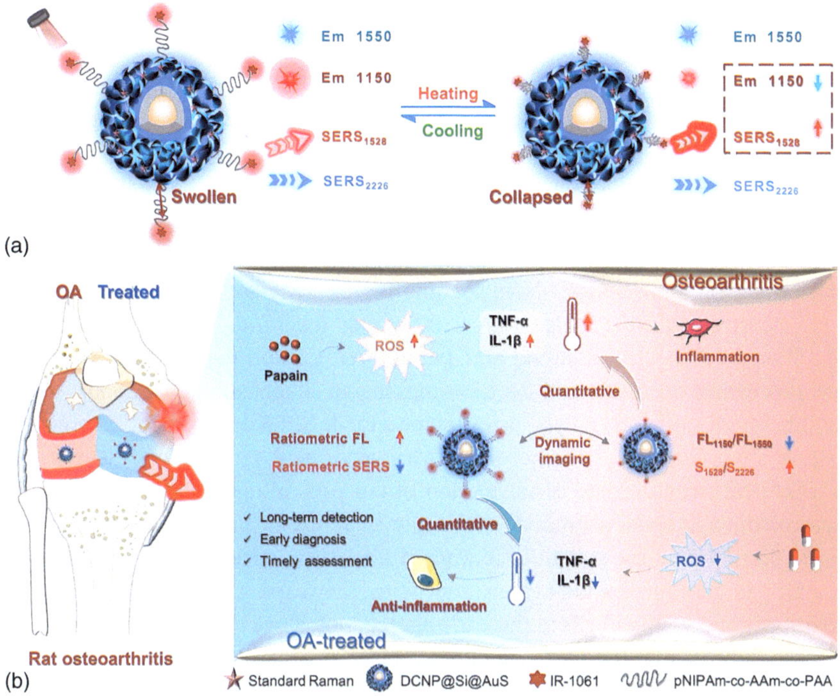

Figure 9.26 Schematic Illustration of fluorescence-guided SERS imaging for hepatic fibrosis staging. The construction of the SERS/fluorescence peptide probes SF-I, SF-IV, SF-D (a) and their fluorescence-guided SERS imaging of multiplex collagen in fibrotic liver tissues across various stages (b). *Source:* Reproduced with permission from Ref. [76] / American Chemical Society.

collagen, respectively, in both solution and tissues. The fluorescence-guided SERS imaging method has suggested exceptional accuracy in determining changes in the type, localization, and quantity of collagen, thus revealing collagen's role in the development of hepatic fibrosis. This high-throughput collagen fingerprinting strategy is suitable for the simultaneous detection of multiple markers in complex biological systems and has a great potential in histopathology staining [76] (Figure 9.26).

These integrated platforms have found applications in single-molecule detection, cellular imaging, and real-time monitoring of biological processes. Ongoing research focuses on developing standardized calibration methods, improving signal processing algorithms for dual-mode data analysis, and expanding multiplexing capabilities through spectral optimization. The continued evolution of these hybrid detection systems promises to advance our understanding of complex molecular systems while providing robust analytical tools for practical applications.

9.3.4 Advanced SERS Instrumentation and AI Analytics

Modern SERS instrumentation has evolved to incorporate advanced optical systems that enhance measurement capabilities. Confocal microscopy integration has improved spatial resolution, enabling detailed mapping of sample surfaces. The development of tip-enhanced Raman spectroscopy (TERS) has pushed the boundaries of spatial resolution to the nanoscale, allowing for precise molecular imaging. Ultrafast spectroscopy capabilities have added temporal resolution to SERS measurements, enabling the study of dynamic processes at unprecedented timescales. Data processing and analysis have become increasingly sophisticated in modern SERS systems. Advanced algorithms, including machine learning approaches, have improved spectral analysis and pattern recognition capabilities. Real-time data processing enables rapid decision-making in diagnostic applications, while cloud-based storage and analysis facilities have made SERS data more accessible and easier to share across research networks. Mrđenović et al. reported on research into the chemical structure and conformation of tau protein aggregates at the growth phase of the aggregation process. While the structure of tau fibrils is dominated by the cross-β pattern, still little is known about the structural rearrangements occurring in the course of tau aggregation. Thus, they captured the TERS spectra for protofibrils and young fibrils at the early aggregation stage to reveal their structure and gain insights into the TERS signature related to structural transitions resulting from the ongoing aggregation. They applied multivariate data analysis to treat the TERS data acquired from protofibrils and young fibrils to reveal discrepancies in the structure of tau aggregates in time [77] (Figure 9.27).

Ho et al. compiled an extensive dataset of bacterial Raman spectra and applied deep learning techniques to precisely identify 30 prevalent bacterial pathogens. Their methodology holds promise for culture-independent pathogen identification and antibiotic susceptibility testing, with potential for broad application in diagnostics involving blood, urine, and sputum samples [78] (Figure 9.28).

9.3.5 Challenges and Future Horizons

Despite these advances, several challenges remain in the development of integrated SERS systems. Standardization of substrate fabrication remains a critical issue, as does the need for improved signal reproducibility and quantification. Long-term stability of hybrid platforms and cost-effective manufacturing processes are ongoing areas of research. However, these challenges also present opportunities for innovation and improvement. Looking to the future, several promising directions are emerging in SERS technology development. Artificial intelligence-driven system optimization promises to improve performance and reliability, while the development of wearable SERS devices could enable continuous monitoring applications. Integration with the Internet of Things (IoT) could create networks of

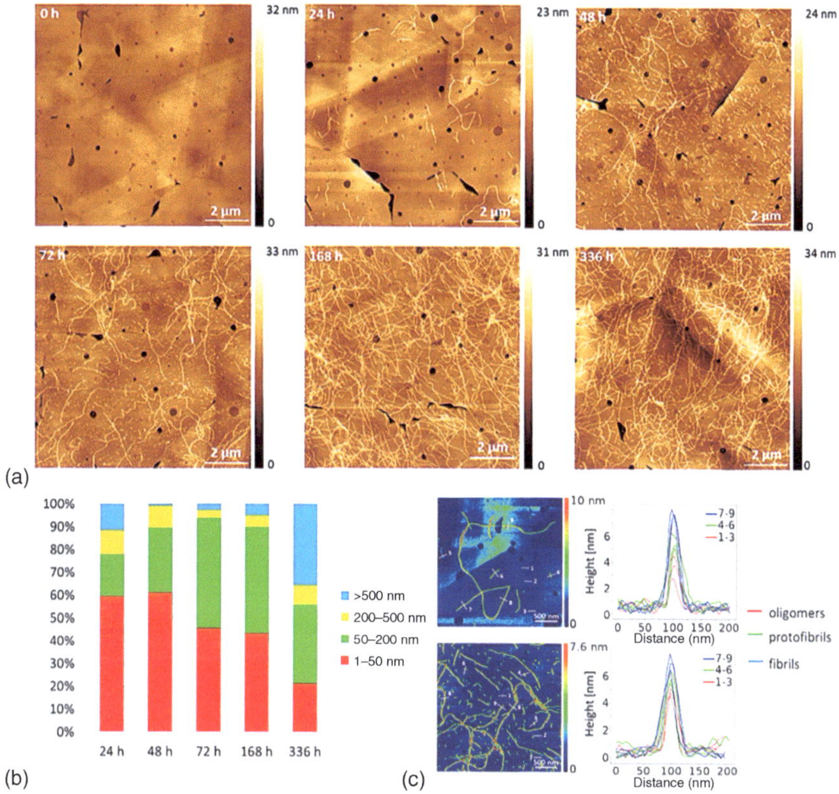

Figure 9.27 Monitoring of morphological changes during the aggregation of tau protein: (a) AFM topography images of tau protein aggregating within 0–336 h, (b) ratios of tau aggregates displaying length of aggregates in the ranges: 1–50 nm, 50–200 nm, 200–500 nm, and above 500 nm, (c) AFM images of tau protein aggregating for 24 and 72 h together with cross-section profiles through oligomers, protofibrils, and fibrils. *Source:* Reproduced with permission from Ref. [77] / The Royal Society of Chemistry.

SERS sensors for distributed sensing applications, while automated calibration and quality control systems could improve reliability and ease of use.

The synthesis of SERS with complementary analytical technologies represents a significant step forward in sensing and diagnostic capabilities. Through continued innovation in substrate design, instrumental integration, and data analysis, these hybrid platforms are poised to revolutionize various fields, from medical diagnostics to environmental monitoring. As these technologies continue to evolve, we can expect to see even more powerful and versatile diagnostic platforms emerging, offering unprecedented insights into molecular processes and enabling more effective decision-making across a wide range of applications.

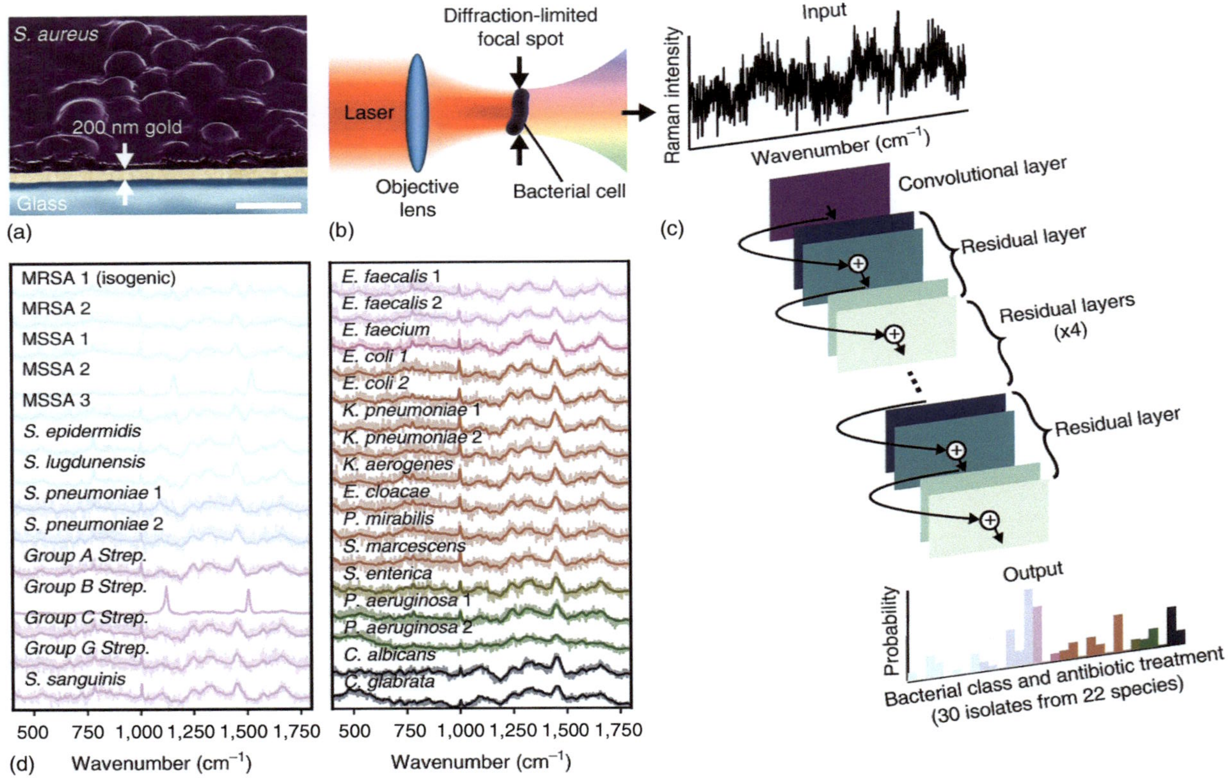

Figure 9.28 A convolutional neural network (CNN) can be used to identify bacteria from the Raman spectra. (a) To build a training dataset of the Raman spectra, we deposit bacterial cells onto gold-coated silica substrates and collect spectra from 2,000 bacteria over monolayer regions for each strain. An SEM cross-section of the sample is shown (gold-coated to allow for visualization of bacteria under electron beam illumination). Scale bar is 1 μm. (b) Conceptual measurement schematic: by focusing the excitation laser source to a diffraction-limited spot size, the Raman signal from single cells can be acquired. (c) Using a one-dimensional residual network with 25 total convolutional layers (see *Methods* for details), the low-signal Raman spectra are classified as one of 30 isolates, which are then grouped by empiric antibiotic treatment. (d) Raman spectra of bacterial species can be difficult to distinguish, and short integration times (1 s) lead to the noisy spectra (SNR = 4.1). Averages of 2,000 spectra from 30 isolates are shown in bold and overlaid on representative examples of noisy single spectra for each isolate. The spectra are color-grouped according to antibiotic treatment. These reference isolates represent over 94% of the most common infections seen at Stanford Hospital in the years 2016–2017. *Source:* Reproduced with permission from Ref. [78] / Springer Nature / CC BY 4.0.

References

1 Xu, C., Wu, F., Yu, P. et al. (2019). In vivo electrochemical sensors for neurochemicals: recent update. *ACS Sensors* 4: 3102–3118.

2 Liu, W., Dong, H., Zhang, L. et al. (2017). Development of an efficient biosensor for the in vivo monitoring of cu^{+} and pH in the brain: rational design and synthesis of recognition molecules. *Angewandte Chemie, International Edition* 56: 16328–16332.

3 Huang, S., Zhang, L., Dai, L. et al. (2021). Nonenzymatic electrochemical sensor with ratiometric signal output for selective determination of superoxide anion in rat brain. *Analytical Chemistry* 93: 5570–5576.

4 Zhu, A., Qu, Q., Shao, X. et al. (2012). Carbon-dot-based dual-emission nanohybrid produces a ratiometric fluorescent sensor for in vivo imaging of cellular copper ions. *Angewandte Chemie, International Edition* 51: 7185–7189.

5 Zhang, L. and Tian, Y. (2018). Designing recognition molecules and tailoring functional surfaces for in vivo monitoring of small molecules in the brain. *Accounts of Chemical Research* 51: 688–696.

6 Chai, X., Zhou, X., Zhu, A. et al. (2013). A two-channel ratiometric electrochemical biosensor for in vivo monitoring of copper ions in a rat brain using gold truncated octahedral microcages. *Angewandte Chemie, International Edition* 52: 8129–8133.

7 Luo, Y., Zhang, L., Liu, W. et al. (2015). A single biosensor for evaluating the levels of copper ion and L-cysteine in a live rat brain with Alzheimer's disease. *Angewandte Chemie, International Edition* 54: 14053–14056.

8 Wang, Y., Zhang, L., and Tian, Y. (2022). Rational design of electrochemical molecular probes for highly selective and long-term measurement in vivo. *Journal of Electrochemistry* 28: 2108451.

9 Dong, H., Zhou, Q., Zhang, L. et al. (2019). Rational design of specific recognition molecules for simultaneously monitoring of endogenous polysulfide and hydrogen sulfide in the mouse brain. *Angewandte Chemie, International Edition* 58: 13948–13953.

10 Yadav, P.K., Martinov, M., Vitvitsky, V. et al. (2016). Biosynthesis and reactivity of cysteine persulfides in signaling. *Journal of the American Chemical Society* 138: 289–299.

11 Liu, Y., Liu, Z., Zhao, F. et al. (2021). Long-term tracking and dynamically quantifying of reversible changes of extracellular Ca^{2+} in multiple brain regions of freely moving animals. *Angewandte Chemie, International Edition* 60: 14429–14437.

12 Li, H., Arroyo-Curras, N., Kang, D. et al. (2016). Dual-reporter drift correction to enhance the performance of electrochemical aptamer-based sensors in whole blood. *Journal of the American Chemical Society* 138: 15809–15812.

13 Liu, F., Dong, H., and Tian, Y. (2019). Real-time monitoring of peroxynitrite ($ONOO^{-}$) in the rat brain by developing a ratiometric electrochemical biosensor. *Analyst* 144: 2150–2157.

14 Zhou, J., Zhang, L., and Tian, Y. (2016). Micro electrochemical pH sensor applicable for real-time ratiometric monitoring of pH values in rat brains. *Analytical Chemistry* 88: 2113–2118.

15 Zhao, F., Zhang, L., Zhu, A. et al. (2016). In vivo monitoring of local pH values in a live rat brain based on the design of a specific electroactive molecular for H^+. *Chemical Communications* 52: 3717–3720.

16 Qian, Y., Zhang, L., and Tian, Y. (2022). Highly stable electrochemical probe with bidentate thiols for ratiometric monitoring of endogenous polysulfide in living mouse brains. *Analytical Chemistry* 94 (2): 1447–1455.

17 Zhang, C., Liu, Z., Zhang, L. et al. (2020). A robust au−C≡C functionalized surface: toward real-time mapping and accurate quantification of Fe^{2+} in the live brains of AD mouse models. *Angewandte Chemie, International Edition* 59: 20499–20507.

18 Feng, T., Ji, W., Zhang, Y. et al. (2020). Zwitterionic polydopamine engineered interface for in vivo sensing with high biocompatibility. *Angewandte Chemie, International Edition* 56: 23445–23449.

19 Wang, X., Xu, T., Zhang, Y. et al. (2021). In vivo detection of redox-inactive neurochemicals in the rat brain with an ion transfer microsensor. *ACS Sensors* 6: 2757.

20 Gu, C., Kong, F., and Liang, S. (2024). In vivo dynamic tracking of cerebral chloride regulation usign molecularly tailored liquid/liquid interfacial ultramicro iontronics. *Science Advances* 10: eadr7218.

21 Robinson, D.L., Hermans, A., Seipel, A.T. et al. (2008). Monitoring rapid chemical communication in the brain. *Chemical Reviews* 108: 2554.

22 ÖNeill, R. (1994). Microvoltammetric techniques and sensors for monitoring neurochemical dynamics in vivo. A review. *Analyst* 119: 767–777.

23 Wu, F., Yu, P., and Mao, L. (2017). Self-powered electrochemical systems as neurochemical sensors: toward self-triggered in vivo analysis of brain chemistry. *Chemical Society Reviews* 46: 2692–2704.

24 Lu, J., Zhuang, X., Wei, H. et al. (2024). Enzymatic galvanic redox potentiometry for in vivo biosensing. *Analytical Chemistry* 96.

25 Pan, C., Wu, F., Mao, J. et al. (2022). Highly stable and selective sensing of hydrogen sulfide in living mouse brain with NiN_4 single-atom catalyst-based galvanic redox potentiometry. *Journal of the American Chemical Society* 144: 14678–14686.

26 Zhao, F., Liu, Y., Dong, H. et al. (2017). An electrochemophysiological microarray for real-time monitoring and quantification of multiple ions in the brain of a freely moving rat. *Angewandte Chemie, International Edition* 59: 10426–10430.

27 Zhu, F., Xue, Y., Ji, W. et al. (2023). Galvanic redox potentiometry for fouling-free and stable serotonin sensing in a living animal brain. *Angewandte Chemie, International Edition* 62: e202212458.

28 Liu, J., Fu, B., and Zhang, Z. (2020). Ionic current rectification triggered photoelectrochemical chiral sensing platform for recognition of amino acid enantiomers on self-standing nanochannel arrays. *Analytical Chemistry* 92: 8670–8674.

29 Fu, B. and Zhang, Z. (2019). Rationally engineered photonic-plasmonic synergistic resonators in second near-infrared window for in vivo photoelectrochemical biodetection. *Nano Letters* 19: 9069–9074.

30 Xu, X., Xu, D., Lu, S. et al. (2024). Atomically dispersed recognition unit for selective in vivo photoelectrochemical medicine detection. *Nature Communications* 15: 8827.

31 Li, G., Alshalalfeh, M., Yang, Y. et al. (2021). Can one measure resonance Raman optical activity? *Angewandte Chemie, International Edition* 60: 22004.

32 Steinke, M., Zunhammer, F., Chatzopoulou, E.I. et al. (2018). Rapid analysis of cell-nanoparticle interactions using single-cell Raman trapping microscopy. *Angewandte Chemie, International Edition* 57: 4946.

33 Feng, E., Zheng, T., He, X. et al. (2023). Plasmon-induced charge transfer-enhanced Raman scattering on a semiconductor: toward amplification-free quantification of SARS-CoV-2. *Angewandte Chemie, International Edition* 62: e202309249.

34 Hess, C. (2021). New advances in using Raman spectroscopy for the characterization of catalysts and catalytic reactions. *Chemical Society Reviews* 50: 3519–3564.

35 Tran, V., Walkenfort, B., König, M. et al. (2019). Rapid, quantitative, and ultrasensitive point-of-care testing: a portable SERS reader for lateral flow assays in clinical chemistry. *Angewandte Chemie, International Edition* 58: 442.

36 Thaveesangsakulthai, I., Chatdarong, K., Somboonna, N. et al. (2024). A large scale study of portable sweat test sensor for accurate, non-invasive and rapid COVID-19 screening based on volatile compound marker detection. *Scientific Reports* 14: 20148.

37 Moothanchery, M., Perumal, J., Mahyuddin, A.P. et al. (2022). Rapid and sensitive detection of ovarian cancer biomarker using a portable single peak Raman detection method. *Scientific Reports* 12459.

38 Murphy, N., Tipping, W.J., Braddick, H.J. et al. (2023). Expanding the range of bioorthogonal tags for multiplex stimulated Raman scattering microscopy. *Angewandte Chemie, International Edition* 62: e202311530.

39 Saar, B.G., Zeng, Y., Freudiger, C.W. et al. (2010). Label-free, real-time monitoring of biomass processing with stimulated Raman scattering microscopy. *Angewandte Chemie, International Edition* 49: 5476–5479.

40 Hu, F., Shi, L., and Min, W. (2019). Biological imaging of chemical bonds by stimulated Raman scattering microscopy. *Nature Methods* 16: 830–842.

41 Xiong, H., Qian, N., and Min, W. (2019). Stimulated Raman excited fluorescence spectroscopy of visible dyes. *Journal of Physical Chemistry Letters* 10 (13): 3563–3570.

42 Timea, F., Andreas, K., and Torsten, F. (2020). Recent advances in nano-photonic techniques for pharmaceutical drug monitoring with emphasis on Raman spectroscopy. *Nano* 9: 19–37.

43 Chaudhry, I., Hu, G., Ye, H. et al. (2024). Toward modeling the complexity of the chemical mechanism in SERS. *ACS Nano* 18: 20835–20850.

44 Yang, H., He, L., Hu, Y. et al. (2015). Quantitative detection of photothermal and photoelectrocatalytic effects induced by SPR from au@Pt nanoparticles. *Angewandte Chemie, International Edition* 54: 11462–11466.

45 Kannan, P.K., Shankar, P., Blackman, C. et al. (2019). Recent advances in 2D inorganic nanomaterials for SERS sensing. *Advanced Materials* 31: 1803432.

46 Ding, S., Yi, J., Li, J. et al. (2016). Nanostructure-based plasmon-enhanced Raman spectroscopy for surface analysis of materials. *Nature Reviews Materials* 1: 16021.

47 Valley, N., Greeneltch, N., Van Duyne, R.P. et al. (2013). A look at the origin and magnitude of the chemical contribution to the enhancement mechanism of surface-enhanced Raman spectroscopy (SERS): theory and experiment. *Journal of Physical Chemistry Letters* 4: 2599–2604.

48 Lane, L.A., Qian, X., Nie, S. et al. (2015). SERS nanoparticles in medicine: from label-free detection to spectroscopic tagging. *Chemical Reviews* 115: 10489–10529.

49 Zheng, X., Ye, Z., Akmal, Z. et al. (2024). Recent progress in SERS monitoring of photocatalytic reactions. *Chemical Society Reviews* 53: 656–683.

50 Dong, J., Zhang, X., Briega-Martos, V. et al. (2019). In situ Raman spectroscopic evidence for oxygen reduction reaction intermediates at platinum single-crystal surfaces. *Nature Energy* 4: 60–67.

51 Das, R., Parveen, S., Bora, A. et al. (2020). Origin of high photoluminescence yield and high SERS sensitivity of nitrogen-doped graphene quantum dots. *Carbon* 160: 273–286.

52 Itoh, T., Procházka, M., Dong, Z.-H. et al. (2023). Toward a new era of SERS and TERS at the nanometer scale: from fundamentals to innovative applications. *Chemical Reviews* 123: 1552–1634.

53 Peng, W., Liu, Y., Lu, M. et al. (2024). Advances in surface-enhanced Raman scattering detection of foodborne pathogens: from recognition-based fingerprint to molecular diagnosis. *Coordination Chemistry Reviews* 518: 216083.

54 Granger, J.H., Schlotter, N.E., Crawford, A.C. et al. (2016). Prospects for point-of-care pathogen diagnostics using surface-enhanced Raman scattering (SERS). *The Sociological Review* 14: 3865–3882.

55 Alcantara, L.C.J., Amenga-Etego, L., Andersson, R. et al. (2022). Methods for fighting emerging pathogens. *Nature Methods* 19: 395–397.

56 Payne, T.D., Klawa, S.J., Jian, T. et al. (2021). Catching COVID: engineering peptide-modified SERS sensors for SARS-CoV-2. *ACS Sensors* 6: 3436–3444.

57 Te Velthuis, A.J.W. and E, F. (2016). Influenza virus RNA polymerase: insights into the mechanisms of viral RNA synthesis. *Nature Reviews. Microbiology* 14: 479–493.

58 Zhou, P., Yang, X., Wang, X. et al. (2020). A pneumonia outbreak associated with a new coronavirus of probable bat origin. *Nature* 579: 270–273.

59 Yang, Y., Peng, Y., Lin, C. et al. (2021). Human ACE2-functionalized gold "virus-trap" nanostructures for accurate capture of SARSCoV-2 and single-virus SERS detection. *Nano Micro Letters* 13: 109.

60 Guan, P.-C., Zhang, H., Li, Z.-Y. et al. (2022). Rapid point-of-care assay by SERS detection of SARS-CoV-2 virus and its variants. *Analytical Chemistry* 94: 17795–17802.

61 Shankar, S., Pan, J., Yang, P. et al. (2024). Viral DNA polymerase structures reveal mechanisms of antiviral drug resistance. *Cell* 187: 5572–5586.

62 Chan, Y.K. and Gack, M.U. (2016). Viral evasion of intracellular DNA and RNA sensing. *Nature Reviews. Microbiology* 14: 360–373.

63 Karagoz, A., Tombuloglu, H., Alsaeed, M. et al. (2023). Monkeypox (mpox) virus: classification, origin, transmission, genome organization, antiviral drugs, and molecular diagnosis. *Journal of Infection and Public Health* 16: 531–541.

64 Zhang, Z., Jiang, H., Jiang, S. et al. (2023). Rapid detection of the monkeypox virus genome and antigen proteins based on surface-enhanced Raman spectroscopy. *ACS Applied Materials & Interfaces* 15: 34419–34426.

65 Yu, Q., Li, J., Zheng, S. et al. (2023). Molybdenum disulfide-loaded multilayer AuNPs with colorimetric-SERS dual-signal enhancement activities for flexible immunochromatographic diagnosis of monkeypox virus. *Journal of Hazardous Materials* 459: 132136.

66 Tian, L., Jiang, Q., Liu, K.-K. et al. (2016). Bacterial nanocellulose-based flexible surface enhanced Raman scattering substrate. *Advanced Materials Interfaces* 3: 1600214.

67 Kim, J.A., Wales, D.J., Thompson, A.J. et al. (2020). Fiber-optic SERS probes fabricated using two-photon polymerization for rapid detection of bacteria. *Advanced Optical Materials* 8: 1901934.

68 Lin, C.-C., Yang, Y.-M., Liao, P.-H. et al. (2014). A filter-like AuNPs@MS SERS substrate for Staphylococcus aureus detection. *Biosensors & Bioelectronics* 53: 519.

69 Lin, C.-C., Lin, C.-Y., Kao, C.-J. et al. (2017). High efficiency SERS detection of clinical microorganism by AgNPs-decorated filter membrane and pattern recognition techniques. *Sensors and Actuators B: Chemical* 241: 513.

70 Wang, D., Ni, H., Wang, Z. et al. (2017). Discrimination of nosiheptide sources with plasmonic filters. *ACS Applied Materials & Interfaces* 9: 13049.

71 Jiang, L., Wang, L., Zhan, D.-S. et al. (2021). Electrochemically renewable SERS sensor: a new platform for the detection of metabolites involved in peroxide production. *Biosensors & Bioelectronics* 175: 112918.

72 Wang, C., Wang, C., Lu, W. et al. (2024). Novel SERS signal amplification strategy for ultrasensitive and specific detection of spinal cord injury-related miRNA. *ACS Sensors* 9: 736–744.

73 Zhou, Q., Niu, X., Zhang, Z. et al. (2024). Glycan profiling in small extracellular vesicles with a SERS microfluidic biosensor identifies early malignant development in lung cancer. *Advancement of Science* 11: 2401818.

74 Ho, K.H.W., Lai, H., Zhang, R. et al. (2024). SERS-based droplet microfluidic platform for sensitive and high-throughput detection of cancer exosomes. *ACS Sensors* 9: 4860–4869.

75 Huang, X., Tian, H., Huang, L. et al. (2023). Well-ordered au nanoarray for sensitive and reproducible detection of hepatocellular carcinoma-associated miRNA via CHA-assisted SERS/fluorescence dual-mode sensing. *Analytical Chemistry* 95: 5955–5966.

76 Li, Q., Xiao, S., Ge, X. et al. (2024). Temperature-activated near-infrared-II fluorescence and SERS dynamic-reversible probes for long-term assessment of osteoarthritis in vivo. *Angewandte Chemie, International Edition* 63: e202408792.

77 Mrđenović, D., Ge, W., Kumar, N. et al. (2022). Nanoscale chemical imaging of human cell membranes using tip-enhanced Raman spectroscopy. *Angewandte Chemie, International Edition* 61: e202210288.

78 Chi-Sing Ho, C.-S., Jean, N. et al. (2019). Rapid identification of pathogenic bacteria using Raman spectroscopy and deep learning. *Nature Communication* 10: 4927.

10

Redefining Point-of-Care Testing

Rapid Diagnostic Innovations

Ryan Amini, Jian Xiao Ma, Cheryl Tong, and Yingfu Li

Department of Biochemistry and Biomedical Sciences, McMaster University, 1280 Main Street West, Hamilton, Ontario, L8S 4K1, Canada

CHAPTER MENU

The techniques discussed in the preceding chapters have all demonstrated their efficacy and reliability as diagnostic tests, as well as their potential for future innovations. Currently, however, each of these techniques suffers from the same flaw—an inability to provide rapid and simple results at the point-of-care (POC) delivery (Table 10.1). Polymerase chain reaction (PCR) methods, which are often described as the gold standard for diagnostic testing, are expensive, laborious, and time-consuming [1, 2]. Tests cannot be completed in a home setting, given that they require costly and bulky thermal cycler machinery. The requirement to travel to a local clinic or hospital with laboratory professionals is also inconvenient for the end user, and delays are expected when an increasing number of people require testing. Amidst the height of the COVID-19 pandemic, RT-PCR tests would require four to eight hours to process samples and an additional one to three days to report results [3, 4]. Unfortunately, enzyme-linked immunosorbent assay (ELISA) tests and similar serological assays suffer from the same drawbacks. Although they are inexpensive and provide a faster turnaround time (i.e., three to four hours) compared to PCR [5], they also require specialized equipment (i.e., antigen-coated microplates, microplate washers, and microplate readers) that are operated by trained laboratory personnel. While there is some optimism that newer biological discoveries like CRISPR-Cas could emerge as a rapid diagnostic test, the current technology is not yet ready to replace traditional methods in clinical use [7, 8]. Given the scarcity of sequences that CRISPR/Cas proteins can detect, the absence

Diagnostic Techniques in Public Health: Analytical and Chemometric Approaches in Disease Detection and Management, First Edition. Edited by Jinghong Li and Yan Du.

Table 10.1 Common diagnostic tests and their limiting factors in becoming POC assays.

Diagnostic technique	Advantages as a diagnostic test	Limiting factors to become POC and general limitations	Refs.
RT-PCR	• Highly sensitive and specific • Real-time monitoring of sample amplification	• Long sample to answer time (several hours—several days) • Thermal cycler machine is expensive and requires skilled technicians to operate • Susceptibility to false-positive results in nonsterile environments	[1–4]
ELISA	• Low cost • Less machinery than gold-standard PCR tests • Accurately detects IgG/IgM after several days of infection onset	• Requires specialized laboratory equipment to perform incubation, washing, and visualization steps • Requires skilled technicians • Susceptible to low sensitivity with lower viral loads	[5, 6]
CRISPR-Cas	• Simple • Inexpensive • Does not require special instrumentation	• Requires additional isothermal amplification methods to scale up targets • Requires skilled technicians • Challenges with multiplexing • Lack of standardized procedures	[7, 8]

of standardized procedures, challenges with multiplexing, and the common need to pair with amplification methods to decrease the limit of detection, more optimization is required before CRISPR/Cas can be the foundation for future diagnostic tests [9].

These limitations highlight the critical need for point-of-care testing (POCT). POCT refers to diagnostic testing performed near the patient in a home setting, offering immediate results without the need for sophisticated laboratory equipment. The ideal POC test, as denoted by the World Health Organization, should meet the "ASSURED" criteria: affordable, sensitive, specific, user-friendly, rapid and robust, equipment-free, and deliverable to the end user (Figure 10.1) [10]. Additionally, the test should ideally be safe to use, and the results should be concordant with established laboratory methods.

In comparison to conventional diagnostic methods, POCT provides a plethora of improved clinical outcomes, such as faster decision-making, improved adherence to treatment, reduced incidence of complications, quicker optimization of treatment, and heightened patient satisfaction [11]. In the context of infectious diseases, POCT

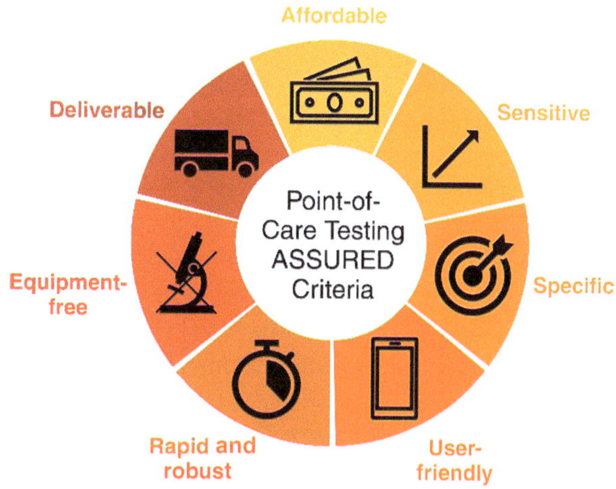

Figure 10.1 Schematic summary of POCT ASSURED criteria.

is imperative for allowing contact tracing and isolation [12]. Timely identification of diseases like COVID-19 is crucial for monitoring and preventing contagious outbreaks. Using the recent pandemic as an example, estimates suggest that a reduction in transmission could be as high as 80% with effective and reliable POCT [13]. The importance of POCT also extends to resource-limited settings where access to centralized laboratories may be restricted. By decentralizing testing and bringing it closer to the patient rather than the provider, POCT ensures timely intervention and management of viral and bacterial infections, thereby improving health outcomes and reducing healthcare disparities.

Given the recent challenges posed by the COVID-19 pandemic, the importance of POCT has become increasingly apparent and only continues to grow. The demand for POCT is evidenced by the growth in the global market, which is expected to increase from $50 billion USD in 2023 to $78 billion in 2028 [14]. Clearly, there is a need for rapid, accurate, and accessible diagnostic tools to manage and mitigate the spread of infectious diseases. This growing popularity underscores the necessity to continually refine, improve, and innovate POCT technologies to meet the evolving healthcare demands. In this chapter, we will highlight these advancements and innovations. We begin with a current overview of the past and present technology (i.e., devices and testing methods) that have grown in popularity over the last century. Then, we will review the integration of these platforms for the detection of deadly viruses such as SARS-CoV-2 and human immunodeficiency virus (HIV). We finally conclude with some perspective on the challenges that common POCT systems need to overcome, while also taking a brief look at the expected role that artificial intelligence (AI) may play in the coming years.

10.1 A New Wave in POCT: Tracking the Advances in Instruments and Methods

As biotechnology evolves and discovers new tools to diagnose diseases, there is always a follow-up pressure to simplify these tools into something more accessible and closer to the patient, hence the term point-of-care. POCT was explored as early as the 1950s, when on-site testing for blood analysis was being conducted and was referred to at the time as "near-patient testing" [15]. The term "point-of-care testing" was first coined by Dr. Gerald J. Kost after the development of a biosensor that was capable of monitoring ionized calcium levels in whole blood during hepatic transplantation [16]. This biosensor's development allowed Dr. Kost to perform whole-blood analysis in the operating room and immediately report the results to the anesthesiology and surgery team, who then used the information in surgery [17]. Another key breakthrough that occurred in the field of POCT was in the 1970s with the development of the first blood glucose meter called the Ames Reflectance Meter [18]. The device would require a large drop of blood to be placed on a paper strip known as the Dextrostix [18]. The Dextrostix was a smaller version of the laboratory method of measuring glucose known as the glucose oxidase-peroxidase method (GOD-POD) [19]. The reaction mechanism first oxidizes glucose to gluconic acid while simultaneously reducing oxygen to hydrogen peroxide by glucose oxidase [19]. The hydrogen peroxide is then split, forming water and an oxygen molecule. The oxygen will then react with 4-aminoantipyrine, which, in the presence of phenol, will produce quinoneimine, which is a colored compound that can be analyzed via colorimetric analysis [19]. Next, the glucometer would then capture reflected light from the surface of the strip using a photoelectric cell in the device. Finally, this would produce a signal that would be displayed on the surface of the meter to show the concentration of glucose in the blood. Creating a POCT for diabetes revolutionized diabetic care for patients, allowing them the information needed to make dietary changes to adapt to their body's glucose requirements. These examples highlight the value of timely results in times of patient need and the value of POCT development as a field. As highlighted by this example, for a device or method to be considered a POCT, it should generally be near the patient, have a short turnaround time, involve no pipetting, and be user-friendly, with no need for previous experience in lab sample analysis techniques.

10.1.1 Chemometric and Analytical Diagnostic Detection Methods

Currently, many chemometric and analytical diagnostic detection methods are involved with POCT. These detection methods are critical to ensuring that POCT can efficiently diagnose diseases and provide correct information to healthcare providers. While there are many different detection methods across the literature, this chapter will focus on the main methods commonly used in POCT, including colorimetric, electrochemical, fluorescence, chemiluminescence, surface plasmon resonance-based (SPR), and surface-enhanced Raman spectroscopy-based (SERS) methods.

10.1.1.1 Colorimetric Detection Methods

A colorimetric detection method uses an established relationship between the sample content and color change produced by a chromogenic reaction to detect an unknown amount of analyte (Figure 10.2a) [21]. These tests can be both qualitative,

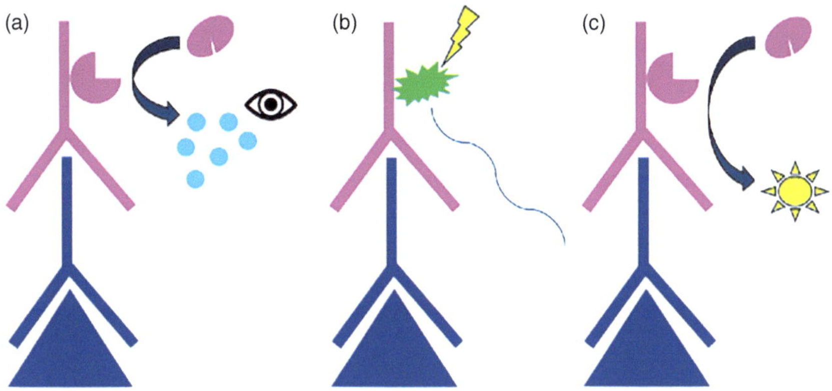

Figure 10.2 Schematics of colorimetric, chemiluminescent, and fluorescent methods. (a) Colorimetric detection works through a reaction that generates products that can create a color that can be physically observed. (b) Fluorescent detection works through incoming light that excites electrons; once the electrons relax, light is emitted and detected. (c) Chemiluminescent reactions generate luminescence through a high-energy intermediate that releases light from the reaction. *Source:* Adapted from Ref. [20].

in the case of technologies that use the human naked eye to determine if the color change has occurred, and quantitative, such as devices like spectrophotometers that can capture light and compute the shift of wavelength to determine the quantity of the analyte [21]. An example of a colorimetric test would include a colorimetric COVID-19 lateral flow device (LFD) developed by Lee et al. [22]. The test uses a lateral flow assay as the device. A CBP31 linker, prepared by genetically fusing the cellulose-binding modules of family 3 [CBM3] and family 1 [CBM1] with antibody-binding B and C domains of Protein A), is used to immobilize the SARS-CoV-2 antibodies in an orientated manner, which serve as the detection method [22, 23]. The test functioned in a way similar to many LFDs developed for COVID-19, in which a sample would be loaded onto a lateral flow paper test strip. The sample would then pass through a conjugate pad, and SARS-CoV-2 antigens would bind to the mouse antibody-conjugated gold nanoparticles. The sample solution would then flow over the test line and bind to the CPB31-linked human antibody, which functions as a capture antibody. The result is a colorimetric change in the test line of the lateral flow paper test strip, indicating a positive or negative diagnosis of COVID-19.

10.1.1.2 Fluorescence Detection Methods

Fluorescence detection methods are very similar to colorimetric detection methods in that they require an established relationship with the sample content. However, instead of a color change produced by a reaction, a fluorescence detection method will have an emitted fluorescence or luminescence that must be measured by an instrument such as a luminometer or a spectrometer (Figure 10.2b). The reaction requires an external light source to excite electrons to a higher energy state, and the emission of fluorescence occurs when the electrons relax to their ground state. An example of fluorescence detection in a POCT is a smartphone-based fluorescent diagnostic system for H5N1 viruses [24]. This device operates through the use of parabolic mirrors between the lateral flow strip and a smartphone camera. The test starts off with an LFD, where the specimen is loaded onto the paper strip. The LFD has an anti-influenza A nucleocapsid antibody on the test line and an anti-mouse IgG on the control line further down the strip. The lateral flow strip is loaded with a bioconjugate that is conjugated with an antibody and a coumarin-derived dendrimer, which is responsible for fluorescence. The sample is then loaded on the test and left to incubate. After allowing the test to incubate, the test is then loaded onto the device. Excitation light (460/560 nm) is used, and fluorescence intensity is measured via a smartphone reader. Overall, the device was able to show a specificity of 98.55% and a sensitivity of 96.55%.

10.1.1.3 Chemiluminescence Detection Methods

Chemiluminescence detection measures the content of an analyte the same way as a fluorescence detection method by tracking the emission of light or luminescence. However, the key difference is how the emission of fluorescence is produced. In chemiluminescence reactions, the emitted fluorescence is generated as a by-product from an unstable intermediate in a chemical reaction (Figure 10.2c). As a result, there is no need for an external light source. An example of a chemiluminescence detection method is a continuous glow chemiluminescent bacterial detection test developed by Chen et al. [25] The device uses 4-mercaptophenylboronic acid (MPBA)-functionalized CuSe nanoprobes to detect *Staphylococcus aureus* and *Escherichia coli*. The MPBA catalyzes the chemiluminescence of luminol at a slower rate than other reactions using horseradish peroxidase (HRP), creating a "flash" type reaction. The MPBA functions as the recognition element for reacting with glycan in a cis-diol confirmation on the bacterial surface [26]. The luminescence can be detected using a smartphone as a detector and analyzer [25]. The LOD of the device is as low as 1.25 cfu/mL. Overall, the device has high sensitivity, which makes it a powerful tool for bacterial detection.

10.1.1.4 Electrochemical Detection Methods

Electrochemical detection methods monitor changes in the electrical signals due to electrochemical reactions that happen in the presence of an electrode [27]. Typically, an electrochemical detection method requires three components. First, a biorecognition element is responsible for monitoring the reaction and generating changes in electrical signals based on different results. Second, a converter is required to convert this signal into data. The last component is an interface to perform data analysis on the results of the method.

Electrochemical detection methods can then further be split into four methods of detection, which are voltammetry, electrochemical impedance spectroscopy, chronoamperometry, and electrochemiluminescence. Voltammetry is the most common method among the four methods of electrochemical detection [27]. This method works through the analysis of electrode current fluctuations with applied potential in the presence of an electrochemical reaction. Electrochemical impedance spectroscopy works by feeding the system with a perturbation signal and

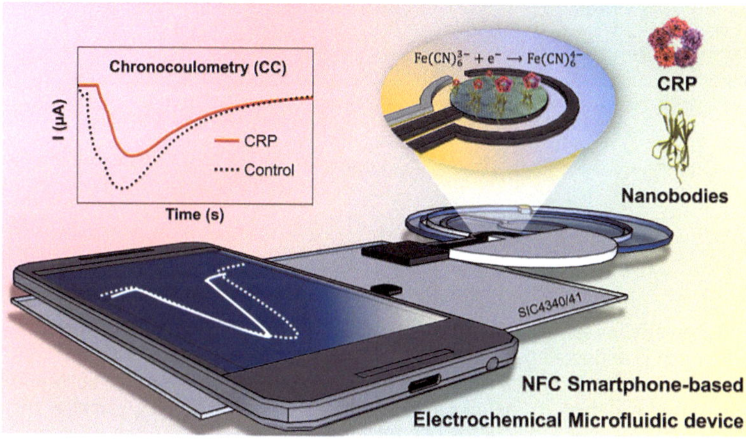

Figure 10.3 An NFC smartphone-based electrochemical microfluidic device. Through the CRP-binding nanobodies on the surface of the electrode, electrochemical signals can be read through the use of a cellphone. The electrochemical detection method was chronocoulometry, which is similar to chronoamperometry, but measures charge instead of current as a function of time. *Source:* Ref. [28] / with permission of American Chemical Society.

observing the corresponding response signal. Chronoamperometry works through the stepping of the electric potential of the working electrode, and the resulting response waveform is measured as current as a function of time. Electrochemiluminescence works through the generation of reactants that are used to react with other substances to form a chemiluminescent reaction.

An example of an electrochemical detection method that is used in a POCT device is an NFC smartphone-based electrochemical microfluidic device developed by Boonkaew et al. for C-Reactive protein (CRP) detection [28]. The device analyzes levels of CRP, which is a known biomarker for systemic inflammation, infection, and other human pathologies. The device performs this through the use of nanobodies binding CRP on the surface of the electrode, which happens to be an NFC potentiostat (Figure 10.3). The electrochemical detection method was chronocoulometry, which is similar to chronoamperometry, but measures charge instead of current as a function of time. The potentiostat was integrated into a Redmi Note 10S smartphone, which would monitor the course of the reaction (Figure 10.3). Overall, the device has a limit of detection of 7.6 pg/mL and a wide linear range of detection (10 ng/mL-100 μg/mL), as well as high specificity for CRP despite other proteins that are routinely present in blood serum samples.

10.1.1.5 Surface Plasmon Resonance and Surface-Enhanced Raman Spectroscopy-Based Detection Methods

SPR-based methods operate through the phenomenon that occurs when electrons on a thin metal sheet are excited by light that is directed at the sheet at a set angle of incidence (Figure 10.4) [30]. The electrons in the sheet will absorb light, begin to

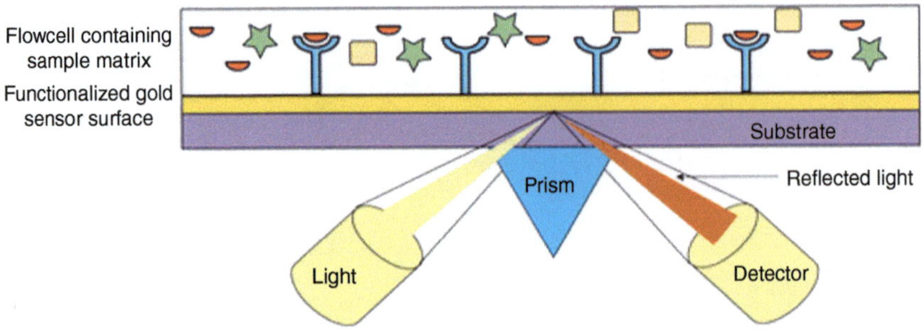

Figure 10.4 Schematic for surface plasmon resonance mechanism. Light enters at an incident angle, which is deflected by a functionalized gold sensor surface. The light is reflected into a detector. Based on the resonance from the surface, there is some absorption of light, which is what is detected. Changes in the surface via protein binding can alter this absorption of light. *Source:* Ref. [29] / MDPI / CC BY 4.0.

oscillate, and form a plasmon. At the same time, the light is refracted to a detector, and the absorbed light due to the SPR will not be detected. Changes that occur on the sheet surface, such as those caused by the binding of ligands to receptors coated onto the surface, can alter the refractive index (RI), which will cause the angle of the light absorbed by SPR to change, thus making it useful to detect interactions between ligands and receptors. This process has been used to detect stroke biomarkers, as shown by Harpaz et al. The authors have developed an SPR chip that can detect stroke markers NT-proBNP and S100β [30]. The device has a capture antibody, and when a sample with one of the biomarkers present is loaded, it will bind to the capture antibody. The RI of the sensor surface will change once binding has occurred, which will be interpreted as a shift in RI. This shift is picked up by the camera in the device. In terms of the value of the device as a POCT, it was still relatively portable at around 5 cm in height and weighing around 1 kg. However, the device still needs to be connected to a pump to work.

SERS-based detection is another surface-sensitive technique that operates through a beam of photons being fired at a surface, which enhances Raman scattering by molecules that are adsorbed to a surface (Figure 10.5). This then allows an SERS signal to be detected and used to determine the presence of low-abundance biomolecules of interest. An example of an SERS POCT device was made by Wu et al. [32]. The device used polyethylene glycol diacrylate (PEGDA)/sodium alginate (SA) composite hydrogels to isolate neurotransmitters (in the case of this study, it was dopamine) for a portable SERS analysis (Figure 10.6). The overall device was separated into two layers. The top layer held the hydrogels to isolate the sample. The second layer was prepared with an SERS substrate assembled with gold@silver core nanorods, which were used to detect neurotransmitters. The steps for operating the device would include adding blood to the top of the device. Then, a syringe was injected into the side of the device, which would create a negative pressure vacuum. This would then cause the analyte to flow down into the hydrogel, separating the blood (Figure 10.6). The net result would be smaller molecules falling through and ending up on the surface, which is embedded with

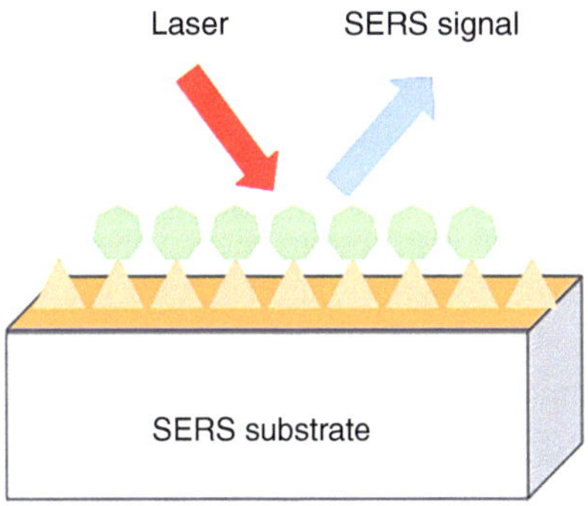

Figure 10.5 Schematic of surface-enhanced Raman spectroscopy mechanism. First, a beam of photons comes from a laser source. The laser photons excite the metal nanostructures, creating a localized plasmonic field. A molecule near the field that is adsorbed will lead to a Raman signal (SERS Signal), which can be detected. *Source:* Ref. [31] / with permission of IDEX Health & Science LLC.

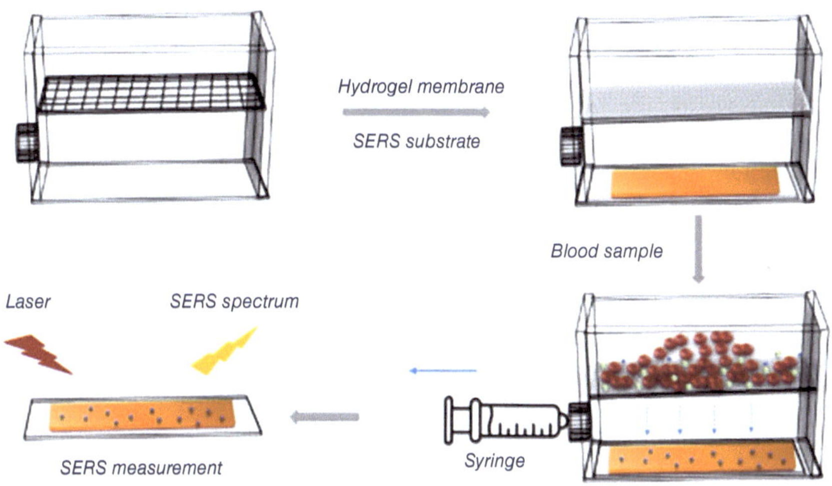

Figure 10.6 A surface-enhanced Raman spectroscopy device. A 3D printer was used to create the device. Hydrogel is then added to the top layer, and the SERS substrate is added to the bottom layer. The blood sample is then added to the top layer and separated by the hydrogel. The syringe helps by creating a negative vacuum. The SERS substrate is then removed from the bottom and read by SERS. *Source:* Ref. [32] / MDPI / CC BY 4.0.

the SERS substrate. The substrate would then be taken out of the device and subjected to an SERS laser for measurement. The end result is a portable device that incorporates sample prep and a POCT device into one device [32]. The whole readout can be performed in five minutes, which makes it valuable for quickly reacting to medical emergencies at a patient's bedside. Thus, due to the device's simplicity, portability, and speed, it is an excellent POCT device for monitoring neurological and cardiovascular disorders.

10.1.2 Chemometric and Analytical Diagnostic Detection Instruments and Devices

Although a plethora of detection methods have been developed, the next challenge is seamlessly integrating these methods into practical devices that achieve rapid and simple detection of infectious diseases. Foundational instruments such as the LFD, which typically uses a fluorescent or colorimetric detection method, have been a popular choice among POC rapid diagnostics [33]. However, despite its widespread use, the LFD involves the simple adoption of conventional steps and is accompanied by its own set of limitations, such as low sensitivity [34–36]. The recent research boom in POCT caused by the COVID-19 pandemic has accordingly pushed for newer innovations. Instruments involving nanotechnology, microfluidics, and microelectrochemical systems (MEMS) have particularly grown in interest as device alternatives. In the following section, we will explore the various diagnostic instruments and devices that have been developed and recently innovated for the detection of infectious diseases, such as LFDs, miniaturized PCR devices, and microfluidic devices. These instruments and devices are summarized in Table 10.2.

10.1.2.1 Lateral Flow Devices

LFDs (also known as lateral flow assays, lateral flow immunoassays, and rapid detection tests) are the most widely used POCT instruments to date. The concept of lateral flow technology originated in the 1950s with the development of the first radioimmunoassays and latex agglutination assays [49, 50]. Progressive refinements in the 1980s then led to the creation of the first commercially successful LFD for home pregnancy testing, providing a simple and rapid method for detecting human chorionic gonadotropin (hCG) in urine [51, 52]. This innovation paved the way for the development of a wide range of lateral flow assays for various applications in disease diagnostics, food safety, and environmental monitoring [53–57].

LFDs operate on the principle of immunochromatography, where the target analyte (such as a virus, bacterium, or biomolecule) in the sample interacts with specific antibodies or antigens embedded in the test strip. A typical test strip consists of several components: a sample pad, a conjugate pad, a nitrocellulose membrane, and an absorbent pad [58]. The testing process starts when a liquid sample is applied to the sample pad, which migrates through the strip by capillary action. As the sample moves through the conjugate pad, it encounters labeled detection antibodies or antigens that bind to the target analyte. This complex then migrates to the nitrocellulose membrane, where it encounters a series of capture antibodies or antigens immobilized in specific zones. The presence or absence of a visible line in the test zone indicates whether the target analyte is present [59].

Generally, LFDs exist in two formats: sandwich assays and competitive assays [60]. In sandwich assays, the target analyte binds to two different antibodies: a labeled detection antibody and a capture antibody. This format is most suitable for larger analytes with multiple epitopes, such as proteins or cells. When the sample containing the target analyte is applied, the analyte forms a “sandwich” between the detection antibody and the capture antibody, and a visible line indicates the presence of an analyte. In contrast, competitive assays are used for smaller analytes with

Table 10.2 Summary of diagnostic instruments.

Instrument	Description	Advantages	Disadvantages	Refs.
Lateral flow devices	Strip-based devices for rapid detection of analytes in samples	• Rapid sample answer time • Inexpensive • Easy-to-use • Highly portable • No equipment required	• Lower sensitivity and specificity • Difficult to integrate quantitative results • Capture receptors can take months to develop • Lack digital connectivity • Difficult to multiplex	[37–39]
LOC devices	Miniaturized devices integrating multiple laboratory functions on a single chip	• Highly portable • Minimal reagent volume • Low power consumption	• Difficult to fabricate and manufacture	[40, 41]
LOAD devices	Disc-shaped devices that use centrifugal force to manage fluid movement	• Easy to multiplex • Easy to control with centrifugal force	• Difficult to fabricate and manufacture • Short shelf-life	[42–44]
Mini-PCR	Portable PCR devices for nucleic acid amplification	• Accurate detection • High specificity and sensitivity	• Machinery is still complex and expensive	[45, 46]
Isothermal nucleic acid amplification devices	Amplification of nucleic acids at constant temperatures	• Rapid sample answer time • Easy-to-use	• Complicated primer design • Expensive reagents	[47, 48]

a single epitope, such as small molecules or haptens. In this format, the sample analyte competes with a labeled analyte for binding sites on the capture antibody. If the analyte is present in the sample, it prevents the labeled analyte from binding to the capture antibody, resulting in no visible line or a weaker line in the test zone. Thus, as opposed to sandwich assays, a visible line is produced when the sample analyte is absent, and the labeled analyte binds to the antibody.

One pathogen for which LFDs are commonly used is the group A *Streptococcus* (GAS) bacteria, which are the root cause of strep throat disease (pharyngitis) [61]. Currently, more than 70 LFDs have been approved and waived by the Clinical Laboratory Improvement Amendments (CLIA) to detect and diagnose GAS [33]. Most of these tests detect the presence of the Lancefield Group A carbohydrate antigen, which is specific to GAS bacteria. A meta-analysis of the diagnostic accuracy of LFDs for GAS pharyngitis was performed by Lean et al. in 2014 [62]. They found a sensitivity and specificity of 86% and 96%, respectively, for the LFDs. These tests play an important role in preventing further health complications that may arise from strep throat, such as rheumatic fever and poststreptococcal glomerulonephritis [63, 64].

Overall, LFDs are highly favored for several reasons—they are cost-effective, easy to miniaturize, highly portable, and independent of equipment. However, their simplicity inherently comes with many technical limitations. The major drawback of LFDs is their lower sensitivity and specificity compared to laboratory-based methods. This is mainly associated with the properties (i.e., size, morphology, and affinity) of the receptor and the imperfect binding kinetics between the receptor and the flowing analyte [37]. In the previously highlighted example of GAS pharyngitis LFDs, the sensitivity of 86% leaves room for many false-negative cases. For this reason, a negative LFD result for streptococcal pharyngitis is typically followed up by culture-based techniques by physicians [65]. Other disadvantages of LFDs include their limited multiplexing capability, the absence of quantitative results, and the lack of digital connectivity. Enhancing the sensitivity and specificity of LFDs through the use of sensitive nanomaterials or coupling with CRISPR-based methods, and integrating these devices with digital technologies for better data recording, are crucial areas for improvement.

10.1.2.2 Lab-on-a-Chip (LOC) Devices

Lab-on-a-chip (LOC) devices represent a significant leap forward in miniaturized diagnostic and analytical technology. These devices integrate multisample biological analyses onto a single chip, typically ranging from a few millimeters to several square centimeters in size [66–68]. The miniaturization allows complex biochemical processes such as sample preparation, reaction, separation, and detection to be automated and streamlined in a highly compact and portable format. Given its potential, LOC research has been explored for several different applications, including chemical synthesis, DNA analysis, and the diagnosis of human diseases [69–71].

The history of LOC devices is intrinsically linked to the origins of microfluidics (the precise control and manipulation of fluids at the microscale level), microelectronics (the precise control of electronics at the microscale level), and photolithography (the process of using light to fabricate transistors in silicon) [72]. One of the pivotal advancements in these fields came about in 1964 with the creation of the first integrated circuit, in which resistors, capacitors, and transistors were placed on a single piece of semiconductor material [73]. Using these methods, the first LOC device, which served as a gas chromatograph, was developed in 1979 by S.C. Terry at Stanford University [74]. This innovation was soon followed by the development of micro total analysis systems (μTAS), which aimed to integrate sample collection and final analysis on a single chip. Over the years, advances in materials science, microfabrication techniques, and biological detection methods have significantly expanded the capabilities of LOC devices, making them more robust, versatile, and accessible for various applications.

Today, the typical LOC device consists of a network of microchannels, chambers, and valves etched or molded into substrates made of materials such as glass, silicon, or polymers [75]. Fluids are driven through these microchannels using methods like capillary action, pressure-driven flow, or electrokinetic flow. Sensors and detectors integrated within the microfluidic network enable real-time monitoring and analysis of small volumes of fluids. This automation of processes reduces the need for manual intervention and enhances the speed and efficiency of biochemical assays.

A recent example of LOC technology in detecting viral infections is the development of an LOC device for the concurrent electrochemical detection of SARS-CoV-2 RNA and antibodies in saliva and plasma [76]. This device, described in a 2022 study by Najjar, integrates Cas12a-based enzymatic detection of SARS-CoV-2 RNA with a sandwich-based ELISA on electrodes functionalized with SARS-CoV-2 antigens (Figure 10.7). The device automates the extraction, concentration, and amplification of viral RNA, providing rapid and accurate results within two hours.

LOC devices offer several advantages, including rapid detection, easy multiplexing capabilities, minimal reagent volume consumption, high accuracy, and excellent portability [40]. Despite their apparent advantages, LOC devices face several notable challenges. The fabrication of LOC devices is complex and costly, often requiring advanced microfabrication techniques and specialized facilities. Efficient integration of sample preparation processes remains challenging as well, as many LOC devices need preprocessed samples to function accurately. Additionally, the lack of standardization in LOC device design and fabrication can lead to variability in performance and difficulties in widespread adoption. To address these challenges, advancements in materials science, microfabrication techniques, and integrated systems design are crucial. Developing more cost-effective and scalable fabrication methods can make LOC devices more accessible and practical for broader use. Nonetheless, because of its unique advantage of miniaturizing complex biochemical operations into a singular chip, LOC devices will play a key role in the future of POCT.

10.1.2.3 Lab-on-a-Disc (LOAD) Devices

Lab-on-a-disc (LOAD) devices are a specialized type of LOC technology that uses centrifugal force to manipulate fluids within a disc-shaped platform [77]. LOAD and LOC devices differ primarily in their fluid manipulation and structural design. LOC devices employ microfluidics, manipulating fluids through microchannels using capillary action, pressure, or electrokinetic flow. In contrast, LOAD devices use centrifugal forces, Euler forces, and Coriolis forces generated by spinning the disc to move fluids through microchannels and chambers [78].

Using this centrifugal method, LOAD devices offer the ability to integrate sample preparation, reagent mixing, and separation in a single instrument [79, 80]. Additionally, by adjusting the rotational speed, different fluidic processes can be precisely managed, enabling complex analytical procedures to be performed on a single disc [81–83]. This method allows for both continuous and batch processing [84]. Continuous processing involves the disc spinning continuously for ongoing fluid movement and processing, suitable for real-time analysis. Batch processing

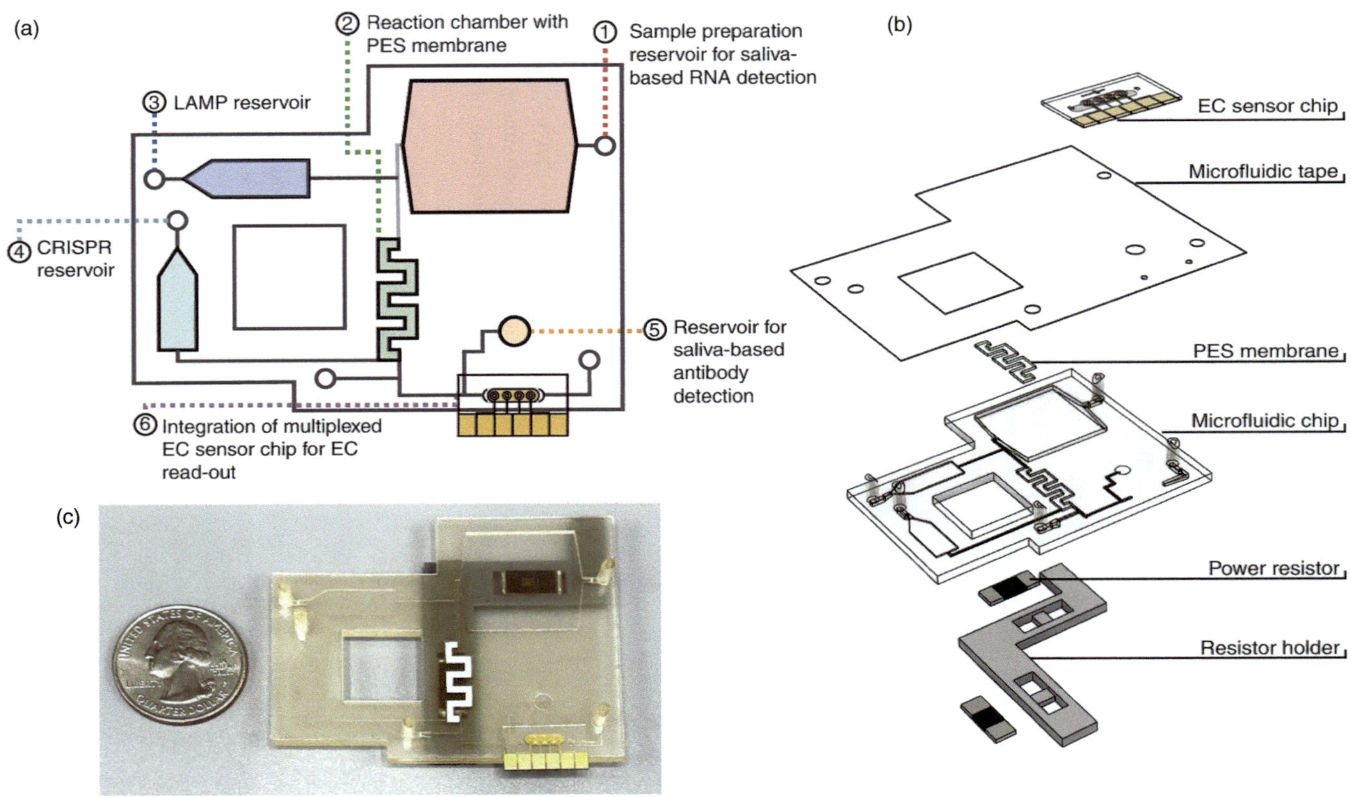

Figure 10.7 (a) Overview of the microfluidic chip engineered for a lab-on-a-chip (LOC) system that identifies the SARS-CoV-2 RNA and antibodies in saliva samples. (1) The user deposits saliva into the antibody detection reservoir and a mixture of saliva and proteinase K into the sample preparation reservoir for incubation. (2) The saliva is then transported over a PES membrane within the reaction chamber to capture RNA and heated to neutralize inhibitors. (3) The LAMP solution is introduced into the reaction chamber for incubation. (4) The CRISPR mixture is added, incubated, and subsequently pumped over the EC sensor chip. (5) The saliva intended for antibody detection is also pumped over the EC sensor chip. (6) After the addition of polystreptavidin-HRP and TMB, the results are obtained from the EC sensor chip using a potentiostat. (b) An enlarged view of the multiplexed system, comprising a heating system, a sealed microfluidic chip, and a multiplexed EC sensor chip. (c) A photograph of the microfluidic system alongside a quarter dollar for size reference. *Source:* Reproduced from Ref. [76] / with permission of Springer Nature.

involves intermittent spinning, allowing specific steps to complete before the next cycle, which is ideal for multistep assays requiring precise timing and control.

A notable example of a LOAD device used for viral detection is presented in the paper by Yao et al., in which they developed a high-throughput centrifugal microfluidic disc for the rapid detection of influenza virus subtypes using an isothermal nucleic acid amplification (INAA) method (Figure 10.8) [85]. The detection platform was capable of identifying six highly pathogenic influenza viruses, including influenza A subtypes H1, H3, H5, H7, and H9, as well as the influenza B virus. The described design features a disc with two parallel detection units, each containing reservoirs for sample loading and storing the reaction mixture, enzymes, and reverse transcriptase. It uses three types of channels—normal liquid, siphon, and pressure-stabilizing—to mix reagents and balance pressure, aided by two types of MembR valves. The process involves rotating the disc at various speeds

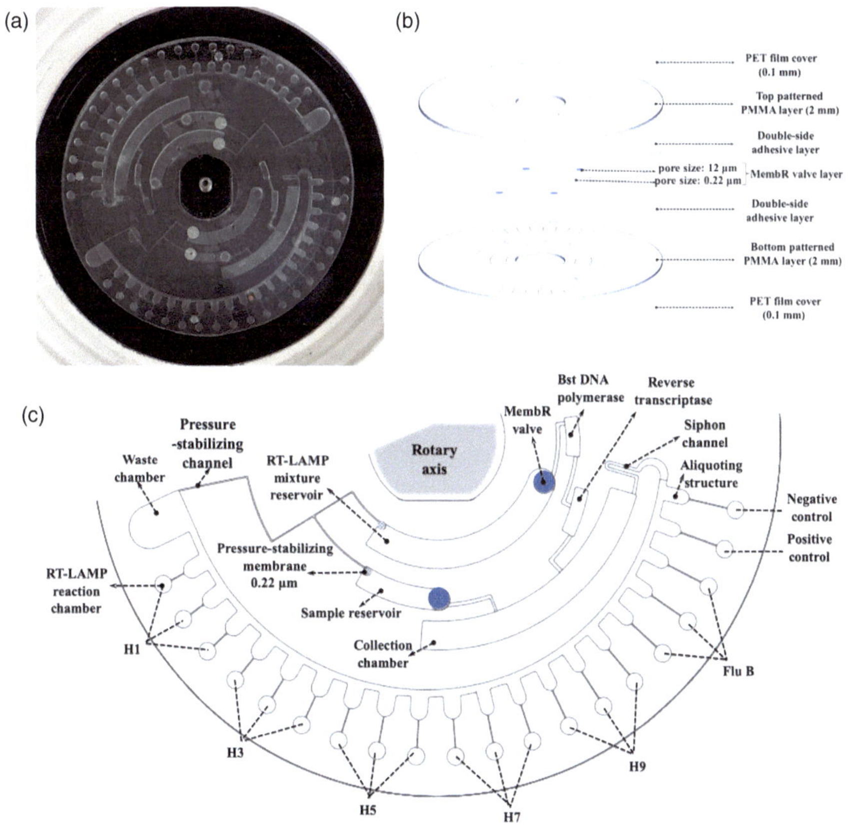

Figure 10.8 Overview of a disc device for RT-LAMP. (a) Photograph showing the disc mounted on the rotary axis. (b) Diagram of the layered disc, featuring a film cover (PET, 0.1-mm thick), a top patterned layer (PMMA, 2-mm thick), a double-sided adhesive layer, a MembR valve layer (PET), another double-sided adhesive layer, a bottom patterned layer, and a final patterned layer. (c) Diagram illustrating a disc with four reservoirs, three types of channels, 20 RT-LAMP reaction chambers, and two types of MembR valves. *Source:* Reproduced from Ref. [85] / with permission of American Chemical Society.

to distribute the reaction buffer and enzymes, ultimately performing the reverse transcription loop-mediated isothermal amplification (RT-LAMP) reaction at 63 °C for 45 minutes, with real-time detection via fluorescence. A specificity of 100% was reached, with a sensitivity as low as 20 copies for certain influenza strains.

LOAD devices offer several key advantages, including precise fluid control through centrifugal force, portability due to their compact design, and large-scale parallelization and multiplexing [86–89]. However, LOAD devices also face challenges such as short shelf life, complexity and cost of designing intricate microfluidic networks, and difficulty working with extremely small volumes [42]. More critical is the need for reliable power sources to control the spinning mechanism, which is in contrast to LOC devices that do not necessitate moving parts. Similar to LOC devices, advances in microfabrication techniques and materials science will be key to overcoming the challenges of complicated design and fabrication. To address the power requirements of LOAD devices, strategies such as using rechargeable batteries, optimizing energy-efficient design, integrating solar power, and implementing mechanical wind-up mechanisms can ensure reliable operation in remote settings [90]. Overall, LOAD devices have attracted great attention in the POCT research field and are considered a viable option as a future POCT device.

10.1.2.4 Miniaturized PCR Devices

PCR methods were previously highlighted in Chapter 5, as well as in this chapter's introduction, as the gold standard of infectious disease diagnostics. PCR is highly regarded because it can exponentially amplify one or a few copies of a particular nucleic acid sequence, making it extremely sensitive and specific [91]. However, the thermal cycler machinery needed for PCR is expensive and difficult for a layperson to operate, negating its ability to perform POC detection [1, 2]. More critically, the equipment is bulky and heavy, restricting its use in centralized laboratories. For PCR to be capable of POC detection, thermal cycler equipment would need to be made portable.

The miniaturization of PCR was therefore realized in the 1990s and early 2000s. Advancements in microfluidics and MEMS have enabled the creation of chip devices capable of precisely controlling temperature changes and performing high-precision genetic analyses [92, 93]. One of the earliest miniature PCR devices described in the literature was developed by Northrup et al. in 1998 [94]. Their instrument used a microfabricated silicon-based chamber that was battery-operated and successfully verified with several biological systems, including HIV and cystic fibrosis. This stationary format, where the reaction chambers are microfabricated on the chip (i.e., chamber-based PCR chips), has been widely replicated since then [95].

One notable paper from González-González et al. reported the validation of the commercially available miniature PCR device sold by Amplyus/miniPCR Bio for the detection of Ebola and Zika virus (Figure 10.9) [96]. Their results demonstrated that the instrument could detect and amplify Ebola and Zika DNA concentrations as low as 50 copies. Overall, using the miniPCR device from Amplyus/miniPCR Bio was

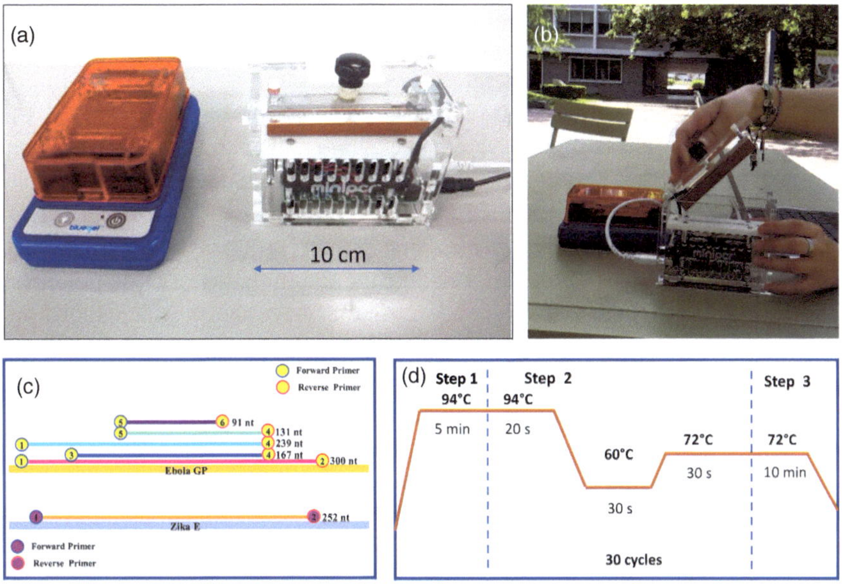

Figure 10.9 A portable PCR system for the detection of Ebola. (a) The miniPCR thermocycler and blueGel electrophoresis chamber. (b) Use of the miniPCR device in an outdoor setting. (c) Various lengths of amplicons can be produced using different sets of primers aimed at the sequence encoding the Ebola virus GP protein. (d) The temperature cycling scheme employed in the PCR protocol. *Source:* Ref. [96] / PLOS / CC BY 4.0.

straightforward and user-friendly. The connection of the device to a laptop allows the user to easily monitor the progress of the temperature cycling.

Miniature PCR is advantageous because it maintains high sensitivity and specificity, similar to conventional PCR, allowing for accurate detection of low levels of nucleic acids [45]. Additionally, each test will only require a minimal sample volume. Despite these benefits, miniature PCR carries two glaring issues from its conventional counterpart. First, the components of miniature PCR are still complex and expensive. Second, miniature PCR does not provide a significant speed advantage compared to conventional PCR. In the aforementioned study by González-González et al., for example, the full amplification protocol using the miniPCR was just six minutes faster than the conventional thermocycler. Although impressive for its size, further innovations will be required for miniature PCR to be practical for testing at points of entry, workplaces, or congregate settings. Current efforts are focused on speeding the process by increasing the rate of heat transfer, specifically by using more efficient heaters and shrinking sample volume [97, 98]. If both these limitations are addressed, miniature PCR will become a valuable POCT technology moving forward.

10.1.2.5 Isothermal Nucleic Acid Amplification Devices

While miniature PCR is not currently suited to circumvent the arduous heating/cooling steps of PCR, implementing an INAA strategy presents a potentially more effective alternative. INAA was similarly discussed in a previous chapter (Chapter 5);

to recap, INAA involves the exponential amplification of nucleic acid sequences at a constant temperature using enzymes, usually strand-displacing polymerases, rather than temperature changes [99].

The first INAA method was LAMP, developed by Notomi et al. in 2000 [100]. LAMP relies on a strand-displacing DNA polymerase that amplifies DNA with high efficiency [101]. It uses a set of four to six primers designed to recognize multiple regions on the target DNA, including two outer primers, two inner primers that initiate DNA synthesis, and sometimes loop primers to accelerate the reaction further. The amplification results in a significant accumulation of DNA products, visible as a turbidity change, color change, or fluorescence signal, which allows for easy detection of the target sequence. Notomi and his group's pioneering work laid the foundation for this field by demonstrating that nucleic acid sequences could be amplified rapidly and efficiently at a single temperature, eliminating the need for complex thermal cycling equipment. Today, there are a plethora of methods outside of LAMP, including rolling circle amplification (RCA) [102, 103], recombinase polymerase amplification (RPA) [104, 105], strand displacement amplification (SDA) [106, 107], helicase-dependent amplification [108], and so on. To implement INAA in POCT platforms, apparatuses will typically include components such as a reaction chamber, a heating element to maintain the required temperature, and a detection system, often involving fluorescence or colorimetric indicators to signal the presence of the target nucleic acid.

A recent application of INAA was demonstrated by our group [109]. The method harnesses the cleaving capabilities of catalytic DNA molecules known as DNAzymes and uses them to cut the genomic SARS-CoV-2 RNA. The cleaved fragments were then used as primers, which bind to a circular DNA template to initiate RCA and fluorescent signal generation (Figure 10.10). This strategy could achieve a detection limit of 500 aM of RNA, as well as a sensitivity and specificity of 86% and 100%, respectively, with clinical saliva samples.

Similar to the miniaturized PCR devices, INAA would be a suitable option for POCT because it packages the powerful capabilities of nucleic acid amplification in a more rapid, inexpensive, and portable format that requires minimal user training [110]. However, there are still challenges to be addressed to enhance the performance and accessibility of INAA devices. Each technique comes with its own advantages and limitations. A common theme among the different methods is that primer design is often complicated. Additionally, the reagents can be sensitive to varying environmental conditions and are often difficult to scale due to their cost [47, 48]. With LAMP, for example, the technique is not ideal for amplifying small nucleic acid fragments; several primers (four to six) are often required, and two of the primer sets are typically long, incurring more cost in synthesis [111]. Further advancements in enzyme engineering and pairing with microfluidic technologies can help overcome these limitations, making INAA devices even more reliable and versatile.

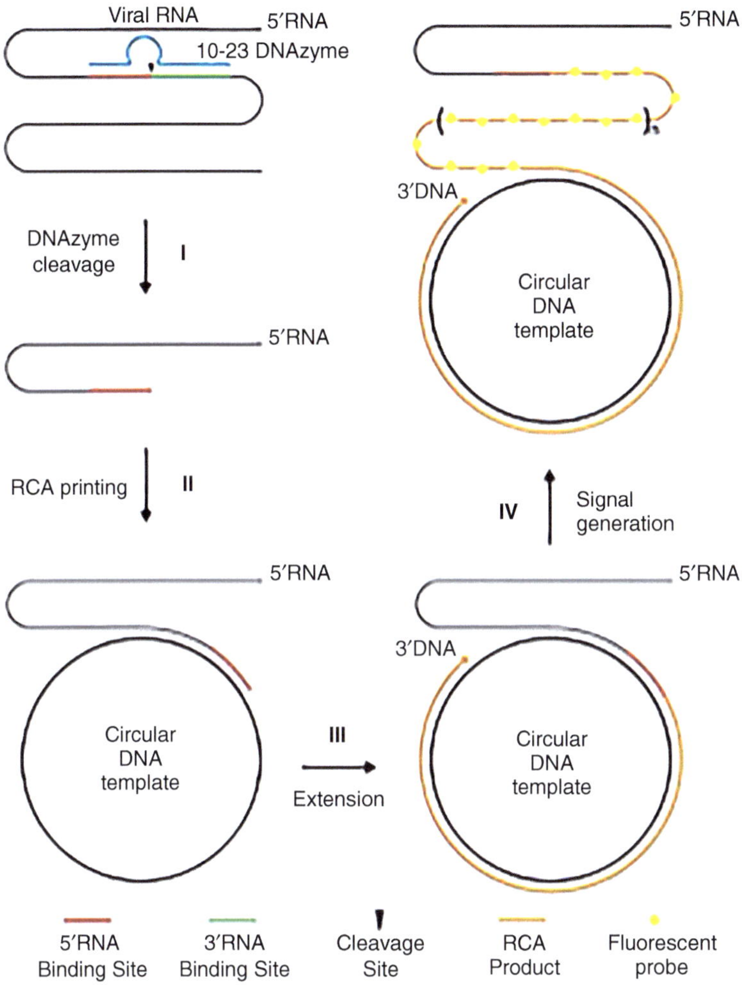

Figure 10.10 Summary of DNAzyme-guided primer production linked with RCA for identifying SARS-CoV-2 genomic RNA. The 10–23 DNAzyme precisely cleaves the viral genomic RNA at a specific site, generating a new 3′ RNA terminus. The terminus then binds to a circular DNA template and functions as a primer for rolling circle amplification. The generated product produces a readable fluorescent signal. *Source:* Ref. [109] / with permission of John Wiley & Sons.

10.1.3 Integration of Artificial Intelligence (AI) into POCT

AI and other machine-learning strategies are powerful information technology tools that have only begun being integrated into the field of POC testing and will continue to play a bigger role in the future. AI is a discipline that harnesses the intelligence of machines and encompasses machine learning (ML) and robotics [112]. However, only information relevant to POCT testing will be further discussed. One of the current challenges of POC testing is that there is often a need for trained users to analyze and interpret the results of the POC test [112]. AI systems have demonstrated the ability to achieve pattern recognition, which can be applied to many different POC tests [113]. In one case, an AI software that uses deep neural networks outperformed 21 board-certified dermatologists at diagnosing skin cancers from biopsy-proven clinical images [114]. Similarly, a deep neural network was used to create software called EasyScanGo that could assist with diagnosing Malaria [112]. The gold-standard test for diagnosing Malaria is a peripheral blood smear, which is then read or diagnosed by healthcare professionals [115]. The AI software demonstrated a diagnostic sensitivity of 91.1% and a specificity of 75.6% (which went up to 85.1% with good-quality slides) [116]. These examples demonstrate the potential value that integration of AI brings through the reduction in resources needed to conduct POC testing, such as not requiring trained experts or physicians on site. As AI develops and technology improves, AI software may become far better at diagnosing diseases than human-trained physicians. AI software is not prone to biases that humans can have, and computerized detection is far more sensitive than human sight and interpretation. Furthermore, AI also has the potential to analyze and model data on outbreaks of diseases and communicate this information to healthcare policymakers. In the context of future pandemics, this could lead to faster response times for diagnosing new disease outbreaks, leading to improved coordination and response efficiency from the health system. Thus, the future design of POC tests potentially should consider how they can be integrated into AI systems and networks to facilitate improvements in the way POC tests are conducted, resources used in POC tests, and the epidemiology information that a POC test connected to cloud networks could provide. In conclusion, AI has the potential to rapidly speed up current POC testing strategies and allow for better allocation of current healthcare resources.

10.2 Enhancing the Existing: Integrating Viral Detection in Current POCT Platforms

Now that we have discussed the various technologies available for POCT, it is time to bring everything together and look at POCT devices for viral diagnostics in practice. Many challenges need to be overcome when detecting viral infections in clinical samples. First, viral loads are low in the earliest stage of infection since it takes several days for the virus to replicate to a detectable level [117, 118]. Tests are prone to giving false negative results during this time, and higher sensitivity and/or an amplification step are needed to improve test accuracy. Second, there are components in biological fluids that can interfere with the detection of analytes, such as immunoglobulin G in plasma and proteases and RNases in saliva [119, 120]. Traditional methods of isolating and purifying the desired analyte require specialized equipment, such as a centrifuge, which is unsuitable for POCT. Third, many assays require multiple steps, such as cell lysis, wash steps, and test reactions, to occur in separate compartments. For POCT, these steps should be automated or circumvented, if possible, to make it user-friendly. Over the years, researchers have come up with creative solutions to enable the sensitive, rapid detection of viral infections. In this section, we will examine some of the methods that have been applied to the detection of HIV and COVID-19 infection, as well as their strengths and challenges.

10.2.1 Human Immunodeficiency Virus (HIV)

10.2.1.1 Background

According to the World Health Organization estimates, there were 39 million people globally living with HIV and 630,000 HIV-related deaths in 2022 [121]. HIV is a retrovirus that primarily targets $CD4^+$ T lymphocytes [122]. Upon infection, HIV reverse-transcribes its single-stranded RNA into DNA, which will be integrated into the host genome. The host cell then transcribes and translates the HIV DNA to produce mature virions. As the virus replicates, the host immune cells are destroyed, leaving the patient susceptible to other infections [122]. If left untreated, HIV infection will eventually progress to the development of acquired immunodeficiency syndrome (AIDS), which is defined as having a $CD4^+$ T cell count of less than 200 cells/μL [123]. There are two HIV subtypes: HIV-1 and HIV-2. HIV-1 is characterized by high transmissibility and faster progression to AIDS compared to HIV-2 [124]. HIV-1 infections make up the majority of HIV cases and can be found worldwide, whereas HIV-2 is confined to West Africa [124]. Although there is no cure for HIV/AIDS, it can be treated with antiretroviral therapy (ART) [121]. ART medicines prevent HIV from replicating and decrease viral load, thereby reducing the immunodeficiency effects of HIV and increasing the life expectancy of the patient, while also reducing the risk of transmission [125]. Treatment should be started as soon as possible, since early treatment is associated with better health outcomes and reduced healthcare costs [126]. This is where rapid tests play an important role, as it allows those potentially infected with HIV to receive an accurate diagnosis early. Patients in Canada seeking HIV tests reported a preference for POCT over laboratory testing because they found POCT to be less stressful and less invasive [127]. A study in South Africa showed that those who were offered POC CD4 tests, which are done after the initial diagnosis to guide treatment, were more likely to return to the clinic for additional HIV care, further highlighting the importance of POCT [128].

10.2.1.2 Current Methods for Diagnosing HIV and Monitoring Disease Progression

Currently, the only approved POCT for HIV diagnosis in North America is LFAs [129, 130]. All of these tests detect the presence of anti-HIV antibodies in whole blood, plasma, serum, or oral fluid. While these tests have high sensitivity (>95%) and specificity (>99%) [131–136], there is one major limitation—antibodies take time to develop and are not detectable until at least three weeks and up to three months after infection [137]. In an effort to diagnose HIV earlier, some newer tests also detect p24 antigen, which is the most abundant HIV protein [138]. p24 will usually be present at detectable levels approximately two to three weeks after infection [137]. However, as antibodies are being produced, they form immune complexes with p24, masking their presence and rendering them undetectable again [139]. As such, reports on the value of antigen detection have shown mixed results—sensitivities ranged from 0% to 29.4% [140–142]. Unfortunately, the risk of transmission is highest during the three-month period when antibodies and antigens cannot be detected consistently [143, 144].

After the initial diagnosis of HIV infection, the patient would ideally receive treatment and should monitor their viral load and/or CD4 cell count. Viral load monitoring is useful for indicating whether treatment is working and if the patient is adhering to the treatment regimen [145]. However, some patients may respond well to treatment and have a low viral load, but still have a low CD4 cell count and are immunosuppressed [146]. The current gold-standard method for measuring viral load is RT-qPCR, and the gold standard for measuring CD4 cell count is flow cytometry. Neither of these methods is suitable at the POC, and there are currently no FDA-approved POCTs for viral load monitoring and CD4 cell count.

10.2.1.3 Improving Antigen Detection

As previously mentioned, p24 antigen detection has shown limited sensitivity on lateral flow assays. One of the main reasons for this is the formation of immune complexes, which mask the antigen [139]. Pretreatment of the sample to dissociate the immune complexes can significantly increase the sensitivity of HIV antigen tests. This can be done using heat or acid. Using an ELISA assay, Steindl et al. compared heat pretreatment, acid pretreatment, and heat pretreatment with the sample diluted in sodium dodecyl sulfate (SDS) and diethylenetriamine-pentaacetic acid (DTPA), and achieved a sensitivity of 62.5%, 60.7%, and 80.4%, respectively, compared to 21.4% without pretreatment [147]. More recently, Parpia et al. heat-shocked samples at 88 °C for four minutes before running the sample on a lateral flow assay similar to commercially available LFDs [148]. They achieved a sensitivity of 95.8%, a significant increase compared to 0 to 29.4% reported from evaluations of other LFDs [140–142, 148]. Pretreatment of samples can potentially be combined with current approved lateral flow assays for improved results. However, this will increase the complexity of the test, and considerations for a portable heating element or a simple acid pretreatment step will have to be made.

10.2.1.4 Isothermal Nucleic Acid Amplification Tests

Nucleic acid amplification tests can be used for both the diagnosis of HIV infection and for quantifying viral load to monitor disease progression, depending on whether the test readout is quantitative or qualitative. While nucleic acid amplification is not typically used for HIV diagnosis due to the complicated machinery required, it is an option worth exploring, since HIV RNA is the first marker to be detectable and will allow for earlier diagnosis compared to current antigen/antibody tests [137]. Since traditional thermal cycling is difficult to conduct at the POC, researchers have turned to isothermal amplification and combined it with various detection methods to create sensitive POC devices for measuring viral load.

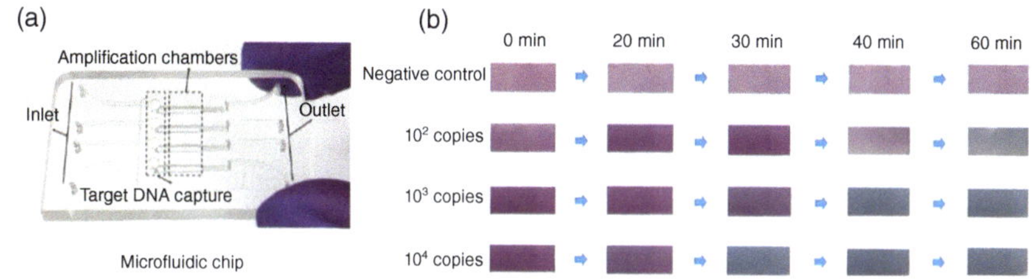

Figure 10.11 (a) A microfluidic chip for RT-LAMP reaction for the quantification of HIV RNA. Four separate reactors allow for the simultaneous testing of four different samples. (b) Color change of RT-LAMP reactions with varying starting RNA concentrations and at various time points. The higher the starting concentration, the quicker a color change is observed. *Source:* Ref. [149] / Springer Nature / CC BY 4.0.

Although real-time tracking of amplification reactions is typically measured with fluorescence, other methods can be used, which include colorimetric and potentiometric detection methods. An example of real-time colorimetric detection of HIV using RT-LAMP is demonstrated by Yin et al. (Figure 10.11) [149]. They used Eriochrome Black T (EBT), which is blue in its natural state and red when it is complexed with Mg^{2+}. At the start of the reaction, EBT is complexed with Mg^{2+}. As the reaction progresses, pyrophosphate ions (PPi^{4-}) and H^+ ions are generated. The negative pyrophosphate ions react with Mg^{2+}, and the H^+ ions further accelerate the dissociation of the EBT-Mg^{2+} complex. Both of these result in free EBT in solution, turning the solution blue. The reaction can be monitored with a smartphone to detect changes in hue over time, which can be used to calculate threshold time [149]. One benefit of using color change instead of fluorescence or absorbance is that the hue of the solution can be measured without a set pathlength and is, therefore, independent of volume and chamber size [149]. This is important, especially for microfluidic chips, as the production process is not perfect, and on such a small scale, a small difference in the size of the chamber can have large effects on the measurement. Additionally, fluorescence detection often requires special filters and light sources, whereas this device does not [149]. The authors reported a limit of detection of 100 RNA copies per 1 μL sample, with a sample-to-answer time of ~1 hour [149]. The main limitation of this study was that the device was only tested with purified RNA samples spiked in plasma. A simple lysis and purification step will need to be integrated with the device and tested with clinical samples before the device can be used in clinical settings. One example of how this can be done is shown by Liu et al. [150]. They first mixed the test samples with a lysis buffer, then inserted the sample into their microfluidic device, which contains a filter membrane. The sample flows through the filter membrane, which filters out any potential inhibitors, before reaching the isothermal amplification reaction chamber [150].

Another method of tracking amplification reactions is by measuring pH changes using potentiometry. During amplification, a hydronium ion is released every time a nucleotide is added onto a strand [151]. This causes a decrease in pH. In a device designed by Kong et al., H^+ ions pass through an ion-selective membrane to interact with a carbon nanotube electrolyte layer, which causes changes in the open-circuit voltage of the sensor [151]. The change in voltage can be measured in real time with the use of a handheld sensor, and viral load can be calculated from the data. A limit of detection of 1 RNA copy/μL is reported, with a sensitivity and specificity of 96.5% and 94.1%, respectively [151]. However, the RNA was once again purified in a separate step before adding the sample to the device, and modifications to the workflow and platform to incorporate lysis and purification will have to be made [151].

Both of the above methods used LAMP reactions, which have an ideal reaction temperature of 60–65 °C. This means that a heating element is required, which usually uses electricity and can be inconvenient for POC use. As such, other groups have explored the use of RPA, which has an ideal temperature of 37–42 °C and can work at room temperature, albeit slower. Heat for this reaction can be provided by body temperature, as demonstrated by Kong et al. (Figure 10.12) [152]. They designed a microfluidic device that is worn on the back of the wrist. With a thermal insulation wristband to help reduce the effects of differences in ambient temperature, the device will reach temperatures of 30–34 °C during the reaction [152]. RPA reagents are immobilized on the chip. Users simply need to insert the sample and wear the device for a 24-minute incubation time [152]. This reaction generates a fluorescent readout, which can be detected with a smartphone. However, since the chip needs to be worn on the wrist during the reaction, fluorescence cannot be measured in real time, and this test is therefore only suitable for diagnosis. A limit of detection of 100 DNA copies/mL is reported. Once again, the major limitation of this test is

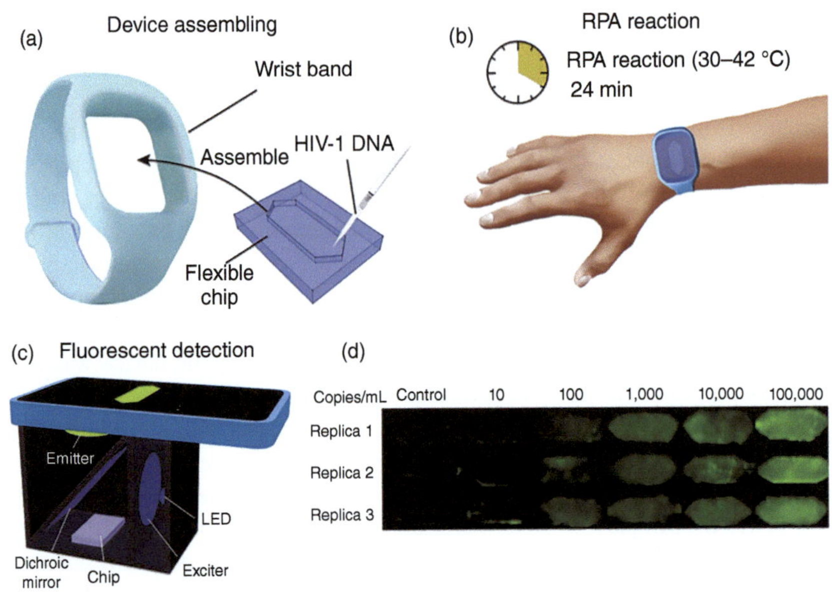

Figure 10.12 Operation of a wearable chip for the isothermal amplification of HIV DNA. (a) Schematic of the device components. (b) Schematic of the device being worn during the RPA reaction to provide the device with heat. (c) Apparatus to measure fluorescence signals generated by the RPA reaction using a smartphone. (d) Smartphone image of fluorescence generated by the RPA reaction with starting HIV DNA concentrations ranging from 10 to 10^5 copies/mL. The limit of detection is 100 copies/mL. *Source:* Ref. [152] / with permission of Elsevier.

that it has not been tested with clinical samples and requires a separate lysis and purification step. Another limitation is that it can be costly. The RPA reagents alone were estimated to cost $8 per reaction [152]. The fluorescence detection setup costs approximately $400 for the excitation filter, dichroic mirror, emission filter, and LED light [152]. Although this setup has its advantages, including high sensitivity and less concern for contamination, the RPA reaction could potentially be coupled with a different detection method, such as a colorimetric readout, in order to reduce cost.

10.2.1.5 CD4 Cell Count

While the gold-standard method for CD4 cell count, flow cytometry, is typically restricted to laboratory use, portable flow cytometers have been developed for POC use. One example of such a device is the CyFlow miniPOC by Partec. The operation of this device is simple—users will have to mix a blood sample with lyophilized antibodies and two buffers, then insert the sample into the device [153]. Results are generated within three minutes, and its performance is comparable to laboratory flow cytometers [154]. The device can also run on battery power for increased portability. Another advantage of this device is that it reports both absolute CD4 cell count (the number of CD4 cells in the sample) and CD4% (the percentage of lymphocytes that are CD4 cells). While absolute CD4 cell count is an overall good indicator of disease progression, CD4 levels can fluctuate [155]. CD4% tends to be more stable and provides additional information as to whether changes in CD4 count are due to fluctuations or if they reflect a trend [156]. Additionally, CD4% is used to measure disease progression in children, since children naturally have higher levels of CD4 cells and other lymphocytes [157]. CD4 count and overall lymphocyte count decrease with age, so absolute cell counts would have to be adjusted for age, whereas CD4% would be a better indicator of ART effectiveness [157, 158]. The biggest limitation of portable cytometers is that they are expensive. As such, other methods of counting CD4 cells are being investigated.

Wang et al. designed a device that uses a method they called micro-a-fluidic ELISA to produce a colorimetric readout (Figure 10.13) [159]. In typical microfluidics, the fluid is moving through microchannels and chambers, while for this device, it is the substrate that moves through stationary fluids. Anti-CD4 antibodies conjugated to magnetic beads are mixed with the sample and bind with CD4 cells. A magnet under the chip causes the magnetic beads carrying CD4 cells to migrate along the chip to a chamber containing antibodies conjugated to HRP, which bind to the CD4 cells. The complex then migrates to a chamber with 3,3′,5,5′-Tetramethylbenzidine (TMB), where the HRP oxidizes the TMB to cause a color change. The intensity of the color is measured using a cell phone application [159]. This device had a 97% accuracy in identifying whether patients had a CD4 count above or below 350 cells/μL, the WHO-recommended cutoff below which ART should be prioritized [159]. This device has several advantages. First, the migration of the capture antibodies is controlled precisely by a magnetic actuator. This allows sufficient time for binding to CD4 and the tagged antibodies, as opposed to lateral flow assays, where the sample flows by the capture antibodies only once. Additionally, the device works with whole blood without the need for any sample processing. Lastly, this

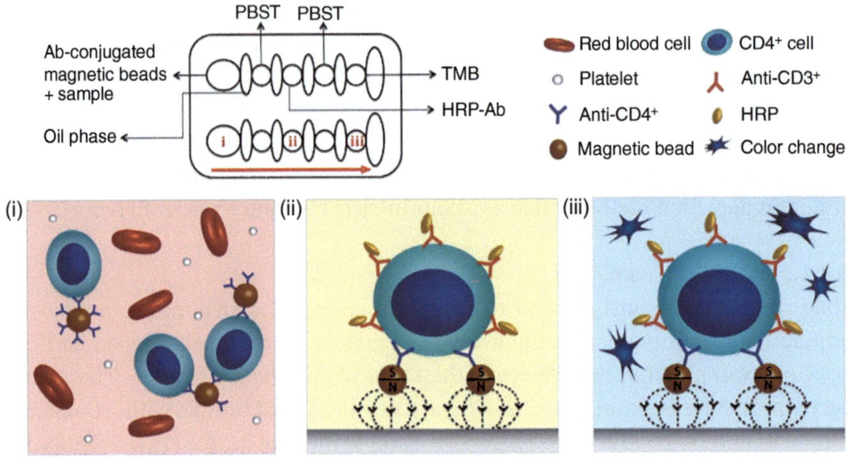

Figure 10.13 Design of a microfluidic chip-based assay. In chamber (i), the sample is mixed with anti-CD4 antibody bound to magnetic beads. The magnetic bead-CD4 cell complexes migrate through a wash chamber to chamber (ii), where anti-CD3 antibodies conjugated to horseradish peroxidase (HRP) bind to the CD4 cells. The complexes then migrate through another wash chamber to chamber (iii), where 3,3′,5,5′-tetramethylbenzidine (TMB) interacts with HRP to cause a color change. *Source:* Reproduced from Ref. [159] / with permission of Springer Nature.

device is easy to use, as it is completely automated, and results can be read within nine minutes, making it suitable for POC use [159].

10.2.2 COVID-19

10.2.2.1 Background

COVID-19 disease is caused by an infection of the SARS-CoV-2 virus. Symptoms can vary greatly between patients, and some examples include fever, cough, weakness, and taste and smell disorders [160]. In the most serious cases, COVID-19 can be fatal. Some also experience long-term effects, a condition known as long COVID, where symptoms can include fatigue, difficulty breathing, and confusion, which can last several months after infection [161]. As of 2025, there have been nearly 705 million cases and over 7 million deaths globally since the emergence of the virus in December 2019 [162]. At its peak, there were over 3.5 million new cases and 16,000 deaths daily [162]. Public health measures, such as frequent testing, isolation of positive cases, and contact tracing, were introduced to reduce the spread of the virus [163]. Current COVID-19 tests can be divided into three categories: nucleic acid tests, antigen tests, and antibody tests.

10.2.2.2 Current Commercial Methods for the Detection of SARS-CoV-2

Nucleic Acid Amplification Tests In the category of nucleic acid tests, laboratory-based RT-qPCR is the gold-standard diagnostic method for COVID-19. However, due to its long turnaround times and high costs, this method could not be used as frequently as would be optimal for limiting the spread of the pathogen. Thus, portable RT-PCR devices for COVID-19 diagnosis were developed. One such example is the Visby Medical Respiratory Health Test. This handheld device uses RT-qPCR to amplify the target RNA and uses a lateral flow assay with a colorimetric readout to display the results [64]. The device has a simple workflow: a patient's nasopharyngeal swab is mixed with buffer, which is added to a chamber in the device. The device needs to be plugged into a power source before closing the chamber and turning on the device. The device then automatically lyses the cells and viruses in the sample and completes an RT-PCR reaction. The biotinylated PCR product binds capture probes and streptavidin-bound HRP, and a color change is produced after an enzymatic reaction [64]. This assay tests for influenza A, influenza B, and COVID-19 simultaneously. The sample to answer time is 30 minutes and has a sensitivity and specificity of 95% and 100%, respectively, which is comparable to laboratory-based RT-qPCR tests [37].

Rapid Antigen Tests Despite the excellent performance of portable RT-PCR devices, they are not widely used, possibly due to their relatively high cost compared to rapid antigen tests (RATs), which were used almost exclusively for POC diagnosis of COVID-19. RATs were LFDs that used a sandwich assay to indicate the presence of antigens in the sample, which was typically a nasal or nasopharyngeal swab [164]. They typically detected nucleocapsid proteins, which formed the capsid around the genetic material, or spike proteins, which were responsible for host cell entry [165, 166]. RATs were especially popular as they could be used for self-testing at home. Although they were widely used, they were far from perfect. Sensitivities were as low as 33.7%, depending on the variant and viral load [167]. For this reason, alternative methods for rapid detection of antigens are needed.

Antibody Tests Rapid antibody tests work in much the same way as RATs, with a sandwich assay on an LFD. These typically detected the presence of IgA, IgM, and/or IgG antibodies against SARS-CoV-2 [168]. In contrast to HIV, however, antibody tests for COVID-19 were mainly used for research purposes rather than diagnosis. This is because the presence of anti-SARS-CoV-2 antibodies did not necessarily indicate a current infection—antibodies could be present in the body from past infections or from vaccinations [169].

10.2.2.3 New Methods for the Detection of SARS-CoV-2

Aptamer-based Detection of Antigens One way to improve the sensitivity of RATs is to improve the binding affinity of the molecular recognition element. Because antibodies are somewhat difficult to modify, one alternative is to use aptamers. Aptamers are single-stranded DNA or RNA molecules that can fold into complex shapes to form binding pockets, allowing them to bind to a variety of targets with high specificity and high affinity [170, 171]. Aptamers are discovered through a process known as Systematic Evolution of Ligands by Exponential Enrichment (SELEX), also referred to as in vitro selection [172]. Briefly, a large library of random nucleotide sequences is incubated with the target to allow binding. After incubation, all unbound sequences are washed away, while bound sequences are amplified by PCR. The PCR product is incubated with the target again, and the process is repeated for several rounds with increased stringency to push for the selection of high-affinity aptamers [173]. This process is advantageous as it allows researchers to have a higher degree of control over sensitivity and specificity issues. For example, to select for an aptamer that can bind to all variants of SARS-CoV-2, the DNA library can be incubated with a different variant at each round, and any aptamers that do not bind to all variants will be washed away [174]. Other advantages of aptamers are that they have high thermal stability and they are easy to produce and modify with low batch-to-batch variation [175–178].

Aptamers can be used with various detection methods and devices in the same way as antibodies. Zhang et al. designed an electrochemical aptamer-based sensor (aptasensor) to detect SARS-CoV-2 spike protein (Figure 10.14) [179]. First, to create a high-affinity aptamer, they took truncated versions of two of the top aptamers from a previous selection for SARS-CoV-2 spike protein-binding aptamers and joined them with a polythymidine linker. Because aptamers are small, there is little steric hindrance, and several aptamers can often bind to the same target, especially larger targets with multiple epitopes, such as the spike protein. This new dimeric aptamer had a 28- to 99-fold higher binding affinity to the spike protein compared to each monomeric aptamer individually [179]. To use this aptamer in an electrochemical sensor, the aptamers were modified to have a thiol group

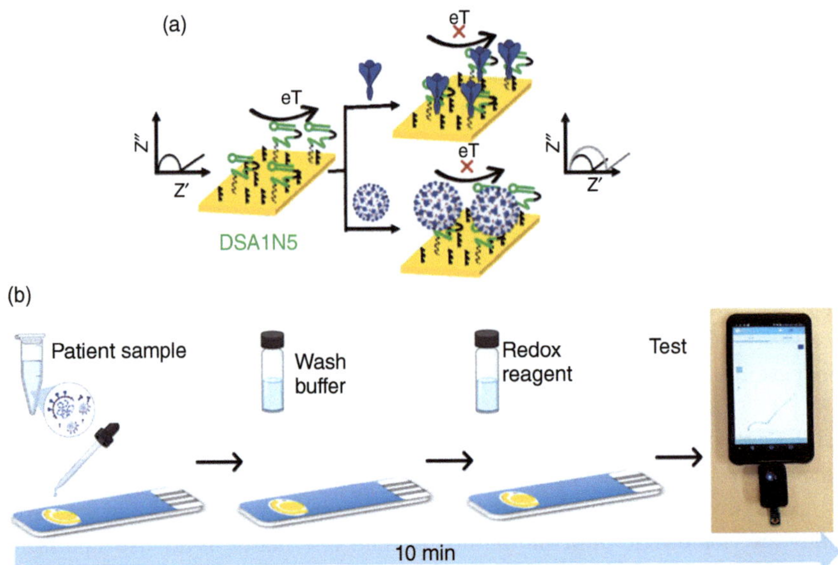

Figure 10.14 An electrochemical assay for SARS-CoV-2. (a) Schematic of the electrode in the electrochemical aptasensor. The binding of purified SARS-CoV-2 spike protein or whole virus to aptamers on the gold electrode causes a change in charge transfer resistance, and the redox reaction of Fe^{2+}/Fe^{3+} ions is blocked. (b) Workflow of the device. A patient sample is added to the electrode, followed by a wash buffer and redox reagent, and the chip can be plugged into a potentiostat, which is then plugged into a smartphone to view the results. The sample-to-answer time is 10 min. *Source:* Ref. [179] / John Wiley & Sons / CC BY-NC 4.0.

at its terminal, and were bound to a gold electrode on a chip that can measure changes in charge transfer resistance. The binding of SARS-CoV-2 spike protein to the aptamers causes an increase in electrochemical impedance, and measurable signals are produced within five minutes. To use the chip, a saliva sample mixed with binding buffer is incubated on the chip for five minutes. Then, an additional binding buffer is added onto the chip for a wash step. Lastly, a redox buffer is added to the chip, and the chip can then be plugged into a potentiostat. The potentiostat can be plugged into a smartphone to read the results. When tested with clinical samples, the test had a sensitivity of 80.5% and a specificity of 100% [179]. While this is better than or comparable to many lateral flow immunoassays, there is room for improvement. For example, the authors later developed an aptamer that binds to all variants with high affinity [174]. This aptasensor can be modified to use the universal aptamer to improve sensitivity for the detection of different SARS-CoV-2 variants.

As previously determined, multimeric aptamers that bind to more than one epitope at a time generally have higher binding affinities compared to individual aptamers. Continuing with this idea, Chauhan et al. designed what they termed a "DNA Net" (Figure 10.15) [180]. Essentially, it is a designer DNA nanostructure shaped as a net, where each vertex contains a cluster of three aptamers spaced in a way that each will bind to one of the three spike protein subunits. The net structure is flexible enough to wrap around the entire virus and have each aptamer cluster bind to a spike protein. Each aptamer is linked to a fluorescence tag and quencher. Upon binding to the target, the quencher is released, and a fluorescence signal is generated. The workflow is simple: all that needs to be done is mix the sample with the DNA Net solution and let it incubate at room temperature for 10 minutes. The signal can then be detected using a portable fluorimeter. The reported limit of detection for this assay is 1 viral copy/μL in artificial saliva and no cross-reactivity with the H1N1 influenza virus, OC43 (common cold) virus, and Zika virus. This test is also inexpensive, with an estimated cost of $1.26 each. The ease of use and high sensitivity make this assay an attractive option for POC diagnostics. However, a drawback is that portable fluorimeters can be expensive. A potential alternative is to use smartphones to detect fluorescence, which would also allow for better connectivity [181].

Aptamers can also be used with SPR, as demonstrated by Cennamo et al. (Figure 10.16) [182]. Their device features a gold film on top of a 1-cm-long plastic optical fiber, with aptamers specific to the receptor binding domain (RBD) of the spike protein immobilized on the gold film [182]. About 50 μL of human serum sample is dropped on the sensor region for an incubation period of 10 minutes, during which time any antigens present in the sample will bind the aptamers. The sensor is illuminated with a halogen light, and the transmission spectrum is recorded once before and once after a wash step using a spectrometer. The spectrometer can be connected to the internet to send and view the results. The authors report a sample-to-answer time of 20 minutes and an estimated limit of detection of 75.3 nM in spiked human serum, with no cross-reactivity to bovine serum albumin (BSA), an H1N1 hemagglutinin protein, and MERS spike protein [182]. One limitation

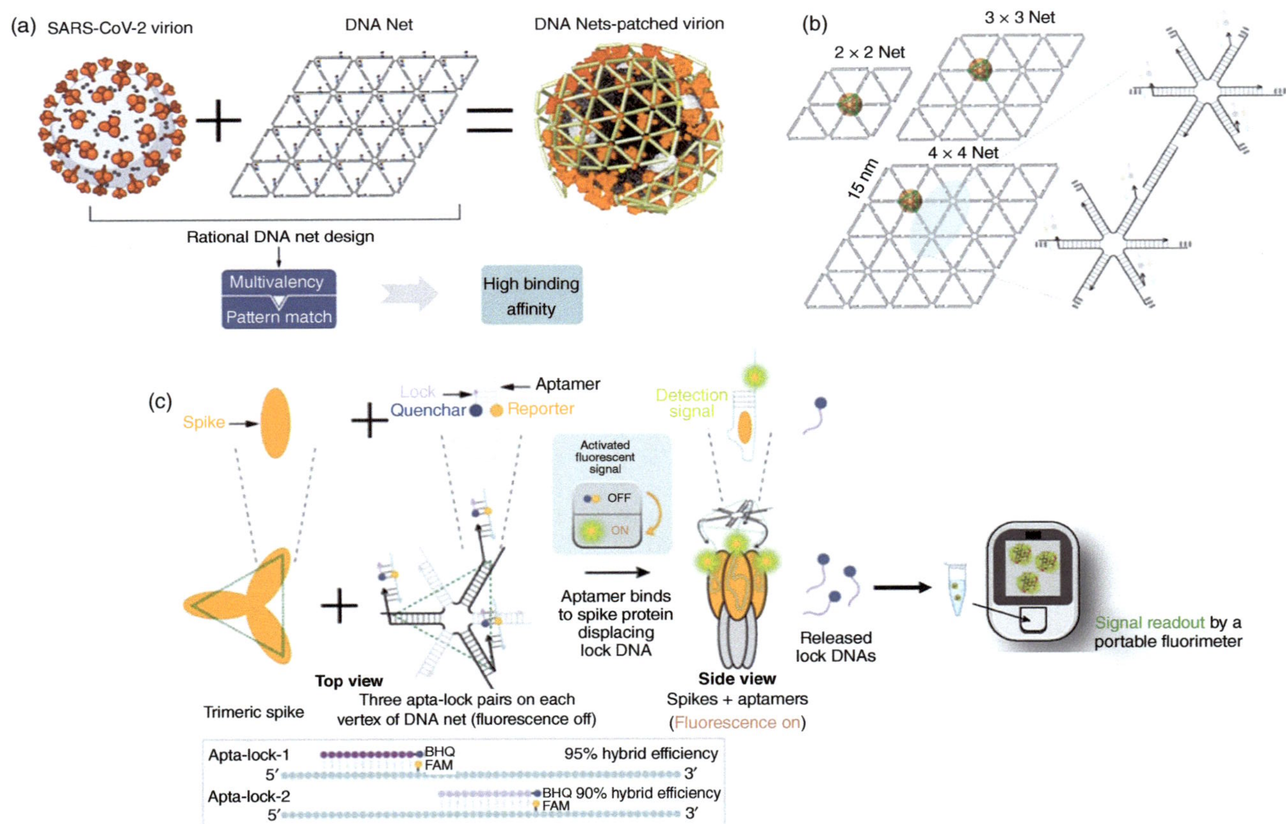

Figure 10.15 Overview of the DNA Net design and assay. (a) A 4 × 4 net is the ideal size to wrap around the SARS-CoV-2 virus and bind with high affinity. (b) A close-up look at two adjacent vertices. Each vertex contains three aptamer-lock pairs. (c) Schematic of fluorescence signal generation upon recognition of the SARS-CoV-2 spike protein. Each of the three aptamer-lock pairs interacts with a subunit of the spike protein. Upon binding, the lock DNA is displaced, and the quencher is released from the fluorescence reporter. The fluorescence signal can be detected with a portable fluorimeter. *Source:* Reproduced from Ref. [180] / with permission of American Chemical Society.

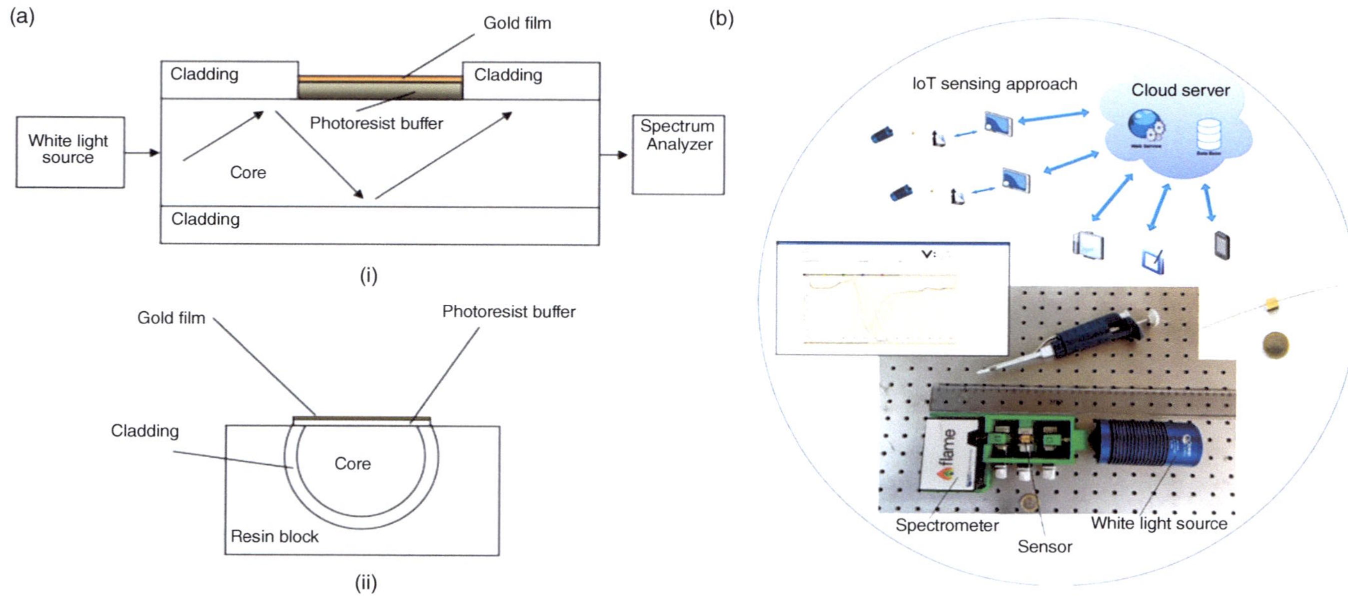

Figure 10.16 Schematic and photograph of the POC diagnostic device that uses surface plasmon resonance. (a) (i) Schematic of the detector components and (ii) cross-sectional view of the sensor. Light travels from the source through the plastic optical fiber core to form surface plasmons at the gold film layer. The reflected light is detected by a spectrometer. (b) Photograph of the device (bottom), an example of a spectrometer readout (left), and a close-up of the sensor with a coin for scale (right). The spectrometer can be connected to the internet for ease of data exchange. *Source:* Adapted from Refs. [182, 206].

of this device is that it was only tested with human serum spiked with protein. Ideally, the device should work with a more readily available sample, such as a nasal swab or saliva, and should be tested with clinical samples. Second, the spectrometer must be connected to the internet in order to view the results, which may not always be possible in resource-limited settings. Lastly, the sensitivity can be improved upon. The authors suggest using a sandwich configuration where an antibody binds the aptamer-antigen complex, as other groups have reported an increase in sensitivity using this method [182]. Nonetheless, this device is portable, sensitive, and rapid, and shows promise as a potential POC device.

Guan et al. described an aptamer-based SERS assay for rapid antigen detection (Figure 10.17) [183]. In this method, the authors used three aptamers against the receptor-binding domain of SARS-CoV-2, each with high binding affinity and specificity and each binding to a different epitope [183]. The three aptamers are conjugated onto magnetic beads and gold nanoparticles with Raman reporters, which form a sandwich with SARS-CoV-2 upon mixing with the sample. The complexes can be separated from the solution using a magnet for a wash step. The shift in the surface plasmon band caused by binding can be read with a handheld Raman spectrometer. The authors estimate a limit of detection of 18 fM of spike protein,

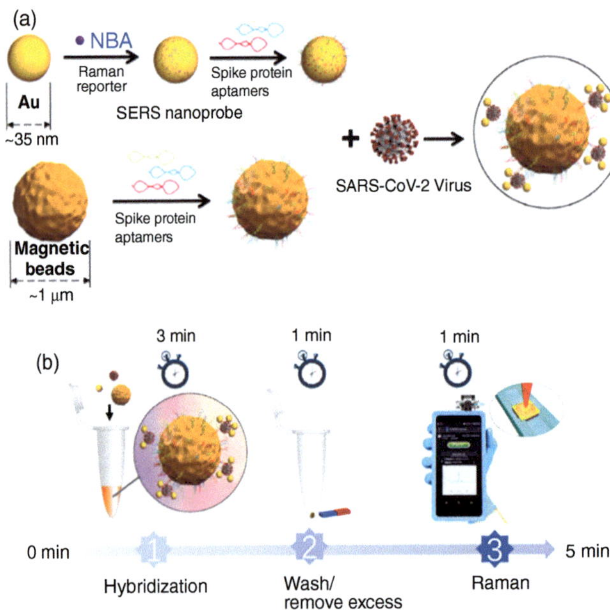

Figure 10.17 Schematic of SERS assay. (a) Gold nanoparticles with Raman reporters and magnetic beads, both conjugated to spike protein-binding aptamers, form a sandwich structure with the SARS-CoV-2 virus. (b) Workflow of the assay. The sample is mixed with magnetic beads and gold nanoparticles conjugated to the aptamer, allowing them to form the sandwich structure. SARS-CoV-2 viruses bound to magnetic beads can be easily separated from the solution and concentrated with a magnet. The Raman signal can be measured with a portable Raman spectrometer. *Source:* Ref. [183] / American Chemical Society.

and a sensitivity and specificity of 95% and 100% respectively, when tested with oropharyngeal swabs spiked with SARS-CoV-2 pseudovirus. The sample-to-answer time is five minutes, and each test costs $0.31 to produce [183]. The high sensitivity and specificity, fast turnaround time, and low cost demonstrate the device's utility as a POC test.

CRISPR-Cas Detection of Nucleic Acids CRISPR-Cas systems can be used to detect target nucleic acid sequences. When coupled with nucleic acid amplification, this system can be a powerful detection method. One example of this is a DNA endonuclease-targeted CRISPR Trans Reporter (DETECTR) assay coupled with RT-LAMP by Broughton et al. [184]. This assay demonstrated the potential of CRISPR-Cas to be used as a POCT. The use of isothermal amplification eliminates the need for a thermal cycler. The results of the assay are visualized on a lateral flow strip, where labeled nucleic acids are captured at either the control or test line depending on whether they have been cleaved by Cas12 [184]. The sample-to-answer time is 30–40 minutes. With a sensitivity of 95% and a specificity of 100%, the performance of this assay is comparable to gold-standard qRT-PCR tests [184, 185]. Additionally, this assay can be developed within two weeks, so it can be rapidly deployed with the emergence of new pathogens. However, there is an RNA extraction step, as well as a need to heat to 62 °C for the RT-LAMP reaction and 37 °C for the Cas12 detection reaction. These steps will either need to be circumvented or automated before it can be used at the POC.

An example of how these challenges can be overcome was demonstrated by Chen et al., who developed a POCT device for SARS-CoV-2 detection that combines RT-RPA with CRISPR-Cas detection (Figure 10.18) [186]. The device is simple to use. Users mix the nasopharyngeal swab sample with a magnetic bead buffer. The buffer contains a detergent that lyses the viral particles. Within this low-pH buffer, the negatively charged viral RNA binds to positively charged magnetic beads. Users load the mixture into the assay cartridge and then mount the cartridge onto a portable droplet magnetofluidics (DM) device. The DM device then takes over the process. The RNA bound to the magnetic beads is magnetically transported away from the swab eluate and through a wash buffer, thereby purifying the RNA. The RNA then gets delivered to the RT-RPA reaction mixture, and the device heats to 45.5 °C using a miniature heating module built into the DM device. At the same time, Cas12 recognizes SARS-CoV-2 RNA and cleaves ssDNA fluorogenic reporters. The portable device automatically measures the fluorescence signal every 10 seconds to give real-time data. Impressively, this device achieved a sensitivity of 100% and a specificity of 100% when tested with unprocessed clinical samples [186].

It is also possible for CRISPR-Cas systems to be used on their own without amplification. In such cases, however, other methods must be used to increase sensitivity. Fozouni et al. found that using more than one CRISPR RNA (crRNA) significantly increases the sensitivity of the assay [187]. By combining three different Cas13-crRNAs, they were able to achieve a limit of detection of 31 copies/μL in purified RNA, compared to a limit of detection of >1,000 copies/μL for each individual crRNA. When tested against patient samples, they were able to correctly

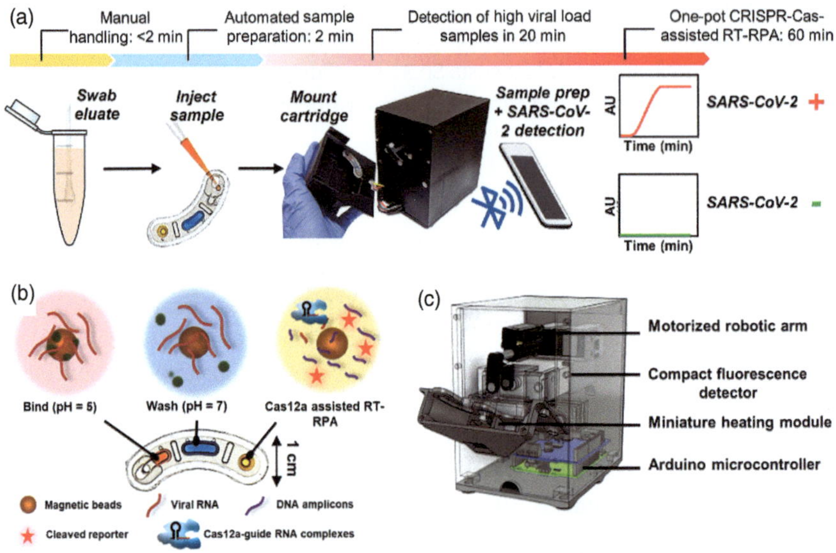

Figure 10.18 Overview of CRISPR-Cas12a-assisted RT-RPA assay. (a) Workflow of the assay. Nasopharyngeal swab samples are mixed with a buffer and injected into the cartridge. The cartridge is mounted onto the droplet magnetofluidic (DM) device, which will automatically carry out the remaining steps of the assay. (b) Schematic of the assay cartridge. The sample is loaded into the first well, where viral RNAs bind to the magnetic beads. The DM device then magnetically transports the beads with bound RNA to the wash buffer to remove any amplification inhibitors. Finally, the beads and RNA are transported to the reaction chamber, where RT-RPA and Cas12a cleavage of fluorescent reporters occur. (c) The DM device automatically transports the magnetic beads through the cartridge wells, heats the RT-RPA and Cas12a cleavage reactions, and detects fluorescence signals. *Source:* Ref. [186] / with permission of Elsevier.

identify 5 out of 5 positive samples, with no cross-reactivity with other respiratory viruses. They also showed that consumer smartphones were able to detect the fluorescence signal that is the readout of the assay. Further tests with clinical samples will have to be done before this test can be approved for use. That being said, this amplification- and heat-free assay makes an attractive POCT for areas that may not have a constant power supply.

While we have only discussed POCT for HIV and COVID-19 infections, many of these tests can be easily adapted to detect other pathogens by changing the molecular recognition element. Each of the discussed detection methods and devices has their own pros and cons. Factors such as the type of pathogen, analyte, and purpose of the test must be considered when designing or selecting a POCT.

10.3 Challenges in POCT

Currently, many different challenges exist across the field for POCT, and this section aims to summarize these issues and provide insights into how they may be overcome.

10.3.1 Sensitivity and Specificity Challenges

One of the metrics for establishing the performance of a POC test is sensitivity and specificity. The sensitivity of the test refers to the ability of the test to determine cases of patients correctly and can be calculated by dividing the true positive patients by the sum of the true positive patients and false negative patients. The specificity of a test refers to the ability of the test to determine the cases of healthy individuals correctly. This can be referred to mathematically as the cases of true negative patients divided by the sum of true negative individuals and false positive patients. It is critical to ensure that POCT tests can perform well and, as a result, there are guidelines in place by healthcare agencies across the world on different specificity and sensitivity requirements for diagnostics across various diseases.

LFDs are the most widely used POCT device type and suffer from sensitivity and specificity challenges. As mentioned previously, these assays have very widespread use across different disease indications and have many benefits, such as being extremely cheap to manufacture and simple to use. However, LFDs are limited to qualitative or semi-quantitative detection of one highly concentrated analyte per test. Several methods that are being researched to improve sensitivity and specificity challenges include improving biorecognition, assay optimization, orientation of capture elements, signal amplification, and extraction and enrichment of the target molecule of the sample.

Adjusting the flow rate in LFDs is critical to improving sensitivity because the flow rate influences the contact time between the biorecognition element and the antigen present in the analyte. This can be contrasted with standard laboratory tests such as ELISA, which have long contact times and allow for better detection of the antigen. This flow rate is primarily caused by the capillary effect of paper fibers known as wicking. Wicking creates a constant flow rate across a single direction when passing across the membrane. An ideal test, however, would aim to reduce the flow rate when passing over the test region to augment the binding reaction between the biorecognition markers and the antigen of interest. Currently, there have been many attempts to remedy this issue. One approach that has been performed is to use flow-interrupting materials, such as salt, sugar, and wax polymer, on the LFD strip to reduce the flow rate [188–190]. One group has demonstrated that just applying pressure to the top of the NC membrane between the test line and control line could lead to a positive test signal improvement of COVID-19 LFD kits by approximately twofold [191].

Improving biorecognition is another key factor to consider when improving the sensitivity and specificity of POCT. This is widely applicable to any type of POCT that uses a biorecognition element. Currently, the most common biorecognition molecule type used is antibodies. However, other recognition elements such as nanobodies and nucleic acid probes such as DNA/RNA aptamers have been experimented with as alternatives to antibodies. Some examples of advantages that an aptamer can have over an antibody include their stability compared to antibodies, which makes them easier to store, or their ability to be easily engineered to enhance binding affinity, such as joining two aptamers together to form a dimeric aptamer [192].

Orientation of the immobilized capture element is another technique that has been experimented with in literature to improve analytical performance [193]. Orientation of the immobilized capture element could improve the antigen-binding capacity in POCT, which would lead to sensitivity and specificity improvements [193]. Typically, LFAs immobilize antibodies in a random orientation, which may cause them to have low sensitivity [194, 195]. In an aforementioned study by Lee et al., the linker CBP31 was used to immobilize capture antibodies in an orientated fashion in the test line of their SARS-CoV-2 LFD. Using 19 clinical samples, the LFD was able to detect SARS-CoV-2 with 100% accuracy [22].

Signal amplification by improving the measurable signal in the test region to provide better color contrast is another area that can be further developed to improve the sensitivity and specificity. One way to improve signal amplification was to make new chemical/physical modifications. One group modified the standard gold nanoparticle label into a polydopamine-coated gold nanoparticle [196]. The resultant limit of detection ended up being 10-fold lower than the standard LFD [196].

Another method to improve the sensitivity and specificity of LFDs is to improve the extraction and enrichment of the sample. This occurs through concentrating the sample before injecting it into the LFD [197]. One group has demonstrated this by using antibody-functionalized magnetic beads to concentrate and isolate a protein of interest before loading it onto an LFD [197]. The result was a 10-fold increase in sensitivity [197]. There are washing steps that are part of this protocol, which have been noted to be less user-friendly [193]. Thus, this type of approach does add additional steps that healthcare teams will need to learn how to perform and could lead to more complicated workflows if implemented more widely.

10.3.2 Multiplexing

Multiplexing is a powerful feature that some POC tests can perform, as it allows for detecting multiple different analytes from a single test, which opens the possibility for more conditions to be diagnosed at a given time [198]. Once again, when considering the most widely used POC test (LFDs), these devices suffer from poor multiplexing compared to other earlier discussed methods. Some of the reasons for challenges in multiplexing LFDs include antibody cross-reactivity, lot-to-lot variability, and dependency on the capillary action between the sample and LFD components [198]. The most limiting issue, however, is the possibility of cross-reactivity, which causes issues such as nonspecific binding, higher background, and false positives. This has been observed in some diseases, such as Flavivirus infections like the dengue virus, which have cross-reactivity with the Zika virus. Some logistical challenges for testing the efficacy of multiplexed tests would include having to find patients with multiple infections in the case of multiplexed tests attempting to test multiple different infections. Other issues include a lack of regulatory guidance when it comes to the development of multiplexed tests. This includes issues such as inter- and intra-assay variability for multiplexed tests that are not clearly defined. Furthermore, it is unclear whether one test in the multiplex failing constitutes all the tests failing.

One of the methods that has been proposed as a remedy to the problem of multiplexing is running individual strips in an "array" format [198]. This would be done by using a holder to distribute the sample to multiple test strips. The key advantage of this strategy is that the tests are performed independently of each other, and thus, no issues with reciprocal interference between assays can occur. However, a disadvantage of this strategy is that it does require relatively large volumes of sample as each strip will usually require 50–100 μL of liquid to complete [198]. Furthermore, the device cost will increase with each strip added, as both the manufacturing of the strip and the reagents needed have to be considered [198].

Another way to solve the issue of cross-reactivity is to consider having alternative biomolecules, such as recombinant proteins, aptamers, or protein scaffolds that generally will have less susceptibility to cross-reactivity [199–201]. However, compared to the catalogs of commercially available antibodies, the acquisition of these alternative biomolecules is not nearly as simple.

10.3.3 Awareness/Distrust of POCT

New developments in POCT will often require extensive education and validation before they can enter mainstream clinical use. In a recent survey conducted by IPSOS among healthcare providers, the top five challenges of POC rapid tests include lack of awareness of POC rapid tests among healthcare providers and lack of trust from the healthcare profession [202]. One of the reasons for distrust of POCT among physicians is due to the differences in specificity and/or sensitivity between at-home tests and laboratory-based tests [203]. Often, the POCT is still sent to the laboratory or supplemented with a laboratory-based test to verify the result. Ways to overcome this distrust include improving the sensitivity and specificity of tests, better training to ensure healthcare providers know the limitations of tests, or simplifying the test so it can be more easily administered.

10.3.4 Mass Production Issues

Scaling up the production of many different types of POCT has proven to be a barrier to making certain POCT devices more widely accessible. While the dominant test device, LFDs, has less mass production issues than other devices, mass production of many POCTs is difficult. For example, LOC devices have been reported to be very challenging to scale up to industrial size [204]. Currently, LOC research is mainly conducted academically with a focus on topics such as added device functionality and increased endpoint detection sensitivity rather than system integration, standardization, and economies of scale to allow for the mass production of devices [204].

In terms of system integration, one must consider the fact that LOC devices often will require many large ancillary devices, such as fluidic pumps, high current power supplies, and signal acquisition devices, such as microscopes or photospectrometers, leading to the analogy of "chip-in-a-lab" to highlight the size and amount of equipment needed to operate these devices [204]. When compared to LFDs, which require very few, if any, extra instrumentation, the amount of ancillary devices required for these microfluidic devices must either be miniaturized or integrated into the devices to make them more applicable as a mainstream POCT device [204].

When researchers develop new technologies, they are often trying to prove a hypothesis rather than design a product for commercial use. Consequently, considerations such as economies of scale are an afterthought rather than a priority in the design of this new POCT. Examples of economies of scale issues in POCT devices include traditional fabrication methods used to create nanoscale structures, such as lithographical techniques that are not able to be scaled up efficiently [204]. Another example is the polymers used in the academic development of microfluidic POCT. Currently, the preferred polymer for developing these tests is polydimethylsiloxane (PDMS) [205]. However, PDMS is challenging to integrate into manufacturing processes due to difficulties in scaling up manufacturing processes and the cost of using PDMS compared to other polymers [204].

Finally, the issue of standardization is very similar to the issues discussed about the economies of scale to produce microfluidic devices. Currently, there is no standard fabrication technique that is commonly used by all commercial and research groups. While in academic research, this may provide flexibility in the development of the device and proving the hypothesis that the research project aims to achieve becomes a far greater challenge once it enters commercial applications. Furthermore, there is a tendency to move from one research question to the next after a hypothesis is proven rather than spending additional time analyzing the reproducibility of the device when it is reconstructed with newer materials [204]. This gap in knowledge presents a fundamental disconnect between academic researchers' goals and the practical research needed to push these devices into the market.

To conclude, as shown using the examples of microfluidic devices, key issues regarding both the peripheral devices required for POCT devices and the scale-up production of POCT devices need to be overcome for these devices to enter mainstream use.

10.4 Conclusion and Future Perspectives

The recent and rapid advancements in POCT have revolutionized the landscape of diagnostic methods, making significant strides in addressing the limitations of conventional laboratory-based diagnostics. The innovations discussed in this chapter highlight the remarkable progress made in developing rapid, accurate, and accessible diagnostic tools. These advancements have crucial implications for the present and future of public health, offering the potential to enhance disease management, improve patient outcomes, and mitigate the spread of infectious diseases.

However, despite these advancements, a multitude of challenges remain. Notably, there is still no POC test for 50% of the WHO priority diseases of epidemic potential [38]. This gap underscores the need for continued research and development in the field. Future perspectives should also focus on overcoming the technical limitations that currently hinder the widespread adoption of POCT. More established tests, such as LFDs, require improvements in sensitivity and specificity, whereas more niche tests, such as LOC and LOAD, need improvements in cost-effectiveness and scalability.

To fully realize the potential of POCT, a concerted effort is required from researchers, healthcare providers, policymakers, and industry stakeholders. If this effort is prioritized in the near future, a potential repeat of the catastrophic conditions of COVID-19 can be mitigated or even prevented (12). Continued investment in research and innovation is essential to develop new diagnostic tools that can meet the evolving demands of global health. The recent advent of AI and nanotechnology will play a major role in enhancing diagnostic precision, enabling miniaturization of devices, and improving the speed and accessibility of tests.

All in all, the future of POCT is promising, with the potential to transform healthcare delivery and improve global health outcomes. As we continue to innovate and address existing challenges, the vision of accessible, rapid, and reliable diagnostics for all becomes increasingly attainable. It is imperative to maintain the momentum in research and development, ensuring that POCT technologies reach their full potential and contribute to a healthier, more equitable world.

References

1 Khan, P., Aufdembrink, L.M., and Engelhart, A.E. (2020). Isothermal SARS-CoV-2 diagnostics: tools for enabling distributed pandemic testing as a means of supporting safe reopenings. *ACS Synthetic Biology* 9: 2861–2880.
2 Mina, M.J. and Andersen, K.G. (2021). COVID-19 testing: one size does not fit all. *Science (New York, N.Y.)* 371 (6525): 126–127.
3 Maia, R., Carvalho, V., Faria, B. et al. (2022). Diagnosis methods for COVID-19: a systematic review. *Micromachines (Basel)* 13 (8): 1349.
4 Kashir, J. and Yaqinuddin, A. (2020). Loop mediated isothermal amplification (LAMP) assays as a rapid diagnostic for COVID-19. *Medical Hypotheses* 141: 109786.
5 Hernández-Pérez, J.M., Martín-González, E., and Pino-Yanes, M. (2020). Strengths and weaknesses of the diagnostic tests for SARS-CoV-2 infection. *Medicina Clinica (English Edition)* 155 (10): 464–465.
6 Padoan, A., Cosma, C., Sciacovelli, L. et al. (2020). Analytical performances of a chemiluminescence immunoassay for SARS-CoV-2 IgM/IgG and antibody kinetics. *Clinical Chemistry and Laboratory Medicine (CCLM)* 58 (7): 1081–1088.
7 Chen, H., Zhou, X., Wang, M., and Ren, L. (2023). Towards point of care CRISPR-based diagnostics: from method to device. *Journal of Functional Biomaterials* 14 (2): 97.
8 van Dongen, J.E., Berendsen, J.T.W., Steenbergen, R.D.M. et al. (2020). Point-of-care CRISPR/Cas nucleic acid detection: recent advances, challenges and opportunities. *Biosensors & Bioelectronics* 15 (166): 112445.

9 Ghouneimy, A., Mahas, A., Marsic, T. et al. (2023). CRISPR-based diagnostics: challenges and potential solutions toward point-of-care applications. *ACS Synthetic Biology* 12 (1): 1–16.

10 Wang, C., Liu, M., Wang, Z. et al. (2021). Point-of-care diagnostics for infectious diseases: from methods to devices. *Nano Today* 37: 101092.

11 Price, C.P. (2001). Point of care testing. *BMJ* 322 (7297): 1285–1288.

12 Simpson, S., Kaufmann, M.C., Glozman, V., and Chakrabarti, A. (2020). Disease X: accelerating the development of medical countermeasures for the next pandemic. *The Lancet Infectious Diseases* 20 (5): 108–115.

13 Kretzschmar, M.E., Rozhnova, G., Bootsma, M.C.J. et al. (2020). Impact of delays on effectiveness of contact tracing strategies for COVID-19: a modelling study. *The Lancet Public Health* 5 (8): e452–e459.

14 MarketsandMarkets [Internet] (2024). Point of Care Diagnostics Market Size, Share, Trends and Revenue Forecast [Latest]. https://www.marketsandmarkets .com/Market-Reports/point-of-care-diagnostic-market-106829185.html (accessed 13 March 2024).

15 England, J.M., Hyde, K., Lewis, S.M. et al. (1995). Guide-lines for near patient testing: haematology. *Clinical and Laboratory Haematology* 17 (4): 301–310.

16 Kost, G.J., Jammal, M.A., Ward, R.E., and Safwat, A.M. (1986). Monitoring of ionized calcium during human hepatic transplantation. Critical values and their relevance to cardiac and hemodynamic management. *American Journal of Clinical Pathology* 86 (1): 61–70.

17 Liu, X., Zhu, X., Kost, G.J. et al. (2019). The creation of point-of-careology. *Point of Care* 18 (3): 77.

18 Kesavadev, J., Das, A.K., Manojan, K.K. et al. (2023). History and evolution of capillary glucose monitoring. *International Journal of Diabetes and Technology* 2 (2): 37.

19 Shaker, G. and Swift, C.J. (2024). Peroxidase-coupled glucose method. In: *StatPearls [Internet]*. Treasure Island (FL): StatPearls Publishing. http://www.ncbi .nlm.nih.gov/books/NBK594277 (accessed 17 June 2024).

20 Choosing The Right Western Blot Detection Method [Internet]. https://www .ptglab.com/support/western-blot-protocol/choosing-the-right-western-blot-detection-method/ (accessed 1 July 2024).

21 Ko, A. and Liao, C. (2023). Paper-based colorimetric sensors for point-of-care testing. *Analytical Methods* 15 (35): 4377–4404.

22 Lee, A.S., Kim, S.M., Kim, K.R. et al. (2023). A colorimetric lateral flow immunoassay based on oriented antibody immobilization for sensitive detection of SARS-CoV-2. *Sensors and Actuators B: Chemical* 15 (379): 133245.

23 Yang, J.M., Kim, K.R., Jeon, S. et al. (2021). A sensitive paper-based lateral flow immunoassay platform using engineered cellulose-binding protein linker fused with antibody-binding domains. *Sensors and Actuators B: Chemical* 15 (329): 129099.

24 Yeo, S.J., Choi, K., Cuc, B.T. et al. (2016). Smartphone-based fluorescent diagnostic system for highly pathogenic H5N1 viruses. *Theranostics* 6 (2): 231–242.

25 Chen, X., Wang, X., Fang, Y. et al. (2022). Long-lasting chemiluminescence-based POCT for portable and visual pathogenic detection and in situ inactivation. *Analytical Chemistry* 94 (23): 8382–8391.

26 Mo, G., He, X., Zhou, C. et al. (2019). A novel ECL sensor based on a boronate affinity molecular imprinting technique and functionalized SiO_2@CQDs/AuNPs/MPBA nanocomposites for sensitive determination of alpha-fetoprotein. *Biosensors & Bioelectronics* 1 (126): 558–564.

27 Ning, Q., Feng, S., Cheng, Y. et al. (2022). Point-of-care biochemical assays using electrochemical technologies: approaches, applications, and opportunities. *Mikrochimica Acta* 189 (8): 310.

28 Boonkaew, S., Szot-Karpińska, K., Niedziółka-Jönsson, J. et al. (2024). NFC smartphone-based electrochemical microfluidic device integrated with nanobody recognition for C-reactive protein. *ACS Sensors [Internet].* https://doi.org/10.1021/acssensors.4c00249.

29 Puiu, M. and Bala, C. (2016). SPR and SPR imaging: recent trends in developing nanodevices for detection and real-time monitoring of biomolecular events. *Sensors* 16 (6): 870.

30 Harpaz, D., Koh, B., Marks, R.S. et al. (2019). Point-of-care surface plasmon resonance biosensor for stroke biomarkers NT-proBNP and S100β using a functionalized gold chip with specific antibody. *Sensors (Basel)* 19 (11): 2533.

31 IDEX Health & Science [Internet]. Surface-enhanced Raman Scattering - Semrock. https://www.idex-hs.com/capabilities/life-science-optics/applications/sers (accessed 1 July 2024).

32 Wu, L., Liu, X., Zong, S. et al. (2023). A SERS composite hydrogel device for point-of-care analysis of neurotransmitter in whole blood. *Biosensors-Basel* 13 (6): 611.

33 Kozel, T.R. and Burnham-Marusich, A.R. (2017). Point-of-care testing for infectious diseases: past, present, and future. *Journal of Clinical Microbiology* 55 (8): 2313–2320.

34 Parvu, V., Gary, D.S., Mann, J. et al. (2021). Factors that influence the reported sensitivity of rapid antigen testing for SARS-CoV-2. *Frontiers in Microbiology* 12: 2611–2611.

35 Dinnes, J., Deeks, J.J., Adriano, A. et al. (2021). Rapid, point-of-care antigen and molecular-based tests for diagnosis of SARS-CoV-2 infection. *Cochrane Database of Systematic Reviews [Internet]* 3. https://doi.org//10.1002/14651858.CD013705.pub2.

36 Stokes, W., Berenger, B.M., Portnoy, D. et al. (2021). Clinical performance of the Abbott Panbio with nasopharyngeal, throat, and saliva swabs among symptomatic individuals with COVID-19. *European Journal of Clinical Microbiology & Infectious Diseases: Official Publication of the European Society of Clinical Microbiology.* 40 (8): 1721–1726.

37 Gasperino, D., Baughman, T., Hsieh, H.V. et al. (2018). Improving lateral flow assay performance using computational modeling. *Annual Review of Analytical Chemistry* 11: 219–244.

38 Budd, J., Miller, B.S., Weckman, N.E. et al. (2023). Lateral flow test engineering and lessons learned from COVID-19. *Nature Reviews Bioengineering* 1 (1): 13–31.

39 Boehringer, H.R. and O'Farrell, B.J. (2021). Lateral flow assays in infectious disease diagnosis. *Clinical Chemistry* 68 (1): 52–58.

40 Arshavsky-Graham, S. and Segal, E. (2022). Lab-on-a-chip devices for point-of-care medical diagnostics. In: *Microfluidics in Biotechnology [Internet]* (ed. J. Bahnemann and A. Grünberger), 247–265. Cham: Springer International Publishing. https://doi.org/10.1007/10_2020_127.

41 Temiz, Y., Lovchik, R.D., Kaigala, G.V., and Delamarche, E. (2015 Jan). Lab-on-a-chip devices: How to close and plug the lab? *Microelectronic Engineering* 25 (132): 156–175.

42 Gorkin, R. III, Nwankire, C.E., Gaughran, J. et al. (2012). Centrifugo-pneumatic valving utilizing dissolvable films. *Lab on a Chip* 12 (16): 2894–2902.

43 Li, L., Wang, C., Nie, Y. et al. (2020 Jun). Nanofabrication enabled lab-on-a-chip technology for the manipulation and detection of bacteria. *TrAC, Trends in Analytical Chemistry* 1 (127): 115905.

44 Kong, L.X., Perebikovsky, A., Moebius, J. et al. (2016). Lab-on-a-CD: a fully integrated molecular diagnostic system. *SLAS Technology* 21 (3): 323–355.

45 González-González, E., Trujillo-de Santiago, G., Lara-Mayorga, I.M. et al. (2020). Portable and accurate diagnostics for COVID-19: Combined use of the miniPCR thermocycler and a well-plate reader for SARS-CoV-2 virus detection. *PLoS One* 15 (8): e0237418.

46 Petralia, S. and Conoci, S. (2017). PCR technologies for point of care testing: progress and perspectives. *ACS Sensors* 2 (7): 876–891.

47 Asiello, P.J. and Baeumner, A.J. (2011). Miniaturized isothermal nucleic acid amplification, a review. *Lab on a Chip* 11 (8): 1420–1430.

48 Chang, C.C., Chen, C.C., Wei, S.C. et al. (2012). Diagnostic devices for isothermal nucleic acid amplification. *Sensors* 12 (6): 8319–8337.

49 Goldsmith, S.J. (1975). Radioimmunoassay: review of basic principles. *Seminars in Nuclear Medicine* 5 (2): 125–152.

50 Singer, J. and Plotz, C. (1956). The latex fixation test. I. Application to the serologic diagnosis of rheumatoid arthritis. *The American Journal of Medicine* 21: 888–892.

51 Vaitukaitis, J.L., Braunstein, G.D., and Ross, G.T. (1972). A radioimmunoassay which specifically measures human chorionic gonadotropin in the presence of human luteinizing hormone. *American Journal of Obstetrics and Gynecology* 113 (6): 751–758.

52 Kalimuthu, K., Arivalagan, J., Mohan, M. et al. (2022). Point of care diagnosis of plant virus: current trends and prospects. *Molecular and Cellular Probes* 1 (61): 101779.

53 Jarvis, J.N., Tenforde, M.W., Lechiile, K. et al. (2020). Evaluation of a novel semiquantitative cryptococcal antigen lateral flow assay in patients with advanced HIV disease. *Journal of Clinical Microbiology* 58 (9). https://doi.org/10.1128/jcm.00441-20.

54 Lalremruata, A., Nguyen, T.T., McCall, M.B.B. et al. (2020). Recombinase polymerase amplification and lateral flow assay for ultrasensitive detection of low-density plasmodium falciparum infection from controlled human malaria infection studies and naturally acquired infections. *Journal of Clinical Microbiology* 58 (5). https://doi.org/10.1128/jcm.01879-19.

55 Huang, L., Zhang, Y., Su, E. et al. (2020). Eight biomarkers on a novel strip for early diagnosis of acute myocardial infarction. *Nanoscale Advances* 2 (3): 1138–1143.

56 Yin, R., Sun, Y., Wang, K. et al. (2020). Development of a PCR-based lateral flow strip assay for the simple, rapid, and accurate detection of pork in meat and meat products. *Food Chemistry* 15 (318): 126541.

57 Lan, J., Sun, W., Chen, L. et al. (2020). Simultaneous and rapid detection of carbofuran and 3-hydroxy-carbofuran in water samples and pesticide preparations using lateral-flow immunochromatographic assay. *Food and Agricultural Immunology* 31 (1): 165–175.

58 Kasetsirikul, S., Shiddiky, M.J.A., and Nguyen, N.T. (2020). Challenges and perspectives in the development of paper-based lateral flow assays. *Microfluidics and Nanofluidics* 24 (2): 17.

59 Meng, L., Turner, A.P.F., and Mak, W.C. (2020). Soft and flexible material-based affinity sensors. *Biotechnology Advances* 1 (39): 107398.

60 Castillo-León, J., Trebbien, R., Castillo, J.J., and Svendsen, W.E. (2021). Commercially available rapid diagnostic tests for the detection of high priority pathogens: status and challenges. *The Analyst* 146 (12): 3750–3776.

61 Brouwer, S., Rivera-Hernandez, T., Curren, B.F. et al. (2023). Pathogenesis, epidemiology and control of Group A Streptococcus infection. *Nature Reviews. Microbiology* 21 (7): 431–447.

62 Lean, W.L., Arnup, S., Danchin, M., and Steer, A.C. (2014). Rapid diagnostic tests for Group A Streptococcal pharyngitis: a meta-analysis. *Pediatrics* 134 (4): 771–781.

63 Gerber, M.A., Baltimore, R.S., Eaton, C.B. et al. (2009). Prevention of rheumatic fever and diagnosis and treatment of acute Streptococcal pharyngitis. *Circulation* 119 (11): 1541–1551.

64 Alhamoud, M.A., Salloot, I.Z., Mohiuddin, S.S. et al. A comprehensive review study on glomerulonephritis associated with post-streptococcal infection. *Cureus* 13 (12): e20212.

65 Shulman, S.T., Bisno, A.L., Clegg, H.W. et al. (2012). Executive summary: clinical practice guideline for the diagnosis and Management of Group A Streptococcal pharyngitis: 2012 update by the Infectious Diseases Society of America. *Clinical Infectious Diseases* 55 (10): 1279–1282.

66 Foudeh, A.M., Didar, T.F., Veres, T., and Tabrizian, M. (2012). Microfluidic designs and techniques using lab-on-a-chip devices for pathogen detection for point-of-care diagnostics. *Lab on a Chip* 12 (18): 3249–3266.

67 Nasseri, B., Soleimani, N., Rabiee, N. et al. (2018). Point-of-care microfluidic devices for pathogen detection. *Biosensors & Bioelectronics* 15 (117): 112–128.

68 Fang, X., Liu, Y., Kong, J., and Jiang, X. (2010). Loop-mediated isothermal amplification integrated on microfluidic chips for point-of-care quantitative detection of pathogens. *Analytical Chemistry* 82 (7): 3002–3006.

69 Krishna, K.S., Li, Y., Li, S., and Kumar, C.S.S.R. (2013). Lab-on-a-chip synthesis of inorganic nanomaterials and quantum dots for biomedical applications. *Advanced Drug Delivery Reviews* 65 (11): 1470–1495.

70 Marasso, S.L., Giuri, E., Canavese, G. et al. (2011). A multilevel Lab on chip platform for DNA analysis. *Biomedical Microdevices* 13 (1): 19–27.

71 Zhu, H., Fohlerová, Z., Pekárek, J. et al. (2020). Recent advances in lab-on-a-chip technologies for viral diagnosis. *Biosensors & Bioelectronics* 1 (153): 112041.

72 Convery, N. and Gadegaard, N. (2019). 30 years of microfluidics. *Micro and Nano Engineering* 1 (2): 76–91.

73 Kilby, J.S. (2007). Miniaturized electronic circuits [US Patent No. 3,138, 743]. *IEEE Solid-State Circuits Society Newsletter* 12 (2): 44–54.

74 Terry, S.C., Jerman, J.H., and Angell, J.B. (1979). A gas chromatographic air analyzer fabricated on a silicon wafer. *IEEE Transactions on Electron Devices* 26 (12): 1880–1886.

75 Lim, Y.C., Kouzani, A.Z., and Duan, W. (2010). Lab-on-a-chip: a component view. *Microsystem Technologies* 16 (12): 1995–2015.

76 Najjar, D., Rainbow, J., Sharma Timilsina, S. et al. (2022). A lab-on-a-chip for the concurrent electrochemical detection of SARS-CoV-2 RNA and anti-SARS-CoV-2 antibodies in saliva and plasma. *Nature Biomedical Engineering* 6 (8): 968–978.

77 Henderson, B., Kinahan, D.J., and Ducrée, J. (2016). The centrifugal microfluidic: lab-on-a-disc platform. In: *Microfluidics for Biologists: Fundamentals and Applications [Internet]* (ed. C.K. Dixit and A. Kaushik), 115–144. Cham: Springer International Publishing. https://doi.org/10.1007/978-3-319-40036-5_5.

78 Tang, M., Wang, G., Kong, S.K., and Ho, H.P. (2016). A review of biomedical centrifugal microfluidic platforms. *Micromachines* 7 (2): 26.

79 Nguyen, H.V., Nguyen, V.D., Nguyen, H.Q. et al. (2019). Nucleic acid diagnostics on the total integrated lab-on-a-disc for point-of-care testing. *Biosensors & Bioelectronics* 15 (141): 111466.

80 H-pui, H., P-man, L., S-yuen, K. H.-c., W. et al. (2014). Allergen screening bioassays: recent developments in lab-on-a-chip and lab-on-a-disc systems. *Bioanalysis* 6 (14): 2005–2018.

81 Miyazaki, C.M., Carthy, E., and Kinahan, D.J. (2020 Nov). Biosensing on the centrifugal microfluidic lab-on-a-disc platform. *Processes* 8 (11): 1360.

82 Sanger, K., Zór, K., Bille Jendresen, C. et al. (2017). Lab-on-a-disc platform for screening of genetically modified *E. coli* cells via cell-free electrochemical detection of *p*-Coumaric acid. *Sensors and Actuators B: Chemical* 1 (253): 999–1005.

83 Michael, I.J., Kim, T.H., Sunkara, V., and Cho, Y.K. (2016). Challenges and opportunities of centrifugal microfluidics for extreme point-of-care testing. *Micromachines* 7 (2): 32.

84 Strohmeier, O., Keller, M., Schwemmer, F. et al. (2015). Centrifugal microfluidic platforms: advanced unit operations and applications. *Chemical Society Reviews* 44 (17): 6187–6229.

85 Yao, Y., Chen, X., Zhang, X. et al. (2020). Rapid detection of Influenza virus subtypes based on an integrated centrifugal disc. *ACS Sensors* 5 (5): 1354–1362.

86 Madou, M., Zoval, J., Jia, G. et al. (2006). Lab on a CD. *Annual Review of Biomedical Engineering* 8: 601–628.

87 Loo, J., But, G.W.C., Kwok, H.C. et al. (2019). A rapid sample-to-answer analytical detection of genetically modified papaya using loop-mediated isothermal amplification assay on lab-on-a-disc for field use. *Food Chemistry* 15 (274): 822–830.

88 Sayad, A.A., Ibrahim, F., Uddin, S.M. et al. (2016). A microfluidic lab-on-a-disc integrated loop mediated isothermal amplification for foodborne pathogen detection. *Sensors and Actuators B: Chemical* 1 (227): 600–609.

89 Morelli, L., Serioli, L., Centorbi, F.A. et al. (2018). Injection molded lab-on-a-disc platform for screening of genetically modified E. coli using liquid–liquid extraction and surface enhanced Raman scattering. *Lab on a Chip* 18 (6): 869–877.

90 King, D. and Ducrée, J. (2015). Optical detection on centrifugal microfluidic lab-on-a-disc platforms. In: *Encyclopedia of Microfluidics and Nanofluidics [Internet]* (ed. D. Li), 2535–2542. New York, NY: Springer. https://doi.org/10.1007/978-1-4614-5491-5_1785.

91 Vogels, C.B.F., Brito, A.F., Wyllie, A.L. et al. (2020). Analytical sensitivity and efficiency comparisons of SARS-CoV-2 RT–qPCR primer–probe sets. *Nature Microbiology* 5 (10): 1299–1305.

92 Park, S., Zhang, Y., Lin, S. et al. (2011). Advances in microfluidic PCR for point-of-care infectious disease diagnostics. *Biotechnology Advances* 29 (6): 830–839.

93 Zhang, C., Xu, J., Ma, W., and Zheng, W. (2006). PCR microfluidic devices for DNA amplification. *Biotechnology Advances* 24 (3): 243–284.

94 Northrup, M., Ching, M., White, R., and Watson, R. (1993). DNA amplification in a microfabricated reaction chamber. *Transducers*' 93: 924–926.

95 Zhang, C. and Xing, D. (2007). Miniaturized PCR chips for nucleic acid amplification and analysis: latest advances and future trends. *Nucleic Acids Research* 35 (13): 4223–4237.

96 González-González, E., Mendoza-Ramos, J.L., Pedroza, S.C. et al. (2019). Validation of use of the miniPCR thermocycler for Ebola and Zika virus detection. *PLoS One* 14 (5): e0215642.

97 Neuzil, P., Zhang, C., Pipper, J. et al. (2006). Ultra fast miniaturized real-time PCR: 40 cycles in less than six minutes. *Nucleic Acids Research* 34 (11): e77.

98 Ullerich, L., Campbell, S., Krieg-Schneider, F. et al. (2017). Ultra-fast PCR technologies for point-of-care testing. *LaboratoriumsMedizin* 41 (5): 239–244.

99 Zhao, Y., Chen, F., Li, Q. et al. (2015). Isothermal amplification of nucleic acids. *Chemical Reviews* 115 (22): 12491–12545.

100 Notomi, T., Okayama, H., Masubuchi, H. et al. (2000). Loop-mediated isothermal amplification of DNA. *Nucleic Acids Research* 28 (12): e63.

101 Notomi, T., Mori, Y., Tomita, N., and Kanda, H. (2015). Loop-mediated isothermal amplification (LAMP): principle, features, and future prospects. *Journal of Microbiology* 53 (1): 1–5.

102 Yan, Y., Li, J., Li, W. et al. (2018). DNA flower-encapsulated horseradish peroxidase with enhanced biocatalytic activity synthesized by an isothermal one-pot method based on rolling circle amplification. *Nanoscale* 10 (47): 22456–22465.

103 Davari, M., van Diepeningen, A.D., Babai-Ahari, A. et al. (2012). Rapid identification of *Fusarium graminearum* species complex using Rolling Circle Amplification (RCA). *Journal of Microbiological Methods* 89 (1): 63–70.

104 Quyen, T.L., Ngo, T.A., Bang, D.D. et al. (2019). Classification of multiple DNA dyes based on inhibition effects on real-time loop-mediated isothermal amplification (LAMP): prospect for point of care setting. *Frontiers in Microbiology* 10.

105 Nzelu, C.O., Kato, H., and Peters, N.C. (2019). Loop-mediated isothermal amplification (LAMP): an advanced molecular point-of-care technique for the detection of Leishmania infection. *PLoS Neglected Tropical Diseases* 13 (11): e0007698.

106 Dai, Y., Furst, A., and Liu, C.C. (2019). Strand displacement strategies for biosensor applications. *Trends in Biotechnology* 37 (12): 1367–1382.

107 Zhou, W., Hu, L., Ying, L. et al. (2018). A CRISPR–Cas9-triggered strand displacement amplification method for ultrasensitive DNA detection. *Nature Communications* 9 (1).

108 Barreda-García, S., Miranda-Castro, R., de-los-Santos-Álvarez, N. et al. (2016). Comparison of isothermal helicase-dependent amplification and PCR for the detection of Mycobacterium tuberculosis by an electrochemical genomagnetic assay. *Analytical and Bioanalytical Chemistry* 408 (30): 8603–8610.

109 Gu, J., Mathai, A., Nurmi, C. et al. (2023). Detection of large genomic RNA via DNAzyme-mediated RNA cleavage and rolling circle amplification: SARS-CoV-2 as a model. Chemistry – A. *European Journal* 29 (27): e202300075.

110 Lee, S.H., Park, S.-m., Kim, B.N. et al. (2019). Emerging ultrafast nucleic acid amplification technologies for next-generation molecular diagnostics. *Biosensors & Bioelectronics* 15 (141): 111448.

111 Srivastava, P. and Prasad, D. (2023). Isothermal nucleic acid amplification and its uses in modern diagnostic technologies. *3 Biotech* 13 (6): 200.

112 Khan, A.I., Khan, M., and Khan, R. (2023). Artificial intelligence in point-of-care testing. *Annals of Laboratory Medicine* 43 (5): 401–407.

113 Wang, P. and Kricka, L.J. (2018). Current and emerging trends in point-of-care technology and strategies for clinical validation and implementation. *Clinical Chemistry* 64 (10): 1439–1452.

114 Esteva, A., Kuprel, B., Novoa, R.A. et al. (2017). Dermatologist-level classification of skin cancer with deep neural networks. *Nature* 542 (7639): 115–118.

115 Laboratory Diagnosis of Malaria: Conventional and Rapid Diagnostic Methods. (2024). *Archives of Pathology & Laboratory Medicine [Internet].* https://meridian.allenpress.com/aplm/article/137/6/805/132487/Laboratory-Diagnosis-of-Malaria-Conventional-and (accessed 17 April 2024).

116 Das, D., Vongpromek, R., Assawariyathipat, T. et al. (2022). Field evaluation of the diagnostic performance of EasyScan GO: a digital malaria microscopy device based on machine-learning. *Malaria Journal* 21 (1): 122.

117 Puhach, O., Meyer, B., and Eckerle, I. (2023). SARS-CoV-2 viral load and shedding kinetics. *Nature Reviews Microbiology [Internet]* 21 (3): 147–161. https://www.nature.com/articles/s41579-022-00822-w (accessed 27 June 2024).

118 Rolland, M., Tovanabutra, S., Dearlove, B. et al. (2020). Molecular dating and viral load growth rates suggested that the eclipse phase lasted about a week in HIV-1 infected adults in East Africa and Thailand. *PLoS Pathog [Internet]* 16 (2): e1008179. https://www.ncbi.nlm.nih.gov/pmc/articles/PMC7004303 (accessed 27 June 2024).

119 Al-Soud, W.A. and Rådström, P. (2001). Purification and characterization of PCR-inhibitory components in blood cells. *Journal of Clinical Microbiology [Internet]* 39 (2): 485–493. https://www.ncbi.nlm.nih.gov/pmc/articles/PMC87763 (accessed 27 June 2024).

120 Ko, H.Y., Li, Y.T., Li, Y.Y. et al. (2023). Optimization and improvement of qPCR detection sensitivity of SARS-CoV-2 in Saliva. *Microbiology Spectrum [Internet]* 11 (3): e04640–e04622. https://journals.asm.org/doi/10.1128/spectrum.04640-22 (accessed 27 June 2024).

121 HIV and AIDS [Internet]. (2024). https://www.who.int/news-room/fact-sheets/detail/hiv-aids (accessed 29 April 2024).

122 Deeks, S.G., Overbaugh, J., Phillips, A., and Buchbinder, S. (2015). HIV infection. *Nature Reviews Disease Primers [Internet]* 1 (1): 1–22. https://www.nature.com/articles/nrdp201535 (accessed 27 June 2024).

123 Bekker, L.G., Beyrer, C., Mgodi, N. et al. (2023). HIV infection. *Nature Reviews Disease Primers [Internet]* 9 (1): 1–21. https://www.nature.com/articles/s41572-023-00452-3 (accessed 1 May 2024).

124 Nyamweya, S., Hegedus, A., Jaye, A. et al. (2013). Comparing HIV-1 and HIV-2 infection: Lessons for viral immunopathogenesis. *Reviews in Medical Virology* 23 (4): 221–240.

125 Menéndez-Arias, L. and Delgado, R. (2022). Update and latest advances in antiretroviral therapy. *Trends in Pharmacological Sciences [Internet]* 43 (1): 16–29. https://www.sciencedirect.com/science/article/pii/S0165614721001930 (accessed 27 June 2024).

126 Girardi, E.M., CAP, S., Monforte, A., and d'Arminio, M. (2007). Late diagnosis of HIV infection: epidemiological features, consequences and strategies to encourage earlier testing. *JAIDS Journal of Acquired Immune Deficiency Syndromes [Internet]* 46: S3. https://journals.lww.com/jaids/fulltext/2007/09011/Late_Diagnosis_of_HIV_Infection__Epidemiological.2.aspx?casa_token=uHkPz5zG5tIAAAAA:_BWwQEr_Q_RPQHurn7d1QCRv3jLFNrLXMKV8mgRP

RFX4BZBbzXXEO3GMTkm8z8Ua3I1KVJcn6AHK0fkiuUIZdu&casa_token=H-PsvzqJ7KwAAAAA:n4kdRDoHdl0q4B6A9dmnyb8jeKmTjc80gjs3Z1qQikAnW3z0Z35giFoMfrpvXYvtj5_rM_QOW2fUX00q18lNLQbB (accessed 28 June 2024).

127 Minichiello, A., Swab, M., Chongo, M. et al. (2017). HIV point-of-care testing in Canadian settings: a scoping review. *Front Public Health [Internet]* 5. https://www.frontiersin.org/journals/public-health/articles/10.3389/fpubh.2017.00076/full (accessed 29 June 2024).

128 Larson, B.A., Schnippel, K., Ndibongo, B. et al. (2012). Rapid Point-of-Care CD4 Testing at Mobile HIV Testing Sites to Increase Linkage to Care: An Evaluation of a Pilot Program in South Africa. *JAIDS Journal of Acquired Immune Deficiency Syndromes [Internet]* 61 (2): e13. https://journals.lww.com/jaids/fulltext/2012/10010/Rapid_Point_of_Care_CD4_Testing_at_Mobile_HIV.21.aspx?casa_token=LlmWXJb_uKgAAAAA:TyP7oCaeM31oGz4rMrOVC8LfKPn2tUzfQGaDea1a9_vfMBEC4Vst_IhyPcvNNbaEjWYPzToKYae3eQ5iLN5oECI-&casa_token=M4B2DyojjWYAAAAA:jcGLyxkv8h2XSsyYxtJ9FA_3_YcIzLRIJS8jdPZBA6R8aF-m2cZnC_zCGjaiCnkttDXAtbuAiD_Z1_MpQGqf-PzU (acessed 29 June 2024].

129 FDA Approved HIV Tests (2024). HIV Testing | HIV/AIDS | CDC [Internet]. https://www.cdc.gov/hiv/partners/testing/laboratorytests.html (accessed 29 June 2024).

130 CATIE - Canada's source for HIV and hepatitis C information [Internet]. (2023). HIV testing technologies. https://www.catie.ca/hiv-testing-technologies (accessed 29 June 2024).

131 Delaney, K.P., Branson, B.M., Uniyal, A. et al. (2006). Performance of an oral fluid rapid HIV-1/2 test: experience from four CDC studies. *AIDS [Internet]* 20 (12): 1655. https://journals.lww.com/aidsonline/abstract/2006/08010/performance_of_an_oral_fluid_rapid_hiv_1_2_test_.11.aspx (accessed 29 June 2024).

132 Adams, S., Luo, W., Wesolowski, L. et al. (2017). Performance evaluation of the point-of-care INSTI™ HIV-1/2 antibody test in early and established HIV infections. *Journal of Clinical Virology [Internet]* 91: 90–94. https://www.sciencedirect.com/science/article/pii/S1386653217300896 (accessed 29 June 2024).

133 Masciotra, S., Price, K.A., Sprinkle, P. et al. (2015). Performance evaluation of the CHEMBIO DPP® (dual path platform) HIV-1/2 assay in early and established infections. *Journal of Clinical Virology [Internet]* 70: 97–100. https://www.sciencedirect.com/science/article/pii/S1386653215002954 (accessed 29 June 2024).

134 Delaney, K.P., Branson, B.M., Uniyal, A. et al. (2011). Evaluation of the performance characteristics of 6 rapid HIV antibody tests. *Clinical Infectious Diseases [Internet]* 52 (2): 257–263. https://doi.org/10.1093/cid/ciq068 (accessed 29 June 2024).

135 Rossetti, R., Smith, T., Luo, W., and Masciotra, S. (2020). Performance evaluation of the MedMira reveal G4 LAB S/P and POC HIV antibody rapid

screening tests using plasma and whole blood specimens. *Journal of Clinical Virology [Internet]* 127: 104344. https://www.sciencedirect.com/science/article/pii/S138665322030086X (accessed 29 June 2024).

136 Kalou, M.B., Watson, A., Clay, S. et al. (2016). Laboratory evaluation of the Chembio Dual Path Platform HIV-Syphilis Assay: original research. *African Journal of Laboratory Medicine [Internet]* 5 (1): 1–6. https://journals.co.za/doi/abs/10.4102/ajlm.v5i1.433 (accessed 29 June 2024).

137 Cohen, M.S., Gay, C.L., Busch, M.P., and Hecht, F.M. (2010). The detection of acute HIV infection. *The Journal of Infectious Diseases [Internet]* 202 (Supplement_2): S270–S277. https://doi.org/10.1086/655651.

138 Anderson, A.M., Tyor, W.R., Mulligan, M.J. et al. (2018). Measurement of human immunodeficiency virus p24 antigen in human cerebrospinal fluid with digital enzyme-linked immunosorbent assay and association with decreased neuropsychological performance. *Clinical Infectious Disease [Internet]* 67 (1): 137–140. https://www.ncbi.nlm.nih.gov/pmc/articles/PMC6005050/ (accessed 29 June 2024).

139 Nishanian, P., Huskins, K.R., Stehn, S. et al. (1990). A simple method for improved assay demonstrates that HIV p24 antigen is present as immune complexes in most Sera from HIV-infected individuals. *The Journal of Infectious Diseases [Internet]* 162 (1): 21–28. https://doi.org/10.1093/infdis/162.1.21 (accessed 29 June 2024).

140 Faraoni, S., Rocchetti, A., Gotta, F. et al. (2013). Evaluation of a rapid antigen and antibody combination test in acute HIV infection. *Journal of Clinical Virology [Internet]* 57 (1): 84–87. https://www.sciencedirect.com/science/article/pii/S1386653213000140 (accessed 30 June 2024).

141 Lewis, J.M., Macpherson, P., Adams, E.R. et al. (2015). Field accuracy of fourth-generation rapid diagnostic tests for acute HIV-1: a systematic review. *AIDS [Internet].* 29(18):2465: https://journals.lww.com/aidsonline/FullText/2015/11280/Field_accuracy_of_fourth_generation_rapid.10.aspx (accessed 30 June 2024).

142 Rosenberg, N.E., Kamanga, G., Phiri, S. et al. (2012). Detection of acute HIV infection: a field evaluation of the Determine® HIV-1/2 Ag/Ab combo test. *The Journal of Infectious Diseases [Internet]* 205 (4): 528–534. https://doi.org/10.1093/infdis/jir789.

143 Wawer, M.J., Gray, R.H., Sewankambo, N.K. et al. (2005). Rates of HIV-1 Transmission per Coital Act, by Stage of HIV-1 Infection, in Rakai, Uganda. *The Journal of Infectious Diseases [Internet].* 191 (9): 1403–1409. https://doi.org/10.1086/429411.

144 Pilcher, C.D., Tien, H.C., Eron, J.J. et al. Brief but efficient: acute HIV infection and the sexual transmission of HIV. *The Journal of Infectious Diseases [Internet]* 189 (10): 1785–1792. https://doi.org/10.1086/386333.

145 Calmy, A., Ford, N., Hirschel, B. et al. (2007). HIV viral load monitoring in resource-limited regions: optional or necessary? *Clinical Infectious Diseases [Internet]* 44 (1): 128–134. https://doi.org/10.1086/510073.

146 Hoffman, J., van Griensven, J., Colebunders, R., and McKellar, M. (2010). Role of the Cd4 Count in HIV Management. *HIV Therapy [Internet].* 4 (1): 27–39. https://doi.org/10.2217/hiv.09.58.

147 Steindl, F., Armbruster, C., Pierer, K. et al. (1998). A simple and robust method for the complete dissociation of HIV-1 p24 and other antigens from immune complexes in serum and plasma samples. *Journal of Immunological Methods* 217 (1–2): 143–151.

148 Parpia, Z.A., Elghanian, R., Nabatiyan, A. et al. (2010). p24 antigen rapid test for diagnosis of acute pediatric HIV infection. *JAIDS Journal of Acquired Immune Deficiency Syndromes [Internet]* 55 (4): 413–419. https://journals.lww.com/00126334-201012010-00001 (accessed 14 June 2024).

149 Yin, K., Pandian, V., Kadimisetty, K. et al. (2020). Real-time colorimetric quantitative molecular detection of infectious diseases on smartphone-based diagnostic platform. *Scientific Reports [Internet].* 10 (1): 9009. https://www.nature.com/articles/s41598-020-65899-w (accessed 27 June 2024).

150 Liu, C., Geva, E., Mauk, M. et al. (2011). An isothermal amplification reactor with an integrated isolation membrane for point-of-care detection of infectious diseases. *Analyst [Internet]* 136 (10): 2069–2076. https://pubs.rsc.org/en/content/articlelanding/2011/an/c1an00007a (accessed 30 June 2024).

151 Kong, H., Zhang, W., Yao, J. et al. (2021). A RT-LAMP based hydrogen ion selective electrode sensing for effective detection HIV-1 RNA with high-sensitivity. *Sensors and Actuators B: Chemical* 329: 129118.

152 Kong, M., Li, Z., Wu, J. et al. (2019). A wearable microfluidic device for rapid detection of HIV-1 DNA using recombinase polymerase amplification. *Talanta [Internet]* 205: 120155. https://www.sciencedirect.com/science/article/pii/S0039914019307817 (accessed 27 June 2024).

153 Wade, D., Diaw, P.A., Daneau, G. et al. (2015). Laboratory and field evaluation of the Partec CyFlow MiniPOC for absolute and relative CD4 T-cell enumeration. *PLoS One [Internet]* 10 (2): e0116663. https://www.ncbi.nlm.nih.gov/pmc/articles/PMC4331543/ (accessed 27 June 2024).

154 Nasi, M., De Biasi, S., Bianchini, E. et al. (2015). Reliable and accurate CD4+ T cell count and percent by the Portable Flow Cytometer CyFlow MiniPOC and "CD4 Easy Count Kit-Dry", as revealed by the comparison with the Gold Standard Dual Platform Technology. *PLoS One [Internet]* 10 (1): e0116848. https://www.ncbi.nlm.nih.gov/pmc/articles/PMC4306486/ (accessed 30 June 2024).

155 Hughes, M.D., Stein, D.S., Gundacker, H.M. et al. (1994). Within-subject variation in CD4 lymphocyte count in asymptomatic human immunodeficiency virus infection: implications for patient monitoring. *The Journal of Infectious Diseases [Internet]* 169 (1): 28–36. https://doi.org/10.1093/infdis/169.1.28.

156 Pirzada, Y., Khuder, S., and Donabedian, H. (2006). Predicting AIDS-related events using CD4 percentage or CD4 absolute counts. *AIDS Research and Therapy [Internet].* 3 (1): 20. https://doi.org/10.1186/1742-6405-3-20.

157 Shearer, W.T., Rosenblatt, H.M., Gelman, R.S. et al. (2003). Lymphocyte subsets in healthy children from birth through 18 years of age: the Pediatric AIDS Clinical Trials Group P1009 study. *Journal of Allergy and Clinical Immunology* 112 (5): 973–980.

158 Yin, D.E., Warshaw, M.G., Miller, W.C. et al. (2014). Using CD4 percentage and age to optimize pediatric antiretroviral therapy initiation. *Pediatrics [Internet]* 134 (4): e1104–e1116. https://doi.org/10.1542/peds.2014-0527 (accessed 30 June 2024).

159 Wang, S., Tasoglu, S., Chen, P.Z. et al. (2014). Micro-a-fluidics ELISA for rapid CD4 cell count at the point-of-care. *Scientific Reports [Internet]* 4 (1): 3796. https://www.nature.com/articles/srep03796 (accessed 30 June 2024).

160 Çalıca Utku, A., Budak, G., Karabay, O. et al. Main symptoms in patients presenting in the COVID-19 period. *Scottish Medical Journal [Internet]* 65 (4): 127–132. https://doi.org/10.1177/0036933020949253.

161 Raveendran, A.V., Jayadevan, R., and Sashidharan, S. (2021). Long COVID: an overview. *Diabetes & Metabolic Syndrome: Clinical Research & Reviews [Internet]* 15 (3): 869–875. https://www.sciencedirect.com/science/article/pii/S1871402121001193 (accessed 30 June 2024).

162 COVID – Coronavirus Statistics – Worldometer [Internet]. https://www.worldometers.info/coronavirus/ (accessed 30 June 2024).

163 Ayouni, I., Maatoug, J., Dhouib, W. et al. (2021). Effective public health measures to mitigate the spread of COVID-19: a systematic review. *BMC Public Health [Internet]* 21 (1): 1015. https://doi.org/10.1186/s12889-021-11111-1 (accessed 30 June 2024).

164 Yamayoshi, S., Sakai-Tagawa, Y., Koga, M. et al. (2020). Comparison of rapid antigen tests for COVID-19. *Viruses [Internet]* 12 (12): 1420. https://www.mdpi.com/1999-4915/12/12/1420 (accessed 30 June 2024).

165 Barlev-Gross, M., Weiss, S., Ben-Shmuel, A. et al. (2021). Spike vs nucleocapsid SARS-CoV-2 antigen detection: application in nasopharyngeal swab specimens. *Analytical and Bioanalytical Chemistry [Internet].* 413 (13): 3501–3510. https://doi.org/10.1007/s00216-021-03298-4.

166 Wang, M.Y., Zhao, R., Gao, L.J. et al. (2020). SARS-CoV-2: structure, biology, and structure-based therapeutics development. *Frontiers in Cellular and Infection Microbiology [Internet]* 10. https://www.frontiersin.org/journals/cellular-and-infection-microbiology/articles/10.3389/fcimb.2020.587269/full (accessed 30 June 2024).

167 Wagenhäuser, I., Knies, K., Hofmann, D. et al. (2023). Virus variant–specific clinical performance of SARS coronavirus two rapid antigen tests in point-of-care use, from November 2020 to January 2022. *Clinical Microbiology and Infection [Internet]* 29 (2): 225–232. https://www.sciencedirect.com/science/article/pii/S1198743X22004220 (accessed 30 June 2024).

168 Kopel, J., Goyal, H., and Perisetti, A. (2021). Antibody tests for COVID-19. *Baylor University Medical Center Proceedings [Internet]* 34 (1): 63–72. https://doi.org/10.1080/08998280.2020.1829261 (accessed 30 June 2024).

169 Jacofsky, D., Jacofsky, E.M., and Jacofsky, M. (2020). Understanding antibody testing for COVID-19. *The Journal of Arthroplasty [Internet]* 35 (7, Supplement): S74–S81. https://www.sciencedirect.com/science/article/pii/S0883540320304423 (accessed 30 June 2024).

170 Ellington, A.D. and Szostak, J.W. (1990). In vitro selection of RNA molecules that bind specific ligands. *Nature [Internet]* 346 (6287): 818–822. https://www.nature.com/articles/346818a0 (accessed 30 June 2024).

171 Gelinas, A.D., Davies, D.R., and Janjic, N. (2016). Embracing proteins: structural themes in aptamer–protein complexes. *Current Opinion in Structural Biology [Internet]* 36 (Complete): 122–132. https://journals.scholarsportal.info/details/0959440x/v36icomplete/122_epstiac.xml (accessed 2 April 2024).

172 Tuerk, C. and Gold, L. (1990). Systematic Evolution of Ligands by Exponential Enrichment: RNA Ligands to Bacteriophage T4 DNA Polymerase. *Science [Internet]* 249 (4968): 505–510. https://www.jstor.org/stable/2874490 (accessed 31 March 2024).

173 Wilson, D.S. and Szostak, J.W. (1999). In vitro selection of functional nucleic acids. *Annual Review of Biochemistry* 68: 611–647.

174 Zhang, Z., Li, J., Gu, J. et al. (2022). A universal DNA Aptamer that recognizes spike proteins of diverse SARS-CoV-2 variants of concern. *Chemistry – A European Journal [Internet]* 28 (15): e202200078. https://onlinelibrary.wiley.com/doi/abs/10.1002/chem.202200078 (accessed 15 September 2023).

175 Sun, H., Zhu, X., Lu, P.Y. et al. (2014). Oligonucleotide Aptamers: New Tools for Targeted Cancer Therapy. *Molecular Therapy Nucleic Acids [Internet]* 3 (8): e182. https://www.ncbi.nlm.nih.gov/pmc/articles/PMC4221593/ (accessed 29 March 2024).

176 Song, K.M., Lee, S., and Ban, C. (2012). Aptamers and their biological applications. *Sensors (Basel) [Internet]* 12 (1): 612–631.

177 Ito, T., Ueno, Y., Komatsu, Y., and Matsuda, A. (2003). Synthesis, thermal stability and resistance to enzymatic hydrolysis of the oligonucleotides containing 5-(N-aminohexyl)carbamoyl-2′-O-methyluridines. *Nucleic Acids Research [Internet]* 31 (10): 2514. https://www.ncbi.nlm.nih.gov/pmc/articles/PMC156055/ (accessed 29 March 2024).

178 Thiviyanathan, V. and Gorenstein, D.G. (2012). Aptamers and the next generation of diagnostic reagents. *Proteomics Clinical Applications [Internet]* 6 (11–12): 563–573. https://www.ncbi.nlm.nih.gov/pmc/articles/PMC4021847/ (accessed 29 March 2024).

179 Zhang, Z., Pandey, R., Li, J. et al. (2021). High-affinity dimeric aptamers enable the rapid electrochemical detection of wild-type and B.1.1.7 SARS-CoV-2 in unprocessed saliva. *Angewandte Chemie International Edition [Internet]* 60 (45): 24266–24274. https://onlinelibrary.wiley.com/doi/abs/10.1002/anie.202110819 (accessed 1 April 2024).

180 Chauhan, N., Xiong, Y., Ren, S. et al. (2023). Net-shaped DNA nanostructures designed for rapid/sensitive detection and potential inhibition of the SARS-CoV-2 virus. *Journal of the American Chemical Society [Internet]* 145 (37): 20214–20228. https://doi.org/10.1021/jacs.2c04835 (accessed 21 June).

181 Damhorst, G.L., Duarte-Guevara, C., Chen, W. et al. (2015). Smartphone-imaged HIV-1 Reverse-Transcription Loop-Mediated Isothermal Amplification (RT-LAMP) on a chip from whole blood. *Engineering (Beijing).* 1 (3): 324–335.

182 Cennamo, N., Pasquardini, L., Arcadio, F. et al. (2021). SARS-CoV-2 spike protein detection through a plasmonic D-shaped plastic optical fiber aptasensor. *Talanta [Internet]* 233: 122532. https://www.sciencedirect.com/science/article/pii/S0039914021004537 (accessed 1 July 2024).

183 Guan, P.C., Zhang, H., Li, Z.Y. et al. (2022). Rapid Point-of-Care Assay by SERS detection of SARS-CoV-2 virus and its variants. *Analytical Chemistry* 94 (51): 17795–17802.

184 Broughton, J.P., Deng, X., Yu, G. et al. (2020). CRISPR-Cas12–based detection of SARS-CoV-2. *Nature Biotechnology [Internet]* 38 (7): 870–874. https://www.ncbi.nlm.nih.gov/pmc/articles/PMC9107629/ (accessed 19 June 2024).

185 Wang, M., Chen, D., Wu, W. et al. (2021). Analytical performance evaluation of five RT-PCR kits for severe acute respiratory syndrome coronavirus 2. *Journal of Clinical Laboratory Analysis [Internet].* 35 (1): e23643. https://onlinelibrary.wiley.com/doi/abs/10.1002/jcla.23643 (accessed 1 July 2024).

186 Chen, F.E., Lee, P.W., Trick, A.Y. et al. (2021). Point-of-care CRISPR-Cas-assisted SARS-CoV-2 detection in an automated and portable droplet magnetofluidic device. *Biosensors and Bioelectronics [Internet]* 190: 113390. https://www.ncbi.nlm.nih.gov/pmc/articles/PMC8170879/ (accessed 18 June 2024).

187 Fozouni, P., Son, S., de Derby, L, M.D. et al. (2021). Amplification-free detection of SARS-CoV-2 with CRISPR-Cas13a and mobile phone microscopy. *Cell [Internet]* 184 (2): 323–333.e9. https://www.cell.com/cell/abstract/S0092-8674(20)31623-8 (accessed 19 June 2024).

188 He, X., Liu, Z., Yang, Y. et al. (2019). Sensitivity enhancement of nucleic acid lateral flow assays through a physical-chemical coupling method: dissoluble saline barriers. *ACS Sensors* 4 (6): 1691–1700.

189 Rivas, L., Medina-Sánchez, M., de la Escosura-Muñiz, A., and Merkoçi, A. (2014). Improving sensitivity of gold nanoparticle-based lateral flow assays by using wax-printed pillars as delay barriers of microfluidics. *Lab on a Chip* 14 (22): 4406–4414.

190 Roy, L., Buragohain, P., and Borse, V. (2022). Strategies for sensitivity enhancement of point-of-care devices. *Biosensors and Bioelectronics: X* 10: 100098.

191 Park, S.B. and Shin, J.H. (2022). Pressed lateral flow assay strips for flow delay-induced signal enhancement in lateral flow assay strips. *BioChip Journal* 16 (4): 480–489.

192 Amini, R., Zhang, Z., Li, J. et al. (2022 Sep). Aptamers for SARS-CoV-2: isolation, characterization, and diagnostic and therapeutic developments. *Analysis & Sensing* 2 (5): e202200012.

193 Omidfar, K., Riahi, F., and Kashanian, S. (2023). Lateral flow assay: a summary of recent progress for improving assay performance. *Biosensors (Basel).* 13 (9): 837.

194 Han, H., Wang, C., Yang, X. et al. (2022). Rapid field determination of SARS-CoV-2 by a colorimetric and fluorescent dual-functional lateral flow immunoassay biosensor. *Sensors and Actuators B Chemical* 351: 130897.

195 Liu, D., Ju, C., Han, C. et al. (2021). Nanozyme chemiluminescence paper test for rapid and sensitive detection of SARS-CoV-2 antigen. *Biosensors and Bioelectronics* 173: 112817.

196 Xu, S., Zhang, G., Fang, B. et al. (2019). Lateral flow immunoassay based on polydopamine-coated gold nanoparticles for the sensitive detection of zearalenone in Maize. *ACS Applied Materials & Interfaces* 11 (34): 31283–31290.

197 Sharma, A., Tok, A.I.Y., Lee, C. et al. (2019). Magnetic field assisted preconcentration of biomolecules for lateral flow assaying. *Sensors and Actuators B: Chemical* 285: 431–437.

198 Anfossi, L., Di Nardo, F., Cavalera, S. et al. (2018). Multiplex lateral flow immunoassay: an overview of strategies towards high-throughput point-of-need testing. *Biosensors-Basel* 9 (1): 2.

199 Lee, J.H., Seo, H.S., Kwon, J.H. et al. (2015). Multiplex diagnosis of viral infectious diseases (AIDS, hepatitis C, and hepatitis A) based on point of care lateral flow assay using engineered proteinticles. *Biosensors and Bioelectronics* 69: 213–225.

200 Le, T.T., Chang, P., Benton, D.J. et al. (2017). Dual recognition element lateral flow assay toward multiplex strain specific influenza virus detection. *Analytical Chemistry.* 89 (12): 6781–6786.

201 Saxena, S.K. (2016). *Proof and Concepts in Rapid Diagnostic Tests and Technologies; BoD – Books on Demand*, 174.

202 IPSOS. FUTURE OF POINT OF CARE & RAPID TESTING [Internet]. (2022). U.S, EU, China. https://www.ipsos.com/sites/default/files/ct/publication/documents/2022-03/Ipsos_Future%20of%20POC%20and%20Rapid%20Testing_March%202022.pdf (accessed 28 June 2024).

203 Rasti, R., Nanjebe, D., Karlström, J. et al. (2017). Health care workers' perceptions of point-of-care testing in a low-income country—a qualitative study in Southwestern Uganda. *PLoS One.* 12 (7): e0182005.

204 Mohammed, M.I., Haswell, S., and Gibson, I. (2015). Lab-on-a-chip or chip-in-a-lab: challenges of commercialization lost in translation. *Procedia Technology.* 20: 54–59.

205 Berthier, E., Young, E.W.K., and Beebe, D. (2012). Engineers are from PDMS-land, biologists are from polystyrenia. *Lab on a Chip* 12 (7): 1224–1237.

206 Cennamo, N., Massarotti, D., Conte, L., and Zeni, L. (2011). Low cost sensors based on SPR in a plastic optical fiber for biosensor implementation. *Sensors (Basel) [Internet]* 11 (12): 11752–11760. https://www.ncbi.nlm.nih.gov/pmc/articles/PMC3252008/ (accessed 1 July 2024).

11

Looking Forward

Perspectives on the Future of Diagnostic Techniques

Yan Du[1,2] *and Jinghong Li*[3,4,5,6]

[1] *Chinese Academy of Sciences, Changchun Institute of Applied Chemistry, State Key Laboratory of Electroanalytical Chemistry, Changchun 130000, China*
[2] *University of Science and Technology of China, School of Applied Chemistry and Engineering, Hefei 230026, China*
[3] *Beijing Life Science Academy, Beijing 102209, China*
[4] *Tsinghua University, Key Laboratory of Bioorganic Phosphorus Chemistry & Chemical Biology, Department of Chemistry, Beijing 100084, China*
[5] *New Cornerstone Science Laboratory, Shenzhen 518054, China*
[6] *University of Science and Technology of China, Hefei National Laboratory of Physical Science at Microscale, Center for BioAnalytical Chemistry, Hefei 230026, China*

CHAPTER MENU

The convergence of molecular diagnostics, microfluidics, plasmonic biosensors, and genomic sequencing is paving the way for faster, more precise, and more accessible diagnostic solutions. Looking forward, three overarching trends will define the future of diagnostics:

(I) Decentralized, point-of-care testing driven by portable and AI-enhanced technologies.
(II) Genomic and multi-omic data integration, enabling precision diagnostics and predictive modeling.
(III) Global standardization of diagnostic methodologies to enhance disease surveillance and outbreak preparedness.

At the core of these advancements lies chemometric and analytical methodologies, which are indispensable for enhancing diagnostic accuracy, reducing false positives/negatives, and improving biological data interpretability. The ability to leverage large-scale diagnostic datasets, extract patterns through machine learning,

Diagnostic Techniques in Public Health: Analytical and Chemometric Approaches in Disease Detection and Management, First Edition. Edited by Jinghong Li and Yan Du.

and refine decision-making in real time will be pivotal in shaping the next phase of diagnostic innovation. Moreover, as digital health ecosystems expand, integrating diagnostic data with electronic health records (EHRs) and mobile health applications will facilitate personalized, predictive, and preventive healthcare strategies.

The following sections examine key technological breakthroughs that will shape the future of diagnostic techniques, including molecular diagnostics, biosensing technologies, and AI-driven predictive analytics. We also address the challenges and ethical considerations in deploying these innovations on a global scale, emphasizing the need for equitable access, regulatory oversight, and responsible data governance. Ultimately, this chapter outlines a roadmap for next-generation diagnostics, prioritizing precision, accessibility, and global health security.

11.1 Introduction: The Evolution of Diagnostic Techniques

11.1.1 The Demand for Precision and Accessibility in Diagnostics

The global demand for rapid, accurate, and accessible diagnostic technologies has increased due to emerging infectious diseases, antimicrobial resistance, and the expansion of decentralized healthcare systems. The COVID-19 pandemic highlighted the urgent need for efficient diagnostic tools, where delays in pathogen detection directly impacted containment efforts and healthcare responses (Chapter 1, No. 7 Ref) [1].

In parallel with the growing medical demand during the COVID-19, the diagnostic industry has experienced substantial market expansion. The global in vitro diagnostics (IVD) market was valued at approximately USD 100 billion in 2022 and is projected to surpass USD 120 billion by 2027, driven by technological advancements in molecular diagnostics, point-of-care (POC) testing, and AI-assisted analysis. The increasing prevalence of chronic diseases, aging populations, and the push for personalized medicine has further fueled this market growth. Companies such as Roche Diagnostics, Abbott, and Thermo Fisher Scientific have heavily invested in automated and AI-driven diagnostic solutions, recognizing the economic potential of precision diagnostics in clinical and public health settings.

While polymerase chain reaction (PCR)-based molecular diagnostics have revolutionized disease detection, their dependency on sophisticated laboratory infrastructure, trained personnel, and extended turnaround times limits their deployment in decentralized settings (Chapter 5, No. 1 Ref) [2]. Consequently, chemometric approaches, which integrate multivariate statistical analysis, signal processing, and machine learning, have emerged as a critical enabler of real-time, high-accuracy diagnostics (Chapter 5, No. 5 Ref) [3]. In nucleic acid-based assays, chemometric techniques enhance signal-to-noise ratios, optimize reaction conditions, and minimize variability, making molecular diagnostics more adaptable for POC applications (Chapter 5, No. 6 Ref) [4].

The commercial viability of next-generation diagnostics is also influenced by healthcare policies, reimbursement frameworks, and the cost-effectiveness of these novel technologies. The shift toward decentralized testing, particularly in resource-limited settings, has driven significant investments in portable biosensing devices and AI-powered diagnostic platforms. The expanding integration of diagnostics with telemedicine and EHRs further reinforces their economic relevance. It is estimated that AI-driven diagnostic solutions will contribute more than USD 20 billion in annual market growth by 2030, particularly in predictive analytics and automated disease detection.

Furthermore, biosensing technologies coupled with chemometric modeling facilitate automated disease surveillance, remote diagnostics, and noninvasive biomarker analysis, significantly improving their clinical and public health applications. The continuous refinement of chemometric-assisted diagnostic models is expected to drive the evolution of cost-effective, portable, and AI-driven diagnostic platforms, ensuring widespread accessibility and clinical reliability. As technological advancements continue to align with market demands, the diagnostic sector is poised to experience unprecedented economic expansion, shaping the future of precision medicine and population health management.

11.1.2 Transforming Public Health Surveillance and Clinical Diagnostics

Recent advancements in molecular biology, microfluidics, and AI have significantly transformed public health surveillance and clinical diagnostics. Traditional diagnostic techniques—including enzyme-linked immunosorbent assays (ELISA) (Chapter 3, No. 6 Ref) [5], immunofluorescence assays, and microbial culture-based methods—are increasingly being replaced or augmented by high-throughput sequencing, digital PCR, and electrochemical biosensing (Chapter 5, No. 3 Ref) [6].

Despite increasing adoption of high-throughput sequencing and molecular diagnostics, ELISA remains indispensable in public health due to its simplicity, affordability, and adaptability (Chapter 3, No. 6 Ref) [5]. Although conventional ELISA faces limitations in sensitivity, throughput, and multiplexing, recent innovations—such as nanobody-, aptamer- (ELASA), and MIP-based assays—have enhanced selectivity and robustness (Chapter 3, No. 8, 10, 12 Ref) [7–9]. Advances in catalytic labels, including nanozymes, and improved enzyme-antibody conjugation (e.g., avidin-biotin systems) further amplify signal transduction (Chapter 3, No. 30, 32 Ref) [10, 11]. Integration with microfluidic and biosensor technologies has led to paper-based, electrochemical, and automated ELISA formats, enabling low-cost, real-time, POC diagnostics (Chapter 3, No. 71, 73, 79 Ref) [12–14]. These developments position ELISA as a vital complement to nucleic acid-based diagnostics, particularly where decentralized, rapid testing is required but sequencing technologies may be impractical.

One of the most revolutionary developments in public health diagnostics is the integration of metagenomic sequencing into pathogen surveillance. The use of metagenomic next-generation sequencing (mNGS) has enabled high-resolution, unbiased pathogen identification, thereby improving early outbreak detection, enhancing antimicrobial resistance monitoring, and strengthening epidemiological surveillance (Chapter 4, No. 2 Ref) [15]. Genomic surveillance played an essential role in variant tracking, vaccine adaptation strategies, and containment measures during the COVID-19 pandemic (Chapter 4, No. 13 Ref) [16].

Recent advances in point-of-care testing (POCT) instrumentation, particularly in LOC and LOAD platforms, have significantly improved diagnostic scalability by miniaturizing laboratory-based processes into compact and portable formats (Chapter 9, No. 51 Ref) [17]. These microfluidic-based systems integrate sample preparation, amplification, and detection into a single platform, reducing reliance on external equipment and trained personnel (Chapter 9, No. 52 Ref) [18]. Additionally, the use of nanomaterial-based biosensors in electrochemical and colorimetric assays has enhanced signal sensitivity, making POCT devices more reliable even in resource-limited settings (Chapter 9, No. 58 Ref) [19].

Beyond infectious disease surveillance, wearable biosensing and bioelectronic innovations have enabled continuous, non-invasive physiological monitoring. The development of flexible, skin-adhering sensors, microfluidic diagnostic platforms, and smartphone-integrated biomarker assays has improved early disease detection, personalized health monitoring, and real-time therapeutic interventions. However, as these technologies generate increasingly complex, multidimensional datasets, chemometric approaches play an indispensable role in ensuring data accuracy, standardization, and integration across different diagnostic modalities (Chapter 1, No. 7 Ref) [1].

11.1.3 The Role of Chemometric Approaches in Optimizing Diagnostic Accuracy and Automation

Despite rapid technological advancements, modern diagnostic tools still face challenges related to signal variability, cross-reactivity, and data reproducibility. Chemometric methodologies provide robust solutions by integrating machine learning, mathematical modeling, and statistical data fusion, thereby enhancing diagnostic accuracy, automation, and reliability (Chapter 5, No. 5 Ref) [3].

Key contributions of chemometrics in diagnostics:

(I) **Pattern Recognition and Machine Learning for Biomarker Identification**
By analyzing high-dimensional multi-omics datasets from genomics, proteomics, and metabolomics, chemometric models extract disease-specific molecular signatures to enhance specificity and reduce false positives (Chapter 4, No. 2 Ref) [15].

(II) **Noise Reduction and Signal Enhancement in Biosensing Technologies**
Real-world biosensors are often affected by environmental fluctuations, signal drift, and biological interference. Chemometric algorithms dynamically adjust sensor outputs, ensuring high-fidelity disease detection even in complex biological matrices (Chapter 8, No. 6 Ref) [20].

(III) **Data Fusion for Multimodal and Multianalyte Diagnostics**
Advanced diagnostic platforms increasingly integrate electrochemical, optical, and mechanical biosensors into a single system (Chapter 9, No. 4 Ref) [21].

Chemometric-driven data fusion techniques allow seamless integration of heterogeneous sensor data, ensuring cross-validated and multidimensional diagnostic outputs (Chapter 4, No. 2 Ref) [15].

(IV) **Predictive Modeling for Disease Risk Assessment and Early Intervention**
Beyond immediate diagnostics, chemometric-driven predictive models analyze longitudinal health data to forecast disease onset, progression, and response to treatment, thus enabling proactive, preventive medicine (Chapter 4, No. 13 Ref) [16].

As next-generation diagnostic platforms continue to evolve, the integration of chemometric modeling with AI-enhanced biosensing technologies is expected to accelerate the transition toward fully automated, real-time disease monitoring, thereby revolutionizing public health preparedness and clinical decision-making.

11.2 The Paradigm Shift in Molecular Diagnostics

11.2.1 Beyond PCR: The Rise of Isothermal and CRISPR-Based Methods

11.2.1.1 Limitations of Traditional PCR-Based Diagnostics

PCR has long been regarded as the gold standard in nucleic acid amplification due to its high sensitivity and specificity. However, its dependence on complex thermal cycling, specialized laboratory equipment, and trained personnel limits its application in POC and resource-limited settings (Chapter 5, No. 19 Ref) [22]. The COVID-19 pandemic further exposed the bottlenecks of PCR-based diagnostics, highlighting slow turnaround times and logistical challenges in decentralized testing (Chapter 2, No. 13 Ref) [23].

Beyond infrastructure constraints, PCR is prone to inhibition from biological samples, leading to false negatives in clinical diagnostics. Additionally, its reliance on predefined target sequences limits its applicability for detecting emerging pathogens or highly mutated viral strains. To overcome these challenges, researchers have increasingly turned to isothermal amplification techniques and CRISPR-based diagnostics, which offer faster, field-deployable alternatives.

11.2.1.2 Isothermal Amplification Methods: Overcoming PCR Limitations

Isothermal amplification techniques have gained traction due to their ability to amplify nucleic acids at constant temperatures, eliminating the need for complex thermal cycling. These methods offer faster reaction times, simplified workflows, and compatibility with portable diagnostic devices, making them particularly suited for on-site and low-resource testing environments (Chapter 5, No. 6 Ref) [4].

Key isothermal amplification methods include the following:

(I) Loop-mediated Isothermal Amplification (LAMP): Uses strand-displacing DNA polymerase and multiple primers to achieve rapid, high-sensitivity nucleic acid detection within 30 minutes (Chapter 5, No. 7 Ref) [24].

(II) Recombinase Polymerase Amplification (RPA): Facilitates ultra-fast, low-temperature amplification, enabling DNA/RNA detection within 10–20 minutes at 37–42 °C (Chapter 5, No. 7 Ref) [24].
(III) Recombinase-aided Amplification (RAA): Developed with independent intellectual property rights in China, RAA provides high specificity and sensitivity, requiring minimal instrumentation for field diagnostics (Chapter 2, No. 1 Ref) [25].

These technologies are increasingly integrated with biosensors and microfluidic platforms to enhance real-time, high-throughput molecular diagnostics, further bridging the gap between laboratory testing and POC applications (Chapter 5, No. 6 Ref) [4].

11.2.1.3 CRISPR-based Diagnostics: A Disruptive Technology for POCT

The emergence of CRISPR-based molecular diagnostics has revolutionized nucleic acid detection by leveraging the programmable nature of CRISPR/Cas systems for rapid, sequence-specific target recognition (Chapter 6, No. 17 Ref) [26]. Unlike PCR, CRISPR-based assays offer single-molecule sensitivity, eliminating the need for complex amplification steps while enabling visual or fluorescence-based readouts (Chapter 6, No. 3 Ref) [27].

Key CRISPR-based diagnostic approaches include the following:

(I) Cas12 and Cas13 Biosensing: Cas12 cleaves single-stranded DNA (ssDNA), while Cas13 targets RNA, enabling highly specific detection of viral and bacterial pathogens (Chapter 6, No. 3 Ref) [27].
(II) SHERLOCK (Specific High-sensitivity Enzymatic Reporter Unlocking): Developed for ultra-sensitive, portable nucleic acid detection, SHERLOCK has been widely applied in COVID-19 diagnostics (Chapter 6, No. 17 Ref) [26].
(III) DETECTR (DNA Endonuclease-Targeted CRISPR Trans Reporter): Uses Cas12 to directly detect viral DNA, offering rapid POC applications for infections such as influenza and HPV (Chapter 6, No. 3 Ref) [27].

By integrating CRISPR technology with microfluidic systems, lateral flow assays, and smartphone-based platforms, researchers are advancing toward real-time, at-home disease detection, paving the way for next-generation, user-friendly molecular diagnostics (Chapter 6, No. 17 Ref) [26].

11.2.2 Next-Generation Sequencing (NGS) and Genomic Surveillance

11.2.2.1 The Increasing Role of Metagenomic Sequencing in Pathogen Detection

Metagenomic next-generation sequencing (mNGS) has revolutionized pathogen detection and infectious disease surveillance by enabling unbiased, high-throughput nucleic acid sequencing from complex biological samples. Unlike PCR-based diagnostics, which require prespecified primers, mNGS provides a culture-independent, hypothesis-free approach for detecting emerging, unknown, and highly mutated pathogens, making it a critical tool in rapid outbreak response and epidemiological surveillance (Chapter 4, No. 10 Ref) [28].

Key advancements in mNGS-driven diagnostics include:

1) Ultra-comprehensive pathogen profiling, simultaneously detecting bacteria, viruses, fungi, and parasites in a single test (Chapter 4, No. 5 Ref) [29].
2) Antimicrobial resistance (AMR) surveillance, enabling the rapid identification of drug resistance genes to inform targeted antimicrobial therapy (Chapter 4, No. 16 Ref) [30].
3) Pathogen evolution tracking, allowing real-time monitoring of genetic mutations in circulating infectious agents to predict future outbreaks.

Beyond infectious disease detection, mNGS is being integrated with host transcriptomics to differentiate between viral, bacterial, and immune-mediated diseases, further expanding its utility in precision medicine. However, its high cost, computational intensity, and data complexity remain challenges, necessitating the development of more efficient bioinformatics pipelines and AI-driven analytics for real-time pathogen identification (Chapter 4, No. 12 Ref) [31].

11.2.2.2 Real-Time Sequencing and AI-Driven Genomic Analysis

The emergence of real-time sequencing technologies, such as nanopore sequencing and single-molecule real-time (SMRT) sequencing, has further accelerated genomic surveillance. Unlike traditional short-read sequencing platforms, these technologies enable long-read, portable sequencing, making them particularly valuable for on-site disease diagnostics and rapid outbreak response (Chapter 4, No. 14 Ref) [32].

Key applications of nanopore sequencing in public health include the following:

1) Rapid viral genome characterization during pandemics (e.g., COVID-19, Ebola) (Chapter 4, No. 14 Ref) [32].
2) On-site antimicrobial resistance profiling, reducing reliance on centralized laboratories (Chapter 4, No. 14 Ref) [32].
3) Portable sequencing for resource-limited settings, improving diagnostic accessibility in remote areas (Chapter 4, No. 14 Ref) [32].

To further improve the scalability, accuracy, and speed of genomic surveillance, researchers are integrating AI and deep learning models for automated pathogen classification, outbreak prediction, and epidemiological forecasting (Chapter 4, No. 12 Ref) [31]. These AI-powered innovations play a pivotal role in the following:

1) Detecting rare and novel pathogens with minimal sequencing coverage, reducing false negatives in diagnostics [31].
2) Predicting disease progression and transmission dynamics through genomic epidemiology models, enhancing preparedness for future pandemics (Chapter 4, No. 11 Ref) [33].
3) Optimizing data interpretation in real-time sequencing, allowing for the rapid identification of clinically significant mutations (Chapter 4, No. 16 Ref) [30].

As NGS technologies continue to evolve, their integration with high-throughput sequencing automation, portable diagnostic platforms, and AI-enhanced analytics is expected to redefine global genomic surveillance. Future innovations will focus on scalable, decentralized sequencing networks, allowing real-time pathogen tracking across international borders while maintaining data interoperability and biosecurity. These advancements will not only enhance the speed and accuracy of disease detection but also enable precision-targeted public health interventions on a global scale.

11.3 Advancements in Biosensing and Bioelectronics

11.3.1 Plasmonic and Nano-Enabled Biosensors

Nanomaterial-based biosensors have transformed high-sensitivity, real-time molecular detection by leveraging the optical, electrical, and chemical properties of nanoscale materials. Among these, plasmonic biosensors utilizing localized surface plasmon resonance (LSPR) in metallic nanostructures—such as gold, silver, and hybrid metal-dielectric nanocomposites—have enabled label-free, ultra-sensitive detection with minimal sample preparation. These biosensors provide rapid disease diagnostics, environmental monitoring, and multianalyte screening, particularly in high-throughput clinical applications (Chapter 8, No. 5 Ref) [34].

A major technological advancement in plasmonic biosensors is the integration of surface-enhanced Raman scattering (SERS), which amplifies Raman signals by several orders of magnitude, facilitating single-molecule-level detection of biomarkers, pathogens, and trace contaminants in biological fluids. The enhancement mechanisms involve electromagnetic field amplification, charge transfer effects, and localized molecular interactions between engineered metallic nanostructures and analytes. These properties make SERS-based diagnostics a powerful platform for early disease detection, antimicrobial resistance profiling, and rapid forensic screening (Chapter 8, No. 3 Ref) [35].

Future developments in plasmonic biosensing technologies will focus on multiplexed, high-throughput analysis. The emergence of:

1) Multimodal plasmonic biosensors combining LSPR with fluorescence, electrochemical, or photothermal effects, enhancing detection sensitivity and specificity.
2) Hybrid nanomaterial-based plasmonic substrates, which integrate semiconductor and quantum-enhanced surfaces for superior signal stability and extended spectral range.
3) 3D plasmonic nanoarrays, designed for high-density biomarker detection, enabling parallel analysis of disease signatures from complex biofluids.

In addition, the integration of AI-driven spectral analysis with plasmonic biosensors will enable automated, real-time data interpretation, significantly improving diagnostic accuracy and reliability. Future platforms will leverage cloud-based AI models for spectral deconvolution and digital signal enhancement, allowing biosensors to adaptively optimize detection parameters based on sample variability.

To further expand real-world applicability, researchers are focusing on scalable, low-cost fabrication of plasmonic biosensors through techniques such as nanoimprint lithography and inkjet-printed plasmonic arrays. These approaches will ensure greater accessibility and commercial scalability, bridging the gap between cutting-edge diagnostic tools and POC healthcare solutions (Chapter 8, No. 5 Ref) [34].

11.3.2 Microfluidic Platforms for High-Throughput Analysis

Microfluidic technology has revolutionized diagnostics by enabling high-throughput automation, miniaturization, and real-time biochemical processing. Next-generation LOC devices now integrate multianalyte detection, AI-assisted data processing, and ultra-fast reaction kinetics, significantly enhancing diagnostic efficiency. By incorporating advanced microfluidic architectures that optimize fluid dynamics, reagent mixing, and single-cell isolation, these platforms are redefining the landscape of precision medicine and decentralized diagnostics (Chapter 7, No. 2 Ref) [36].

One of the most transformative aspects of microfluidic systems is their ability to process complex biological samples with minimal reagent consumption while maintaining high analytical throughput. These technologies are now widely adopted for high-speed pathogen screening, automated blood chemistry analysis, and multiplexed molecular diagnostics, substantially reducing the time-to-result for clinical decision-making (Chapter 7, No. 2 Ref) [36].

Recent advancements have expanded the scope of high-throughput microfluidics, leading to the development of

1) Digital droplet microfluidics, which enables ultra-sensitive single-molecule and single-cell analysis, crucial for early cancer detection and infectious disease monitoring.
2) Self-powered capillary-driven microfluidic chips, eliminating the need for external pumps and making them ideal for field-deployable, cost-effective diagnostics.
3) Multiphase microfluidic networks, designed for high-throughput screening of drug candidates and personalized therapeutic profiling.
4) 3D-printed microfluidic biosensors, which allow for rapid prototyping and scalable production of customizable diagnostic devices.

Looking forward, AI-enhanced microfluidic platforms will drive next-generation high-throughput screening (HTS), allowing real-time pattern recognition, adaptive fluid control, and cloud-integrated disease surveillance. AI integration into POCT is expected to transform diagnostic accuracy and accessibility, as evidenced by the successful deployment of deep learning-assisted image analysis for malaria and skin cancer detection (Chapter 10, No. 64 Ref) [37]. AI-based smartphone biosensing platforms have also been developed for real-time viral detection, reducing diagnostic turnaround time and improving outbreak response (Chapter 10, No. 67 Ref) [38]. Future innovations will focus on microfluidic arrays for parallelized multi-biomarker analysis, expanding the capabilities of POC and remote healthcare applications. These advancements will be instrumental in bridging the gap between high-end laboratory diagnostics and widely accessible healthcare solutions (Chapter 7, No. 2 Ref) [36]. Combined with AI-driven analysis, these emerging technologies have the potential to reshape infectious disease diagnostics by providing rapid, on-site results that match the accuracy of traditional laboratory methods (Chapter 10, No. 80 Ref) [39]. However, these advances raise concerns regarding algorithmic biases, as current training datasets may not adequately represent diverse patient populations (Chapter 10, No. 70 Ref) [40]. Ensuring equitable access to AI-assisted diagnostics requires standardized validation protocols and international collaboration on regulatory frameworks.

11.3.3 Flexible Bioelectronics and Wearable Diagnostics

The transition from laboratory-based diagnostics to wearable biosensors has been facilitated by advancements in flexible bioelectronics, which enable continuous health monitoring through noninvasive, real-time sensing. Wearable biosensors utilize stretchable and biocompatible materials integrated with electrochemical and optical transducers to detect physiological biomarkers in sweat, saliva, and interstitial fluids (Chapter 8, No. 6 Ref) [20].

Recent developments in flexible bioelectronics have led to the fabrication of highly sensitive, multifunctional wearable sensors capable of detecting chemical and physical biomarkers simultaneously. For instance, wearable electrochemical biosensors have been designed to monitor glucose levels, lactate concentration, and pH in sweat, providing valuable insights into metabolic health and hydration status (Chapter 8, No. 6 Ref) [20].

Furthermore, motion-sensitive bioelectronics, such as strain sensors and pressure-sensitive conductive films, have been incorporated into wearable devices to track biomechanical movements, muscle activity, and cardiovascular signals. These technologies enable early detection of neuromuscular disorders, cardiovascular diseases, and stress-related conditions, enhancing personalized healthcare and preventive medicine (Chapter 8, No. 6 Ref) [20].

Looking ahead, the integration of flexible bioelectronics with wireless communication and data analytics will pave the way for next-generation smart healthcare systems. The development of battery-free, self-powered biosensors using energy-harvesting technologies will further enhance the feasibility of long-term continuous monitoring, reducing the dependency on external power sources and bulky equipment (Chapter 8, No. 6 Ref) [20].

By bridging the gap between conventional clinical diagnostics and wearable health monitoring, flexible bioelectronics are set to transform personalized medicine, enabling real-time, decentralized disease management and preventive healthcare solutions.

11.4 The Fusion of Chemometric Approaches with AI-Driven Approaches in Future Diagnostics

11.4.1 Data-Driven Precision Diagnostics

As diagnostic techniques advance, the integration of AI and machine learning (ML) has become instrumental in transforming data interpretation, predictive modeling, and real-time disease monitoring. AI-driven algorithms can analyze vast amounts of diagnostic data, including genomic sequences, imaging results, and biosensor outputs, to detect subtle disease signatures that might be overlooked by conventional methods. This shift toward data-driven precision diagnostics enhances early disease detection, personalized treatment strategies, and outbreak prediction models (Chapter 7, No. 6 Ref) [41].

One of the most impactful applications of AI in diagnostics is its role in predictive modeling. AI-powered systems can process historical health data, identifying correlations and risk factors for disease onset. These insights enable healthcare professionals to implement targeted prevention measures, thereby reducing the burden of chronic and infectious diseases. For example, machine learning models trained on genomic data have demonstrated the ability to predict antimicrobial resistance patterns, guiding clinicians in selecting the most effective treatments (Chapter 4, No. 14 Ref) [32].

AI-assisted microfluidic systems are another promising frontier in diagnostic technology. Microfluidic devices generate large datasets from single-cell profiling and high-throughput molecular diagnostics, which AI algorithms can process in real time. This capability enables precise biomarker quantification and rapid disease classification, facilitating more efficient clinical decision-making (Chapter 7, No. 101 Ref) [42].

Furthermore, deep learning algorithms have enhanced the capabilities of imaging-based diagnostics, including AI-assisted pathology and radiology. By analyzing high-resolution images, AI can detect morphological changes in cells and tissues indicative of disease, improving diagnostic accuracy and reducing human error. These systems are particularly valuable in screening for cancers, neurological disorders, and infectious diseases (Chapter 7, No. 7 Ref) [43].

11.4.2 Standardization, Interoperability, and Global Data Sharing

The rapid expansion of diagnostic technologies has underscored the necessity for standardization and interoperability across different platforms. Disparities in diagnostic protocols, variations in data formats, and lack of uniform quality controls hinder the global adoption of advanced diagnostic tools. Establishing standardized methodologies ensures consistency in test performance, enhances reproducibility, and facilitates regulatory approval for emerging diagnostic technologies (Chapter 7, No. 12 Ref) [44].

One major challenge in global health diagnostics is the harmonization of pathogen surveillance systems. Different regions employ varying methodologies for genetic sequencing, disease reporting, and epidemiological tracking, creating barriers to international collaboration. To address this, initiatives such as the development of open-access global pathogen surveillance databases have been proposed. These databases aggregate genomic and epidemiological data, allowing researchers worldwide to monitor emerging infectious diseases, track viral mutations, and assess antimicrobial resistance trends in real time (Chapter 1, No. 1 Ref) [45].

The integration of AI-driven analytics with global health data networks offers a transformative solution to these challenges. By leveraging cloud computing and big data analytics, AI can process real-time diagnostic data from multiple sources, enabling rapid response to public health threats. This capability was demonstrated during the COVID-19 pandemic, where AI-assisted genomic sequencing helped track viral evolution and inform public health policies (Chapter 4, No. 17 Ref) [46].

Another critical aspect of global diagnostics is ensuring equitable access to high-quality diagnostic tools. Many low-resource settings lack the infrastructure for advanced molecular diagnostics, exacerbating health disparities. Standardization of POCT devices and mobile health technologies can bridge this gap, enabling decentralized diagnostic solutions that are scalable and cost-effective (Chapter 7, No. 12 Ref; Chapter 10, No. 1 Ref) [44, 47].

Looking ahead, the fusion of chemometric approaches with AI-driven diagnostics will redefine the landscape of disease detection and surveillance. By integrating machine learning models with standardized diagnostic platforms, researchers can develop more accurate, rapid, and globally accessible testing solutions. This paradigm shift will be instrumental in shaping the next generation of public health strategies, ensuring a more robust and interconnected diagnostic infrastructure worldwide.

11.5 Outlook: Challenges and Future Directions in Diagnostic Techniques

Diagnostic science is on the cusp of a paradigm shift fueled by the convergence of advanced biosensing, AI-driven data analytics, genomic sequencing, and high-throughput screening technologies. These innovations are moving diagnostics beyond centralized laboratories toward real-time, decentralized, and intelligent systems, enabling more predictive and personalized healthcare. However, realizing this future requires overcoming persistent challenges related to accessibility, scalability, and governance. High costs, specialized infrastructure needs, and gaps in regulatory oversight currently limit the reach of next-generation tools. Tackling these issues—while strategically harnessing new technologies—is critical to ensure that emerging diagnostic techniques deliver equitable improvements in global health outcomes.

11.5.1 Overcoming Barriers to Accessibility and Scalability

Many cutting-edge diagnostic technologies remain financially and logistically impractical for widespread use, especially in low-resource settings. Advanced molecular tests (e.g., CRISPR-based assays and microfluidic platforms) often rely on expensive reagents, precision fabrication, and controlled lab conditions, restricting their availability to well-funded institutions. In addition, many novel diagnostics require skilled personnel and complex equipment, creating a reliance on centralized facilities that is impractical in rural or underserved areas. Supply chain inefficiencies further compound the problem, as essential components produced in limited locations lead to high costs and frequent shortages when distributed globally. Developing cost-effective manufacturing methods and utilizing inexpensive materials are crucial for scaling these innovations, as is designing systems that can operate with minimal infrastructure or training.

Key strategies to improve accessibility and scalability include the following:

1) Low-cost, Scalable Platforms: Transition to polymer- and paper-based diagnostic devices that are cheap to produce and easy to mass-manufacture. These disposable POC tests can provide reliable results without the need for elaborate laboratory setups.
2) User-friendly, Automated Devices: Engineer diagnostics with built-in sample processing and result interpretation (e.g., precalibrated cartridges and integrated readouts) to minimize the need for trained operators. Simplifying workflows in this way allows frontline healthcare workers to use advanced tests with minimal training.
3) Mobile and Connected Diagnostics: Leverage smartphones and Internet of Things (IoT)-enabled biosensors to decentralize testing. For example, camera-based smartphone assays and portable reader devices can perform molecular tests on-site and transmit results for remote analysis. Such connected tools enable real-time disease detection in both urban and remote settings, dramatically expanding reach.
4) Localized Production and Supply Chains: Establish regional manufacturing hubs for diagnostic reagents and kits to reduce dependence on a single supply source. Strengthening local production and distribution networks helps ensure that tests and materials remain available and affordable even in hard-to-reach areas.

By focusing on affordability, ease of use, and decentralized production, these approaches aim to bridge the gap between high-end innovations and the realities of resource-limited healthcare systems. While technological advances can greatly expand diagnostic reach, their successful deployment must be coupled with practical design and distribution strategies. Without thoughtful adaptation to real-world constraints, even the most powerful diagnostic tools may fail to achieve widespread impact. Ultimately, improving accessibility and scalability through these measures will lay the groundwork for next-generation diagnostics to benefit all populations.

Importantly, as these solutions roll out, they must be accompanied by robust oversight to ensure they are used effectively and ethically in diverse settings.

11.5.2 Addressing Ethical and Regulatory Challenges in AI and Genomic Diagnostics

The rapid adoption of AI-driven diagnostics and advanced genomic techniques (such as real-time sequencing for pathogen surveillance) introduce complex ethical and regulatory challenges. While these innovations can dramatically improve diagnostic accuracy and speed, they also bring new risks related to data privacy, algorithmic bias, and unequal access. Without appropriate safeguards and policies, next-generation diagnostics could inadvertently widen healthcare disparities or erode public trust. Several key considerations must be addressed to ensure an equitable and responsible implementation of these technologies:

1) Data Privacy and Security: AI-powered diagnostic platforms often rely on large-scale data collection and cloud-based analysis, raising concerns about who controls sensitive health data and how it is protected. Strong data governance policies—including encryption, secure data storage, and transparent data-sharing agreements—are essential to maintain patient confidentiality and prevent misuse of personal genomic information.
2) Algorithmic Bias and Fairness: Many machine learning models in diagnostics are trained on datasets that do not fully represent the diversity of global populations. This can lead to biased algorithms that underperform for underrepresented groups, exacerbating health inequalities. To prevent such outcomes, developers and regulators need to mandate bias audits, use diverse training data, and require continuous validation of AI diagnostic tools across different demographics.
3) Disparities in Genomic Surveillance: Cutting-edge genomic diagnostics (like pathogen genome sequencing for outbreak tracking) are predominantly available in high-income regions, creating an imbalance in global health security. Low-resource countries may lack access to sequencing technology or the capacity to analyze genomic data, limiting their ability to benefit from or contribute to these advances. International data-sharing collaborations and investments in global sequencing infrastructure are needed to democratize genomic surveillance, ensuring that insights (e.g., emerging disease variants) are shared equitably.
4) Regulatory Framework Gaps: Existing medical device regulations and approval processes often lag behind the pace of innovation in AI and genomics. Current frameworks may not adequately address issues like the continuous learning nature of AI algorithms or the cross-border flow of digital health data. Proactive regulatory updates are required—for instance, establishing guidelines for algorithm transparency and accountability (such as explainable AI in diagnostics) and creating oversight mechanisms for genomic data use. Harmonizing standards across jurisdictions will facilitate the safe adoption of AI-driven and digital diagnostics worldwide.

Addressing these ethical and regulatory challenges is as crucial as the technological development itself. Ensuring patient data privacy and algorithmic fairness will maintain public trust in new diagnostic methods, while international cooperation on standards will prevent regulatory mismatches that could delay life-saving tools. In essence, strong ethical guidelines and forward-looking policies must evolve in tandem with diagnostic innovations. By instituting rigorous governance (from data protection laws to AI auditing requirements) and fostering global consensus on best practices, stakeholders can ensure that advanced diagnostics are deployed responsibly and inclusively, protecting individual rights and promoting health equity.

11.5.3 Strategic Implementation for Global Health Impact

Even as technological breakthroughs and policy frameworks advance, a strategic approach is required to integrate these developments into health systems for maximum global impact. Collaboration between innovators, healthcare providers, and policymakers will be pivotal in translating laboratory innovations into practical solutions on the ground. This begins with aligning new diagnostic tools with existing healthcare infrastructure and workflows. For example, developers should work with local clinics to ensure that POC devices are compatible with the clinic's capabilities (power supply, training level, etc.) and with digital health record systems for seamless data integration. Early involvement of end-users in design and deployment planning can greatly increase the uptake and effectiveness of novel diagnostics in real-world settings.

Equally important is scalability in manufacturing and distribution. Global health agencies and industry partners must invest in streamlined production pipelines that can rapidly scale output of tests (especially during disease outbreaks) without compromising quality. Strengthening supply chains—through diversified manufacturing sites and stockpiles of critical materials—will make diagnostic tools more resilient to disruptions and available where they are needed most. In parallel, international regulatory harmonization can smooth the path for new diagnostics to be approved and adopted across different countries. Cooperative frameworks (for instance, mutual recognition of diagnostic approvals or unified guidelines for digital health) can reduce delays in bringing innovations to underserved regions. Additionally, building data interoperability standards will allow diagnostic devices and surveillance systems around the world to share information in real time, aiding in early outbreak detection and coordinated responses.

A critical enabler of these advancements is the role of chemometrics in diagnostic science. As diagnostic platforms integrate increasingly complex biosensors, AI-driven analytics, and multimodal detection systems, the need for robust, quantitative measurement frameworks becomes paramount. Chemometric techniques—encompassing signal processing, multivariate analysis, and data standardization—play a pivotal role in ensuring the accuracy, reproducibility, and cross-platform comparability of diagnostic results. In particular, chemometrics enhances spectral deconvolution in optical biosensing, refines electrochemical signal interpretation, and optimizes algorithmic decision-making in AI-assisted diagnostics. By embedding chemometric methodologies within future diagnostic workflows, researchers can mitigate signal artifacts, reduce variability, and maximize the reliability of real-time disease detection. As global diagnostic networks expand, chemometrics will serve as a crucial bridge between raw sensor outputs and clinically actionable insights, reinforcing the integrity and predictive power of next-generation healthcare technologies.

A strategic, global perspective must also prioritize equity. This means proactively directing resources and training to build diagnostic capacity in low- and middle-income countries, rather than introducing new technologies only in wealthier healthcare systems. Initiatives such as technology transfer programs, open-access diagnostic software, and subsidized pricing for low-income regions can help level the playing field. Moreover, continuous monitoring and evaluation should guide course corrections: collecting data on how new diagnostic tools perform across different populations will identify remaining gaps and drive improvements.

In conclusion, the future of diagnostic techniques will not be defined by technology alone, but by how intelligently and inclusively these innovations are implemented. By coupling cutting-edge diagnostics with thoughtful investment in infrastructure, education, and governance, the global community can usher in an era of more agile and proactive healthcare. Early disease detection, personalized treatment selection, and real-time public health surveillance could become standard practice worldwide. Achieving this vision will require steadfast commitment to scaling up access, enforcing ethical standards, and fostering international collaboration. With strategic action, the next generation of diagnostics can fundamentally improve global health-delivering timely, precise, and accessible care to populations everywhere, and transforming disease management into a truly preventive and equitable endeavor.

References

1 Gushulak, B. and Weekers, J. (2009). Macpherson D migrants and emerging public health issues in a globalized world: threats, risks and challenges, an evidence-based framework. *Emerging Health Threats Journal* 2: e10.
2 Wang, C., Horby, P.W., Hayden, F.G., and Gao, G.F. (2020). A novel coronavirus outbreak of global health concern. *Lancet* 395 (10223): 470–473.
3 Fire, A. and Xu, S.Q. (1995). Rolling replication of short DNA circles. *Proceedings of the National Academy of Sciences of the United States of America* 92 (10): 4641–4645.
4 Notomi, T., Okayama, H., Masubuchi, H. et al. (2000). Loop-mediated isothermal amplification of DNA. *Nucleic Acids Research* 28 (12): e63.
5 Reen, D.J. (1994). Enzyme-linked Immunosorbent assay (ELISA). In: *Basic Protein and Peptide Protocols* (ed. J.M. Walker), 461–466. Totowa, NJ: Humana Press.
6 Vermeesch, J.R., Voet, T., and Devriendt, K. (2016). Prenatal and pre-implantation genetic diagnosis. *Nature Reviews Genetics* 17 (10): 643–656.

7 Li, D., Morisseau, C., McReynolds, C.B. et al. (2020). Development of improved double-Nanobody Sandwich ELISAs for human soluble epoxide hydrolase detection in peripheral blood mononuclear cells of diabetic patients and the prefrontal cortex of multiple sclerosis patients. *Analytical Chemistry* 92 (10): 7334–7342.

8 Drolet, D.W., Moon-McDermott, L., and Romig, T.S. (1996). An enzyme-linked oligonucleotide assay. *Nature Biotechnology* 14 (8): 1021–1025.

9 Chen, C., Luo, J., Li, C. et al. (2018). Molecularly imprinted polymer as an antibody substitution in pseudo-immunoassays for chemical contaminants in food and environmental samples. *Journal of Agricultural and Food Chemistry* 66 (11): 2561–2571.

10 Tao, X., Wang, X., Liu, B., and Liu, J. (2020). Conjugation of antibodies and aptamers on nanozymes for developing biosensors. *Biosensors and Bioelectronics* 168: 112537.

11 Wei, H. and Wang, E. (2013). Nanomaterials with enzyme-like characteristics (nanozymes): next-generation artificial enzymes. *Chemical Society Reviews* 42 (14): 6060–6093.

12 Shang, L., Cheng, Y., and Zhao, Y. (2017). Emerging droplet microfluidics. *Chemical Reviews* 117 (12): 7964–8040.

13 Lizama, J.H., Chen, C.-J., Chang, W.-C. et al. (2024). Planarized through-hole valves enabling multilayered microfluidic architecture towards pipette-free ELISA. *Microchemical Journal* 200: 110373.

14 Gong, F., Wei, H.-x., Qi, J. et al. (2021). Pulling-force spinning top for serum separation combined with paper-based microfluidic devices in COVID-19 ELISA diagnosis. *ACS Sensors* 6 (7): 2709–2719.

15 Scott, J., Heike, S., Brittany, G. et al. (2015). *In Standards for Pathogen Detection Via Next Generation Sequencing*. MD: NIST Gaithersburg.

16 Wang, S.W., Tan, Y., Li, S.J., and Zhu, T.H. (2022). Structural and dynamic analysis of leaf-associated fungal Community of Walnut Leaves Infected by leaf spot disease based Illumina high-throughput sequencing technology. *Polish Journal of Microbiology* 71 (3): 429–441.

17 Das, R., Parveen, S., Bora, A., and Giri, P.K. (2020). Origin of high photoluminescence yield and high SERS sensitivity of nitrogen-doped graphene quantum dots. *Carbon* 160: 273–286.

18 Itoh, T., Procházka, M., Dong, Z.-C. et al. (2023). Toward a new era of SERS and TERS at the Nanometer scale: from fundamentals to innovative applications. *Chemical Reviews* 123 (4): 1552–1634.

19 Zhou, P., Yang, X.-L., Wang, X.-G. et al. (2020). A pneumonia outbreak associated with a new coronavirus of probable bat origin. *Nature* 579 (7798): 270–273.

20 Amirjani, A., Kamani, P., Hosseini, H.R.M., and Sadrnezhaad, S.K. (2022). SPR-based assay kit for rapid determination of Pb2+. *Analytica Chimica Acta* 1220: 340030.

21 Zhang, L. and Tian, Y. (2018). Designing recognition molecules and tailoring functional surfaces for in vivo monitoring of small molecules in the brain. *Accounts of Chemical Research* 51 (3): 688–696.

22 Liu, Y., Yang, Y., Wang, G. et al. (2023). Multiplexed discrimination of SARS-CoV-2 variants via plasmonic-enhanced fluorescence in a portable and automated device. *Nature Biomedical Engineering* 7 (12): 1636–1648.
23 Liu, Y.Y., Ren, W.C., Xue, Z.T. et al. (2023). Real-time recombinase-aided amplification assay for rapid amplification of the IS1081 gene of Mycobacterium tuberculosis. *European Journal of Clinical Microbiology & Infectious Diseases* 42 (8): 963–972.
24 Piepenburg, O., Williams, C.H., Stemple, D.L., and Armes, N. (2006). A DNA detection using recombination proteins. *PLOS Biology* 4 (7): e204.
25 Caetano-Anollés, G. (1996). Scanning of nucleic acids by in vitro amplification: new developments and applications. *Nature Biotechnology* 14 (13): 1668–1674.
26 Rollie, C., Schneider, S., Brinkmann, A.S. et al. (2015). Intrinsic sequence specificity of the Cas1 integrase directs new spacer acquisition. *Elife* 4.
27 Li, Y., Li, S.Y., and Wang, J. (2019). Liu G CRISPR/Cas systems towards next-generation biosensing. *Trends in Biotechnology* 37 (7): 730–743.
28 Bentley, D.R., Balasubramanian, S., Swerdlow, H.P. et al. (2008). Accurate whole human genome sequencing using reversible terminator chemistry. *Nature* 456 (7218): 53–59.
29 Tang, W.J., Zhang, Y., Luo, C. et al. (2021). Clinical application of metagenomic next-generation sequencing for suspected infections in patients with primary immunodeficiency disease. *Frontiers in Immunology* 12.
30 Chen, M.J., Qin, Y.Y., Deng, F.M. et al. (2021). Illumina MiSeq sequencing reveals microbial community succession in salted peppers with different salinity during preservation. *Food Research International* 143.
31 Lang, J.D., Zhu, R.R., Sun, X. et al. (2021). Evaluation of the MGISEQ-2000 sequencing platform for Illumina target capture sequencing libraries. *Frontiers in Genetics* 12.
32 Lin, L.J., Du, F.M., Zeng, J. et al. (2020). Deep insights into fungal diversity in traditional Chinese sour soup by Illumina MiSeq sequencing. *Food Research International* 137.
33 Hu, Z., Cheng, L., and Wang, H. (2015). The Illumina-solexa sequencing protocol for bacterial genomes. *Methods in Molecular Biology* 1231: 91–97.
34 Amirjani, A., Salehi, K., and Sadrnezhaad, S.K. (2022). Simple SPR-based colorimetric sensor to differentiate Mg2+ and Ca2+ in aqueous solutions. *Spectrochimica Acta Part A: Molecular and Biomolecular Spectroscopy* 268: 120692.
35 Sun, W., Huang, X., and Wang, X. (2023). CRISPR-based molecular diagnostics: a review. *Shengwu Gongcheng Xuebao/Chin. J. Biotechnol* 39 (1): 60–73.
36 Cassedy, A., Parle-McDermott, A., and R, O.'k. (2021). Virus detection: a review of the current and emerging molecular and immunological methods. *Frontiers in Molecular Biosciences* 8.
37 Tang, M., Wang, G., Kong, S.-K., and Ho, H.-P. (2016). A review of biomedical centrifugal microfluidic platforms. *Micromachines* 7 (2): 26.
38 Miyazaki, C.M., Carthy, E., and Kinahan, D.J. (2020). Biosensing on the centrifugal microfluidic lab-on-a-disc platform. *Processes* 8 (11): 1360.

39 Zhang, C., Xu, J., Ma, W., and Zheng, W. (2006). PCR microfluidic devices for DNA amplification. *Biotechnology Advances* 24 (3): 243–284.

40 Strohmeier, O., Keller, M., Schwemmer, F. et al. (2015). Centrifugal microfluidic platforms: advanced unit operations and applications. *Chemical Society Reviews* 44 (17): 6187–6229.

41 Whitesides, G.M. (2006). The origins and the future of microfluidics. *Nature* 442 (7101): 368–373.

42 Zhou, J., Dong, J., Hou, H. et al. (2024). High-throughput microfluidic systems accelerated by artificial intelligence for biomedical applications. *Lab on a Chip* 24 (5): 1307–1326.

43 Toumey, C. (2009). Plenty of room, plenty of history. *Nature Nanotechnology* 4 (12): 783–784.

44 Ramakrishnan, J., Gaurav, P.T.R., Chandar, N.S., and Sudharsan, N.M. (2021). Structural design, analysis and DOE of MEMS-based capacitive accelerometer for automotive airbag application. *Microsystem Technologies* 27 (3): 763–777.

45 Holroyd, T.A., Oloko, O.K., Salmon, D.A. et al. (2020). Communicating recommendations in public health emergencies: the role of public health authorities. *Health Security* 18 (1): 21–28.

46 Langelier, C., Kalantar, K.L., Moazed, F. et al. (2018). Integrating host response and unbiased microbe detection for lower respiratory tract infection diagnosis in critically ill adults. *Proceedings of the National Academy of Sciences of the United States of America* 115 (52): E12353–E12362.

47 Khan, P., Aufdembrink, L.M., and Engelhart, A.E. (2020). Isothermal SARS-CoV-2 diagnostics: tools for enabling distributed pandemic testing as a means of supporting safe Reopenings. *ACS Synthetic Biology* 9 (11): 2861–2880.

Index

Diagnostic Techniques in Public Health: Analytical and Chemometric Approaches in Disease Detection and Management, First Edition. Edited by Jinghong Li and Yan Du.

d